Astrophysics: Theories and Applications

Astrophysics: Theories and Applications

Editor: Audria Baldwin

NY RESEARCH PRESS
New York

Published by NY Research Press
118-35 Queens Blvd., Suite 400,
Forest Hills, NY 11375, USA
www.nyresearchpress.com

Astrophysics: Theories and Applications
Edited by Audria Baldwin

International Standard Book Number: 978-1-63238-733-2 (Hardback)

Cataloging-in-Publication Data

Astrophysics : theories and applications / edited by Audria Baldwin.
p. cm.
Includes bibliographical references and index.
ISBN 978-1-63238-733-2
1. Astrophysics. 2. Cosmic physics. 3. Physics. 4. Astronomy. I. Baldwin, Audria.
QB461 .A88 2020
523.01--dc23

Contents

Preface

The purpose of the book is to provide a glimpse into the dynamics and to present opinions and studies of some of the scientists engaged in the development of new ideas in the field from very different standpoints. This book will prove useful to students and researchers owing to its high content quality.

Astrophysics is the sub-field of astronomy that uses the principles of physics and chemistry to ascertain the nature of various astronomical objects instead of their positions in space. The objects that are studied under astrophysics are sun, galaxies, stars, extrasoler planets, the cosmic microwave background, etc. The properties that are studied in this field include luminosity, temperature, density and chemical composition. However the modern astronomical studies often involve a substantial amount of work in the domains of theoretical and observational physics. Observational astrophysics includes radio astronomy, infrared astronomy, optical astronomy, ultraviolet, x-ray and gamma ray astronomy, etc. Some other areas of study in this discipline include the properties of dark energy, black holes, dark matter and other celestial bodies. This book unravels the recent studies in the field of astrophysics. Also included herein is a detailed explanation of the various theories and applications of astrophysics. It is appropriate for students seeking detailed information in this area as well as for experts.

At the end, I would like to appreciate all the efforts made by the authors in completing their chapters professionally. I express my deepest gratitude to all of them for contributing to this book by sharing their valuable works. A special thanks to my family and friends for their constant support in this journey.

Editor

1

Thermodynamics of Charged AdS Black Holes in Rainbow Gravity

Ping Li, Miao He, Jia-Cheng Ding, Xian-Ru Hu, and Jian-Bo Deng

Institute of Theoretical Physics, Lanzhou University, Lanzhou 730000, China

Correspondence should be addressed to Jian-Bo Deng; dengjb@lzu.edu.cn

Guest Editor: Farook Rahaman

In this paper, the thermodynamic property of charged AdS black holes is studied in rainbow gravity. By the Heisenberg Uncertainty Principle and the modified dispersion relation, we obtain deformed temperature. Moreover, in rainbow gravity we calculate the heat capacity in a fixed charge and discuss the thermal stability. We also obtain a similar behaviour with the liquid-gas system in extending phase space (including P and r) and study its critical behavior with the pressure given by the cosmological constant and with a fixed black hole charge Q. Furthermore, we study the Gibbs function and find its characteristic swallow tail behavior, which indicates the phase transition. We also find that there is a special value about the mass of test particle which would lead the black hole to zero temperature and a diverging heat capacity with a fixed charge.

1. Introduction

It is known Lorentz symmetry which is one of most important symmetries in nature; however, some researches indicate that the Lorentz symmetry might be violated in the ultraviolet limit [1–5]. Since the standard energy-momentum dispersion relation relates to the Lorentz symmetry, the deformation of Lorentz symmetry would lead to the modification of energy-momentum dispersion relation. In fact, some calculations in loop quantum gravity have showed the dispersion relations may be deformed. Meanwhile, based on the deformed energy-momentum dispersion relation the double special relativity has arisen [6, 7]. In this theory, in addition to the velocity of light being the maximum velocity attainable there is another constant for maximum energy scale in nature which is the Planck energy E_P. It gives different picture for the special relativity in microcosmic physics. The theory has been generalized to curved spacetime by Joao Magueijo and Lee Smolin, called gravity's rainbow [8]. In their theory, the geometry of spacetime depends on the energy of the test particle and observers of different energy would see different geometry of spacetime. Hence, a family of energy-dependent metrics named rainbow metrics will describe the geometry of spacetime, which is different from general gravity theory. Based on the nonlinear of Lorentz transformation, the energy-momentum dispersion relation can be rewritten as

$$E^2 f^2\left(\frac{E}{E_P}\right) - p^2 g^2\left(\frac{E}{E_P}\right) = m^2, \quad (1)$$

where E_P is the Planck energy. The rainbow functions $f(E/E_P)$ and $g(E/E_P)$ are required to satisfy

$$\lim_{E/E_P\longrightarrow 0} f\left(\frac{E}{E_P}\right) = 1,$$
$$\lim_{E/E_P\longrightarrow 0} g\left(\frac{E}{E_P}\right) = 1. \quad (2)$$

In this case, the deformed energy-momentum dispersion relation equation (1) will go back to classical one when the energy of the test particle is much lower than E_p. Due to this energy-dependent modification to the dispersion relation, the metric $h(E)$ in gravity's rainbow could be rewritten as [9]

$$h(E) = \eta^{ab} e_a(E) \otimes e_b(E), \quad (3)$$

where the energy dependence of the frame fields is

$$e_0(E) = \frac{1}{f(E/E_P)}\widetilde{e}_0, \qquad e_i(E) = \frac{1}{g(E/E_P)}\widetilde{e}_i, \tag{4}$$

and here the tilde quantities refer to the energy-independent frame fields. This leads to a one-parameter Einstein equation

$$G_{\mu\nu}\left(\frac{E}{E_P}\right) + \Lambda\left(\frac{E}{E_P}\right) g_{\mu\nu}\left(\frac{E}{E_P}\right) = 8\pi G\left(\frac{E}{E_P}\right) T_{\mu\nu}\left(\frac{E}{E_P}\right), \tag{5}$$

where $G_{\mu\nu}(E/E_P)$ and $T_{\mu\nu}(E/E_P)$ are energy-dependent Einstein tensor and energy-momentum tensor and $\Lambda(E/E_P)$ and $G(E/E_P)$ are energy-dependent cosmological constant and Newton constant. Generally, many forms of rainbow functions have been discussed in literatures; in this paper we will mainly employ the following rainbow functions:

$$f\left(\frac{E}{E_P}\right) = 1, \qquad g\left(\frac{E}{E_P}\right) = \sqrt{1 - \eta\left(\frac{E}{E_P}\right)^n}, \tag{6}$$

which has been widely used in [10–18].

Recently, Schwarzschild black holes, Schwarzschild AdS black holes, and Reissner-Nordstrom black holes in rainbow gravity [19–21] have been studied. Ahmed Farag Alia, Mir Faizald, and Mohammed M. Khalile [15] studied the deformed temperature about charged AdS black holes in rainbow gravity based on Heisenberg Uncertainty Principle (HUP), $E = \Delta p \sim 1/r_+$. In this paper, we study the thermodynamical property about the charged AdS black holes in rainbow gravity based on the usual HUP, $p = \Delta p \sim 1/r_+$. Moreover, we study how the mass of test particle influences thermodynamical property for charged AdS black holes.

The paper is organized as follows. In the next section, by using the HUP and the modified dispersion relation, we obtain deformed temperature, and we also calculate heat capacity with a fixed charge and discuss the thermal stability. In Section 3, we find the charged AdS black holes have similar behaviour with the liquid-gas system with the pressure given by the cosmological constant while we treat the black holes charge Q as a fixed external parameter, not a thermodynamic variable. We also calculate the Gibbs free energy and find characteristic swallow tail behavior. Finally, the conclusion and discussion will be offered in Section 4.

2. The Thermal Stability

In rainbow gravity the line element of the modified charged AdS black holes can be described as [15]

$$ds^2 = -\frac{N}{f^2}dt^2 + \frac{1}{Ng^2}dr^2 + \frac{r^2}{g^2}d\Omega^2, \tag{7}$$

where

$$N = 1 - \frac{2M}{r} + \frac{Q^2}{r^2} + \frac{r^2}{l^2}. \tag{8}$$

Generally, $-3/l^2 = \Lambda$ which is cosmological constant. Because all energy dependence in the energy-independent coordinates must be in the rainbow functions f and g, N is independent on the energy of test particle [8]. In gravity's rainbow, the deformed temperature related to the standard temperature T_0 was [15]

$$T = -\frac{1}{4\pi}\lim_{r\to r_+}\sqrt{\frac{-g^{11}}{g^{00}}}\frac{(g^{00})'}{g^{00}} = \frac{g(E/E_P)}{f(E/E_P)}T_0, \tag{9}$$

where r_+ is horizon radius.

In gravity's rainbow, although the metric depends on the energy of test particle, the usual HUP can be still used [19]. For simplicity we take $n = 2$ in the following discussion, by combining (1) with (6) we can get

$$g = \sqrt{1 - \eta G_0 m^2}\sqrt{\frac{r_+^2}{r_+^2 + \eta G_0}}, \tag{10}$$

where $G_0 = 1/E_P^2$, m is the mass of test particle, and η is a constant parameter.

Generally, the standard temperature was given by [22]

$$T_0 = \frac{1}{4\pi}\left(\frac{1}{r_+} + \frac{3r_+}{l^2} - \frac{Q^2}{r_+^3}\right). \tag{11}$$

When using (6) and (10), we can get the temperature of charged AdS black holes in rainbow gravity

$$T = gT_0 = \frac{1}{4\pi k}\sqrt{\frac{r_+^2}{r_+^2 + \eta G_0}}\left(\frac{1}{r_+} + \frac{3r_+}{l^2} - \frac{Q^2}{r_+^3}\right), \tag{12}$$

where $k = 1/\sqrt{1 - \eta G_0 m^2}$. It is easy to find $T = T_0$ when $\eta = 0$. Equation (12) shows that there are two solutions when $T = 0$, one corresponds to extreme black hole and the other to $m^2 = 1/\eta G$. The second solution indicates the temperature of black holes completely depends on the mass of test particle when the black holes keep with fixed mass, charge, and anti-de Sitter radius. The bigger the mass of test particle is, the smaller the temperature of black holes is. When $m^2 = 1/\eta G$, the temperature keeps zero. Generally, due to gravity's rainbow, a minimum radius with respect to the black hole is given and is related to a radius of black hole remnant when the temperature tends to zero [15]. However, our paper shows all black holes can keep zero temperature when the test particle mass approaches a value, such as $m^2 = 1/\eta G$. But due to $m \ll M_P$ in general condition, it may be difficult to test the phenomenon with zero temperature about black holes.

In general, the thermal stability can be determined by the heat capacity, which is also used to the systems of black holes [17, 23–25]. In other words, the positive heat capacity corresponds to a stable state and the negative heat capacity corresponds to unstable state. In following discussions, we will focus on the heat capacity to discuss the stability of black holes. When $N = 0$, the mass of charged AdS black holes can be calculated as

$$M = \frac{1}{2}\left(r_+ + \frac{Q^2}{r_+} + \frac{r_+^3}{l^2}\right). \tag{13}$$

Based on the first law $dM = TdS$ with the deformed temperature [20], the modified entropy can be computed

$$S = \int \frac{dM}{T} = \pi k r_+ \sqrt{r_+^2 + \eta G_0} + \pi k \eta G_0 \ln\left(r_+ + \sqrt{r_+^2 + \eta G_0}\right). \tag{14}$$

Note that the next leading order is logarithmic as $S \approx \pi r_+^2 + (1/2)\pi\eta G_0 \ln(4r_+^2)$, which is similar to the quantum correction in [26–31]. With $A = 4\pi r_+^2$ we can get $S \approx A/4 + (1/2)\pi\eta G_0 \ln(A/\pi)$. We can find the result is in agreement with the standard entropy $S = A/4$ when $\eta = 0$, which is standard condition.

The heat capacity with a fixed charge can be calculated as

$$C_Q = T\frac{dS}{dT} = \left(\frac{\partial M/\partial r_+}{\partial T/\partial r_+}\right) = 2\pi k \frac{\left(-Q^2l^2r_+^2 + l^2r_+^4 + 3r_+^6\right)\left(r_+^2 + \eta G_0\right)^{3/2}}{3r_+^7 + \left(6\eta G_0 - l^2\right)r_+^5 + 3Q^2l^2r_+^3 + 2\eta G_0 Q^2 l^2 r_+}, \tag{15}$$

which shows that C_Q reduces to standard condition [22] with $\eta = 0$. Obviously, the heat capacity is diverging when $m^2 = 1/\eta G$. Generally, when the temperature vanishes, the heat capacity also tends to zero. However, our paper shows a different and anomalous phenomenon. Fortunately, the phenomenon is just an observation effect; the result gives us a way to test the theory of rainbow gravity. Some of the conditions above indicate that the mass of test particle does not influence the forms of temperature, entropy, and heat capacity but only changes their amplitudes.

The numerical methods indicate there are three situations corresponding to zero, one, and two diverging points of heat capacity respectively, which have been described in Figures 1, 2, and 3. Figure 1 shows a continuous phase and does not appear phase transition with $l < l_c$. In Figure 2, there are one diverging point and two stable phases for $C_Q > 0$ with $l = l_c$, phase 1 and phase 2, which individually represent a phase of large black hole (LBH) and a small black hole (SBH). In Figure 3, one can find there are three phases and two diverging points with $l > l_c$. Phase 1 experiences a continuous process from a unstable phase $C_Q < 0$ to a stable phase $C_Q > 0$; phase 2 is a pure unstable phase with $C_Q < 0$; phase 3 is a stable phase with $C_Q > 0$. It is easy to see phase 1 represents the phase of SBH and phase 3 represents the phase of LBH. However, there is a special unstable phase 2 between phase 1 and phase 3. This indicates when the system evolutes from phase 3 to phase 1, the system must experience a medium unstable state which could be explained as an exotic quark-gluon plasma with negative heat capacity [32, 33].

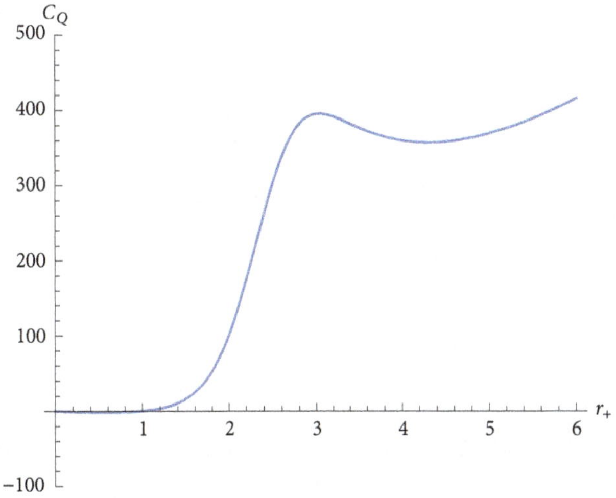

FIGURE 1: C_Q–r_+ diagram of charged AdS black holes in the rainbow gravity. It corresponds to $l = 6$. We have set $Q = 1, \eta = 1, m = 0$.

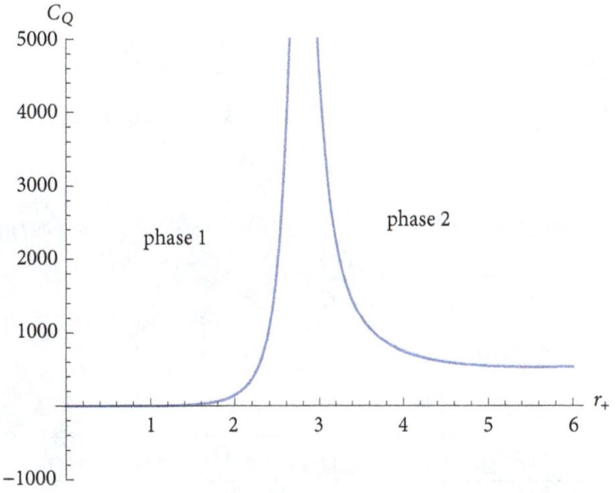

FIGURE 2: C_Q–r_+ diagram of charged AdS black holes in the rainbow gravity. It corresponds to $l = 7.05$. We have set $Q = 1, \eta = 1, m = 0$.

3. The Phase Transition of Charged AdS Black Holes in Extending Phase Space

Surprisingly, although rainbow functions modify the $\Lambda(E/E_P)$ term, they do not affect thermodynamical pressure related to the cosmological constant [17, 34]. So we can take the following relation:

$$P = -\frac{\Lambda(0)}{8\pi} = \frac{3}{8\pi l^2}. \tag{16}$$

Since David Kubiznak and Robert B. Mann have showed the critical behaviour of charged AdS black holes and completed the analogy of this system with the liquid-gas system [22], in what follows we will study whether the critical behavior of the charged AdS black holes system in rainbow

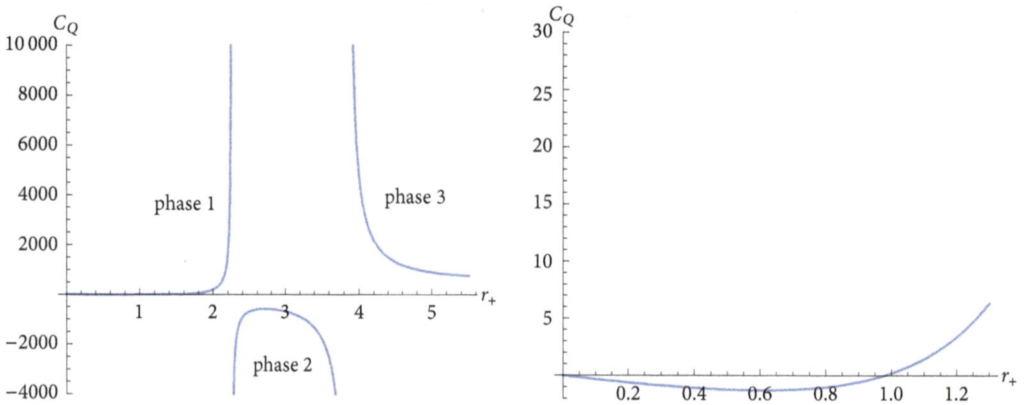

FIGURE 3: $C_Q - r_+$ diagram of charged AdS black holes in the rainbow gravity. It corresponds to $l = 8$, right one corresponds to an extending part of near $r_+ = 1$. We have set $\eta = 1, Q = 1, m = 0$.

gravity is kept. Using (12) and (16) in extended phase space, we can get

$$P = \frac{k}{2}\sqrt{\frac{r_+^2 + \eta G_0}{r_+^4}}T - \frac{1}{8\pi}\frac{1}{r_+^2} + \frac{1}{8\pi}\frac{Q^2}{r_+^4}. \tag{17}$$

Similarly with [17], the critical point is obtained from

$$\frac{\partial P}{\partial r_+} = 0,$$
$$\frac{\partial^2 P}{\partial r_+^2} = 0, \tag{18}$$

which leads to

$$r_c = \sqrt{\frac{2^{4/3}\eta G_0 Q^2 + 2^{4/3}Q^4 + 2Q^2(x+y)^{1/3} + 2^{2/3}(x+y)^{2/3}}{(x+y)^{1/3}}},$$
$$T_c = \frac{1}{2\pi k}\frac{r_c^2 - 2Q^2}{r_c^4 + 2\eta G_0 r_c^2}\sqrt{r_c^2 + \eta G_0}, \tag{19}$$
$$P_c = \frac{r_c^4 - 3Q^2 r_c^2 - 2\eta G_0 Q^2}{8\pi r_c^4 (r_c^2 + 2\eta G_0)},$$

where $x = \eta G_0 Q^2(\eta G_0 + Q^2)$ and $y = Q^2(\eta G_0 + Q^2)(\eta G_0 + 2Q^2)$. We can obtain

$$\frac{P_c r_c}{T_c} = k\frac{r_c^4 - 3Q^2 r_c^2 - 2\eta G_0 Q^2}{4r_c(r_c^2 - 2Q^2)\sqrt{r_c^2 + \eta G}}, \tag{20}$$

which shows the critical ratio is deformed due to the existence of rainbow gravity. It is notable that (20) will back to the usual ratio with $\eta = 0$. Generally, for charged AdS black holes, the pressure and temperature are demanded as positive real value. From (19), when $P_c > 0$ and $T_c > 0$, we have

$$r_c^4 - 3Q^2 r_c^2 - 2\eta G_0 Q^2 > 0,$$
$$r_c > \sqrt{2}Q, \tag{21}$$

which indicates a restriction between Q and η. The $P - r_+$ diagram has been described in Figure 4. From Figure 4, we can find that charged AdS black holes in rainbow gravity have an analogy with the Van-der-Waals system and have a first-order phase transition with $T < T_c$. Namely, when considering rainbow gravity with the form of (6), the behavior like Van-der-Waals system can also be obtained.

Based on [14, 35] the black hole mass is identified with the enthalpy, rather than the internal energy, so the Gibbs free energy for fixed charge in the rainbow gravity will be

$$G = H - TS$$
$$= \frac{1}{2}\left(r_+ + \frac{Q^2}{r_+} + \frac{r_+^3}{l^2}\right) - kT(\pi r_+\sqrt{r_+^2 + \eta G_0} \tag{22}$$
$$+ \pi\eta G_0 \ln\left(r_+ + \sqrt{r_+^2 + \eta G_0}\right),$$

which has been showed in Figure 5. Because the picture of G demonstrates the characteristic swallow tail behaviour, there is a first-order transition in the system.

4. Conclusion

In this paper, we have studied the thermodynamic behavior of charged AdS black holes in rainbow gravity. By the modified dispersion relation and HUP, we got deformed temperature in charged AdS black holes using no-zero mass of test particle. We have discussed the divergence about the heat capacity with a fixed charge. Our result shows that the phase structure has a relationship with AdS radius l. When $l = l_c$, there is only one diverging point about heat capacity; when $l > l_c$, we have found there are two diverging points and three phases including two stable phases and one unstable phase. In particular, an analogy between the charged AdS black holes in the rainbow gravity and the liquid-gas system is discussed. We have also showed $P - r_+$ critical behavior about the charged AdS black holes in the rainbow gravity. The consequence shows there is the Van-der-Waals like behavior in the rainbow gravity when η and Q coincide with (21). The rainbow functions deform the forms of critical pressure,

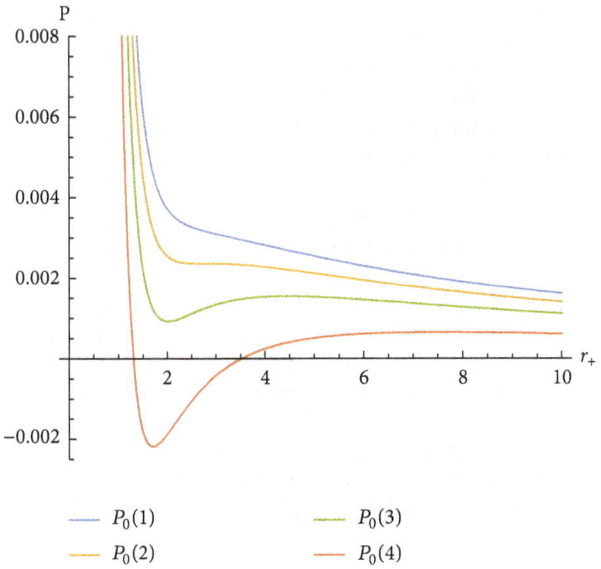

FIGURE 4: $P - r_+$ diagram of charged AdS black holes in the rainbow gravity. The temperature of isotherms decreases from top to bottom. The $P_0(1)$ line corresponds to one-phase for $T > T_c$. The critical state, $T_c = 0.0358$, is denoted by the $P_0(2)$ line. The lowest two lines correspond to the smaller temperature than the critical temperature. We have set $Q = 1, \eta = 1, m = 0$.

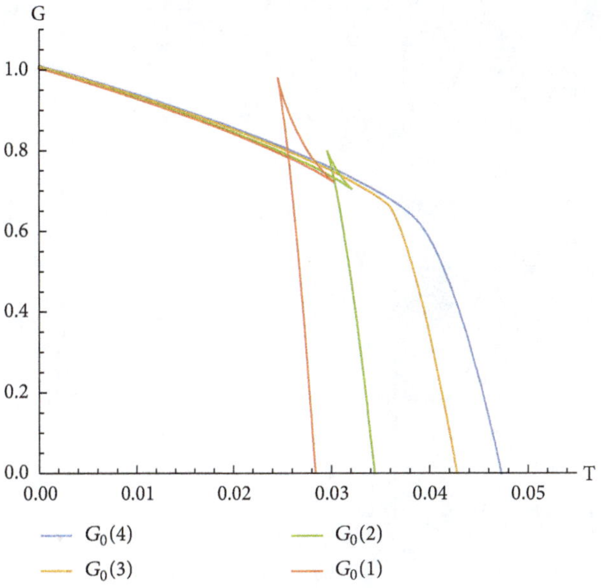

FIGURE 5: Gibbs free energy of charged AdS black holes in rainbow gravity. The blue line $G_0(3)$ corresponds to the critical pressure $P_c \approx 0.0024$, the line $G_0(4)$ corresponds to pressure $P > P_c$, and the others corresponds to pressure $P < P_c$. We have set $Q = 1, \eta = 1, m = 0$.

temperature, and radius. At last, we have discussed the Gibbs free energy and have obtained characteristic "swallow tail" behaviour which can be the explanation of first-order phase transition.

We find the mass of test particle does not influence the forms of temperature, entropy, and heat capacity but only changes their amplitudes. Moreover, there is a special value about the mass of test particle encountered $m^2 = 1/\eta G$, which would lead to zero temperature and diverging heat capacity for charged AdS black holes in rainbow gravity.

Conflicts of Interest

The authors declare that there are no conflicts of interest regarding the publication of this paper.

Acknowledgments

We would like to thank the National Natural Science Foundation of China (Grant no. 11571342) for supporting us on this work.

References

[1] R. Iengo, J. G. Russo, and M. Serone, "Renormalization group in Lifshitz-type theories," *Journal of High Energy Physics*, vol. 2009, no. 11, 2009.

[2] A. Adams, N. Arkani-Hamed, S. Dubovsky, A. Nicolis, and R. Rattazzi, "Causality, analyticity and an IR obstruction to UV completion," *Journal of High Energy Physics*, vol. 2006, no. 10, 2006.

[3] B. M. Gripaios, "Modified gravity via spontaneous symmetry breaking," *Journal of High Energy Physics*, vol. 2004, no. 10, 2004.

[4] J. Alfaro, P. González, and Á. Ricardo, "Electroweak standard model with very special relativity," *Physical Review D*, vol. 91, no. 10, Article ID 105007, 2015.

[5] H. Belich and K. Bakke, "Geometric quantum phases from Lorentz symmetry breaking effects in the cosmic string space-time," *Physical Review D: Particles, Fields, Gravitation and Cosmology*, vol. 90, no. 2, 2014.

[6] J. Magueijo and L. Smolin, "String theories with deformed energy-momentum relations, and a possible nontachyonic bosonic string," *Physical Review D: Particles, Fields, Gravitation and Cosmology*, vol. 71, Article ID 026010, 2005.

[7] J. Magueijo and L. Smolin, "Generalized Lorentz invariance with an invariant energy scale," *Physical Review D: Particles, Fields, Gravitation and Cosmology*, vol. 67, Article ID 044017, 2003.

[8] J. Magueijo and L. Smolin, "Gravity's rainbow," *Classical and Quantum Gravity*, vol. 21, pp. 1725–1736, 2004.

[9] J.-J. Peng and S.-Q. Wu, "Covariant anomaly and Hawking radiation from the modified black hole in the rainbow gravity theory," *General Relativity and Gravitation*, vol. 40, no. 12, pp. 2619–2626, 2008.

[10] G. Amelino-Camelia, J. Ellis, N. E. Mavromatos, and D. V. Nanopoulos, "Distance measurement and wave dispersion in a Liouville-string approach to quantum gravity," *International Journal of Modern Physics A*, vol. 12, no. 3, pp. 607–623, 1997.

[11] U. Jacob, F. Mercati, G. Amelino-Camelia, and T. Piran, "Modifications to Lorentz invariant dispersion in relatively boosted frames," *Physical Review D: Particles, Fields, Gravitation and Cosmology*, vol. 82, no. 8, Article ID 084021, 2010.

[12] G. Amelino-Camelia, "Quantum-spacetime phenomenology," *Living Reviews in Relativity*, vol. 16, no. 1, p. 5, 2013.

[13] A. F. Ali, "Black hole remnant from gravity's rainbow," *Physical Review D: Particles, Fields, Gravitation and Cosmology*, vol. 89, no. 10, Article ID 104040, 2014.

[14] A. F. Ali, M. Faizal, and M. M. Khalil, "Absence of black holes at LHC due to gravity's rainbow," *Physics Letters. B. Particle Physics, Nuclear Physics and Cosmology*, vol. 743, pp. 295–300, 2015.

[15] A. F. Ali, M. Faizal, and M. M. Khalil, "Remnant for all black objects due to gravity's rainbow," *Nuclear Physics B*, vol. 894, pp. 341–360, 2015.

[16] S. H. Hendi and M. Faizal, "Black holes in Gauss-Bonnet gravity's rainbow," *Physical Review D: Particles, Fields, Gravitation and Cosmology*, vol. 92, no. 4, Article ID 044027, 2015.

[17] S. H. Hendi, B. Eslam Panah, and S. Panahiyan, "Three-dimensional dilatonic gravity's rainbow: exact solutions," *PTEP. Progress of Theoretical and Experimental Physics*, vol. 76, no. 5, pp. 1–15, 2016.

[18] S. Gangopadhyay and A. Dutta, "Constraints on rainbow gravity functions from black-hole thermodynamics," *EPL (Europhysics Letters)*, vol. 115, no. 5, 2016.

[19] Y. Gim and W. Kim, "Thermodynamic phase transition in the rainbow Schwarzschild black hole," *Journal of Cosmology and Astroparticle Physics*, vol. 2014, no. 10, 2014.

[20] Y.-W. Kim, S. K. Kim, and Y.-J. Park, "Thermodynamic stability of modified Schwarzschild-AdS black hole in rainbow gravity," *The European Physical Journal C*, vol. 76, no. 10, p. 557, 2016.

[21] C. Liu, "Charged Particle's Tunneling in a Modified Reissner-Nordstrom Black Hole," *International Journal of Theoretical Physics*, vol. 53, no. 1, pp. 60–71, 2014.

[22] D. Kubiznak and R. B. Mann, "P-V criticality of charged AdS black holes," *Journal of High Energy Physics*, vol. 2012, no. 7, pp. 1–25, 2012.

[23] G.-M. Deng, J. Fan, X. Li, and Y.-C. Huang, "Thermodynamics and phase transition of charged AdS black holes with a global monopole," *International Journal of Modern Physics A*, vol. 33, no. 3, Article ID 1850022, 2018.

[24] G. Deng and Y. Huang, "Q-Φ criticality and microstructure of charged AdS black holes in f(R) gravity," *International Journal of Modern Physics A*, vol. 32, no. 35, Article ID 1750204, 2017.

[25] S. H. Hendi and M. Momennia, "AdS charged black holes in Einstein–Yang–Mills gravity's rainbow: Thermal stability and p-v criticality," *Physics Letters B*, vol. 777, pp. 222–234, 2018.

[26] D. V. Fursaev, "Temperature and entropy of a quantum black hole and conformal anomaly," *Physical Review D: Particles, Fields, Gravitation and Cosmology*, vol. 51, no. 10, Article ID R5352, 1995.

[27] R. K. Kaul and P. Majumdar, "Logarithmic correction to the Bekenstein-Hawking entropy," *Physical Review Letters*, vol. 84, no. 23, pp. 5255–5257, 2000.

[28] S. Das, P. Majumdar, and R. K. Bhaduri, "General logarithmic corrections to black-hole entropy," *Classical and Quantum Gravity*, vol. 19, no. 9, pp. 2355–2367, 2002.

[29] A. Chatterjee and P. Majumdar, "Universal canonical black hole entropy," *Physical Review Letters*, vol. 92, no. 14, Article ID 141301, 2004.

[30] F. J. Wang, Y. X. Gui, and C. R. Ma, "Entropy corrections for Schwarzschild black holes," *Physics Letters. B. Particle Physics, Nuclear Physics and Cosmology*, vol. 660, no. 3, pp. 144–146, 2008.

[31] B. Eslam Panah, "Effects of energy dependent spacetime on geometrical thermodynamics and heat engine of black holes: Gravity's rainbow," *Physics Letters. B. Particle Physics, Nuclear Physics and Cosmology*, vol. 787, pp. 45–55, 2018.

[32] P. Burikham and T. Chullaphan, "Comments on holographic star and the dual QGP," *Journal of High Energy Physics*, vol. 2014, no. 5, 2014.

[33] P. Burikham and C. Promsiri, "The Mixed Phase of Charged AdS Black Holes," *Advances in High Energy Physics*, vol. 2016, 2016.

[34] S. H. Hendi, S. Panahiyan, B. Eslam Panah, M. Faizal, and M. Momennia, "Critical behavior of charged black holes in Gauss-Bonnet gravity's rainbow," *Physical Review D: Particles, Fields, Gravitation and Cosmology*, vol. 94, no. 2, Article ID 024028, 2016.

[35] D. Kastor, S. Ray, and J. Traschen, "Enthalpy and the mechanics of AdS black holes," *Classical and Quantum Gravity*, vol. 26, no. 19, Article ID 195011, 16 pages, 2009.

2

Thermodynamic Prescription of Cosmological Constant in the Randall-Sundrum II Brane

Tanwi Bandyopadhyay

Adani Institute of Infrastructure Engineering, Ahmedabad-382421, India

Correspondence should be addressed to Tanwi Bandyopadhyay; tanwib@gmail.com

Academic Editor: Elias C. Vagenas

In this work, we apply the quantum corrected entropy function derived from the Generalized Uncertainty Principle (GUP) to the holographic equipartition law to study the cosmological scenario in the Randall-Sundrum (RS) II brane. An extra driving term has come up in the effective Friedmann equation for a homogeneous, isotropic, and spatially flat universe. Further, thermodynamic prescription of the universe constraints this term eventually with an order equivalent to that of the cosmological constant.

1. Introduction

In order to give an explanation of higher-dimensional theory, Randall and Sundrum [1, 2] proposed an idea of a bulk-brane model, where the four-dimensional world in which we live is called the 3-brane (a domain wall) that is embedded in a higher-dimensional space-time (bulk). According to the theory, the brane confines all the matter field; only gravity propagates in the bulk. Moreover the extra fifth dimension need not be finite; it can extend to infinity in either side of the brane. The concept of brane world scenarios shows a possibility to resolve the problem of unification of all forces and particles in nature. The main equations governing the cosmological evolutions of the brane differ from the corresponding Friedmann equations in standard cosmology [3–6]. The difference lies in the fact that the energy density of the brane appears to be in a quadratic form whereas, in standard cosmology, the energy density appears linearly in the field equations. This model is also consistent with string theory and may resolve the so-called hierarchy problem or the source of dark energy and dark matter [7, 8]. The later theory is one of the overwhelming theories of the current era. The concept of dark matter had been first proposed [9, 10] in the context of studying galaxy clusters. The dark energy, on the other hand, is a completely new component which produces sufficient negative pressure. This drives the cosmic acceleration which has also been substantiated by the observational evidence over the years. The observational data clearly states that the current universe is flat having an approximate cosmic content of 21% dark matter, 72% dark energy, and the rest in the form of visible matter and radiation. All these imply that the standard cosmological models need to be modified with the models of dark matter and dark energy. Unfortunately, very little is known about dark energy. Hence there exist many prospective candidates for this cosmic component. Among them, cosmological constant Λ is the most popular having an equation of state $p_\Lambda = -\rho_\Lambda$. This model is known as the ΛCDM model (cold dark matter) [11–14]. This theory has a major drawback in terms of order of measurement. The observed value of Λ is many orders of magnitude smaller than its theoretical value predicted in quantum field theory. This is termed as the cosmological constant problem and, to resolve this, one of the many proposed cosmological models is the varying cosmological constant ($\Lambda(t)$CDM) model [15–20].

On the other hand, one of the key features of quantum theory of gravity is called the holographic principle. This states that, in a bounded system, the number of degrees of freedom is associated with entropy and scales with the area enclosed [21–23]. Under this principle, gravity is shown to be an entropic force derived from the changes in the Bekenstein-Hawking entropy [24–26]. Further, many studies focussed on derivation and investigation of the Friedmann and acceleration equations in the background of entropic cosmology [27–29]. Various forms of entropy have been applied in these studies [30–37]. In some of them, an extra driving term is derived from entropic forces on the horizon of the universe in order to explain its accelerated expansion. Intrigued by the holographic principle, very recently Padmanabhan [38] has proposed a different approach saying that the cosmic space is emergent as the cosmic time progresses. It has been termed the holographic equipartition law. According to this, the rate of expansion of the universe is related to the difference between the surface degrees of freedom on the holographic horizon and the bulk degrees of freedom inside. Keeping this in the background, the cosmological equations were derived and examined both in classical and in modified theories of gravity [39–45]. For most of these studies, the Bekenstein-Hawking entropy played the major role.

Very recently, a similar study has been carried out in [46], where a modified Rényi entropy was chosen instead of the Bekenstein-Hawking entropy and a constant-like term was obtained in the field equations. Imposing an analytical constraint, this term showed behavior similar to the varying cosmological constant. Further, the power-law corrected entropy was also tested in the same mechanism and similar results were found in [47]. This surely necessitates more investigation into the alternative studies of dark energy and the cosmological constant in modified gravity theories. We have followed this novel approach to study the underlying cosmological scenario in the RS II brane model considering the quantum corrected form of the entropy function derived from the Generalized Uncertainty Principle (GUP) [48]. A similar study has been carried out in [44] in Einstein's gravity but our entropy function is unique in its $\sqrt{Area}$ form. The necessity and motivation for choosing this entropy function was discussed later in detail. The GUP corrected entropy was applied to the holographic equipartition law in a four-dimensional universe embedded in a conformally flat five-dimensional space-time. Consequently, an analogous extra driving term is derived in the modified Friedmann equations. Further thermodynamical investigations showed that this extra term is of an order identical to the order of cosmological constant.

The paper is organized in the following way: In Section 2, we briefly review the $\Lambda(t)$CDM model and the modified field equations in the context of brane world gravity. In Section 3, the expansion of the cosmic space is treated as an emergent process and the modified Friedmann equations are retrieved from the holographic equipartition law in the absence of any dark energy component. Section 4 presents a brief review of GUP corrected entropy. Section 4.1 discusses the results of application of GUP corrected entropy into the holographic equipartition law. In Section 4.2, the validity of the Generalized Second Law of Thermodynamics (GSLT) is assumed and the behavior of the extra driving term is analyzed. Finally, a brief discussion on our study is made in Section 5.

2. Main Equations: $\Lambda(t)$CDM Model in Brane World

A homogeneous, isotropic, spatially flat Friedmann-Robertson-Walker (FRW) universe in the natural unit system ($G = c = \hbar = k_B = 1$) is given by

$$ds^2 = dt^2 - a^2(t)\left[dr^2 + r^2\left(d\theta^2 + \sin^2\theta d\phi^2\right)\right] \tag{1}$$

which is considered to be embedded in a conformally flat five-dimensional space-time. The form of the energy momentum tensor for a combination of dark matter and dark energy is

$$T_\mu{}^\nu = (\rho_m + p_m + \rho_\Lambda + p_\Lambda)\,u_\mu u^\nu - (p_m + p_\Lambda)\,\delta_\mu{}^\nu. \tag{2}$$

Generally a barotropic equation of state $p_m = \omega_m \rho_m$ is chosen for the matter part on the brane having energy density ρ_m and pressure p_m and a variable cosmological constant is chosen as the component of dark energy having energy density ρ_Λ and pressure $p_\Lambda(= -\rho_\Lambda)$. The four-velocity u_μ in the comoving coordinate system takes the form $u_\mu = \delta_\mu{}^t$. Thus the effective Einstein equations on the brane are [49]

$$\frac{\dot{a}^2}{a^2} = H^2 = \frac{8\pi}{3}\left[\rho_T\left(1 + \frac{\rho_T}{2\lambda}\right)\right] \tag{3}$$

and

$$\begin{aligned}\frac{\ddot{a}}{a} &= \dot{H} + H^2 \\ &= -\frac{4\pi}{3}\left[\rho_T\left(1 + \frac{2\rho_T}{\lambda}\right) + 3p_\Lambda\left(1 + \frac{\rho_T}{\lambda}\right)\right],\end{aligned} \tag{4}$$

where $\rho_T = \rho_m + \rho_\Lambda$ is the total energy density, $p_T = p_m + p_\Lambda$ is the total pressure, λ is the positive brane tension, the Hubble parameter is given by $H(t) = \dot{a}/a$, and a(t) is the scale factor in the flat FRW brane model.

Equations (3) and (4) can be explicitly written as

$$\frac{\dot{a}^2}{a^2} = \frac{8\pi}{3}\rho_{m\text{eff}} + \frac{1}{3}\Lambda(t)_{\text{eff}} \tag{5}$$

and

$$\begin{aligned}\frac{\ddot{a}}{a} = -\frac{4\pi}{3}&\Big[(\rho_m + 3p_m) \\ &+ \frac{1}{\lambda}\left(2\rho_m + 3p_m + 2\rho_m\rho_\Lambda + 3\rho_m p_m\right)\Big] + \frac{1}{3}\Lambda(t)_{\text{eff}},\end{aligned} \tag{6}$$

where $\rho_{m\text{eff}} = \rho_m(1 + \rho_m/2\lambda)$ and $\Lambda(t)_{\text{eff}} = 8\pi[\rho_\Lambda(1 + \rho_\Lambda/2\lambda + \rho_m/\lambda)]$.

For the present brane model with matter field given by (2), the explicit form of the energy momentum conservation relation ($T_\mu{}^\nu{}_{;\nu} = 0$) is

$$\dot{\rho_m} + 3\frac{\dot{a}}{a}(\rho_m + p_m) = -\dot{\rho_\Lambda} \simeq -\frac{\Lambda(\dot{t})_{\text{eff}}}{8\pi}. \tag{7}$$

Instead of a variable ρ_Λ, if we choose a constant ρ_Λ, then the field equations together with the continuity equation will be identical to the corresponding equations in the standard ΛCDM model.

3. Field Equations Derived from the Holographic Equipartition Law

For a pure de Sitter universe with Hubble parameter H, the holographic principle can be described by the relation [38]

$$N_{\text{sur}} = N_{\text{bulk}}, \tag{8}$$

where N_{sur} denotes the number of the degrees of freedom on the holographic screen with Hubble radius $r_H = 1/H$:

$$N_{\text{sur}} = \frac{4\pi}{H^2} = 4S_H. \tag{9}$$

Here S_H is the entropy on the Hubble horizon. The number of degrees of freedom in bulk is said to obey the equipartition law of energy

$$N_{\text{bulk}} = \frac{2|E|}{T}. \tag{10}$$

In the context of brane world models, the induced active gravitational mass on the brane $|M| = |E|$ has the form [50]

$$\begin{aligned}|M| &= \frac{4\pi}{3H^3}\left|\left(\rho_T + 3p_T + \frac{3\rho_T p_T}{\lambda} + \frac{2\rho_T^2}{\lambda}\right)\right| = -\epsilon \\ &\cdot \frac{4\pi}{3H^3}\left\{\left[(\rho_m + 3p_m) + \frac{1}{\lambda}(2\rho_m + 3p_m + 2\rho_m\rho_\Lambda + 3\rho_m p_m)\right] + \frac{1}{4\pi} \cdot \Lambda(t)_{\text{eff}}\right\}\end{aligned} \tag{11}$$

for the choice of the matter field (2). The parameter ϵ is defined later. Using the above expression of $|M|$ and the horizon temperature $T = H/2\pi$, we get the expression of N_{bulk} as

$$\begin{aligned}N_{\text{bulk}} &= -\epsilon\frac{16\pi^2}{3H^4}\left\{\left[(\rho_m + 3p_m) + \frac{1}{\lambda}(2\rho_m + 3p_m + 2\rho_m\rho_\Lambda + 3\rho_m p_m)\right] + \frac{1}{4\pi} \cdot \Lambda(t)_{\text{eff}}\right\}.\end{aligned} \tag{12}$$

Since the real world is not purely but asymptotically de Sitter, therefore one may propose that the expansion rate of the cosmic volume is related to the difference of these two degrees of freedom. The analytical form of this is described as [38]

$$\frac{dV}{dt} = l_p^2\left(N_{\text{sur}} - \epsilon N_{\text{bulk}}\right). \tag{13}$$

Equation (13) is known as the holographic equipartition law. Here $V = 4\pi/3H^3$ is the cosmic volume and the parameter ϵ is defined by [38, 51]

$$\epsilon \equiv \begin{cases} +1, & \text{when } \left[(\rho_m + 3p_m) + \frac{1}{\lambda}(3p_m + 2\rho_m + 3\rho_m p_m)\right] < 0 \\ -1, & \text{when } \left[(\rho_m + 3p_m) + \frac{1}{\lambda}(3p_m + 2\rho_m + 3\rho_m p_m)\right] > 0. \end{cases} \tag{14}$$

Here, we have considered that there is no dark energy component in the 3-brane, i.e., $\Lambda(t)_{\text{eff}} \sim \rho_\Lambda = 0$. In this case $[(\rho_m + 3p_m) + (1/\lambda)(3p_m + 2\rho_m + 3\rho_m p_m)] < 0$ for the acceleration of the universe. Hence from (11) and (14), the definition of the parameter ϵ is well justified.

One can write from (9), (12), and (13)

$$-4\pi\frac{\dot{H}}{H^4} = \left\{4S_H + \frac{16\pi^2}{3H^4}\left[(\rho_m + 3p_m) + \frac{1}{\lambda}(3p_m + 2\rho_m + 3\rho_m p_m)\right]\right\} \tag{15}$$

or equivalently

$$\dot{H} = -\frac{4\pi}{3}\left[(\rho_m + 3p_m) + \frac{1}{\lambda}(3p_m + 2\rho_m + 3\rho_m p_m)\right] - \frac{H^4 S_H}{\pi}. \tag{16}$$

The acceleration equation is therefore read as

$$\frac{\ddot{a}}{a} = -\frac{4\pi}{3}\left[(\rho_m + 3p_m) + \frac{1}{\lambda}(3p_m + 2\rho_m + 3\rho_m p_m)\right] + H^2\left(1 - \frac{H^2 S_H}{\pi}\right). \tag{17}$$

Thus we have derived the acceleration equation from the holographic equipartition law and an extra driving term appears on the right side of the equation. This term vanishes when one chooses the Bekenstein-Hawking entropy for S_H. The acceleration equation will then be

$$\frac{\ddot{a}}{a} = -\frac{4\pi}{3}\left[(\rho_m + 3p_m) + \frac{1}{\lambda}(3p_m + 2\rho_m + 3\rho_m p_m)\right] \tag{18}$$

which is identical to (6) with $\Lambda(t)_{\text{eff}} \sim \rho_\Lambda = 0$. Hence in this case, the field equation and the corresponding energy conservation equation become

$$\frac{\dot{a}^2}{a^2} = \frac{8\pi G}{3}\rho_{m\text{eff}} \tag{19}$$

and

$$\dot{\rho}_m + 3\frac{\dot{a}}{a}(\rho_m + p_m) = 0. \tag{20}$$

However, any other form of S_H will not result in the above set of equations and the cosmological implications will definitely be something else.

4. GUP Corrected Entropy on the Horizon

In recent years, a number of studies in general relativity and modified gravity theories came to surface due to the discovery of different aspects of black hole solutions. Black holes are thermodynamic objects with well-defined entropy. Generally, the Bekenstein-Hawking entropy [52–54]

$$S_{BH} = \frac{A}{4l_p^2} \tag{21}$$

is chosen for the same. Here A is the surface area of the sphere with the Hubble horizon $r_H = 1/H$ and $l_p = \sqrt{G\hbar/c^3} \simeq 10^{-35}$m is the Planck length. With $A = 4\pi r_H^2$, we can write

$$S_{BH} = \frac{\pi r_H^2}{l_p^2}. \tag{22}$$

Instead of a flat universe, if we choose a nonflat universe, then the apparent horizon $r_A = 1/\sqrt{H^2 + k/a^2}$ should be used as the horizon radius instead of the Hubble horizon. Corrections in this entropy formula were needed to accommodate the newly emerging physics from string theory and loop quantum gravity (LQG). Several of these theories predicted quantum corrections to the entropy-area relation [55–64]

$$S_{QG} = \frac{A}{4l_p^2} + C_0 \ln\left(\frac{A}{4l_p^2}\right) + \sum_{n=1}^{\infty} C_n \left(\frac{A}{4l_p^2}\right)^{-n}, \tag{23}$$

where the coefficients C_n are model dependent parameters. Recent rigorous calculations from LQG have fixed the value of $C_0 = -1/2$ [59]. On the other hand, Mead [65] first pointed out that the Heisenberg uncertainty principle could be affected by gravity. Later, a considerable amount of effort had been put to the modified commutation relations between position and momenta commonly known as the Generalized Uncertainty Principle (GUP) from different perspectives of quantum aspects of gravity. All these studies eventually led to the GUP corrected entropy form [66–69]

$$S_{GUP} = \frac{A}{4l_p^2} + \frac{\sqrt{\pi}\alpha_0}{4}\sqrt{\frac{A}{4l_p^2}} - \frac{\pi\alpha_0^2}{64}\ln\left(\frac{A}{4l_p^2}\right) + O\left(l_p^3\right). \tag{24}$$

Here α_0 is a dimensionless constant prescribed in the deformed commutation relations [70]. The leading contribution of this new entropy function lies in its second term $\sim \sqrt{Area}$. This is an extra term to the already existing logarithmic correction to entropy derived from the quantum gravity effects. Due to the difference in the leading order correction term, the underlying nature of such model needs to be investigated in four-dimensional Einstein's gravity as well as in higher-dimensional modified theories of gravity. Based on many similarities between the black hole horizon and cosmological horizon and on the assumption that the universe should be described by the quantum language, we employ this newly obtained GUP corrected entropy of the black hole horizon as the entropy of the cosmological horizon in the natural unit system

$$S_Q = \frac{A}{4} + \frac{\sqrt{\pi}\alpha_0}{4}\sqrt{\frac{A}{4}} - \frac{\pi\alpha_0^2}{64}\ln\left(\frac{A}{4}\right) \tag{25}$$

which on further calculation becomes

$$S_Q = S_{BH}\left[1 + \frac{\alpha_0 H}{4} - \frac{\alpha_0^2 H^2}{64}\ln\left(\frac{\pi}{H^2}\right)\right]. \tag{26}$$

Here $S_{BH} = \pi/H^2$. The novelty of this expression is that when $\alpha_0 = 0$, then S_Q becomes S_{BH}.

4.1. Consequences of GUP Corrected Entropy into the Holographic Equipartition Law. Here, we apply the GUP corrected entropy function S_Q into the holographic equipartition law; i.e., we consider that

$$S_H = S_Q = S_{BH}\left[1 + \frac{\alpha_0 H}{4} - \frac{\alpha_0^2 H^2}{64}\ln\left(\frac{\pi}{H^2}\right)\right]. \tag{27}$$

Substituting this new form of S_H in (17), we have

$$\frac{\ddot{a}}{a} = -\frac{4\pi}{3}\left[\left(\rho_m + 3p_m\right) + \frac{1}{\lambda}\left(3p_m + 2\rho_m + 3\rho_m p_m\right)\right] + \left[\frac{\alpha_0^2 H^4}{64}\ln\left(\frac{\pi}{H^2}\right) - \frac{\alpha_0 H^3}{4}\right]. \tag{28}$$

The extra driving term appearing on the right side of the equation needs to be positive for the current cosmic acceleration.

In the brane world gravity, the field equations together with the continuity equation then become

$$\frac{\dot{a}^2}{a^2} = \frac{8\pi}{3}\rho_{m\text{eff}} + f_\alpha(H) \tag{29}$$

$$\frac{\ddot{a}}{a} = -\frac{4\pi}{3}\left[\left(\rho_m + 3p_m\right) + \frac{1}{\lambda}\left(2\rho_m + 3p_m + 3\rho_m p_m\right)\right] + f_\alpha(H) \tag{30}$$

and

$$\dot{\rho_m} + 3\frac{\dot{a}}{a}\left(\rho_m + p_m\right) = -\frac{3\dot{f_\alpha}(H)}{8\pi}, \tag{31}$$

where the extra term $f_\alpha(H)$ is given by

$$f_\alpha(H) = \frac{\alpha_0^2 H^4}{64}\ln\left(\frac{\pi}{H^2}\right) - \frac{\alpha_0 H^3}{4}. \tag{32}$$

Let us now discuss the evolution of this extra driving term from entropy function (27) and acceleration equation (30). Equation (30) is the final equation incorporating all three corrections. As S_{BH} is positive, hence the following restriction is to be obeyed by the parameters for S_Q to be positive:

$$\left[\frac{\alpha_0 H}{16}\ln\left(\frac{\pi}{H^2}\right) - 1\right] < \frac{4}{\alpha_0 H}. \tag{33}$$

Again for the current cosmic acceleration

$$0 < f_\alpha(H) = \frac{\alpha_0 H^3}{4}\left[\frac{\alpha_0 H}{16}\ln\left(\frac{\pi}{H^2}\right) - 1\right]. \tag{34}$$

Hence it is clear from (33) and (34) that

$$f_\alpha(H) < H^2. \tag{35}$$

A similar constraint can be derived from the study of the Generalized Second Law of Thermodynamics (GSLT) as presented in the following subsection.

4.2. Generalized Second Law of Thermodynamics (GSLT). Here we shall discuss the GSLT in the current prescription. Considering S_T as the total entropy of the universe, one can write

$$\dot{S_T} = \dot{S_Q} + \dot{S_I}, \quad (36)$$

where S_I is the entropy of matter inside the horizon. From (27), we can write

$$\dot{S_Q} = \dot{S_{BH}}\left[1 - \left(\frac{\alpha_0{}^2H^2}{64} - \frac{\alpha_0 H}{8}\right)\right], \quad (37)$$

where

$$\dot{S_{BH}} = \frac{d}{dt}\left(\frac{\pi}{H^2}\right) = -\frac{2\pi\dot{H}}{H^3}. \quad (38)$$

Since $\dot{S_{BH}} > 0$, to satisfy $\dot{S_Q} > 0$, the following restriction needs to be obeyed:

$$\left(\frac{\alpha_0{}^2H^2}{64} - \frac{\alpha_0 H}{8}\right) < 1. \quad (39)$$

In order to obtain the rate of change of entropy of the matter inside the horizon, we consider the Gibbs equation [71, 72]

$$T_I dS_I = dE_I + p_T dV, \quad (40)$$

where V is the volume inside the horizon and $E_I = \rho_T dV$ stands for the internal energy. The temperature of the matter T_I inside the horizon has been assumed to be equivalent to the horizon temperature $T = H/2\pi$. In absence of any dark energy component, this equation takes the form

$$\begin{aligned} T\dot{S_I} &= \left[\dot{\rho_m} + 3\frac{\dot{a}}{a}\left(\rho_m + p_m\right)\right]V \\ &= -\frac{3f_\alpha(\dot{H})V}{8\pi}, \end{aligned} \quad (41)$$

where we have used modified continuity equation (31) to obtain the expression of $\dot{S_I}$. Taking time derivative of (32) and using the expression of horizon temperature T, one can yield

$$\dot{S_I} = \dot{S_{BH}}\left[\frac{\alpha_0{}^2H^2}{32}\ln\left(\frac{\pi}{H^2}\right) - \frac{3\alpha_0 H}{8} - \frac{\alpha_0{}^2H^2}{64}\right]. \quad (42)$$

Thus from (37) and (42), the rate of change of total entropy of the universe becomes

$$\begin{aligned} &\dot{S_T} \\ &= \dot{S_{BH}}\left[1 - \left\{\frac{\alpha_0 H}{4} + \frac{\alpha_0{}^2H^2}{32} - \frac{\alpha_0{}^2H^2}{32}\ln\left(\frac{\pi}{H^2}\right)\right\}\right]. \end{aligned} \quad (43)$$

Again as $\dot{S_{BH}} > 0$, to satisfy $\dot{S_T} > 0$, the following condition must be attained:

$$\left[\frac{\alpha_0 H}{4} + \frac{\alpha_0{}^2H^2}{32} - \frac{\alpha_0{}^2H^2}{32}\ln\left(\frac{\pi}{H^2}\right)\right] < 1. \quad (44)$$

From (39) and (44), one can easily derive

$$f_\alpha(H) > \frac{H^2}{2}. \quad (45)$$

Thus, we attain a very interesting result from (35) and (45)

$$\frac{H^2}{2} < f_\alpha(H) < H^2. \quad (46)$$

Following the arguments of [47] as for the observational constraint $\dot{H} < 0$ [73], one can assume H_0 to be the minimum value for H and arrive at a stricter constraint

$$\begin{aligned} &\frac{H_0{}^2}{2} < f_\alpha(H) < H_0{}^2 \\ &\Longrightarrow O\left(f_\alpha(H)\right) \lesssim O\left(H_0{}^2\right). \end{aligned} \quad (47)$$

This result is analogous to the one presented in both [46, 47], though, in the former study, a mathematical condition was imposed to obtain similar restriction while, in the latter, it evolved through the validity of the GSLT. Further probing into the standard ΛCDM model, we obtain $\Lambda = 3H_0{}^2\Omega_\Lambda$. This implies that

$$O\left(\frac{\Lambda}{3}\right) = O\left(H_0{}^2\Omega_\Lambda\right). \quad (48)$$

As from the Planck (2015) results [14], $\Omega_\Lambda = 0.692$, which is of order one. This yields to

$$O\left(\frac{\Lambda}{3}\right) \simeq O\left(H_0{}^2\right). \quad (49)$$

Thus the order of the extra driving term in the acceleration equation becomes equivalent to the order of the cosmological constant term. This result however seems to be model-independent as the positive brane tension did not play any significant role in deriving the analogy.

5. Discussions

In the present work, our aim was to study the cosmic evolution in the brane world gravity with the help of the holographic equipartition law. We have applied the quantum corrected form of the entropy function derived from the Generalized Uncertainty Principle in the holographic equipartition law to derive the modified cosmological equations in a homogeneous, isotropic, and spatially flat 3-brane embedded in a five-dimensional bulk. The novelty of the study lies in the $\sqrt{Area}$ form of the entropy function. It was noticed that the acceleration equation contains an extra driving term of an order consistent with the order of the cosmological constant. A similar constraint was obtained assuming the validity of GSLT. The study remained to be model-independent and the positive brane tension did not play any crucial role for the attained result. However, it should be understood that our aim was not to verify the GSLT in the modified gravity theory. Rather we were interested in the evolution of the extra driving term appearing in the acceleration equation due to imposition of the holographic equipartition law for a specific GUP corrected entropy function whose leading order term is different from the existing forms. This may shed new light on the studies of the cosmological constant problem in modified gravity theories.

Conflicts of Interest

The author declares that they have no conflicts of interest.

Acknowledgments

The author is thankful to IUCAA, Pune, for their warm hospitality and excellent research facilities where part of the work has been done during a visit under the Associateship Programme.

References

[1] L. Randall and R. Sundrum, "Large mass hierarchy from a small extra dimension," *Physical Review Letters*, vol. 83, no. 17, pp. 3370–3373, 1999.

[2] L. Randall and R. Sundrum, "An alternative to compactification," *Physical Review Letters*, vol. 83, no. 23, pp. 4690–4693, 1999.

[3] P. Binétruy, C. Deffayet, U. Ellwanger, and D. Langlois, "Brane cosmological evolution in a bulk with cosmological constant," *Physics Letters B*, vol. 477, pp. 285–291, 2000.

[4] P. Binétruy, C. Deffayet, and D. Langlois, "Non-conventional cosmology from a brane universe," *Nuclear Physics, B*, vol. 565, no. 1-2, pp. 269–287, 2000.

[5] J. Ponce De Leon, "Variation of G, Λ(4) and Vacuum Energy from Brane-World Models," *Modern Physics Letters A*, vol. 17, no. 37, pp. 2425–2441, 2002.

[6] J. Ponce De Leon, "Equivalence between space-time-matter and brane-world theories," *Modern Physics Letters A*, vol. 16, no. 35, pp. 2291–2303, 2001.

[7] A. Lukas, B. A. Ovrut, K. S. Stelle, and D. Waldram, "Universe as a domain wall," *Physical Review D: Particles, Fields, Gravitation and Cosmology*, vol. 59, no. 8, 086001, 9 pages, 1999.

[8] A. Lukas, B. A. Ovrut, and D. Waldram, "Cosmological solutions of Hořava-Witten theory," *Physical Review D: Particles, Fields, Gravitation and Cosmology*, vol. 60, no. 8, Article ID 086001, 1999.

[9] F. Zwicky, "Die Rotverschiebung von extragalaktischen Nebeln," *Helvetica Physica Acta*, vol. 6, pp. 110–127, 1933.

[10] F. Zwiecky, "Nebulae as Gravitational Lenses," *Physical Review Journals*, vol. 51, p. 290, 1937.

[11] S. Perlmutter, G. Aldering, M. Della Valle et al., "Discovery of a supernova explosion at half the age of the Universe," *Nature*, vol. 391, no. 6662, pp. 51–54, 1998.

[12] A. G. Riess, A. V. Filippenko, and P. Challis, "Observational evidence from supernovae for an accelerating universe and a cosmological constant," *The Astronomical Journal*, vol. 116, p. 1009, 1998.

[13] A. G. Riess, L.-G. Strolger, and S. Casertano, "New Hubble Space Telescope Discoveries of Type Ia Supernovae at $z \geq 1$: Narrowing Constraints on the Early Behavior of Dark Energy," *The Astrophysical Journal*, vol. 659, no. 1, p. 98, 2007.

[14] P. A. R. Ade et al., "Planck 2015 results - XIII. Cosmological parameters," *Astronomy & Astrophysics (A&A)*, vol. 594, Article ID A13, 63 pages, 2016.

[15] I. L. Shapiro and J. Solà, "The scaling evolution of the cosmological constant," *Journal of High Energy Physics*, no. 02, Article ID 006, 2002.

[16] S. Basilakos, M. Plionis, and J. Solà, "Hubble expansion and structure formation in time varying vacuum models," *Physical Review D: Particles, Fields, Gravitation and Cosmology*, vol. 80, no. 8, Article ID 083511, 2009.

[17] J. P. Mimoso and D. Pavón, "Entropy evolution of universes with initial and final de Sitter eras," *Physical Review D: Particles, Fields, Gravitation and Cosmology*, vol. 87, Article ID 047302, 2013.

[18] M. H. P. M. van Putten, "Accelerated expansion from cosmological holography," *Monthly Notices of the Royal Astronomical Society: Letters*, vol. 450, no. 1, pp. L48–L51, 2015.

[19] A. Gómez-Valent, E. Karimkhani, and J. Solà, "Background history and cosmic perturbations for a general system of self-conserved dynamical dark energy and matter," *Journal of Cosmology and Astroparticle Physics*, vol. 12, Article ID 048, 2015.

[20] J. A. Lima, S. Basilakos, and J. Solà, "Nonsingular decaying vacuum cosmology and entropy production," *General Relativity and Gravitation*, vol. 47, no. 4, article 40, 2015.

[21] G. 't Hooft, "Dimensional Reduction in Quantum Gravity," *Salamfest*, pp. 0284–296, 1993.

[22] L. Susskind, "A predictive Yukawa unified SO(10) model: higgs and sparticle masses," *Journal of Mathematical Physics*, vol. 36, no. 7, article 139, pp. 6377–6396, 1995.

[23] R. Bousso, "The holographic principle," *Reviews of Modern Physics*, vol. 74, no. 3, pp. 825–874, 2002.

[24] T. Padmanabhan, "Equipartition of energy in the horizon degrees of freedom and the emergence of gravity," *Modern Physics Letters A*, vol. 25, no. 14, pp. 1129–1136, 2010.

[25] T. Padmanabhan, "Thermodynamical Aspects of Gravity: New insights," *Reports on Progress in Physics*, vol. 73, Article ID 046901, 2010.

[26] E. Verlinde, "On the origin of gravity and the laws of Newton," *Journal of High Energy Physics*, vol. 2011, 29 pages, 2011.

[27] A. Sheykhi, "Entropic corrections to Friedmann equations," *Physical Review D: Particles, Fields, Gravitation and Cosmology*, vol. 81, no. 10, Article ID 104011, 2010.

[28] A. Sheykhi and S. H. Hendi, "Power-law entropic corrections to Newton's law and Friedmann equations," *Physical Review D: Particles, Fields, Gravitation and Cosmology*, vol. 84, no. 4, 2011.

[29] S. Mitra, S. Saha, and S. Chakraborty, "Modified Hawking temperature and entropic force: a prescription in FRW model," *Modern Physics Letters A*, vol. 30, no. 13, 1550058 pages, 2015.

[30] D. A. Easson, P. H. Frampton, and G. F. Smoot, "Entropic accelerating universe," *Physics Letters B*, vol. 696, no. 3, pp. 273–277, 2011.

[31] D. A. Easson, P. H. Frampton, and G. F. Smoot, "Entropic inflation," *International Journal of Modern Physics A*, vol. 27, Article ID 1250066, 2012.

[32] Y. F. Cai, J. Liu, and H. Li, "Entropic cosmology: a unified model of inflation and late-time acceleration," *Physics Letters B*, vol. 690, no. 3, pp. 213–219, 2010.

[33] T. S. Koivisto, D. F. Motaá, and M. Zumalacárregui, "Constraining entropic cosmology," *Journal of Cosmology and Astroparticle Physics*, vol. 02, p. 027, 2011.

[34] N. Komatsu and S. Kimura, "Non-adiabatic-like accelerated expansion of the late universe in entropic cosmology," *Physical Review D: Particles, Fields, Gravitation and Cosmology*, vol. 87, Article ID 043531, 2013.

[35] N. Komatsu and S. Kimura, "Evolution of the universe in entropic cosmologies via different formulations," *Physical Review D: Particles, Fields, Gravitation and Cosmology*, vol. 89, Article ID 123501, 2014.

[36] M. P. Dabrowski and H. Gohar, "Abolishing the maximum tension principle," *Physics Letters B*, vol. 748, pp. 428–431, 2015.

[37] C. Tsallis and L. J. L. Cirto, "Black hole thermodynamical entropy," *The European Physical Journal C*, vol. 73, no. 7, p. 2487, 2013.

[38] T. Padmanabhan, "Emergence and Expansion of Cosmic Space as due to the Quest for Holographic Equipartition," https://arxiv.org/abs/1206.4916.

[39] R. G. Cai, "Emergence of Space and Spacetime Dynamics of Friedmann-Robertson-Walker Universe," https://arxiv.org/abs/1207.0622.

[40] K. Yang, Y.-X. Liu, and Y.-Q. Wang, "Emergence of Cosmic Space and the Generalized Holographic Equipartition," *Physical Review D*, vol. 86, Article ID 104013, 8 pages, 2012.

[41] Y. Ling and W.-J. Pan, "Note on the emergence of cosmic space in modified gravities," *Physical Review D*, vol. 88, Article ID 043518, 2013.

[42] A. F. Ali, "Emergence of cosmic space and minimal length in quantum gravity," *Physics Letters B*, vol. 732, pp. 335–342, 2014.

[43] T. Padmanabhan, "Emergent gravity paradigm: recent progress," *Modern Physics Letters A*, vol. 30, no. 3-4, 1540007, 21 pages, 2015.

[44] X.-X. Zeng and Y. Chen, "Quantum gravity corrections to fermions' tunnelling radiation in the Taub-NUT spacetime," *General Relativity and Gravitation*, vol. 47, no. 4, article 47, 2015.

[45] H. Moradpour and A. Sheykhi, "From the Komar mass and entropic force scenarios to the Einstein field equations on the Ads brane," *International Journal of Theoretical Physics*, vol. 55, no. 2, pp. 1145–1155, 2016.

[46] N. Komatsu, "Cosmological model from the holographic equipartition law with a modified Rényi entropy," *The European Physical Journal C*, vol. 77, no. 229, 2017.

[47] N. Komatsu, "Thermodynamic constraints on a varying cosmological-constant-like term from the holographic equipartition law with a power-law corrected entropy," *Physical Review D*, vol. 96, Article ID 103507, 2017.

[48] P. Vergueno and E. C. Vagenas, "Semiclassical corrections to black hole entropy and the generalized uncertainty principle," *Physics Letters B*, vol. 742, pp. 15–18, 2015.

[49] T. Bandyopadhyay, A. Baveja, and S. Chakraborty, "Gravitational collapse in the context of brane world scenario with decaying vacuum energy," *Modern Physics Letters A*, vol. 23, no. 9, pp. 685–693, 2008.

[50] Y. Ling and J. P. Wu, "A note on entropic force and brane cosmology," *Journal of Cosmology and Astroparticle Physics*, vol. 017, p. 1008, 2010.

[51] T. Padmanabhan, "Emergent perspective of gravity and dark energy," *Research in Astronomy and Astrophysics*, vol. 12, p. 891, 2012.

[52] J. D. Bekenstein, "Black holes and entropy," *Physical Review D: Particles, Fields, Gravitation and Cosmology*, vol. 7, pp. 2333–2346, 1973.

[53] J. D. Bekenstein, "Generalized second law of thermodynamics in black-hole physics," *Physical Review D: Particles, Fields, Gravitation and Cosmology*, p. 3292, 1974.

[54] J. D. Bekenstein, "Statistical black hole thermodynamics," *Physical Review D: Particles, Fields, Gravitation and Cosmology*, vol. 12, no. 10, pp. 3077–3085, 1975.

[55] R. K. Kaul and P. Majumdar, "Logarithmic correction to the Bekenstein-Hawking entropy," *Physical Review Letters*, vol. 84, no. 23, pp. 5255–5257, 2000.

[56] A. Ghosh and P. Mitra, "An improved estimate of black hole entropy in the quantum geometry approach," *Physics Letters B*, vol. 616, no. 1-2, pp. 114–117, 2005.

[57] A. J. Medved and E. C. Vagenas, "When conceptual worlds collide: the generalized uncertainty principle and the Bekenstein-Hawking entropy," *Physical Review D: Particles, Fields, Gravitation and Cosmology*, vol. 70, no. 12, Article ID 124021, 5 pages, 2004.

[58] G. Amelino-Camelia, M. Arzano, and A. Procaccini, "Severe constraints on the loop-quantum-gravity energy-momentum dispersion relation from the black-hole area-entropy law," *Physical Review D: Particles, Fields, Gravitation and Cosmology*, vol. 70, no. 10, Article ID 107501, 2004.

[59] K. A. Meissner, "Black-hole entropy in loop quantum gravity," *Classical and Quantum Gravity*, vol. 21, no. 22, pp. 5245–5251, 2004.

[60] S. Das, P. Majumdar, and R. K. Bhaduri, "General logarithmic corrections to black-hole entropy," *Classical and Quantum Gravity*, vol. 19, no. 9, pp. 2355–2367, 2002.

[61] Y. S. Myung, "Logarithmic corrections to three-dimensional black holes and de Sitter spaces," *Physics Letters B*, vol. 579, no. 1-2, pp. 205–210, 2004.

[62] M. Domagala and J. Lewandowski, "Black-hole entropy from quantum geometry," *Classical and Quantum Gravity*, vol. 21, no. 22, pp. 5233–5243, 2004.

[63] A. Chatterjee and P. Majumdar, "Universal canonical black hole entropy," *Physical Review Letters*, vol. 92, no. 14, Article ID 141301, 2004.

[64] M. M. Akbar and S. Das, "Entropy corrections for Schwarzschild and Reissner Nordström black holes," *Classical and Quantum Gravity*, vol. 21, no. 6, pp. 1383–1392, 2004.

[65] C. A. Mead, "Possible Connection Between Gravitation and Fundamental Length," *Physical Review D*, vol. 135, Article ID B849, 1964.

[66] B. Majumder, "Black hole entropy and the modified uncertainty principle: a heuristic analysis," *Physics Letters B*, vol. 703, no. 4, pp. 402–405, 2011.

[67] R. J. Adler, P. Chen, and D. I. Santiago, "The generalized uncertainty principle and black hole remnants," *General Relativity and Gravitation*, vol. 33, no. 12, pp. 2101–2108, 2001.

[68] G. Amelino-Camelia, M. Arzano, Y. Ling, and G. Mandanici, "Black-hole thermodynamics with modified dispersion relations and generalized uncertainty principles," *Classical and Quantum Gravity*, vol. 23, no. 7, pp. 2585–2606, 2006.

[69] B. Majumder, "Black hole entropy with minimal length in tunneling formalism," *General Relativity and Gravitation*, vol. 45, no. 11, pp. 2403–2414, 2013.

[70] S. Das and E. C. Vagenas, "Universality of quantum gravity corrections," *Physical Review Letters*, vol. 101, Article ID 221301, 4 pages, 2008.

[71] B. Wang, Y. G. Gong, and E. Abdalla, "Thermodynamics of an accelerated expanding universe," *Physical Review D: Particles, Fields, Gravitation and Cosmology*, vol. 74, Article ID 083520, 2006.

[72] G. Izquierdo and D. Pavon, "Dark energy and the generalized second law," *Physics Letters B*, vol. 633, no. 4-5, pp. 420–426, 2006.

[73] P. B. Krishna and T. Mathew, "Holographic equipartition and the maximization of entropy," *Physical Review D*, vol. 96, Article ID 063513, 2017.

3

Ghost Dark Energy in the DGP Braneworld

M. Abdollahi Zadeh[1] and A. Sheykhi[1,2]

[1] *Physics Department and Biruni Observatory, College of Sciences, Shiraz University, Shiraz 71454, Iran*
[2] *Research Institute for Astronomy and Astrophysics of Maragha (RIAAM), P.O. Box 55134-441, Maragha, Iran*

Correspondence should be addressed to A. Sheykhi; asheykhi@shirazu.ac.ir

Academic Editor: Ricardo G. Felipe

We investigate the ghost model of dark energy in the framework of DGP braneworld. We explore the cosmological consequences of this model by determining the equation of state parameter, ω_D, the deceleration, and the density parameters. We also examine the stability of this model by studying the squared of the sound speed in the presence/absence of interaction term between dark energy and dark matter. We find out that in the absence of interaction between two dark sectors of the universe we have $\omega_D \longrightarrow -1$ in the late time, while in the presence of interaction ω_D can cross the phantom line -1. In both cases the squared of sound speed v_s^2 does not show any signal of stability. We also determine the statefinder diagnosis of this model as well as the $\omega_D - \omega_D'$ plane and compare the results with the ΛCDM model. We find that $\omega_D - \omega_D'$ plane meets the freezing region in the absence of interaction between two dark sectors, while it meets both the thawing and the freezing regions in the interacting case.

1. Introduction

The current acceleration of the universe expansion is strongly confirmed by the type Ia supernova observations [1] and also supported by the astrophysical data from WMAP [2]. A component which is responsible for this acceleration is called dark energy (DE) with negative pressure, which can overcome the gravity force between the galaxies and push them to accelerate, though its nature and origin is still an open question in the modern cosmology. There are two approaches for explanation of the cosmic acceleration. The first one is the modified gravity models such as $f(R)$ gravity [3] and scalar-tensor theories [4], and the second approach is the idea of the existence of a strange type of energy whose gravity is repulsive such as the cosmological constant Λ [5] and the dynamical DE models [6, 7]. Against the cosmological constant Λ which has constant equation of state (EoS) parameter, $\omega_D = -1$, further observations detect a small variation in the EoS parameter of DE in favor of a dynamical DE with $\omega_D > -1$ in the past and even $\omega_D < -1$ in the late time [8].

An interesting model for probing the dynamical DE model is the ghost dark energy (GDE) model proposed in [9]. The advantages of this model are that it does not introduce any new degree of freedom in contrast to the most DE models which explain the accelerated expansion by introducing new degree(s) of freedom or by modifying the underlying theory of gravity. This is important because, with introducing new degrees of freedom, one needs to investigate the nature and new consequences in the universe so it seems to be impressive and economic if we can explain DE puzzle by using currently known fluids and fields of nature. Actually, GDE model which is based on the Veneziano ghost in Quantum Chromodynamics (QCD) can act as the source of DE [10] and its existence are required for resolution of the U(1) problem in QCD [11]. Indeed, the ghosts are decoupled from the physical states and make no contribution in flat Minkowski space, but it produces a small vacuum energy density in a dynamic background or a curved spacetime proportional to $\Lambda_{QCD}^3 H$, where H is the Hubble parameter and Λ_{QCD} is QCD mass scale of order a $100MeV$ [12]. Different features of GDE have been studied in ample details [13].

Independent of the DE puzzle, for explanation of the cosmic acceleration, special attention is also paid to extra dimensional theories, in which our universe is realized as a 3-brane embedded in a higher dimensional spacetime. Based on the braneworld model, all the particle fields in the standard model are confined to a four-dimensional brane, while gravity is free to propagate in all dimension. One of the original models of braneworld is introduced by Dvali-Gabadadze-Porrati (DGP) [14], which describes our universe as a 4D brane embedded in a 5D Minkowskian bulk with infinite size. In this model the recovery of the usual gravitational laws on the brane is obtained by adding an Einstein-Hilbert term to the action of the brane computed with the brane intrinsic curvature. It is a well-known that the DGP model has two branches of solutions. The self-accelerating branch of DGP model can explain the late time cosmic speed-up without recourse to DE or other components of energy [15, 16]. However, the self-accelerating DGP branch has ghost instabilities and it cannot realize phantom divide crossing by itself. To realize phantom divide crossing it is necessary to add at least a component of energy on the brane. On the other hand, the normal DGP branch cannot explain acceleration but it has the potential to realize a phantom-like phase by dynamical screening on the brane. Adding a DE component to the normal branch solution brings new facilities to explain late time acceleration and also better matching with observations. These are the motivations to add DE to this braneworld setup [17, 18]. In this work we would like to investigate the GDE model in the framework of the DGP braneworld. This study is of great importance, since we can incorporate and disclose the effects of the extra dimension on the evolution of the cosmological parameters on the brane when the DE source is in the form of GDE.

This paper is organized as follows. In Section 2 we formulate the GDE model in the context of the DGP braneworld. We also consider both interacting and noninteracting cases and explore various cosmological parameters as well as cosmological planes. Besides the discussion of instability analysis, we study the $\omega_D - \omega_D'$ plane and properties of statefinder parameters. We finish with closing remarks in Section 3.

2. The GDE in the DGP Model

In the DGP cosmology, a homogeneous, spatially flat, and isotropic 3-dimensional brane which is embedded in a 5-dimensional Minkowskian bulk can be described by the following Friedmann equation [19]

$$H^2 = \left(\sqrt{\frac{\rho_m + \rho_D}{3m_p^2} + \frac{1}{4r_c^2}} + \frac{\epsilon}{2r_c}\right)^2, \quad (1)$$

or equivalently

$$H^2 - \frac{\epsilon}{r_c}H = \frac{1}{3m_p^2}\left(\rho_m + \rho_D\right), \quad (2)$$

where $H = \dot{a}/a$ is the Hubble parameter, $r_c = m_{\mathrm{pl}}^2/(2m_5^3)$ [20] is the crossover length scale reflecting the competition between 4D and 5D effects of gravity, and $\epsilon = \pm 1$ corresponds to the two branches of solutions of the DGP model. Before going any further, it is worth noting that if $H^{-1} \ll r_c$ (early times) the 4D general relativity is recovered; otherwise the 5D effect becomes significant. Also $\epsilon = +1$ corresponds to the self-accelerating solution where the universe may accelerate in the late time purely due to modification of gravity [15, 16], while $\epsilon = -1$ can produce the acceleration only if a DE component is included on the brane. Here, to accommodate GDE into the formalism we take $\epsilon = -1$.

The fractional energy density parameters are defined as

$$\Omega_m = \frac{\rho_m}{3m_p^2H^2}, \quad \Omega_D = \frac{\rho_D}{3m_p^2H^2}, \quad \Omega_{r_c} = \frac{1}{4r_c^2H_0^2}, \quad (3)$$

where H_0 is the Hubble parameter at redshift $z = 0$. The Friedmann equation (2) can be rewritten in terms (3) as

$$\Omega_m + \Omega_D + 2\epsilon\frac{H_0}{H}\sqrt{\Omega_{r_c}} = 1. \quad (4)$$

We introduce $\Omega_{DGP} = 2\epsilon\sqrt{\Omega_{r_c}}H_0/H$, which comes from the extra dimension. Thus the Friedmann equation (4) can be reexpressed as

$$\Omega_m + \Omega_D + \Omega_{DGP} = 1. \quad (5)$$

For the GDE density we have

$$\rho_D = \alpha H, \quad (6)$$

where α is a constant of order Λ_{QCD}^3 and Λ_{QCD} is the QCD mass scale [12]. Taking the time derivative of the energy density ρ_D and using (6) we obtain

$$\dot{\rho}_D = \rho_D\frac{\dot{H}}{H}. \quad (7)$$

For the FRW universe filled with DE and DM, with mutual interaction, the energy-momentum conservation law can be written as

$$\dot{\rho}_m + 3H\rho_m = Q, \quad (8)$$

$$\dot{\rho}_D + 3H\left(1 + \omega_D\right)\rho_D = -Q, \quad (9)$$

where $Q = 3b^2H(\rho_D + \rho_m)$ is considered as the interaction term between DE and DM also b^2 is the coupling constant of interaction Q.

We know that (i) our universe is in a DE dominated phase and (ii) our universe that is our habitat is stable. These imply that any variable DE model should result in a stable DE dominated universe. So it is worth investigating the stability of the GDE in DGP braneworld against perturbation. The intended indicator for checking the stability of a proposed DE model is to study the behavior of the squared sound speed ($v_s^2 = dP/d\rho$) [21]. If $v_s^2 < 0$ we have the classical instability of a given perturbation because the perturbation of the background energy density is an oscillatory function and may grow or decay with time. When $v_s^2 > 0$, we expect a stable universe against perturbations because the perturbation in the energy density propagates in the environment. We continue discussion of stability in the linear perturbation regime where the perturbed energy density of the background can be written as

$$\rho(t,x) = \rho(t) + \delta\rho(t,x), \tag{10}$$

where $\rho(t)$ is unperturbed background energy density. For the energy conservation equation ($\nabla_\mu T^{\mu\nu} = 0$) which yields [21]

$$\delta\ddot{\rho} = v_s^2 \nabla^2 \delta\rho(t,x), \tag{11}$$

we encounter two cases. In the first case where $v_s^2 > 0$, we observe an ordinary wave equation which has a wave solution in the form $\delta\rho = \delta\rho_0 e^{-i\omega t + i\vec{k}.\vec{x}}$ (stable universe). In the second case where $v_s^2 < 0$, the frequency of the oscillations becomes pure imaginary and the density perturbations will grow with time as $\delta\rho = \delta\rho_0 e^{\omega t + i\vec{k}.\vec{x}}$ (unstable universe). Since v_s^2 plays a crucial role in determining the stability of DE model, we rewrite it in terms of EoS parameter as

$$v_s^2 = \frac{\dot{P}}{\dot{\rho}} = \frac{\dot{\rho}_D w_D + \rho_D \dot{w}_D}{\dot{\rho}_D (1+u) + \rho_D \dot{u}}, \tag{12}$$

where $P = P_D$ is the pressure of DE, $\rho = \rho_m + \rho_D$ is the total energy density of DE and DM, and $u = \Omega_m/\Omega_D$ is the energy density ration.

On the other sides, Sahni et al. [22] proposed new geometrical diagnostic pair parameter $\{r, s\}$, known as statefinder parameter, for checking the viability of newly introduced DE models. Unlike the physical variables which depend on the properties of physical fields describing DE models, the statefinder pair primarily depends on the scale factor and hence it depends on the metric of the spacetime. The r and s parameters are defined as [22]

$$r = \frac{\dddot{a}}{aH^3}, \qquad s = \frac{r-1}{3(q-1/2)}, \tag{13}$$

where r can rewrite as

$$r = 1 + 3\frac{\dot{H}}{H^2} + \frac{\ddot{H}}{H^3}. \tag{14}$$

and then

$$r = 2q^2 + q - \frac{\dot{q}}{H}. \tag{15}$$

Let us note that in the $\{r, s\}$ plane, $s > 0$ corresponds to a quintessence-like model of DE and $s < 0$ corresponds to a phantom-like model of DE. Also the studies on a flat ΛCDM model and matter dominated universe (SCDM) show that for these models $\{r, s\} = \{1, 0\}$ and $\{r, s\} = \{1, 1\}$, respectively. In above equations q is the deceleration parameter which is given by

$$q = -1 - \frac{\dot{H}}{H^2}. \tag{16}$$

In what follows we discuss the $\omega_D - \omega_D'$ plane which introduced by Caldwell and Linder [23] for analyzing the dynamical property of various DE models and distinguish these models (ω_D' represents the evolution of ω_D). The models can be categorized into two different classes: (i) $\omega_D' > 0$ and $\omega_D < 0$ which present the thawing region. (ii) $\omega_D' < 0$ and $\omega_D < 0$ which present the freezing region. It should be noted that the ΛCDM model corresponds to a fixed point $\{\omega_D = -1, \omega_D' = 0\}$ in the $\omega_D - \omega_D'$ plane. We shall consider the noninteracting and interacting cases, separatively. **It deserves to mention here that in all figures we selected values for the fixed parameters, which is custom and the observational evidence confirms it, as $\Omega_{r_c} = 0.0003$ [24] or $\Omega_D(z = 0)$ that can be between 0.7 and .75 and $H(z = 0)$ that can be between 60 and 100. It should be noted that the results and conclusions are not sensitive to these choices and also do not do any observational work on this title; thus we by helping papers and experience selected these values for the fixed parameters.**

2.1. Noninteracting Case. We start to obtain the cosmological parameters for GDE in the DGP braneworld by ignoring the interaction term ($Q = 0$). The deceleration parameter q can be obtained by taking the time derivative of (2), which lead to

$$\frac{\dot{H}}{H^2} = \frac{-3(1-\Omega_{DGP}) + 3\Omega_D}{2 - \Omega_{DGP} - \Omega_D}. \tag{17}$$

Using relation (16), we find

$$q = -1 - \frac{-3(1-\Omega_{DGP}) + 3\Omega_D}{2 - \Omega_{DGP} - \Omega_D}. \tag{18}$$

Inserting (7) in (9) we have

$$\omega_D = -1 - \frac{1}{3}\frac{\dot{H}}{H^2}, \tag{19}$$

where by replacing (17) in it, we get

$$\omega_D = -\frac{1}{2 - \Omega_{DGP} - \Omega_D}. \tag{20}$$

Also we obtain ω_D' from the above equation as

$$\omega_D' = -\frac{3(-1 + \Omega_{DGP} + \Omega_D)(\Omega_{DGP} + \Omega_D)}{(-2 + \Omega_{DGP} + \Omega_D)^3}. \tag{21}$$

Note that, in order to find the evolution of density parameter Ω_D, we take the time derivative of relation $\Omega_D = \rho_D/(3m_p^2H^2)$, after combining the result with (7) and (17), which yields

$$\Omega_D' = \Omega_D(1+q). \tag{22}$$

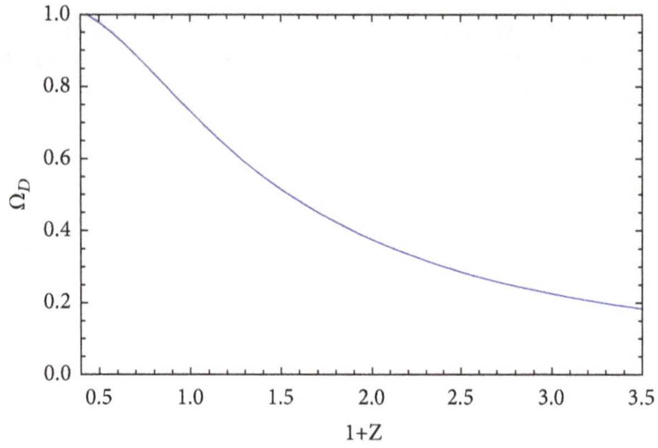

FIGURE 1: The evolution of Ω_D versus redshift parameter z for noninteracting GDE in DGP model. Here, we have taken $\Omega_D(z = 0) = 0.73$, $H(z = 0) = 67$, and $\Omega_{r_c} = 0.0003$.

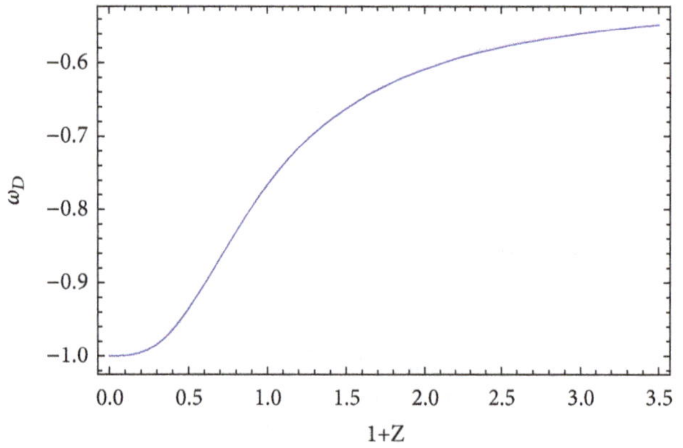

FIGURE 2: The evolution of ω_D versus redshift parameter z for noninteracting GDE in DGP model. Here, we have taken $\Omega_D(z = 0) = 0.73$, $H(z = 0) = 67$, and $\Omega_{r_c} = 0.0003$.

In Figure 1, we plot the evolution of Ω_D versus redshift parameter z. It is obvious that Ω_D tends to 0 in the early universe where $1 + z \longrightarrow \infty$, while at the late time where $1 + z \longrightarrow 0$, we have $\Omega_D \longrightarrow 1$. Clearly, (20) for the EoS parameter shows that, at the late time where $\Omega_D \longrightarrow 1$, the EoS parameter mimics the cosmological constant, namely, $\omega_D \longrightarrow -1$, **which is showed in** Figure 2. In Figure 3, the behavior of the deceleration parameter q is plotted and indicates that indeed there is a decelerated expansion at the early stage of the universe followed by an accelerated expansion. The energy density ratio is defined as $u = \Omega_m/\Omega_D$, which by using (5) can be written

$$u = -1 + \frac{1}{\Omega_D}\left(1 - \Omega_{DGP}\right). \tag{23}$$

Differentiating (23) and (20) and then substituting the results in (12) we get the squared of sound speed as

$$v_s^2 = -\frac{2\Omega_D\left(-1 + \Omega_{DGP} + \Omega_D\right)}{\left(-2 + \Omega_{DGP}\right)\left(-2 + \Omega_{DGP} + \Omega_D\right)^2}. \tag{24}$$

The evolution of v_s^2 against z for the noninteracting GDE in the framework of DGP braneworld is plotted in Figure 4. From graphical analysis of v_s^2 one concludes that this model does not indicate any signal of stability, that is, $v_s^2 < 0$ during the history of the universe. We can also find the statefinder parameters r and s by taking derivative of (18) and using (13) and (15). The results are

$$r = 10 + \frac{18}{\left(-2 + \Omega_{DGP} + \Omega_D\right)^3} + \frac{45}{\left(-2 + \Omega_{DGP} + \Omega_D\right)^2} + \frac{36}{-2 + \Omega_{DGP} + \Omega_D}, \tag{25}$$

$$s = \frac{2\left(-1 + \Omega_{DGP} + \Omega_D\right)^2}{\left(-2 + \Omega_{DGP} + \Omega_D\right)^2}. \tag{26}$$

The graphical behavior of the statefinder parameters $\{r, s\}$ given in (25) and (26) is plotted in Figures 5 and 6, showing that, at late time where $\Omega_D \longrightarrow 1$, we have $\{r, s\} = \{1, 0\}$ which implies that GDE mimics the cosmological constant at the late time, as expected.

Let us study the trajectory in the statefinder plane and analyze this model from the statefinder viewpoint. For this purpose, we plot the statefinder diagram in the $r - s$ in Figure 7, which shows the cure gets to the point $\{r, s\} = \{1, 0\}$ in the end, which implies that the model corresponds to the ΛCDM model at the late time. For complementarity of the diagnostic, we also plot the trajectories of statefinder pair $r - q$ in Figure 8 which ends in the future to $r = 1$, $q = -1$ corresponding to the de Sitter expansion.

The ω_D–ω_D' plane for the noninteracting GDE in the DGP scenario is shown in Figure 9. Again, we see that this plane corresponds to ΛCDM model, i.e., ($\omega_D = -1$, $\omega_D' = 0$), and the trajectory meets the freezing region as well.

2.2. Interacting Case. Differentiating the modified Friedmann equation (2) and using (7) and (8) we reach

$$\frac{\dot{H}}{H^2} = \frac{3\left(b^2 - 1\right)\left(1 - \Omega_{DGP}\right) + 3\Omega_D}{2 - \Omega_{DGP} - \Omega_D}, \tag{27}$$

$$q = -1 - \frac{3\left(b^2 - 1\right)\left(1 - \Omega_{DGP}\right) + 3\Omega_D}{2 - \Omega_{DGP} - \Omega_D}. \tag{28}$$

Next, the EoS parameter can be determined by substituting (7) into the semiconservation law equation (9) and using (27). We find

$$\omega_D = \frac{b^2\left(\Omega_{DGP} - 2\right)\left(\Omega_{DGP} - 1\right) + \Omega_D}{\Omega_D\left(-2 + \Omega_{DGP} + \Omega_D\right)}. \tag{29}$$

Taking differentiation with respect to $x = \ln a$ from above equation we get

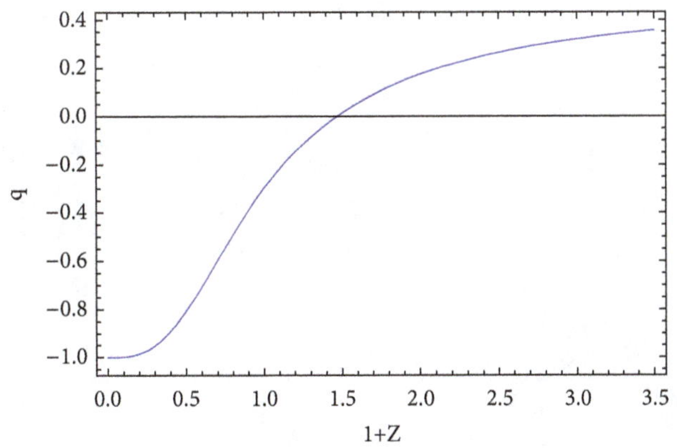

FIGURE 3: The evolution of the deceleration parameter q versus redshift parameter z for noninteracting GDE in DGP model. Here, we have taken $\Omega_D(z=0)=0.73$, $H(z=0)=67$, and $\Omega_{r_c}=0.0003$.

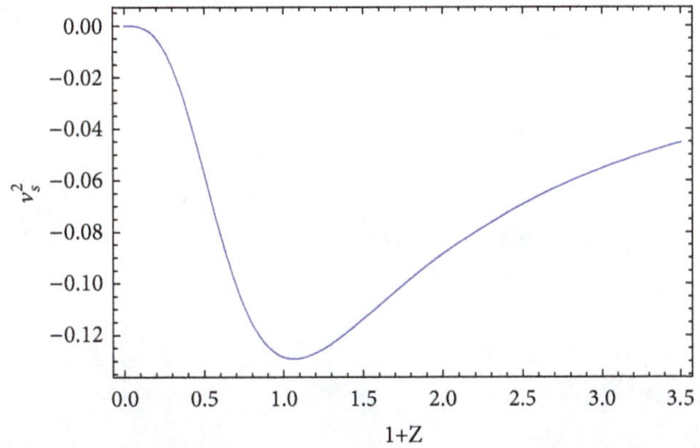

FIGURE 4: The evolution of the squared of sound speed v_s^2 versus redshift parameter z for noninteracting GDE in DGP model. Here, we have taken $\Omega_D(z=0)=0.73$, $H(z=0)=67$, and $\Omega_{r_c}=0.0003$.

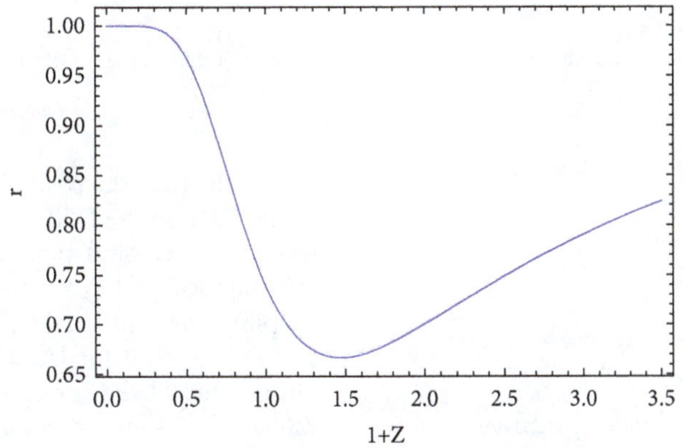

FIGURE 5: The evolution of the statefinder parameter r versus the redshift parameter z for noninteracting GDE in DGP model. Here, we have taken $\Omega_D(z=0)=0.73$, $H(z=0)=67$, and $\Omega_{r_c}=0.0003$.

$$\omega_D' = -\frac{3\left[\left(-1+b^2\right)\left(-1+\Omega_{DGP}\right)-\Omega_D\right]\left[b^2\left(-2+\Omega_{DGP}\right)^2+\left(-\Omega_{DGP}+b^2\left(-4+3\Omega_{DGP}\right)-\Omega_D\right)\Omega_D\right]}{\Omega_D\left(-2+\Omega_{DGP}+\Omega_D\right)^3}, \tag{30}$$

where the prime indicates derivative with respect to $x=\ln a$. We can obtain the equation of motion for Ω_D as

$$\Omega_D' = \frac{3\Omega_D\left(-1+\Omega_D+\Omega_{DGP}+b^2\left(1-\Omega_{DGP}\right)\right)}{-2+\Omega_{DGP}+\Omega_D}, \tag{31}$$

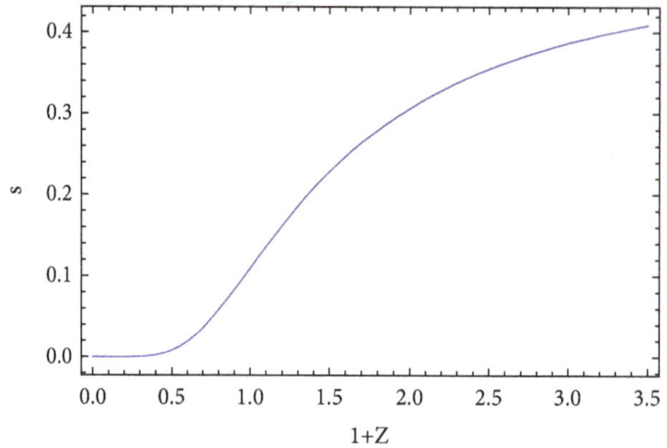

FIGURE 6: The evolution of the statefinder parameter s versus the redshift parameter z for noninteracting GDE in DGP model. Here, we have taken $\Omega_D(z=0) = 0.73$, $H(z=0) = 67$, and $\Omega_{r_c} = 0.0003$.

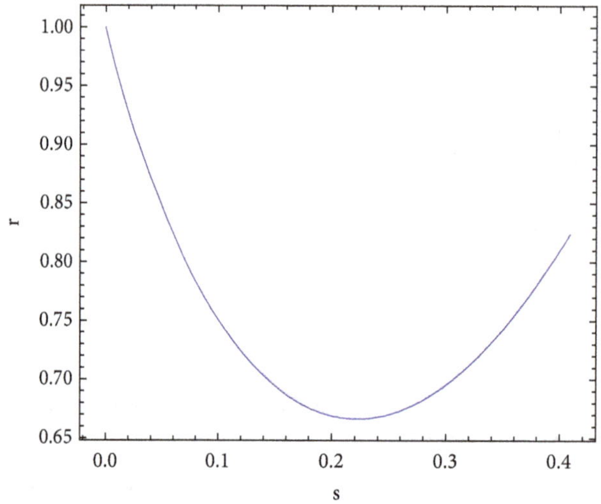

FIGURE 7: The evolution of the statefinder parameter r versus s for noninteracting GDE in the DGP model. Here, we have taken $\Omega_D(z=0) = 0.73$, $H(z=0) = 67$, and $\Omega_{r_c} = 0.0003$.

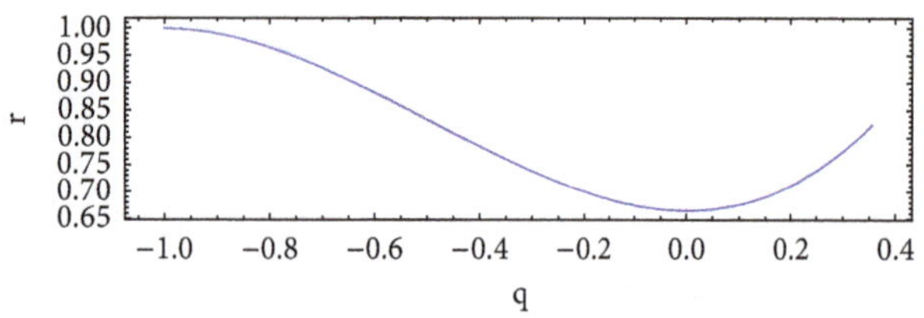

FIGURE 8: The evolution of the statefinder parameter r versus the deceleration parameter q for noninteracting GDE in the DGP model. Here, we have taken $\Omega_D(z=0) = 0.73$, $H(z=0) = 67$, and $\Omega_{r_c} = 0.0003$.

To illustrate the cosmological consequences of the interacting GDE in the DGP braneworld, we plot their evolution in terms of redshift parameter z. In Figure 10, we present the graphical of Ω_D versus z for the different values of the coupling constant b^2. As expected, we see both $\Omega_D \longrightarrow 1$ and $\Omega_D \longrightarrow 0$ for late time and early time, respectively. The graphical behavior of the EoS parameter for the different values of b^2 shows crossing of phantom line as plotted in Figure 11. The stability of interacting GDE in DGP model can be obtained by differentiating with respect to time of (23) and (29)

$$v_s^2 = -\frac{b^2\left[\left(-2+\Omega_{DGP}\right)^3 + \Omega_D\left(6+\left(-6+\Omega_{DGP}\right)\Omega_{DGP}\right)\right] + \Omega_D\left(-2+2\Omega_{DGP}+2\Omega_D\right)}{\left(-2+\Omega_{DGP}\right)\left(-2+\Omega_{DGP}+\Omega_D\right)^2}. \quad (32)$$

The evolution of the deceleration parameter q and the squared of sound speed v_s^2 versus redshift parameter z are plotted in Figures 12 and 13, respectively. In Figure 12, we see for different values of b^2 with the interacting GDE in DGP

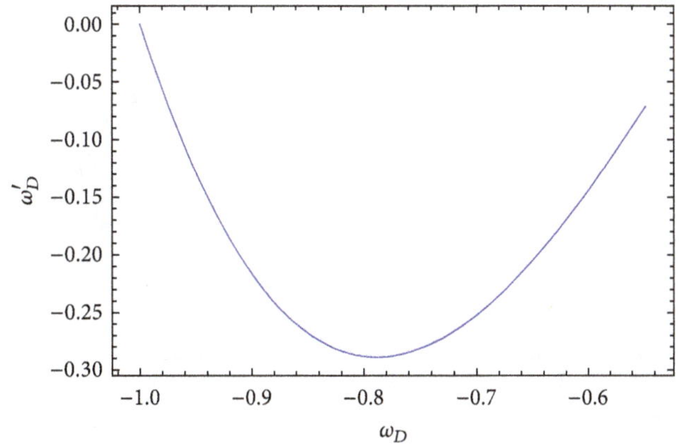

FIGURE 9: The $\omega_D - \omega'_D$ diagram for noninteracting GDE in the DGP model. Here, we have taken $\Omega_D(z = 0) = 0.73$, $H(z = 0) = 67$, and $\Omega_{r_c} = 0.0003$.

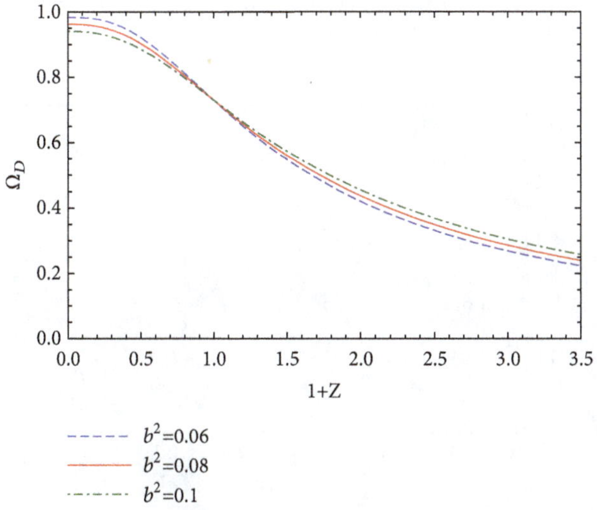

FIGURE 10: The evolution of Ω_D versus redshift parameter z for interacting GDE in the DGP model. Here, we have taken $\Omega_D(z = 0) = 0.73$, $H(z = 0) = 67$, and $\Omega_{r_c} = 0.0003$.

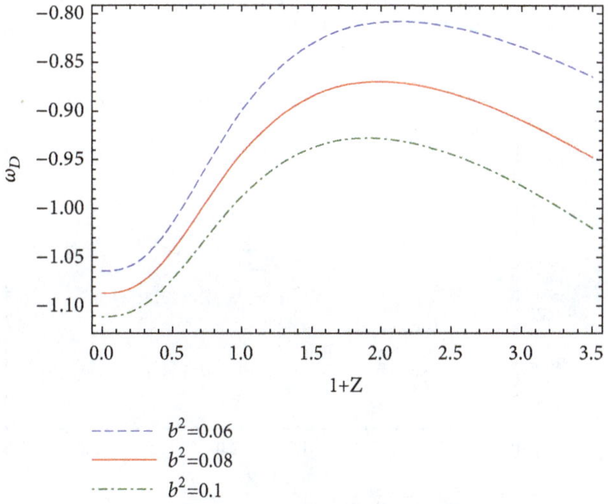

FIGURE 11: The evolution of ω_D versus redshift parameter z for interacting GDE in the DGP model. Here, we have taken $\Omega_D(z = 0) = 0.73$, $H(z = 0) = 67$, and $\Omega_{r_c} = .0003$.

model, our universe has a phase transition from deceleration to an acceleration, while by keeping the same situation in Figure 13, this universe cannot be stable. As the value of b^2 decreases the severity of instability also decreases. Like previous section, the statefinder parameters are obtained by taking derivative of (28) and using (13) and (15)

$$r = 10 + \frac{18\left(1+b^2-b^2\Omega_{DGP}\right)}{\left(-2+\Omega_{DGP}+\Omega_D\right)^3} + \frac{9\left(-1+b^2\left(-1+\Omega_{DGP}\right)\right)\left(-5+b^2\left(-3+2\Omega_{DGP}\right)\right)}{\left(-2+\Omega_{DGP}+\Omega_D\right)^2} + \frac{9\left(4+b^2\left(4-3\Omega_{DGP}\right)\right)}{-2+\Omega_{DGP}+\Omega_D}, \tag{33}$$

$$s = 2 + \frac{2-2b^2\left(-1+\Omega_{DGP}\right)}{\left(-2+\Omega_{DGP}+\Omega_D\right)^2} + \frac{4+b^2\left(3-2\Omega_{DGP}\right)}{-2+\Omega_{DGP}+\Omega_D} + \frac{b^2}{2b^2\left(1-\Omega_{DGp}\right)+\Omega_{DGP}+\Omega_D}. \tag{34}$$

We obtain $\{r, s\} = \{1, 0\}$ for ΛCDM model from (33) and (34) in the limiting case where $b^2 = 0$, $\Omega_{DGP} = 0$, and $\Omega_D \longrightarrow 1$ (in the late time). Also, Figures 14 and 15 show that r and s are positive through the entire life of the universe and turn to 1 and 0 at the late time, respectively. The $\{r, s\}$ evolutionary trajectories for the interacting GDE in the framework of the DGP braneworld for different values of b^2 are shown in Figure 16. From Figure 16, we can see that at the late time all curves tend to the ΛCDM fixed point $\{r = 1, s = 0\}$; also different b^2 results in different evolution trajectories of statefinder which states that r is smaller when b^2 is larger. The $r-q$ diagrams are plotted for different values of b^2 in Figure 17

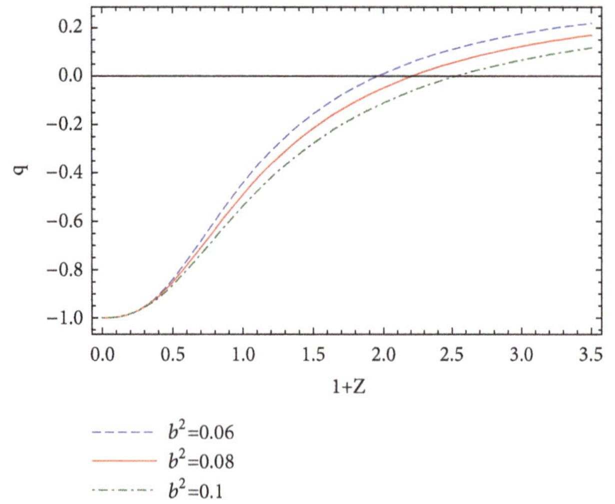

FIGURE 12: The evolution of the deceleration parameter q versus redshift parameter z for interacting GDE in the DGP model. Here, we have taken $\Omega_D(z = 0) = 0.73$, $H(z = 0) = 67$, and $\Omega_{r_c} = 0.0003$.

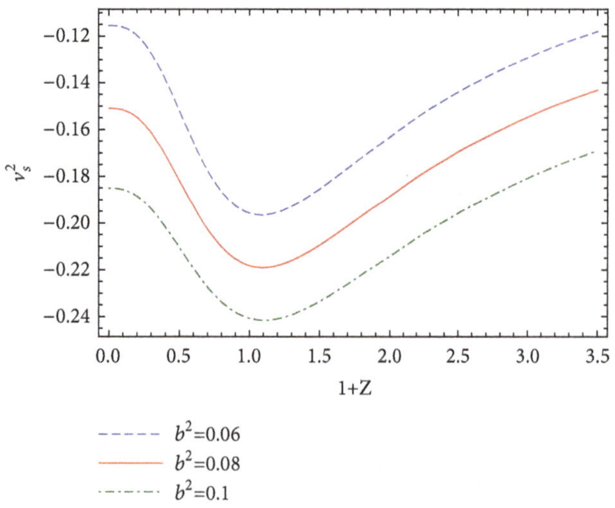

FIGURE 13: The evolution of the squared of sound speed v_s^2 versus redshift parameter z for interacting GDE in the DGP model. Here, we have taken $\Omega_D(z = 0) = 0.73$, $H(z = 0) = 67$, and $\Omega_{r_c} = 0.0003$.

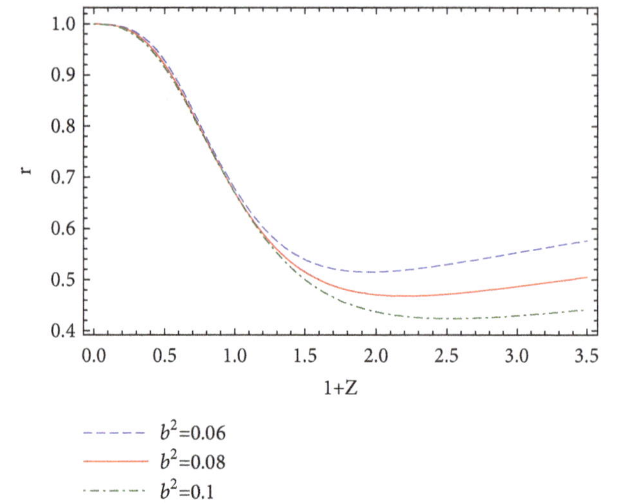

FIGURE 14: The evolution of the statefinder parameter r versus the redshift parameter z for interacting GDE in DGP model. Here, we have taken $\Omega_D(z = 0) = 0.73$, $H(z = 0) = 67$, and $\Omega_{r_c} = .0003$.

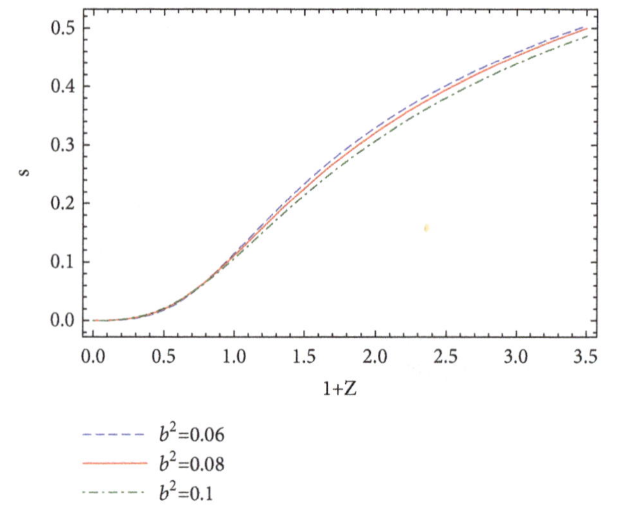

FIGURE 15: The evolution of the statefinder parameter s versus the redshift parameter z for interacting GDE in the DGP model. Here, we have taken $\Omega_D(z = 0) = 0.73$, $H(z = 0) = 67$, and $\Omega_{r_c} = .0003$.

which mimics the de Sitter expansion, namely, $r = 1, q = -1$ in the far future where $z \longrightarrow 0$. In Figure 18, we plot the $\omega_D - \omega'_D$ plane for different values of b^2 which show that the trajectories meet both the thawing and the freezing regions as well.

3. Closing Remarks

We have made a versatile study on both noninteracting and interacting GDE in the framework DGP model through well-known cosmological parameters as well as planes. We summarize our results as follows. For noninteracting case, we have found that the density parameter tends to zero at the early universe while at the late time we have $\Omega_D \longrightarrow 1$. Meanwhile the EoS parameter cannot cross the phantom line and mimics the cosmological constant at the late time (Figure 2). We have shown that our universe has a phase transition from deceleration to an acceleration, though we do not receive any signal of stability. The statefinder plane shows that the trajectory corresponds to quintessence model ($s > 0$ and $r < 1$) while at late time we have $\{r, s\} = \{1, 0\}$ for ΛCDM model as expected. The $\omega_D - \omega'_D$ plane in Figure 9 meets the freezing region as well.

For interacting case, we find that the density and the deceleration parameters as well as the EoS parameter are consistent with observational data. We have seem that as the value of b^2 decreases the severity of instability decreases. From $r - s$ plane, we can see that at the late time all cures tend to the ΛCDM fixed point $\{r = 1, \ s = 0\}$. Besides, for different values of b^2, the different evolution trajectories of statefinder

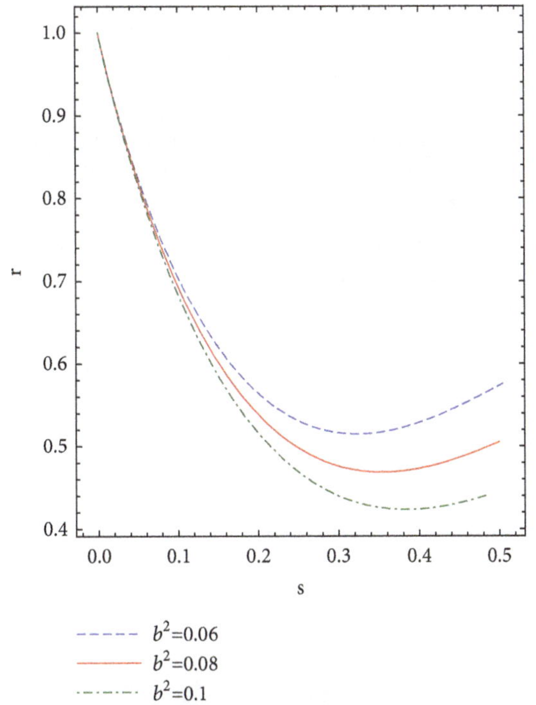

FIGURE 16: The evolution of the statefinder parameter r versus s for interacting GDE in the DGP model. Here, we have taken $\Omega_D(z = 0) = 0.73$, $H(z = 0) = 67$, and $\Omega_{r_c} = .0003$.

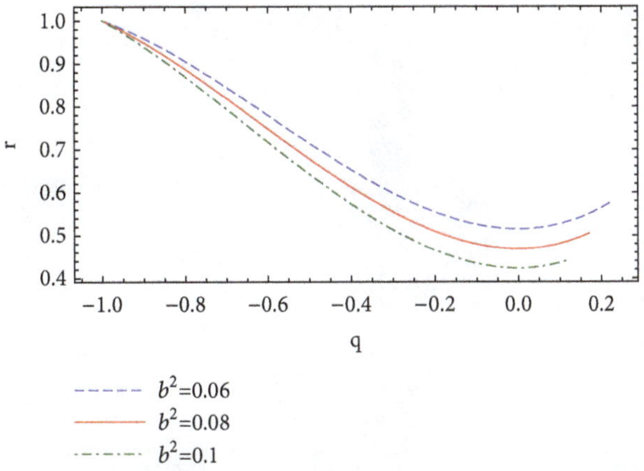

FIGURE 17: The evolution of the statefinder parameter r versus the deceleration parameter q for interacting GDE in the DGP model. Here, we have taken $\Omega_D(z = 0) = 0.73$, $H(z = 0) = 67$, and $\Omega_{r_c} = .0003$.

are shown which indicates that r is smaller when b^2 is larger. The $r - q$ plane is plotted in Figure 17 which mimics the de Sitter expansion, namely, $r = 1, q = -1$ in the far future where $z \longrightarrow 0$. In the end, the $\omega_D - \omega'_D$ plane exhibits both freezing and thawing regions of the universe for all values of b^2. Again, in this case $v_s^2 < 0$ which implies that interacting GDE in the DGP braneworld is not stable against perturbation.

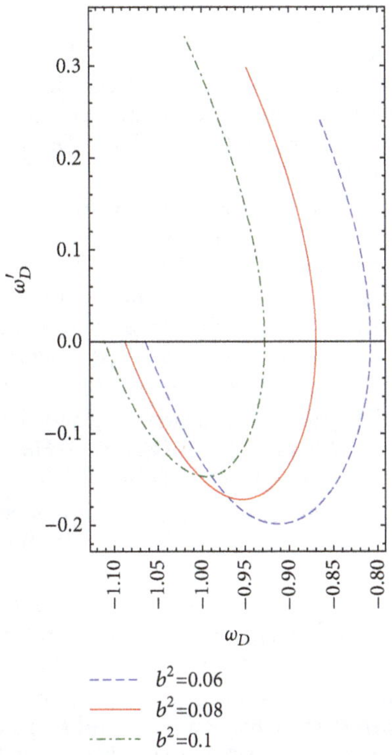

FIGURE 18: The $\omega_D - \omega'_D$ diagram for interacting GDE in the DGP model. Here, we have taken $\Omega_D(z = 0) = 0.73$, $H(z = 0) = 67$, and $\Omega_{r_c} = .0003$.

Conflicts of Interest

The authors declare that they have no conflicts of interest.

Acknowledgments

The authors thank Shiraz University Research Council. This work has been supported financially by Research Institute for Astronomy & Astrophysics of Maragha (RIAAM), Iran.

References

[1] S. Perlmutter, G. Aldering, and G. Goldhaber, "Measurements of Ω and Λ from 42 High-Redshift Supernovae," *The Astrophysical Journal*, vol. 517, no. 2, pp. 565–586, 1999.

[2] A. G. Riess, A. V. Filippenko, P. Challis et al., "Observational evidence from supernovae for an accelerating universe and a cosmological constant," *The Astronomical Journal*, vol. 116, no. 3, pp. 1009–1038, 1998.

[3] G. Hinshaw, D. Larson, and E. Komatsu, "Nine-year Wilkinson Microwave Anisotropy Probe (WMAP) observations: cosmological parameter results," *The Astrophysical Journal Supplement Series*, vol. 208, no. 2, p. 19, 2013.

[4] S. Capozziello, V. F. Cardone, S. Carloni, and A. Troisi, "Curvature quintessence matched with observational data,"

International Journal of Modern Physics D, vol. 12, no. 10, pp. 1969–1982, 2003.

[5] S. Nojiri and S. D. Odintsov, "Introduction to modified gravity and gravitational alternative for dark energy," *International Journal of Geometric Methods in Modern Physics*, vol. 4, no. 1, article 115, 2007.

[6] S. Nojiri and S. D. Odintsov, "Unified cosmic history in modified gravity: from F(R)F(R) theory to Lorentz non-invariant models," *Physics Reports*, vol. 505, pp. 59–144, 2011.

[7] G. Papagiannopoulos, S. Basilakos, J. D. Barrow, and A. Paliathanasis, "New integrable models and analytical solutions in f (R) cosmology with an ideal gas," *Physical Review D: Particles, Fields, Gravitation and Cosmology*, vol. 97, no. 2, 2018.

[8] L. Amendola, "Scaling solutions in general nonminimal coupling theories," *Physical Review D: Particles, Fields, Gravitation and Cosmology*, vol. 60, Article ID 043501, 1999.

[9] J. Uzan, "Cosmological scaling solutions of nonminimally coupled scalar fields," *Physical Review D: Particles, Fields, Gravitation and Cosmology*, vol. 59, no. 12, 1999.

[10] A. Sheykhi, "Interacting new agegraphic dark energy in nonflat Brans-Dicke cosmology," *Physical Review D: Particles, Fields, Gravitation and Cosmology*, vol. 81, no. 2, 2010.

[11] K. Karami, A. Sheykhi, M. Jamil, Z. Azarmi, and M. M. Soltanzadeh, "Interacting entropy-corrected new agegraphic dark energy in Brans-Dicke cosmology," *General Relativity and Gravitation*, vol. 43, no. 1, pp. 27–39, 2011.

[12] A. Sheykhi, E. Ebrahimi, and Y. Yousefi, "Generalized ghost dark energy in Brans-Dicke theory," *Canadian Journal of Physics*, vol. 91, no. 8, pp. 662–667, 2013.

[13] F. Felegary, F. Darabi, and M. R. Setare, "Interacting holographic dark energy model in Brans-Dicke cosmology and coincidence problem," *International Journal of Modern Physics D: Gravitation, Astrophysics, Cosmology*, vol. 27, no. 3, 1850017, 18 pages, 2018.

[14] V. Sahni and A. Starobinsky, "The case for a positive cosmological Λ-term," *International Journal of Modern Physics D*, vol. 9, no. 4, p. 373, 2000.

[15] M. Li, "A model of holographic dark energy," *Physics Letters B*, vol. 603, no. 1-2, pp. 1–5, 2004.

[16] D. Pavon and W. Zimdahl, "Holographic dark energy and cosmic coincidence," *Physics Letters B*, vol. 628, no. 3-4, pp. 206–210, 2005.

[17] M. R. Setare and E. N. Saridakis, "Non-minimally coupled canonical, phantom and quintom models of holographic dark energy," *Physics Letters B*, vol. 671, no. 3, pp. 331–338, 2009.

[18] A. Sheykhi, "Thermodynamics of interacting holographic dark energy with the apparent horizon as an IR cutoff," *Classical and Quantum Gravity*, vol. 27, no. 2, Article ID 025007, 2010.

[19] S. Nojiri and S. D. Odintsov, "Unifying phantom inflation with late-time acceleration: scalar phantom-non-phantom transition model and generalized holographic dark energy," *General Relativity and Gravitation*, vol. 38, pp. 1285–1304, 2006.

[20] R.-G. Cai, "A dark energy model characterized by the age of the universe," *Physics Letters B*, vol. 657, no. 4-5, pp. 228–231, 2007.

[21] A. Sheykhi, "Interacting agegraphic tachyon model of dark energy," *Physics Letters B*, vol. 682, no. 4-5, pp. 329–333, 2010.

[22] A. A. Mamon, "Reconstruction of interaction rate in holographic dark energy model with Hubble horizon as the infrared cut-off," *International Journal of Modern Physics D*, vol. 26, no. 11, 2017.

[23] M. A. Zadeh, A. Sheykhi, and H. Moradpour, "Holographic dark energy with the sign-changeable interaction term," *International Journal of Modern Physics D: Gravitation, Astrophysics, Cosmology*, vol. 26, no. 8, 1750080, 13 pages, 2017.

[24] B. Feng, X. L. Wang, and X. M. Zhang, "Dark energy constraints from the cosmic age and supernova," *Physics Letters B*, vol. 607, no. 1-2, pp. 35–41, 2005.

4

Hawking Radiation of a Single-Partition Black Hole

Youngsub Yoon

Department of Physics and Astronomy, Seoul National University, Seoul 08826, Republic of Korea

Correspondence should be addressed to Youngsub Yoon; youngsub@post.harvard.edu

Academic Editor: Farook Rahaman

Brian Kong and the present author recently presented a new area spectrum and showed that the new area spectrum implies that the decay time of a single-partition black hole (i.e., a black hole with the area not big enough to have two or more partitions) is roughly constant. In this article, we show why the decay time of a single-partition black hole is roughly constant.

1. Introduction

It is now very well-known that black hole emits particles thanks to Hawking's semiclassical treatment of black hole [1]. However, as Rovelli and Smolin discovered loop quantum gravity as a possible method to quantize gravity in late 80s [2], it was destined that Hawking's semiclassical treatment of Hawking radiation needed some modification. Soon, it was discovered that area is discrete according to loop quantum gravity [3–5], and Rovelli found the first connection between the Bekenstein-Hawking entropy and loop quantum gravity in his seminal paper in 1996 [6]. Soon, Hawking radiation spectrum was also approached from the loop quantum gravity framework using the discreteness of area [7–9]. In mid-2010s, the present author approached the problem of Hawking radiation spectrum from the discreteness of area as well and found surprising results such as full discreteness of Hawking radiation spectrum [10] and Maxwell-Boltzmann nature of Hawking radiation spectrum [11]. Also, Brian Kong and the present author suggested a new area spectrum in [12], boldly claiming that the conventional area spectrum known in loop quantum gravity community could be wrong, if one considers that, for area two-form, one needs to use Levi-Civita tensor instead of Levi-Civita symbol as conventionally used in loop quantum gravity community. This consideration predicted that the area spectrum was the square root of the conventional one. As for the evidence of this area spectrum, we showed that the Bekenstein-Hawking entropy was approximately satisfied (we conjectured that it would be exactly satisfied if the effect of extra dimension is considered, which can lead to the prediction of fine structure constant) and found a strange, coincidental numerical relation that implied that the decay time of a single-partition black hole is roughly constant. In this article, we show why the decay time of a single-partition black hole is roughly constant. This new further evidence will give another support the area spectrum proposed by Brian Kong and us. The organization of this paper is as follows. In Section 2, we briefly review loop quantum gravity and its approach to Hawking radiation, particularly focusing on our previous work. In Section 3, we briefly review Brian Kong and the present author's work on new area spectrum focusing on the strange, coincidental numerical relation. In Section 4, we explain our own work on Hawking radiation briefly mentioned in Section 2 in more detail. In Section 5, we explain what is special about single-partition black hole. In Section 6, we derive the constancy of decay time for single-partition black hole. In Section 7, we conclude our paper.

2. Loop Quantum Gravity and Its Approach to Hawking Radiation

According to loop quantum gravity, the eigenvalues of the area operator are quantized [3–5] and the black hole area, as much as any area, is the sum of these eigenvalues. For example, let us say that we have the following area eigenvalues (i.e., the unit areas):

$$A_i = A_1, A_2, A_3, A_4, A_5, A_6 \ldots . \quad (1)$$

Table 1: C(A).

y	A	K(A)	I(A)	C(A)
1	17.8	4	767.4	191.8
2	21.1	14	2740	195.7
3	23.4	32	5552	173.5
4	25.1	50	9276	185.5
5	26.6	72	14000	194.4
6	27.8	110	19814	180.1
7	28.9	154	26817	174.1
8	29.9	204	35109	172.1
9	30.8	262	44797	171.0
10	31.6	326	55990	171.7
11	32.4	388	68803	177.3
12	33.1	474	83353	175.8
13	33.7	584	99761	170.8
14	34.4	684	118155	172.7
15	35.0	804	138664	172.5

Then, the black hole area A must be given by the following formula:

$$A = \sum_i N_i A_i, \tag{2}$$

where N_is are nonnegative integers. Here, we can regard the black hole as having $\sum N_i$ partitions, each of which has one of the A_i as its area.

In [10], the present author showed that the black hole area decrease upon a single emission of photon during Hawking radiation must be given by the unit area. In other words, we have

$$\Delta A = -A_i \tag{3}$$

As the Bekenstein-Hawking entropy is given by $S = kA/4$, [1, 13] and we know $\Delta Q = T\Delta S$, the energy decrease is given by

$$\Delta Q = -\frac{A_i}{4} kT \tag{4}$$

Since this energy must be equal to the energy of photon emitted (i.e., $\Delta Q = -E_{photon}$) the energy of the photon emitted during the Hawking radiation is given by

$$E_{photon} = \frac{A_i}{4} kT \tag{5}$$

In particular, the Hawking radiation spectrum is discrete.

3. Area Spectrum and Single-Partition Black Holes

In [12] (see also [14]), Brian Kong and the present author showed that the area spectrum (i.e., the unit areas) in black hole horizon is given by

$$A = 8\pi \sum_i \sqrt{\frac{1}{2}\sqrt{2j_i^u\left(j_i^u+1\right)+2j_i^d\left(j_i^d+1\right)-j_i^t\left(j_i^t+1\right)}} \tag{6}$$

with the degeneracy $(2j_i^u + 1) + (2j_i^d + 1)$ where j_i^u, j_i^d, and j_i^t satisfy the triangle inequality (i.e., $j_i^1 + j_i^2 \geq j_i^3$) and j_i^t is an integer, and both j_i^u and j_i^d are integers or half-integers at the same time. We want to remark that (6) is not a conventional one and it has an extra square root compared to [14]. However, Brian Kong and the present author argued in [12] that we arrive at this formula, if we use, in the equation for the area two-form, a Levi-Civita tensor instead of a Levi-Civita symbol as conventionally done in loop quantum gravity community.

Using this, we calculated the degeneracy of area spectrum using Java. This yielded $K(A)$ in Table 1 where $K(A)$ is the number of states for area equal to or below A, and $I(A)$ is given by

$$I(A) = \int_{A_{cut}}^{A} \sqrt{A'}\left(e^{A'/4} - 1\right) dA' \tag{7}$$

and y is conveniently defined as follows:

$$y = 2j_i^u\left(j_i^u+1\right)+2j_i^d\left(j_i^d+1\right)-j_i^t\left(j_i^t+1\right) \tag{8}$$

Of course, using y, we have the following relation:

$$A(y) = 8\pi\sqrt{\frac{1}{2}\sqrt{y}} \tag{9}$$

$C(A)$ is given by

$$C(A) = \frac{I(A)}{K(A)} \tag{10}$$

The point of this table is that $C(A)$ is roughly constant. We found this accidentally. Of course $C(A)$ is not exactly but only approximately constant, but the biggest value for $C(A)$ in our result is 195.7, deviating from the "right value" of C, i.e., the value of C for large A by only about 13 percent. As $K(A)$ at this biggest value of $C(A)$ is only 14, it necessarily has a big "statistical" variation. Therefore, 195.7 is not a big deviation.

Then, in Sections X and XI of the same paper, we showed that this constancy can be explained if the decay time of a single-partition black hole is constant. Let us briefly summarize some of the main reasonings.

Let us denote the average number of photons emitted from a black hole during the infinitesimal time Δt by $j_{\Delta t}$. Then, we have

$$j_{\Delta t} = \frac{\Delta t}{\tau} \tag{11}$$

For macroscopic black hole, if dn_{photon} is the number of photons with a frequency between ν and $\nu + d\nu$ emitted during unit time, from Planck's blackbody radiation formula, we have

$$dn_{\text{photon}} = \frac{2\pi\nu^2}{c^2}\frac{A_{BH}d\nu}{e^{h\nu/kT} - 1} \tag{12}$$

where A_{BH} is the area of black hole. Of course, we naturally have

$$\frac{1}{\tau} = \int dn_{photon} \tag{13}$$

Now let us say that, during the time Δt, the number of emitted photons which correspond to a decrease in the black hole area by amount a is given by $x_{a(y),\Delta t}$. Here, y is given by (8), and $a(y)$ is given by (9).

Then, in section XI, by some calculations we showed

$$x_{a,\Delta t} = \frac{3\sqrt{\pi}c}{2\sqrt{A_{BH}}}\frac{1}{e^{a/4} - 1}\Delta t \tag{14}$$

We proceeded by borrowing the idea from a paper by Bekenstein and Mukhanov [7]. The idea is that $x_{a,\Delta t}/j_{\Delta t}$ is proportional to the degeneracy of the black hole after emission divided by the degeneracy of the black hole before emission. Let us denote the degeneracy of the black hole with area A_{BH} by $W(A_{BH})$. Then, from (11) and (14), we have

$$\frac{W(A_{BH} - a)}{W(A_{BH})} \propto \frac{\tau}{\sqrt{A_{BH}}}\frac{1}{e^{a/4} - 1} \tag{15}$$

Now, if a black hole is a single-partition black hole (a black hole not big enough to have two or more partitions), upon emission of a photon, there is no other value for area decrease than $a = A_{BH}$. Plugging this value to the above formula, we get

$$\frac{W(0)}{W(A_{BH})} \propto \frac{\tau}{\sqrt{A_{BH}}}\frac{1}{e^{A_{BH}/4} - 1} \tag{16}$$

which implies

$$W(A_{BH}) \propto \frac{\sqrt{A_{BH}}}{\tau}\left(e^{A_{BH}/4} - 1\right) \tag{17}$$

If τ is constant for single-partition black hole, then it is precisely our earlier relations (7) and (10). Also, one thing very noticeable is that these relations do not hold for multipartitioned black holes. The minimum area a black hole can have is given by $A(y = 1) = 4\pi\sqrt{2}$. For it to be a single-partition black hole, its area should be smaller than $2A(y = 1) = A(y = 16)$. Our relations suddenly do not hold beginning from $y = 16$ even though it holds very well for $y < 16$. This is the reason why the table is presented up to $y = 15$.

However, we later realized formula (14), which we derived in section XI of [12], is due to an error. Nevertheless, we obtained a strong hint that Table 1 is due to Hawking radiation of single-partition black hole.

4. Quantum Corrections to the Hawking Radiation Spectrum

As mentioned in Section 2, the present author showed in [10] that, upon emission of a single photon during Hawking radiation, only single area quanta can decrease. Let us briefly repeat the main argument in that paper. (In an earlier work [11], the author showed that Hawking radiation should follow Maxwell-Boltzmann statistics rather than Bose-Einstein statistics as considered in [10]. We adjust our presentation here considering this earlier work.)

Consider the following problem [15].

Let us say that the unit areas $A_1, A_2, A_3, \cdots$ have degeneracies $d_1, d_2, d_3, \cdots$. Suppose we have a black hole with area A which satisfies $A = \sum_i N_i A_i$ as explained before in (2). For a given configuration ($N_i = N_1, N_2, N_3, \cdots$), how many different ways can this be achieved?"

According to Rovelli [6], the area quanta are distinguishable, as they have fixed location on the black hole horizon, and, only if they are so, the Bekenstein-Hawking entropy can be satisfied. Using this distinguishability, the answer to the above question is given by

$$Q = N!\prod_{i=1}^{\infty}\frac{d_i^{N_i}}{N_i!} \tag{18}$$

where $N = \sum_i N_i$.

By maximizing Q and using (5), one obtains that the Hawking radiation for macroscopic black hole is given by the following Maxwell-Boltzmann distribution:

$$N_i = Nd_i e^{-hf_i/(kT)} = Nd_i e^{-A_i/4} \tag{19}$$

(Note that if we do not follow the present author's earlier work [11], this would be the familiar Einstein-Bose distribution of Hawking radiation, namely, $N_i/N = d_i/(\exp(hf_i/kT) - 1)$.)

We can do better than this. We can express the above expression in terms of the black hole horizon area A. By plugging the above formula to (2), we get

$$N = \frac{A}{\sum_i d_i A_i e^{-A_i/4}} \tag{20}$$

By plugging the above formula back to (19), we get

$$N_i = \frac{Ad_i e^{-A_i/4}}{\sum_i d_i A_i e^{-A_i/4}} \tag{21}$$

Notice that the above expression is proportional to the black hole horizon area A. Therefore, we conclude that the Hawking radiation for macroscopic black hole is proportional to the area of black hole, as much as Planck's blackbody radiation spectrum is proportional to the area of blackbody.

5. Single-Partition Black Hole

Let us see what happens for (2) for a single-partition black hole. In such a case we have $N = 1$. This implies $N_i = 1$ for the concerned i and 0 for other unconcerned values of i. Now, remember what N_i meant in (19). It is exactly Hawking radiation spectrum of given frequency. As the sum of N_i is N, N is the sum of Hawking radiation of every frequency, namely, total Hawking radiation. Thus, we conclude that the decay time of the single-partition black hole is constant as N, which is 1, is constant. In other words, τ is constant as desired.

6. Derivation

In this section, we derive (7) and (10) using the correct method.

At the end of Section 4, we showed that Hawking radiation is proportional to the black hole area A. However, this is only half correct as this holds true only if the temperature of black hole were fixed constant. In reality, the temperature of black hole depends on the black hole area A.

So, let us consider this dependence. The number of black hole area quanta is given by (20), and the number of photons emitted per a second is given by (12) and (13). That is,

$$\frac{1}{\tau} = A_{BH} \frac{2\pi}{c^2} \int_{\pi\sqrt{2}}^{\infty} \frac{u^2 du}{e^u - 1} \left(\frac{kT}{h} \right)^3 \quad (22)$$

where the integration range is not from 0 but from $\pi\sqrt{2}$ because the minimum unit area satisfies $A(y = 1)/4 = \pi\sqrt{2}$ which is nonzero. As $kT = 1/(8\pi M)$ and $A = 16\pi M^2$ (in this article and earlier ones we only consider Schwarzschild black hole for simplicity), we can write

$$\frac{1}{\tau} = N \frac{\alpha_0}{A_{BH}^{3/2}} \quad (23)$$

for some constant α_0 calculable from (20) and (22). Therefore, for macroscopic black hole, we can write

$$\frac{x_{A_i,\Delta t}}{\Delta t} = N_i \frac{\alpha_0}{A_{BH}^{3/2}} \quad (24)$$

which, when plugged in (21), is given by

$$\frac{x_{A_i,\Delta t}}{\Delta t} = \frac{d_i e^{-A_i/4}}{\sqrt{A_{BH}}} \left(\frac{\alpha_0}{\sum_i d_i A_i e^{-A_i/4}} \right) \quad (25)$$

Now, let us assume that this formula holds for single-partition black hole and further assume that the left-hand side of the above equation is constant, which is reasonable as N_i, being 1, is constant. Then, as the factors in the parenthesis in the above equation are merely constant, we immediately see that (7) and (10) are reproduced.

7. Discussions and Conclusions

One may argue that the table is not fitted correctly as it is fitted with Bose-Einstein distribution factor $e^{A/4} - 1$ while the table should be fitted with the Maxwell-Boltzmann distribution factor $e^{A/4}$ if author's earlier work [11] is correct. This is true, but the difference is insignificant as the lowest value for $e^{A/4}$ is about 85, which means that extra "-1" would not significantly affect the fitting.

In this paper, we showed what the thermodynamics of single-partition black hole implies about the degeneracy of area spectrum up from $y = 1$ to $y = 15$. For your information, the number of degeneracies for $A(y)$ up to $y = 65$ is available at (http://youngsubyoon.com/65.csv). If you add them up, you get $K(A)$.

In conclusion, we strongly believe that the extremely strange behaviors of Table 1 (the approximate constancy of C and the relation not holding beginning from exactly $y = 16$) are not a mere coincidence, instead they indicate the correctness of the area spectrum presented in [12].

Conflicts of Interest

The authors declare that they have no conflicts of interest.

References

[1] S. W. Hawking, "Particle creation by black holes," *Communications in Mathematical Physics*, vol. 43, no. 3, pp. 199–220, 1975.

[2] C. Rovelli and L. Smolin, "Loop space representation of quantum general relativity," *Nuclear Physics. B. Theoretical, Phenomenological, and Experimental High Energy Physics. Quantum Field Theory and Statistical Systems*, vol. 331, no. 1, pp. 80–152, 1990.

[3] C. Rovelli and L. Smolin, "Discreteness of area and volume in quantum gravity," *Nuclear Physics B*, vol. 442, no. 3, pp. 593–619, 1995.

[4] S. Frittelli, L. Lehner, and C. Rovelli, "The complete spectrum of the area from recoupling theory in loop quantum gravity," *Classical and Quantum Gravity*, vol. 13, no. 11, pp. 2921–2931, 1996.

[5] A. Ashtekar and J. Lewandowski, "Quantum theory of geometry. I. Area operators," *Classical and Quantum Gravity*, vol. 14, no. 1A, pp. A55–A81, 1997.

[6] C. Rovelli, "Black hole entropy from loop quantum gravity," *Physical Review Letters*, vol. 77, no. 16, pp. 3288–3291, 1996.

[7] J. D. Bekenstein and V. F. Mukhanov, "Spectroscopy of the quantum black hole," *Physics Letters B*, vol. 360, no. 1-2, pp. 7–12, 1995.

[8] M. Barreira, M. Carfora, and C. Rovelli, "Physics with non-perturbative quantum gravity: radiation from a quantum black hole," *General Relativity and Gravitation*, vol. 28, no. 11, pp. 1293–1299, 1996.

[9] K. V. Krasnov, "Quantum geometry and thermal radiation from black holes," *Classical and Quantum Gravity*, vol. 16, no. 2, pp. 563–578, 1999.

[10] Y. Yoon, "Quantum corrections to the Hawking radiation spectrum," *Journal of the Korean Physical Society*, vol. 68, no. 6, pp. 730–734, 2016.

[11] Y. Yoon, "Maxwell-Boltzmann type Hawking radiation," *Modern Physics Letters A*, vol. 32, no. 12, 1750071, 4 pages, 2017.

[12] B. Kong and Y. Yoon, "Black hole entropy predictions without the Immirzi parameter and Hawking radiation of a single-partition black hole," *Journal of the Korean Physical Society*, vol. 68, no. 6, pp. 735–751, 2016.

[13] J. D. Bekenstein, "Black holes and entropy," *Physical Review D: Particles, Fields, Gravitation and Cosmology*, vol. 7, pp. 2333–2346, 1973.

[14] T. Tanaka and T. Tamaki, "Black hole entropy for the general area spectrum," *Journal of Physics: Conference Series*, vol. 229, p. 012080, 2010.

[15] D. J. Griffiths, *Introduction to quantum mechanics*, Prentice Hall, Upper Saddle River, NJ, 2005.

5

Phase Transitions of GUP-Corrected Charged AdS Black Hole

Meng-Sen Ma[1,2] and Yan-Song Liu[2]

[1]Institute of Theoretical Physics, Shanxi Datong University, Datong 037009, China
[2]Department of Physics, Shanxi Datong University, Datong 037009, China

Correspondence should be addressed to Meng-Sen Ma; mengsenma@gmail.com

Academic Editor: Rong-Gen Cai

We study the thermodynamic properties and critical behaviors of the topological charged black hole in AdS space under the consideration of the generalized uncertainty principle (GUP). It is found that only in the spherical horizon case there are Van der Waals-like first-order phase transitions and reentrant phase transitions. From the equation of state we find that the GUP-corrected black hole can have one, two, and three apparent critical points under different conditions. However, it is verified by the Gibbs free energy that in either case there is at most one physical critical point.

1. Introduction

Since Hawking-Page phase transition of Schwarzschild-AdS black hole was explored in [1], phase structures and critical behaviors of various black holes in AdS space have been extensively studied [2–13]. Following [14, 15], the cosmological constant was considered as the thermodynamic pressure and the conjugated quantity was taken as the thermodynamic volume. In this extended phase space, the black hole mass M should be identified with the enthalpy. Although in [2, 3], the Van der Waals (VdW)-like first-order phase transition was first found in the RN-AdS black hole, in the extended phase space it was found that the critical behaviors of the RN-AdS black hole have more similarities to that of the VdW liquid/gas system [16]. This finding aroused many relevant studies on the critical phenomena of various AdS black holes in the extended phase space [17–25]. Furthermore, some special critical behaviors such as the reentrant phase transition (RPT), the triple critical point, the isolated critical point, and even the "critical curve" for several black holes have been explored [26–35].

After considering quantum gravity effects, thermodynamic quantities of black holes may be modified. For example, the generalized uncertainty principle (GUP) will lead to the corrected temperature and entropy [36–43]. Thus, the GUP should also influence the critical behaviors of black holes correspondingly. In [44], the author studied the effects of the GUP to all orders in the Planck length on the thermodynamics and the phase transition of the Schwarzschild black hole. In this paper we consider the usually used simpler form of the GUP

$$\Delta x \geq \frac{\hbar}{\Delta p} + \frac{\alpha^2}{\hbar}\Delta p \geq 2\alpha \sim l_p, \tag{1}$$

where l_p is the Planck length and α is a positive constant with length dimension whose upper limits can be given by the recent discovered gravitational waves [45]. On the basis of this relation, the corrected temperature and entropy for some static and stationary black holes were given in [46]. Using these corrected thermodynamic quantities, we have studied the critical behaviors of the Schwarzschild-AdS black hole and the RN-AdS black hole in [47]. With the GUP corrections, we find that the Hawking-Page phase transition for the AdS black holes no longer always occurs. In this paper, we will further study the critical behaviors and phase transitions of the corrected charged topological AdS black hole in the extended phase space. We find that a combination of α and the electric charge Q can be used to classify the various kinds of critical behaviors.

The plan of this paper is as follows: In Section 2 we introduce the corrected thermodynamic quantities of the charged AdS black hole and simply discuss their properties.

In Section 3 we find the critical points and analyze the numbers of the critical points. In Section 4 we study the critical behaviors of the black hole according to the Gibbs free energy. In Section 5 we summarize our results and discuss the possible future directions.

2. Thermodynamics of the Charged Topological AdS Black Hole with GUP Correction

In Einstein gravity in four-dimensional space-time, we have the charged topological AdS black hole solution,

$$ds^2 = -f(r)\,dt^2 + f(r)^{-1}\,dr^2 + r^2 d\Omega_k^2, \tag{2}$$

with the metric function [48]

$$f(r) = k - \frac{8\pi GM}{\Sigma_k r} + \frac{16\pi^2 G^2 Q^2}{\Sigma_k^2 r^2} + \frac{r^2}{l^2}, \tag{3}$$

where the parameters M, Q are the ADM mass and electric charge of the black hole and l represents the cosmological radius. $d\Omega_k^2$ denotes the line element of a two-dimensional Einstein space with constant scalar curvature $2k$ and volume Σ_k. Without loss of generality, one can take $k = 1$ (spherical horizon), $k = 0$ (planar/toroidal horizon), and $k = -1$ (hyperbolic horizon). Besides, we set $4\pi G/\Sigma_k = 1$ for simplicity. Although Σ_k has different values for different k, this simplification will not affect our physical results.

According to the metric function in (3), the black hole mass is

$$M = \frac{3kr_h^2 + 8\pi P r_h^4 + 3Q^2}{6r_h}, \tag{4}$$

where r_h denotes the position of the event horizon of the black hole. Here P is the thermodynamic pressure and is taken to be $P = -\Lambda/8\pi = 3/8\pi l^2$.

The surface gravity of the black hole is

$$\kappa = \frac{f'(r_h)}{2} = \frac{kr_h^2 + 8\pi P r_h^4 - Q^2}{2r_h^3}. \tag{5}$$

In the semiclassical case, the temperature and entropy for the black hole are

$$T = \frac{\hbar\kappa}{2\pi}, \qquad S = \frac{A}{4\hbar}. \tag{6}$$

As a thermodynamic system, the thermodynamic quantities of the black hole should satisfy the thermodynamic identity:

$$dM = TdS + \Phi dQ + VdP, \tag{7}$$

where the electric potential measured at infinity with reference to the horizon is $\Phi = Q/r_h$ and the thermodynamic volume is $V = 4\pi r_h^3/3$.

Generally, black hole entropy should be a function of the horizon area; namely, $S = S(A)$ [49]. Therefore, the temperature of a black hole can be generally expressed as [46]

$$T = \left.\frac{\partial M}{\partial S}\right|_Q = \frac{dA}{dS} \times \left.\frac{\partial M}{\partial A}\right|_Q = \frac{dA}{dS} \times \frac{\kappa}{8\pi}. \tag{8}$$

According to Heisenberg uncertainty principle, one can derive $dA/dS \simeq \Delta A/\Delta S = const.$ This is just the work of Bekenstein and Hawking, which gives the results in (6).

Considering the effect of GUP, it is shown that [46]

$$\frac{dA}{dS} \simeq \frac{(\Delta A)_{min}}{(\Delta S)_{min}} = 4\hbar', \tag{9}$$

where $\hbar'$ is the effective Planck "constant" and is defined as

$$\hbar' = \frac{2\hbar}{\alpha^2}\left(r_h^2 - r_h\sqrt{r_h^2 - \alpha^2}\right). \tag{10}$$

Thus, the GUP-corrected black hole temperature becomes

$$T' = \frac{\hbar'\kappa}{2\pi} = \frac{\hbar\left(r_h - \sqrt{r_h^2 - \alpha^2}\right)\left(kr_h^2 + 8\pi P r_h^4 - Q^2\right)}{2\pi\alpha^2 r_h^2}. \tag{11}$$

From (5), one can see that the usual temperature of the charged AdS black hole will become negative for very small r_h, while T' give a mandatory requirement $r_h \geq \alpha$, from which we find that the temperature T' can be always positive when the condition $Q^2 < \alpha^2(k + 8\pi P\alpha^2)$ is satisfied. In Figure 1, we compare the behaviors of the usual temperature T and the corrected temperature T' for the charged AdS black hole. For smaller Q, T' is indeed always positive. Besides, with the GUP corrections the $T' - r$ curve exhibits more fruitful structures.

Because GUP only constrains the minimal length, it only influences the temperature and the entropy. The electric charge and the electric potential will remain unchanged. The first law of black hole thermodynamics $dM = T'dS' + \Phi dQ + VdP$ should still be established in this case. Therefore, the GUP-corrected entropy of the black hole can be derived.

$$S' = \int \left.\frac{dM}{T'}\right|_{Q,P} = \int \left.\frac{1}{T'}\frac{\partial M}{\partial r}\right|_{Q,P} dr + S_0 = \frac{\pi}{2\hbar}\left[r_h^2 + r_h\sqrt{r_h^2 - \alpha^2} - \alpha^2 \ln\left(\frac{\sqrt{r_h^2 - \alpha^2} + r_h}{\alpha}\right)\right], \tag{12}$$

$$= \frac{A}{4\hbar} - \frac{\pi\alpha^2}{4\hbar}\ln\frac{A}{\pi\alpha^2} + \cdots \tag{13}$$

Here the effect of GUP leads to a subleading logarithmic term, which also exists in many other quantum corrected entropy. Our entropy is a little different from that in [46], where the authors take an indefinite integral and treat the integral constant as zero. We take the integration constant $S_0 = \alpha^2 \ln\alpha$ to obtain a dimensionless logarithmic term. S_0 cannot be fixed by some physical consideration. To determine S_0

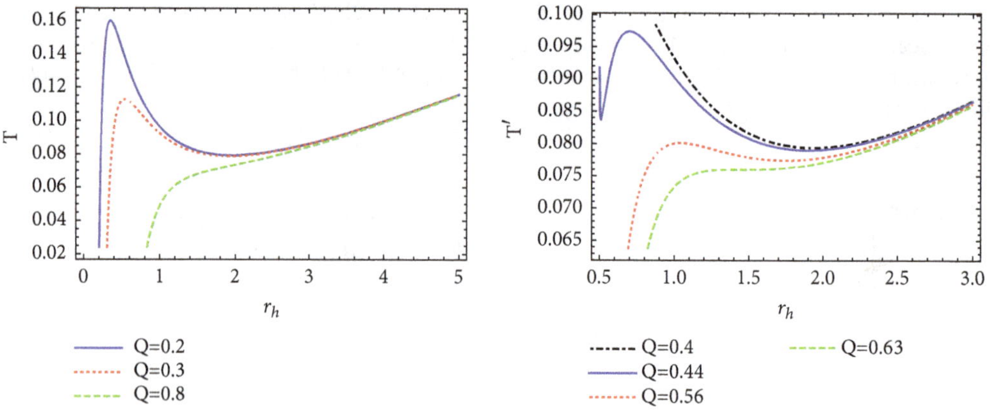

FIGURE 1: The left panel: the standard temperature of the RN-AdS black hole. The right panel: the GUP-corrected temperature with $\alpha = 0.5$. In both cases we take $k = 1$ and $P = 0.01$.

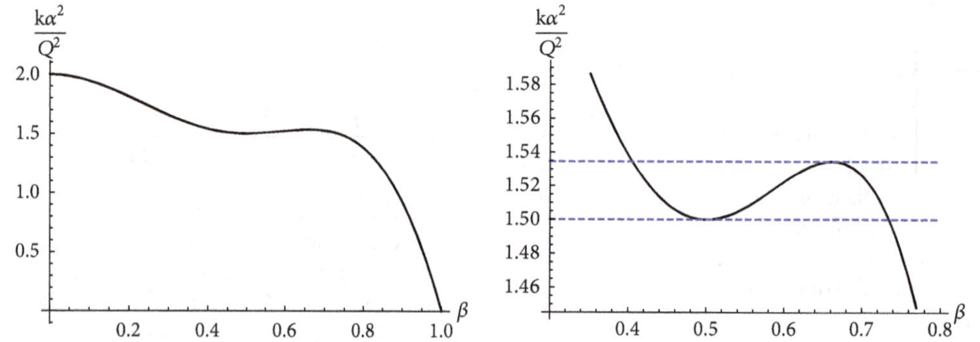

FIGURE 2: Number of the apparent critical points. There are at most three critical points for $1.5 < \alpha^2/Q^2 < 1.535$.

completely, one has to invoke the quantum theory of gravity. It should be noted that the corrected entropy is independent of the parameter k and it is always positive. Moreover, due to the existence of the logarithmic term in the corrected entropy, the Smarr formula no more exists.

3. Multiple Critical Points

In this section, we try to ascertain the number of the critical points. Below we always set $\hbar = 1$ for simplicity. From (11), we can derive the equation of state

$$P = \frac{r_h^2\left[2\pi T'\left(\sqrt{r_h^2 - \alpha^2} + r_h\right) - k\right] + Q^2}{8\pi r_h^4}. \tag{14}$$

To derive the critical points, one should solve the following two equations:

$$\frac{\partial P}{\partial r_h} = \frac{\partial^2 P}{\partial r_h^2} = 0. \tag{15}$$

One can also use another equivalent pair of equations, $\partial T'/\partial r_h = \partial^2 T'/\partial r_h^2 = 0$, to determine the critical points of the system. In either case, the results are the same.

The two expressions are lengthy, so we will not list them here. Combining them, we obtain an equation

$$\begin{aligned} &r_h^4\left(12Q^2 - \alpha^2 k\right) + 2r_h^3\sqrt{r_h^2 - \alpha^2}\left(\alpha^2 k + 3Q^2\right) \\ &\quad - 3kr_h^5\sqrt{r_h^2 - \alpha^2} - 26\alpha^2 Q^2 r_h^2 \\ &\quad - 4\alpha^2 Q^2 r_h\sqrt{r_h^2 - \alpha^2} + 16\alpha^4 Q^2 = 0. \end{aligned} \tag{16}$$

We set $\beta = \sqrt{1 - \alpha^2/r_h^2}$; thus $0 \le \beta \le 1$. Utilizing β, (16) can be simplified to

$$\begin{aligned} &\alpha^4(2\beta + 1)\left[\alpha^2\left(\beta^2 - \beta + 1\right)k \right. \\ &\quad \left. + 2\left(4\beta^5 - \beta^4 - 5\beta^3 + 2\beta^2 + \beta - 1\right)Q^2\right] = 0. \end{aligned} \tag{17}$$

In the case of $\alpha \neq 0$, we obtain a constraint equation

$$\frac{k\alpha^2}{Q^2} = \frac{2\left(4\beta^5 - \beta^4 - 5\beta^3 + 2\beta^2 + \beta - 1\right)}{-\beta^2 + \beta - 1}. \tag{18}$$

As is depicted in Figure 2, the right-hand side of (18) is nonnegative (in fact, when $\beta = 1$, the RHS of (18) is zero; however, because we are only interested in the nontrivial case with $\alpha \neq 0$, this excludes the possibility of $\beta = 1$). This means

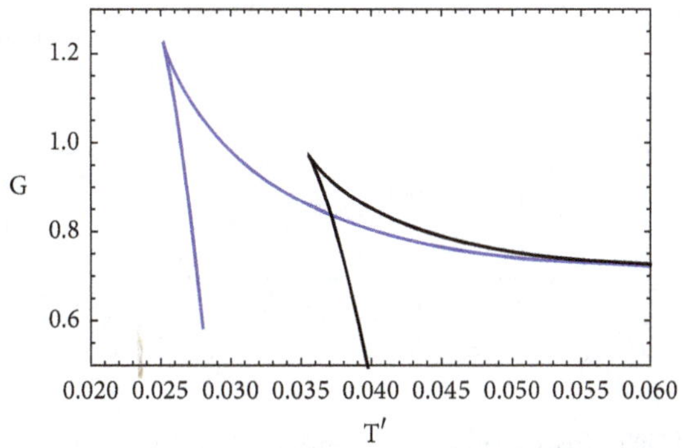

FIGURE 3: The case with $\alpha^2/Q^2 = 1.6$. The blue curve and the black curve correspond to $P = 0.001$ and $P = 0.002$, respectively.

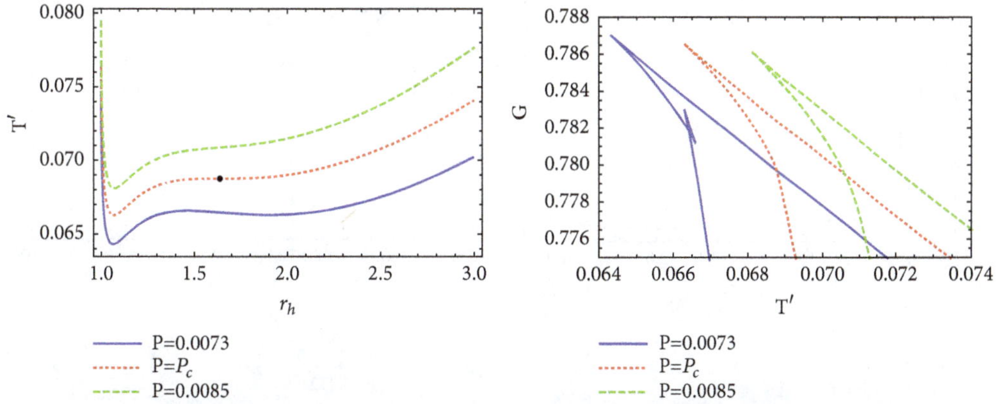

FIGURE 4: The $\alpha^2/Q^2 = 1.4$ case with $\alpha = 1$ and $Q = 0.845$. The critical pressure is $P_c = 0.0079$.

that there is no $P - V$ criticality in the cases with $k = 0$ and $k = -1$. Besides, when the electric charge $Q = 0$, (17) becomes

$$k\alpha^6 (2\beta + 1)\left(\beta^2 - \beta + 1\right) = 0, \tag{19}$$

which also has no real solutions for β in the range $0 \le \beta \le 1$. If $k = 0$, this equation is well satisfied; however one can easily check that there is still no critical point in the ($k = 0$, $Q = 0$) case.

Thus, below we are only concerned with the $k = 1$ case with nonzero electric charge Q. We find that the number of critical points depends on the value of α^2/Q^2. When $\alpha^2/Q^2 > 1.535$ or $\alpha^2/Q^2 < 1.5$, there is only one critical point. When $\alpha^2/Q^2 = 1.535$ or $\alpha^2/Q^2 = 1.5$, there are two critical points. And three critical points occur when $1.5 < \alpha^2/Q^2 < 1.535$.

On the basis of these critical points, we can further discuss the heat capacity at constant pressure,

$$C_p = T' \left.\frac{\partial S'}{\partial T'}\right|_{P,Q} = T' \left.\frac{\partial S'/\partial r_h}{\partial T'/\partial r_h}\right|_{P,Q}, \tag{20}$$

which can reflect the local thermodynamic stability of the black hole. The divergence points of the heat capacity occur at zeros of $\partial T'/\partial r_h$, which are the extremal points in the $T' - r_h$ curve. The sign of C_p is also completely determined by $\partial T'/\partial r_h$ because $\partial S'/\partial r_h = \pi(\sqrt{r_h^2 - \alpha^2} + r_h) > 0$ and the corrected temperature is greater than zero if $Q^2 < \alpha^2(1+8\pi P)$.

4. The Critical Behaviors and Gibbs Free Energy

Below we discuss the critical behaviors of the RN-AdS black hole according to the numbers of the apparent critical points.

4.1. One Critical Point. In this case, we take $\alpha^2/Q^2 = 1.4$ and $\alpha^2/Q^2 = 1.6$, respectively. First, for $\alpha^2/Q^2 = 1.6$ one can easily find that the critical value of the pressure is negative, which means that no second-order phase transition occurs. In fact, there is also no VdW-like first-order phase transition. As is shown in Figure 3, the Gibbs free energy exhibits a cusp for any positive given pressure.

The critical behaviors in the case with $\alpha^2/Q^2 = 1.4$ have been analyzed in [47]. When the pressure is greater than the critical pressure P_c, there is two branches in the $T' - r_h$ curve, which corresponds to a cusp in G. When the pressure is lower than P_c, the $T' - r_h$ curve exhibits four branches. From left to right, we call them the small black hole, the left-intermediate black hole, the right-intermediate black hole, and the large black hole. According to the slope of the $T' - r_h$ curve one can figure out that C_p is negative in the small black hole branch and the right-intermediate black hole branch and it is positive in the left-intermediate black hole branch and the large black hole branch. From the $G - T'$ figure, one can see a standard VdW-like first-order phase transition. We illustrate this in Figure 4. For $P \in (P_t, P_z)$, reentrant phase transition takes

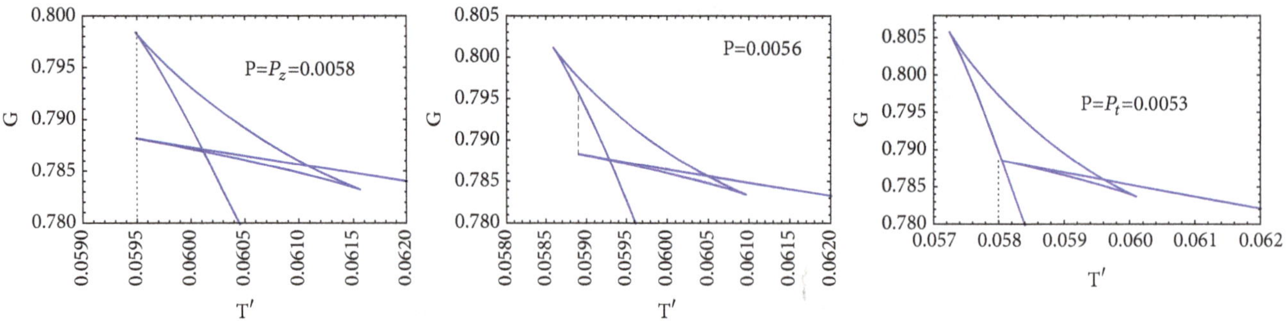

FIGURE 5: Characteristic behavior of Gibbs free energy for the reentrant phase transition. Here we also take $\alpha = 1$ and $Q = 0.845$.

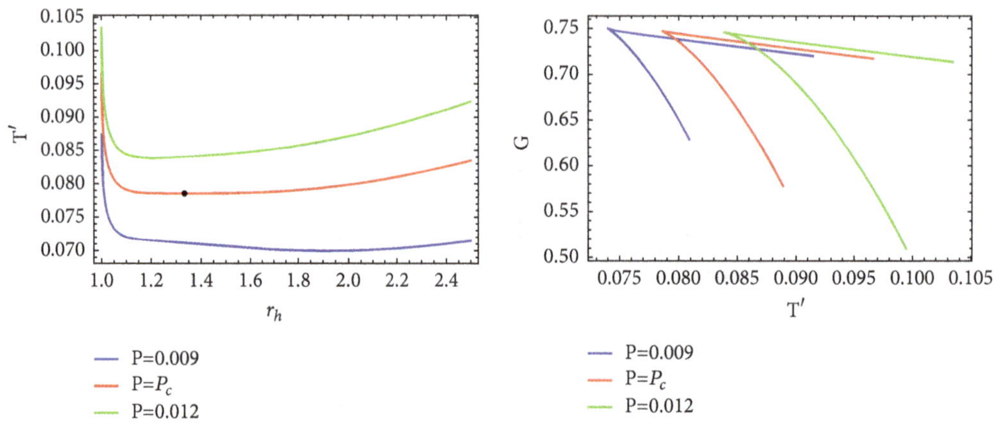

FIGURE 6: The $\alpha^2/Q^2 = 1.535$ case with $\alpha = 1$ and $Q = 0.807$.

place. In this case, starting off from the largest G, the black hole will first evolve along the branch of the large black hole. Then at some point the black hole will undergo a zero-order phase transition and jump to the left-intermediate branch. Finally, undergoing a first-order phase transition, the black hole returns back to the original large black hole. This process has been illustrated in Figure 5. When $P < P_t$, the large black hole is globally thermodynamic stable, and thus no phase transition occurs.

4.2. Two Critical Points. When $\alpha^2/Q^2 = 1.535$, one can easily check that the smaller critical point is false because the critical pressure P_c is negative. The larger critical point is ($P_c = 0.0103$, $T_c = 0.0786$, $r_{hc} = 1.333$). However, as is shown in Figure 6, the black hole only has two branches and exhibits a cusp in G. The lower branch has the positive heat capacity and smaller Gibbs free energy. Thus it is more stable for the large black hole and no phase transition occurs in this case.

For $\alpha^2/Q^2 = 1.5$, the smaller critical point is ($T_{c1} = 0.0613$, $P_{c1} = 0.00497$, $r_{c1} = 1.155$), and the larger critical point is ($T_{c2} = 0.0753$, $P_{c2} = 0.00945$, $r_{c2} = 1.474$). As is illustrated in Figure 7, for $P \in (P_{c1}, P_{c2})$ the black hole has four branches. Similar to the one critical point case ($\alpha^2/Q^2 = 1.4$), for $P \in (P_t, P_{c2})$ the black hole always has a VdW-like first-order phase transition and for $P \in (P_t, P_z)$ there exists the reentrant phase transition. Below a physical critical

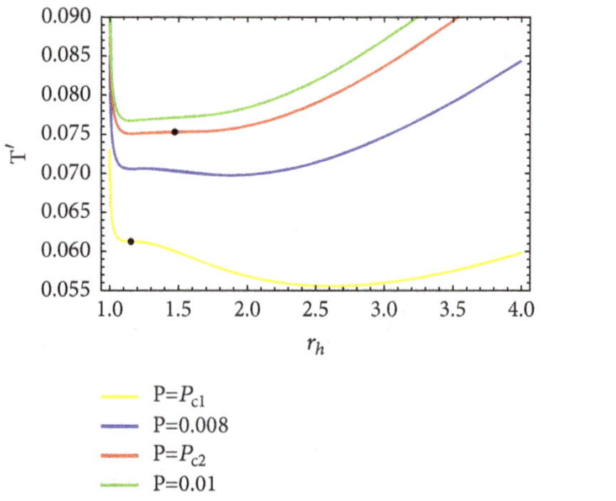

FIGURE 7: The $\alpha^2/Q^2 = 1.5$ case with $\alpha = 1$ and $Q = 0.816$. The reentrant phase transition takes place for $P \in (P_t, P_z)$ with $P_t = 0.00887$, $P_z = 0.009$.

point there should be a VdW-like first-order phase transition. Because $P_{c1} < P_t$, for $P \in (P_{c1}, P_t)$ or $P < P_{c1}$ the large black hole is always globally thermodynamically stable. Therefore, the smaller critical point is indeed an apparent one, which does not correspond to any second-order phase transition. In this case, the Gibbs free energy has similar behaviors to that in Figures 4 and 5.

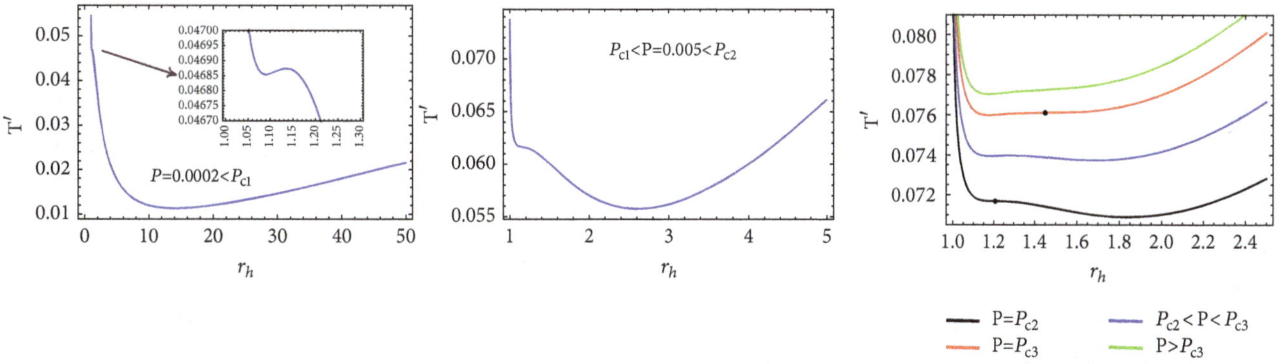

FIGURE 8: $T' - r_h$ curves for $\alpha^2/Q^2 = 1.51$ with $\alpha = 1$ and $Q = 0.814$.

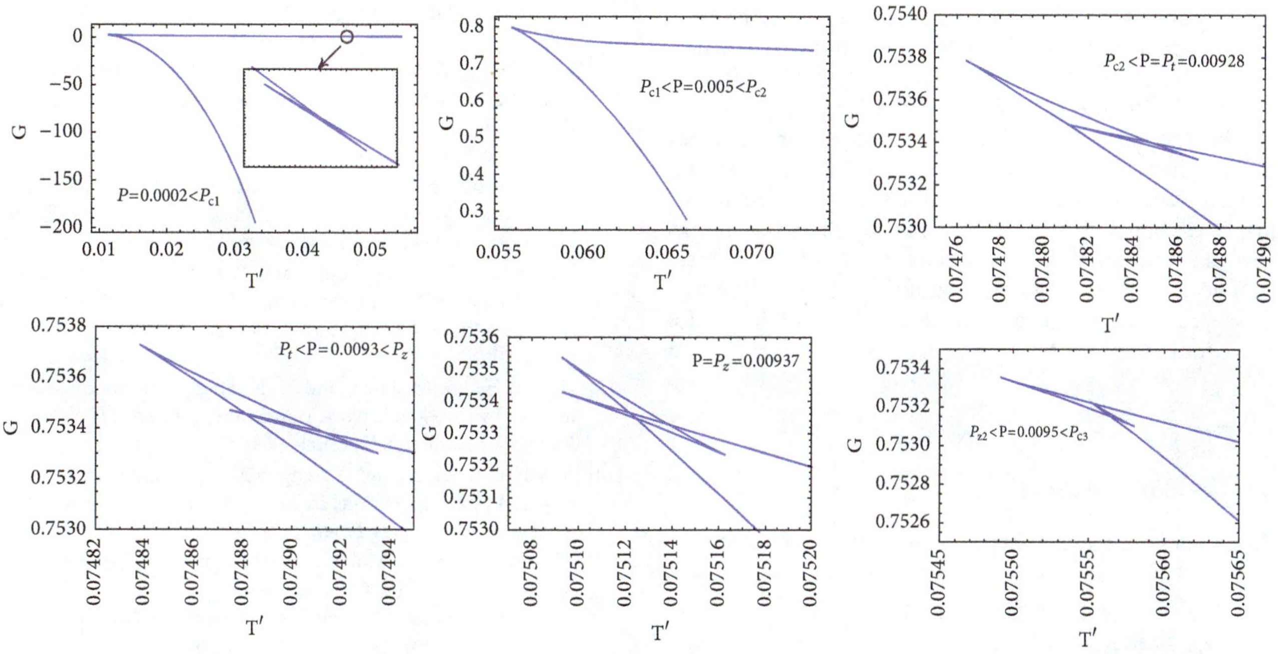

FIGURE 9: The different behaviors of the Gibbs free energy at different pressures. $\alpha^2/Q^2 = 1.51$ with $\alpha = 1$ and $Q = 0.814$.

4.3. Three Critical Points. In this case, we select $\alpha^2/Q^2 = 1.51$. Three apparent critical points are ($T_{c1} = 0.0497$, $P_{c1} = 0.00111$, $r_{c1} = 1.118$), ($T_{c2} = 0.0717$, $P_{c2} = 0.00826$, $r_{c2} = 1.207$), and ($T_{c3} = 0.0761$, $P_{c3} = 0.00966$, $r_{c3} = 1.45$), respectively.

As is shown in Figure 8, for $P < P_{c1}$ and $P \in (P_{c2}, P_{c3})$, there are four branches and there are two branches for $P > P_{c3}$ and $P \in (P_{c1}, P_{c2})$. According to the illustration in Figure 9, when $P < P_{c1}$ and $P_{c2} < P < P_t$, there are swallowtail behaviors; however the swallowtail never intersects with the large black hole branch. When $P \in (P_{c1}, P_{c2})$, it is a cusp in the Gibbs free energy. This means that the large black hole is always globally thermodynamically stable when $P < P_t$ and "$c1$" and "$c2$" are not physical critical points. For $P > P_{c3}$, the Gibbs free energy also exhibits a cusp and for $P \in (P_t, P_{c3})$ there is always a phase transition of first order. Thus, "$c3$" is a physical critical point, which corresponds to a second-order phase transition.

5. Conclusion and Discussion

We begin by considering the charged AdS black hole with general topology ($k = 0, \pm 1$). After considering the GUP we find that the temperature of the RN-AdS black hole, not like its semiclassical counterpart, can be always positive and has more fruitful structures. The GUP-corrected entropy is always positive and is independent of the electric charge Q and the parameter k. By analyzing the equation of state, we find that only in the $k = 1$ case the black hole can have critical behaviors. In particular, one can judge the number of the critical points according to the values of α^2/Q^2. Apparently, the number of the critical points can be one, two, and three. In either case, there is the unique physical critical point. The main results have been summarized in Table 1.

For $P > 1.535$, the apparent critical point has a negative pressure, which is unphysical. At any positive pressure, the Gibbs free energy exhibits a cusp. For $P = 1.535$, the smaller

TABLE 1: The critical behaviors of RN-AdS black hole with GUP corrections in the $k = 1$ case.

α^2/Q^2	# apparent critical points	# physical critical points	behavior
$\left(\frac{1}{1+8\pi P}, 1.5\right)$	1	1	VdW& RPT
1.5	2	1	VdW& RPT
(1.5, 1.535)	3	1	VdW& RPT
1.535	2	0	cusp
$(1.535, \infty)$	1	0	cusp

one of the two apparent critical points also has negative pressure. Below or above the larger critical point, the $T' - r_h$ curve only has two branches, which corresponds to a cusp in the Gibbs free energy. For $1.5 \le P < 1.535$, only the rightmost critical point is the physical one. In the first three cases in Table 1, there is always the VdW-like phase transitions, which occur at the range $P \in (P_t, P_c)$, $P \in (P_t, P_{c2})$, $P \in (P_t, P_{c3})$, respectively, and the RPT takes place for $P \in (P_t, P_z)$.

The Born-Infeld-AdS black hole and the Kerr-AdS black hole can have the reentrant behavior due to their complex horizon structure. GUP only modifies the thermodynamic quantities, but not the geometric structure like the metric. If we take into account the corrections of the GUP in these complicated black holes, there should be more interesting phase structure and critical phenomena. This will be left for future studies.

Conflicts of Interest

The authors declare that they have no conflicts of interest.

Acknowledgments

This work is supported in part by the National Natural Science Foundation of China (Grants No. 11605107 and No. 11475108) and by the Natural Science Foundation of Shanxi (Grant No. 201601D021022).

References

[1] S. W. Hawking and D. N. Page, "Thermodynamics of black holes in anti-de Sitter space," *Communications in Mathematical Physics*, vol. 87, no. 4, pp. 577–588, 1982/83.

[2] A. Chamblin, R. Emparan, C. V. Johnson, and R. C. Myers, "Charged AdS black holes and catastrophic holography," *Physical Review D: Particles, Fields, Gravitation and Cosmology*, vol. 60, no. 6, 1999.

[3] A. Chamblin, R. Emparan, C. V. Johnson, and R. C. Myers, "Holography, thermodynamics, and fluctuations of charged AdS black holes," *Physical Review D: Particles, Fields, Gravitation and Cosmology*, vol. 60, no. 10, Article ID 104026, 1999.

[4] C. S. Peça and J. P. S. Lemos, "Thermodynamics of Reissner-Nordström-anti-de Sitter black holes in the grand canonical ensemble," *Physical Review D: Particles, Fields, Gravitation and Cosmology*, vol. 59, no. 12, Article ID 124007, 1999.

[5] X. N. Wu, "Multicritical phenomena of Reissner-Nordström anti–de Sitter black holes," *Physical Review D: Particles, Fields, Gravitation and Cosmology*, vol. 62, no. 12, Article ID 124023, 2000.

[6] Y. S. Myung, Y.-W. Kim, and Y.-J. Park, "Thermodynamics and phase transitions in the Born-Infeld-anti-de Sitter black holes," *Physical Review D: Particles, Fields, Gravitation and Cosmology*, vol. 78, no. 8, Article ID 084002, 2008.

[7] H. Quevedo and A. Sanchez, "Geometrothermodynamics of asymptotically anti-de Sitter black holes," *Journal of High Energy Physics*, no. 9, 034 pages, 2008.

[8] M. Cadoni, G. D'Appollonio, and P. Pani, "Phase transitions between Reissner-Nordstrom and dilatonic black holes in 4D AdS spacetime," *Journal of High Energy Physics*, no. 3, 100, 27 pages, 2010.

[9] H. Liu, H. Lü, M. Luoa, and K.-N. Shao, "Thermodynamical metrics and black hole phase transitions," *Journal of High Energy Physics*, vol. 2010, no. 12, article 054, 2010.

[10] A. Sahay, T. Sarkar, and G. Sengupta, "Thermodynamic geometry and phase transitions in Kerr-Newman-AdS black holes," *Journal of High Energy Physics*, vol. 2010, article 118, 2010.

[11] R. Banerjee, S. K. Modak, and S. Samanta, "Second order phase transition and thermodynamic geometry in Kerr-AdS black holes," *Physical Review D: Particles, Fields, Gravitation and Cosmology*, vol. 84, Article ID 064024, 2011.

[12] M.-S. Ma, F. Liu, and R. Zhao, "Continuous phase transition and critical behaviors of 3D black hole with torsion," *Classical and Quantum Gravity*, vol. 31, no. 9, Article ID 095001, 2014.

[13] M.-S. Ma and R. Zhao, "Phase transition and entropy spectrum of the BTZ black hole with torsion," *Physical Review D: Particles, Fields, Gravitation and Cosmology*, vol. 89, no. 4, Article ID 044005, 2014.

[14] D. Kastor, S. Ray, and J. Traschen, "Enthalpy and the mechanics of AdS black holes," *Classical and Quantum Gravity*, vol. 26, no. 19, Article ID 195011, 16 pages, 2009.

[15] B. P. Dolan, "Pressure and volume in the first law of black hole thermodynamics," *Classical and Quantum Gravity*, vol. 28, no. 23, Article ID 235017, 2011.

[16] D. Kubiznak and R. B. Mann, "P-V criticality of charged AdS black holes," *Journal of High Energy Physics*, vol. 2012, no. 7, article 033, 2012.

[17] R.-G. Cai, L.-M. Cao, L. Li, and R.-Q. Yang, "P-V criticality in the extended phase space of Gauss-Bonnet black holes in AdS space," *Journal of High Energy Physics*, vol. 2013, no. 9, article 5, 2013.

[18] S. Chen, X. Liu, and C. Liu, "P-V Criticality of an AdS Black Hole in f(R) Gravity," *Chinese Physics Letters*, vol. 30, no. 6, Article ID 060401, 2013.

[19] S. H. Hendi and M. H. Vahidinia, "Extended phase space thermodynamics and P-V criticality of black holes with a nonlinear source," *Physical Review D: Particles, Fields, Gravitation and Cosmology*, vol. 88, no. 8, Article ID 084045, 11 pages, 2013.

[20] N. Altamirano, D. Kubizňák, R. Mann, and Z. Sherkatghanad, "Thermodynamics of rotating black holes and black rings: phase transitions and thermodynamic volume," *Galaxies*, vol. 2, no. 1, pp. 89–159, 2014.

[21] J.-X. Mo and W.-B. Liu, "*P-V* criticality of topological black holes in Lovelock-Born-Infeld gravity," *The European Physical Journal C*, vol. 74, article 2836, 2014.

[22] H. Xu, W. Xu, and L. Zhao, "Extended phase space thermodynamics for third-order Lovelock black holes in diverse dimensions," *The European Physical Journal C*, vol. 74, article 3074, 2014.

[23] M.-S. Ma and Y.-Q. Ma, "Critical behaviors of a black hole in an asymptotically safe gravity with cosmological constant," *Classical and Quantum Gravity*, vol. 32, no. 3, p. 035024, 2015.

[24] M.-S. Ma and R. Zhao, "Stability of black holes based on horizon thermodynamics," *Physics Letters B*, vol. 751, pp. 278–283, 2015.

[25] H.-H. Zhao, L.-C. Zhang, M.-S. Ma, and R. Zhao, "Phase transition and Clapeyron equation of black holes in higher dimensional AdS spacetime," *Classical and Quantum Gravity*, vol. 32, no. 14, Article ID 145007, 2015.

[26] N. Altamirano, D. Kubiznak, and R. Mann, "Reentrant phase transitions in rotating AdS black holes," *Physical Review D: Particles, Fields, Gravitation and Cosmology*, vol. 88, no. 10, Article ID 101502, 5 pages, 2013.

[27] S.-W. Wei and Y.-X. Liu, "Triple points and phase diagrams in the extended phase space of charged Gauss-Bonnet black holes in AdS space," *Physical Review D: Particles, Fields, Gravitation and Cosmology*, vol. 90, no. 4, Article ID 044057, 2014.

[28] B. P. Dolan, A. Kostouki, D. Kubizňák, and R. B. Mann, "Isolated critical point from Lovelock gravity," *Classical and Quantum Gravity*, vol. 31, no. 24, Article ID 242001, 2014.

[29] A. M. Frassino, D. Kubizňák, R. B. Mann, and F. Simovic, "Multiple reentrant phase transitions and triple points in Lovelock thermodynamics," *Journal of High Energy Physics*, vol. 2014, no. 9, article 80, 2014.

[30] M. Zhang, D.-C. Zou, and R.-H. Yue, "Reentrant Phase Transitions and Triple Points of Topological AdS Black Holes in Born-Infeld-Massive Gravity," *Advances in High Energy Physics*, vol. 2017, Article ID 3819246, 11 pages, 2017.

[31] A. Dehyadegari and A. Sheykhi, "Reentrant phase transition of Born-Infeld-AdS black holes," *Physical Review D: Particles, Fields, Gravitation and Cosmology*, vol. 98, no. 2, 2018.

[32] R. A. Hennigar, R. B. Mann, and E. Tjoa, "Superfluid black holes," *Physical Review Letters*, vol. 118, no. 2, Article ID 021301, 2017.

[33] R. A. Hennigar, E. Tjoa, and R. B. Mann, "Thermodynamics of hairy black holes in Lovelock gravity," *Journal of High Energy Physics*, vol. 2017, no. 2, article 70, 2017.

[34] H. Dykaar, R. A. Hennigar, and R. B. Mann, "Hairy black holes in cubic quasi-topological gravity," *Journal of High Energy Physics*, no. 5, 045, front matter+29 pages, 2017.

[35] M.-S. Ma and R.-H. Wang, "Peculiar P-V criticality of topological Hořava-Lifshitz black holes," *Physical Review D: Particles, Fields, Gravitation and Cosmology*, vol. 96, no. 2, Article ID 024052, 2017.

[36] R. J. Adler, P. Chen, and D. I. Santiago, "The generalized uncertainty principle and black hole remnants," *General Relativity and Gravitation*, vol. 33, no. 12, pp. 2101–2108, 2001.

[37] A. J. Medved and E. C. Vagenas, "When conceptual worlds collide: the generalized uncertainty principle and the Bekenstein-Hawking entropy," *Physical Review D: Particles, Fields, Gravitation and Cosmology*, vol. 70, no. 12, Article ID 124021, 5 pages, 2004.

[38] Z. Ren and Z. Sheng-Li, "Generalized uncertainty principle and black hole entropy," *Physics Letters. B. Particle Physics, Nuclear Physics and Cosmology*, vol. 641, no. 2, pp. 208–211, 2006.

[39] K. Nouicer, "Quantum-corrected black hole thermodynamics to all orders in the Planck length," *Physics Letters B*, vol. 646, no. 2-3, pp. 63–71, 2007.

[40] W. Kim, E. J. Son, and M. Yoon, "Thermodynamics of a black hole based on a generalized uncertainty principle," *Journal of High Energy Physics*, vol. 2008, no. 1, article 35, 2008.

[41] B. Majumder, "Quantum black hole and the modified uncertainty principle," *Physics Letters B*, vol. 701, no. 4, pp. 384–387, 2011.

[42] Z. W. Feng, H. L. Li, X. T. Zu, and S. Z. Yang, "Quantum corrections to the thermodynamics of Schwarzschild-Tangherlini black hole and the generalized uncertainty principle," *The European Physical Journal C*, vol. 76, article 212, 2016.

[43] E. C. Vagenas, S. M. Alsaleh, and A. F. Ali, "GUP parameter and black-hole temperature," *EPL (Europhysics Letters)*, vol. 120, no. 4, Article ID 40001, 2017.

[44] Y. Sabri and K. Nouicer, "Phase transitions of a GUP-corrected Schwarzschild black hole within isothermal cavities," *Classical and Quantum Gravity*, vol. 29, no. 21, 215015, 15 pages, 2012.

[45] Z.-W. Feng, S.-Z. Yang, H.-L. Li, and X.-T. Zu, "Constraining the generalized uncertainty principle with the gravitational wave event GW150914," *Physics Letters B*, vol. 768, pp. 81–85, 2017.

[46] L. Xiang and X. Q. Wen, "Black hole thermodynamics with generalized uncertainty principle," *Journal of High Energy Physics*, vol. 2009, no. 10, article 46, 2009.

[47] Z. Sun and M.-S. Ma, "The critical behaviors of the black holes with the generalized uncertainty principle," *EPL (Europhysics Letters)*, vol. 122, no. 6, p. 60002, 2018.

[48] R.-G. Cai and K.-S. Soh, "Topological black holes in the dimensionally continued gravity," *Physical Review D: Particles, Fields, Gravitation and Cosmology*, vol. 59, no. 4, Article ID 044013, 1999.

[49] J. D. Bekenstein, "Black holes and entropy," *Physical Review D: Particles, Fields, Gravitation and Cosmology*, vol. 7, pp. 2333–2346, 1973.

6

Anisotropic Quintessence Strange Stars in $f(T)$ Gravity with Modified Chaplygin Gas

Pameli Saha and Ujjal Debnath

Department of Mathematics, Indian Institute of Engineering Science and Technology, Shibpur, Howrah 711 103, India

Correspondence should be addressed to Pameli Saha; pameli.saha15@gmail.com

Academic Editor: Salvatore Mignemi

In this paper, we study the existence of strange star in the background of $f(T)$ modified gravity where T is a scalar torsion. In KB metric space, we derive the equations of motion using anisotropic property within the spherically strange star with modified Chaplygin gas in the framework of modified $f(T)$ gravity. Then we obtain many physical quantities to describe the physical status such as anisotropic behavior, energy conditions, and stability. By the matching condition, we calculate the unknown parameters to evaluate the numerical values of mass, surface redshift, etc., from our model to make comparison with the observational data.

1. Introduction

In modern cosmology, cosmic acceleration is an interesting discovery. The observation of type Ia supernovae (SNeIa) together with the cosmic microwave background (CMB), large scale structure surveys (LSS), and Wilkinson Microwave Anisotropy Probe (WMAP) [1–4] ensures the presence of an exotic energy component dominating our universe which is entitled as dark energy (DE) **having equation of state** $p = w\rho$ **with strong negative pressure. For accelerating expansion** w **must satisfy the range** $w < -1/3$**. If** $-1 < w < -1/3$ **then it belongs to quintessence phase and if** $w < -1$**, then it belongs to phantom regime. In particular, when** $w = -1 \Longrightarrow p = -\rho$ **then the equation of state of the interior region of a Gravastar (gravitationally vacuum condense star) is described in** [5–10]. There are many investigations of this cosmic expansion and nature of DE based on different ways. These efforts can be classified as follows: (i) **to modify the entire cosmic energy by including new components of DE** and (ii) to modify Einstein-Hilbert action to get different types of modified theories of gravity such as $f(R)$ gravity [11, 12], R being the Ricci scalar; $f(T)$ gravity [13], T being the torsion; $f(R,T)$ gravity [14], Gauss-Bonnet gravity, i.e., $f(G)$ modified gravity [15], etc. Here we assume only $f(T)$ gravity theory.

Since general relativity is similar to $f(T)$, this theory could be a substitute form of the generalized general relativity, named as $f(T)$ theory of gravity. The teleparallel equivalence of gravity (TEGR) gives the concept of this theory. There is defined Riemann-Cartan space-time together with Weitzenbock connections rather than Levi-Civita connections in $f(T)$ theory. **Here**, nonzero torsion and zero curvature appear in the background space-time. Einstein gives this definition of space-time to give an idea of gravitation related to tetrad and torsion. Instead of metric field, tetrad field takes an important role in dynamic field in TEGR.

In $f(T)$ **gravity, equations of motions are second-order differential equations like GR whereas equations of motion are fourth-order in** $f(R)$ **gravity.** So, the former one is more convenient than the latter one. Recently, a wide interest has been seen to study the $f(T)$ gravity [16–20]. There is no doubt of excellence of $f(T)$ theory to explain the cosmic acceleration and analysis on large scale (clustering of galaxies) [21]. But GR must be a fantabulous agreement with solar system test and pulsar observation [22].

In theoretical astrophysics, $f(T)$ version of BTZ black hole solutions has been calculated as $f(T)$ theory was supported for examining the effects of $f(T)$ models in 3 dimensions [23]. **Later on [24], violation of Lorentz invariance made the first violation of black hole thermodynamics**

in $f(T)$ gravity. Recently there are some static solutions which are spherically symmetric with charged source in $f(T)$ theory [25]. The physical conditions have been studied [26] for the existence of astrophysical stars in $f(T)$ theory after obtaining a large group of static perfect fluid solutions [27]. **Capozziello et. al [28] have shown that, instead of $f(R)$ gravity, $f(T)$ removes the singularities for the exact black hole solution in D-Dimensions. Wormhole solution has been studied under $f(T)$ gravity by Sharif and Rani [29]. They have also investigated $f(T)$ gravity for static wormhole solution to verify energy conditions [30]. Again, for charged noncommutative wormhole solutions in f(T) gravity, Sharif and Rani [31, 32] have seen that this solution exists by violating energy conditions.**

Generally, perfect fluid (isotropic fluid) inside the stellar object to study stellar structure and evolution is assumed because there exists isotropic pressure inside the fluid sphere. However, present observation shows that the fluid pressure of the highly compact astrophysical objects like X-ray pulsar, Her-X-1, X-ray buster 4U 1820-30, millisecond pulsar SAXJ1804.4-3658, etc. becomes anisotropy in nature which means the pressure can be rotten into two components such that one is radial pressure (p_r) and the other is transverse pressure (p_t). Now, $\Delta = p_t - p_r$ is known as the anisotropic factor. The anisotropy may arise for the different cases such as the existence of solid core, in presence of type P superfluid, phase transition, rotation, magnetic field, mixture of two fluids, and existence of external field. Generally, strange quark matter contains u, d, and s quarks. There are two ways to classify the formation of strange matter [33]. One way is the transformation of the quark hadron phase in the early universe and the other way is the reformation of neutron stars to strange matter at ultrahigh densities. A strange star is composed of the strange matter. Again the strange star can be classified into two types: Type I strange star with $M/R > 0.3$ and Type II strange star with $0.2 < M/R < 0.3$. Depending on mass, radius, and energy density, the strange star is distinguished from the neutron star [34]. It has been the most interesting topic to study the models of anisotropic stars for the last periods in GR and modified theories of gravity [35]. There have been many discussions about anisotropic star models in [36–41]. It is becoming a scientific tool to discuss the compact star models with Krori-Barua metric [42–44]. **It has been seen in [45] that neutron star solution in $f(T)$ gravity model is possible if $f(T)$ is a linear function of scalar torsion.**

Recently, Abbas and his collaborations [46–50] have discussed the anisotropic compact star models in GR, $f(R)$, $f(G)$, and $f(T)$ theories in diagonal tetrad case with Krori and Barua (KB) metric. **Abbas et al. [49] have studied anisotropic strange star which corresponds to quintessence dark energy model with the help equation of state $p = \alpha\rho$, where $0 < \alpha < 1$. A study of strange star with MIT bag model in the framework of $f(T)$ gravity has been done by Abbas et al. [51]. Here, our** main motivation of this paper is to study the anisotropic strange star models in the framework of $f(T)$ gravity with diagonal tetrad in presence of electric field and modified Chaplygin gas. In Section 2, we give a brief idea of $f(T)$ gravity. **In Section 3, we study anisotropic quintessence strange star in $f(T)$ gravity with the help of modified Chaplygin gas. In Section 4, we analyze many physical phenomenon of this whole system. By matching of two metrics, the unknown constants are found out. We also make stability analysis. In Section 5, we calculate the mass function, compactness, and surface redshift function from our model to compare with observational data and finally, in Section 6, we give the summarization.**

2. $f(T)$ Gravity: Fundamentals

In this section, we briefly overview the basics of $f(T)$ gravity. We define the torsion and the con-torsion tensor as follows [51]:

$$T^{\alpha}_{\mu\nu} = \Gamma^{\alpha}_{\nu\mu} - \Gamma^{\alpha}_{\mu\nu} = e^{\alpha}_{i}\left(\partial_{\mu}e^{i}_{\nu} - \partial_{\nu}e^{i}_{\mu}\right) \quad (1)$$

$$K^{\mu\nu}_{\alpha} = -\frac{1}{2}\left(T^{\mu\nu}_{\alpha} - T^{\nu\mu}_{\alpha} - T^{\mu\nu}_{\alpha}\right) \quad (2)$$

and the components of the tensor $S^{\mu\nu}_{\alpha}$ are defined as

$$S^{\mu\nu}_{\alpha} = \frac{1}{2}\left(K^{\mu\nu}_{\alpha} + \delta^{\mu}_{\alpha}T^{\beta\nu}_{\beta} - \delta^{\nu}_{\alpha}T^{\beta\mu}_{\beta}\right); \quad (3)$$

one can write the torsion scalar as

$$T = T^{\alpha}_{\mu\nu}S^{\mu\nu}_{\alpha} \quad (4)$$

Now, one can define the modified teleparallel action by replacing T with a function of T, in analogy to $f(R)$ gravity [52, 53], as follows:

$$S = \int d^4xe\left[\frac{1}{16\pi}f(T) + L_{Matter}(\Phi_A)\right] \quad (5)$$

where we used $G = c = 1$ and Φ_A is matter fields.

The ordinary matter is an anisotropic fluid so that the energy-momentum tensor is given by

$$T^{\nu}_{\mu} = (\rho + p_t)u_{\mu}u^{\nu} - p_t\delta_{\mu}{}^{\nu} + (p_r - p_t)v_{\mu}v^{\nu} \quad (6)$$

where u^{μ} is the four-velocity, v^{μ} is radial four vectors, ρ is the energy density, p_r is the radial pressure, and p_t is transverse pressure. Further, the energy-momentum tensor for electromagnetic field is given by

$$E^{\nu}_{\mu} = \frac{1}{4\pi}\left(g^{\delta\omega}F_{\mu\delta}F^{\nu}_{\omega} - \frac{1}{4}g^{\nu}_{\mu}F_{\delta\omega}F^{\delta\omega}\right) \quad (7)$$

where $F_{\mu\nu}$ is the Maxwell field tensor defined as

$$F_{\mu\nu} = \Phi_{\nu,\mu} - \Phi_{\mu,\nu} \quad (8)$$

and Φ_{μ} is the four potential.

3. Anisotropic Strange Quintessence Star in $f(T)$ Gravity

We consider the KB metric [42] describing the interior space-time of a strange star

$$ds^2 = -e^{a(r)}dt^2 + e^{b(r)}dr^2 + r^2\left(d\theta^2 + \sin^2\theta d\phi^2\right) \quad (9)$$

where we assume $a(r)$ and $b(r)$ are

$$a(r) = Br^2 + Cr^3, \quad b(r) = Ar^3 \tag{10}$$

where A, B, and C are arbitrary constants. For the charged fluid source with density $\rho(r)$, radial pressure $p_r(r)$, and tangential pressure $p_t(r)$, the Einstein-Maxwell (EM) equations reduce to the form ($G = c = 1$) [51]

$$T(r) = \frac{2e^{-b}}{r}\left(a' + \frac{1}{r}\right) \tag{11}$$

$$T'(r) = \frac{2e^{-b}}{r}\left(a'' + \frac{1}{r^2} - T\left(b' + \frac{1}{r}\right)\right) \tag{12}$$

where the prime $'$ denotes the derivative with respect to the radial coordinate r.

Now the equations of motion for anisotropic fluid are [51]

$$4\pi\rho + E^2 = \frac{f}{4} - \left(T - \frac{1}{r^2} - \frac{e^{-b}}{r}\left(a' + b'\right)\right)\frac{f_T}{2} \tag{13}$$

$$4\pi p_r - E^2 = \left(T - \frac{1}{r^2}\right)\frac{f_T}{2} - \frac{f}{4} \tag{14}$$

$$4\pi p_t + E^2 = \left[\frac{T}{2} + e^{-b}\left(\frac{a''}{2} + \left(\frac{a'}{4} + \frac{1}{2r}\right)\right)\right]\frac{f_T}{2} - \frac{f}{4} \tag{15}$$

$$\frac{\cot\theta}{2r^2}T' f_{TT} = 0 \tag{16}$$

$$E(r) = \frac{1}{r}\int_0^r 4\pi r^2 \sigma e^{\lambda/2} dr = \frac{q(r)}{r^2} \tag{17}$$

where $q(r)$ is the total charge within a sphere of radius r.

We introduce the modified Chaplygin gas (MCG) having equation of state [54]

$$p_r = \xi\rho - \frac{\zeta}{\rho^{\alpha}} \tag{18}$$

where ξ, α, and ζ are free parameters of the model.

From (16) we get

$$f(T) = \beta T + \beta_1 \tag{19}$$

where β and β_1 are integration constants and we assume $\beta_1 = 0$ for simple case.

Now from (10), (11), (13), (14), (18), and (19) we obtain the equation in ρ

$$8\pi(1+\xi)\rho^{\alpha+1} - \beta e^{-Ar^3}(2B + 3Cr + 3Ar)\rho^{\alpha} - 8\pi\zeta = 0 \tag{20}$$

Here we take $\alpha = 1$; then (20) reduces to the quadratic equation in ρ

$$8\pi(1+\xi)\rho^2 - \beta e^{-Ar^3}(2B + 3Cr + 3Ar)\rho - 8\pi\zeta = 0 \tag{21}$$

Solving this equation we get the value of energy density as

$$\rho = \frac{(2B\beta + 3C\beta r + 3A\beta r) + \sqrt{256\zeta\pi^2 e^{2Ar^3}(1+\xi) + (2B\beta + 3C\beta r + 3A\beta r)^2}}{16\pi e^{Ar^3}(1+\xi)} \tag{22}$$

and corresponding components are

$$p_r = \frac{\xi(2B\beta + 3C\beta r + 3A\beta r) + \xi\sqrt{256\zeta\pi^2 e^{2Ar^3}(1+\xi) + (2B\beta + 3C\beta r + 3A\beta r)^2}}{16\pi e^{Ar^3}(1+\xi)} - \frac{\zeta}{\left((2B\beta + 3C\beta r + 3A\beta r) + \sqrt{256\zeta\pi^2 e^{2Ar^3}(1+\xi) + (2B\beta + 3C\beta r + 3A\beta r)^2}\right)/16\pi e^{Ar^3}(1+\xi)} \tag{23}$$

$$\rho + 3p_r = \frac{(1+3\xi)(2B\beta + 3C\beta r + 3A\beta r) + (1+3\xi)\sqrt{256\zeta\pi^2 e^{2Ar^3}(1+\xi) + (2B\beta + 3C\beta r + 3A\beta r)^2}}{16\pi e^{Ar^3}(1+\xi)} - \frac{3\zeta}{\left((2B\beta + 3C\beta r + 3A\beta r) + \sqrt{256\zeta\pi^2 e^{2Ar^3}(1+\xi) + (2B\beta + 3C\beta r + 3A\beta r)^2}\right)/16\pi e^{Ar^3}(1+\xi)} \tag{24}$$

$$E^2 = \frac{\beta}{2r^2}e^{Ar^3}\left(-1 + 3Ar^3\right) + \frac{\beta}{2r^2} - \frac{(2B\beta + 3C\beta r + 3A\beta r) + \sqrt{256\zeta\pi^2 e^{2Ar^3}(1+\xi) + (2B\beta + 3C\beta r + 3A\beta r)^2}}{4e^{Ar^3}(1+\xi)} \tag{25}$$

$$p_t = \frac{\beta e^{-Ar^3}}{8\pi}\left\{2B + 3C - 3Ar + \frac{3}{2}Br^3(2C + A) + \frac{3}{4}r(C - A)\left(3Cr^3 + 2\right) + \frac{1}{r^2}\right\} - \frac{\beta}{8\pi r^2} + \frac{(2B\beta + 3C\beta r + 3A\beta r) + \sqrt{256\zeta\pi^2 e^{2Ar^3}(1 + \xi) + (2B\beta + 3C\beta r + 3A\beta r)^2}}{16\pi e^{Ar^3}(1 + \xi)} \tag{26}$$

Now from Figures 1 and 2, we conclude that anisotropic strange star in $f(T)$ gravity with modified Chaplygin gas acts as a dark energy candidate due to $\rho > 0$, $p_r < 0$. Again with the help of Figures 3 and 4, we notice that the equation of state $w = p_r/\rho$ lies between $-1/3$ and -1; i.e., the corresponding model belongs to quintessence phase not phantom phase.

The amount of net charge inside a sphere having radius r is

$$q = r^2\sqrt{\frac{\beta}{2r^2}e^{Ar^3}\left(-1 + 3Ar^3\right) + \frac{\beta}{2r^2} - \frac{(2B\beta + 3C\beta r + 3A\beta r) + \sqrt{256\zeta\pi^2 e^{2Ar^3}(1 + \xi) + (2B\beta + 3C\beta r + 3A\beta r)^2}}{4e^{Ar^3}(1 + \xi)}} \tag{27}$$

4. Physical Analysis

The central density ρ_0 and central radial pressure p_0 are given by

$$\rho_0 = \rho(r = 0) = \frac{2B\beta + \sqrt{256\zeta\pi^2(1 + \xi) + 4B^2\beta^2}}{16\pi(1 + \xi)} \tag{28}$$

and

$$p_0 = p_r(r = 0) = \frac{2B\beta\xi + \xi\sqrt{256\zeta\pi^2(1 + \xi) + 4B^2\beta^2}}{16\pi(1 + \xi)} - \frac{\zeta}{\left(2B\beta + \sqrt{256\zeta\pi^2(1 + \xi) + 4B^2\beta^2}\right)/16\pi(1 + \xi)} \tag{29}$$

In this section, we investigate the nature of the anisotropic compact star as the following subsection.

Figures 1–5 represent the plots by taking $B = 5$, $C = 1$, $A = 2$, $\xi = 2$, and $\zeta = 1$.

4.1. Anisotropic Behavior. Now we take the derivatives of (22) and (23) with respect to r, given by

$$\frac{d\rho}{dr} = \frac{3\beta(C + A) + \left(768\zeta e^{2Ar^3}A\pi^2(1 + \xi)r^2 + 768A\zeta e^{2Ar^3}\pi^2(1 + \xi)r^2 + 2(3C\beta + 3A\beta)(2B\beta + 3Cr\beta + 3Ar\beta)\right)/2\sqrt{256\zeta e^{2Ar^3}\pi^2(1 + \xi) + (2B\beta + 3Cr\beta + 3Ar\beta)^2}}{16e^{Ar^3}\pi(1 + \xi)} - \frac{24e^{Ar^3}A\pi r^2(1 + \xi)\left\{\beta(2B + 3Cr + 3Ar) + \sqrt{256\zeta e^{2Ar^3}\pi^2(1 + \xi) + (2B\beta + 3Cr\beta + 3Ar\beta)^2}\right\}}{16e^{Ar^3}\pi(1 + \xi)^2} \tag{30}$$

and

$$\frac{dp_r}{dr} = \frac{\xi\left\{3\beta(C + A) + \left(1536\zeta e^{2Ar^3}A\pi^2(1 + \xi)r^2 + 2(3C\beta + 3A\beta)(2B\beta + 3Cr\beta + 3Ar\beta)\right)/2\sqrt{256\zeta e^{2Ar^3}\pi^2(1 + \xi) + (2B\beta + 3Cr\beta + 3Ar\beta)^2}\right\}}{16e^{Ar^3}\pi(1 + \xi)} + \frac{16\zeta e^{Ar^3}\pi(1 + \xi) + \left\{3\beta(C + A) + \left(1536\zeta e^{2Ar^3}A\pi^2(1 + \xi)r^2 + 2(3C\beta + 3A\beta)(2B\beta + 3Cr\beta + 3Ar\beta)\right)/2\sqrt{256\zeta e^{2Ar^3}\pi^2(1 + \xi) + (2B\beta + 3Cr\beta + 3Ar\beta)^2}\right\}}{\left\{\beta(2B + 3Cr + 3Ar) + \sqrt{256\zeta e^{2Ar^3}\pi^2(1 + \xi) + (2B\beta + 3Cr\beta + 3Ar\beta)^2}\right\}^2} - \frac{48\zeta e^{Ar^3}A\pi r^2(1 + \xi)}{\beta(2B + 3Cr + 3Ar) + \sqrt{256\zeta e^{2Ar^3}\pi^2(1 + \xi) + (2B\beta + 3Cr\beta + 3Ar\beta)^2}} - \frac{24\xi e^{Ar^3}A\pi r^2(1 + \xi)\left\{\beta(2B + 3Cr + 3Ar) + \sqrt{256\zeta e^{2Ar^3}\pi^2(1 + \xi) + (2B\beta + 3Cr\beta + 3Ar\beta)^2}\right\}}{128e^{2Ar^3}\pi^2(1 + \xi)^2} \tag{31}$$

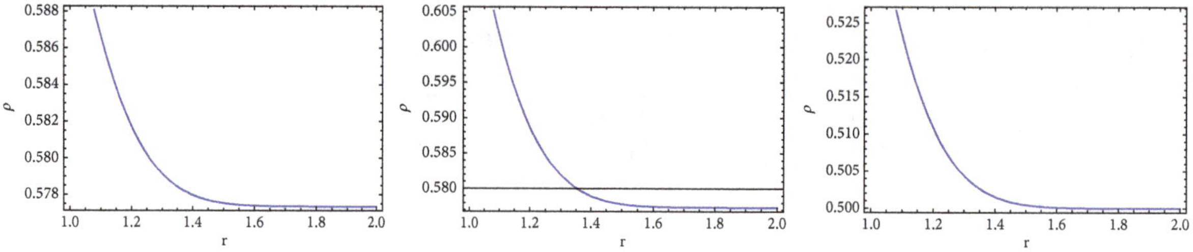

FIGURE 1: This figure represents the variation of ρ versus r (km) for the strange star taking $\beta = 1$, $\beta = 2$, and $\beta = 3$.

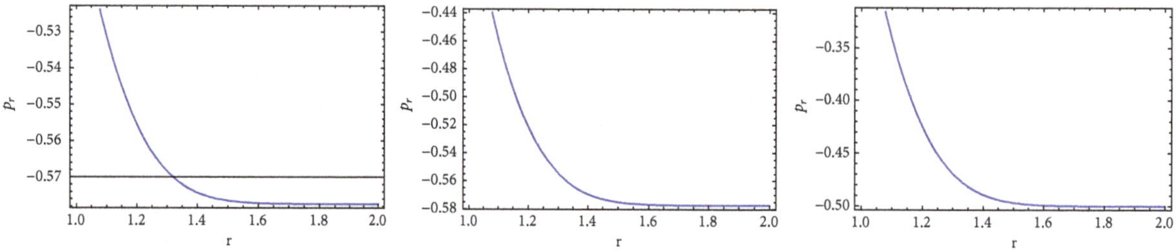

FIGURE 2: This figure represents the variation of p_r versus r (km) for the strange star taking $\beta = 1$, $\beta = 2$, and $\beta = 3$.

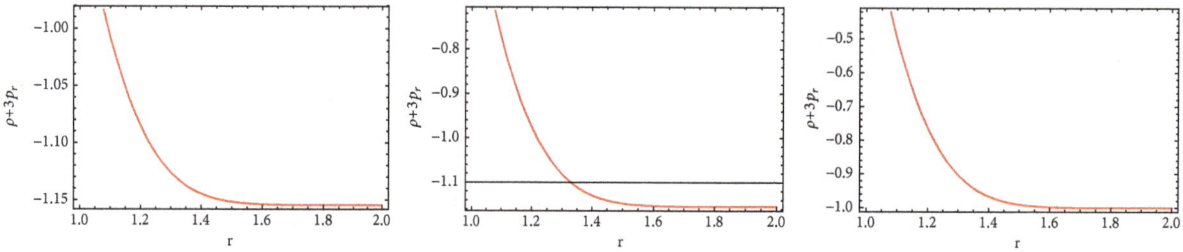

FIGURE 3: This figure represents the variation of $\rho + 3p_r$ versus r (km) for the strange star taking $\beta = 1$, $\beta = 2$, and $\beta = 3$.

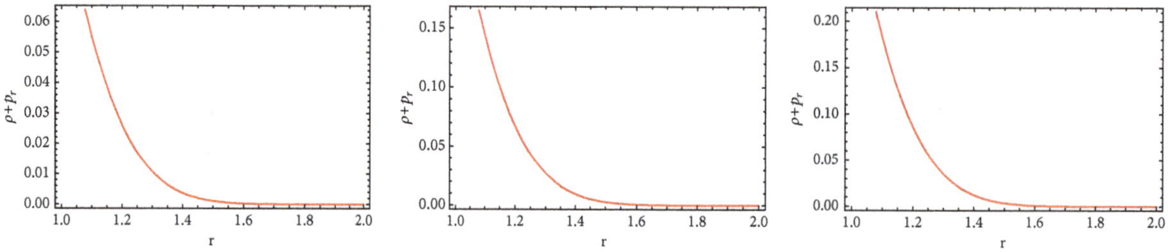

FIGURE 4: This figure represents the variation of $\rho + p_r$ versus r (km) for the strange star taking $\beta = 1$, $\beta = 2$, and $\beta = 3$.

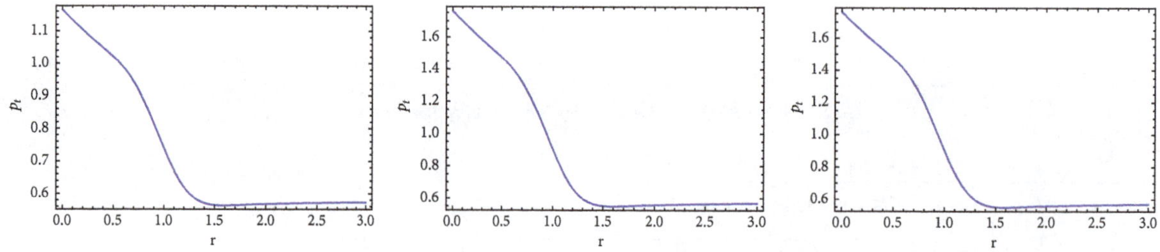

FIGURE 5: This figure represents the variation of p_t versus r (km) for the strange star taking $\beta = 1$, $\beta = 2$, and $\beta = 3$.

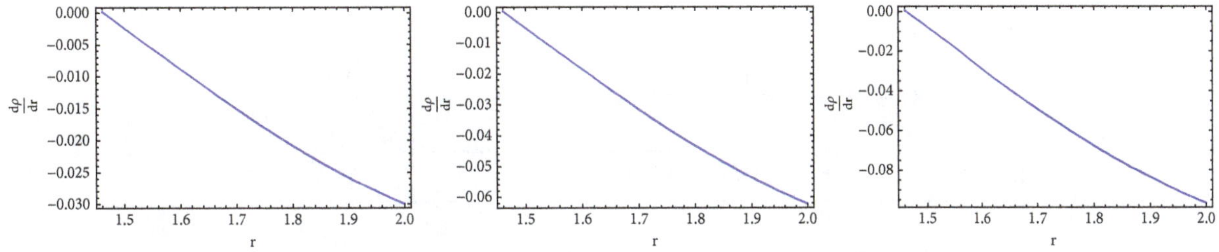

FIGURE 6: This figure represents the variation of $d\rho/dr$ versus r (km) for the strange star taking $\beta = 1$, $\beta = 2$, and $\beta = 3$.

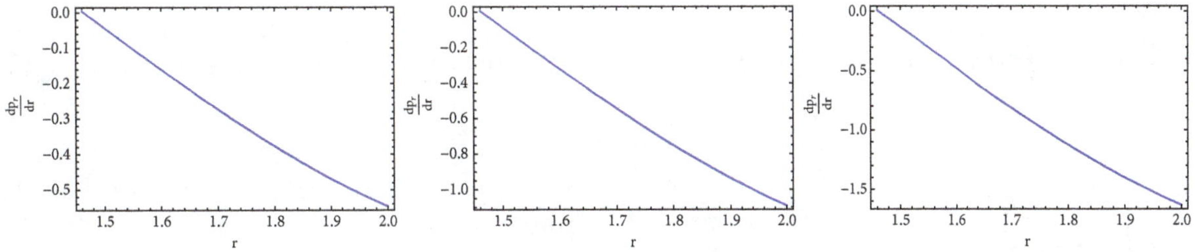

FIGURE 7: This figure represents the variation of dp_r/dr versus r (km) for strange the star taking $\beta = 1$, $\beta = 2$, and $\beta = 3$.

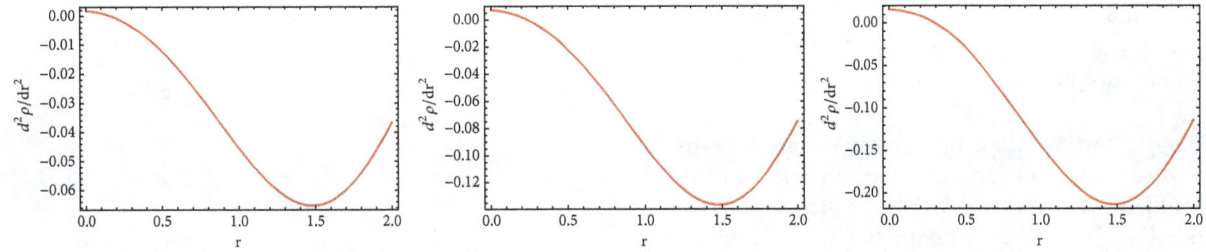

FIGURE 8: This figure represents the variation of $d^2\rho/dr^2$ versus r (km) for the strange star taking $\beta = 1$, $\beta = 2$, and $\beta = 3$.

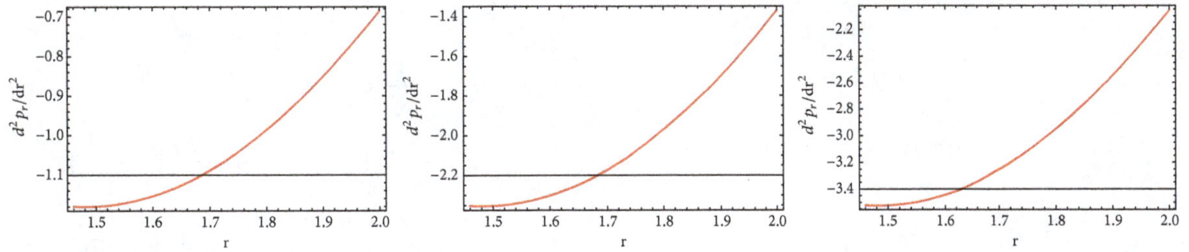

FIGURE 9: This figure represents the variation of d^2p_r/dr^2 versus r (km) for the strange star taking $\beta = 1$, $\beta = 2$, and $\beta = 3$.

Now we present the evolution of $d\rho/dr$ and dp_r/dr by Figures 6 and 7. Figure 6 shows that $d\rho/dr$ decreases keeping $d\rho/dr < 0$ (as energy density decreases) and Figure 7 shows that dp_r/dr decreases keeping $dp_r/dr < 0$ (as for dark energy pressure is negatively very high; i.e., pressure decreases negatively). From Figures 6, 7, 8, and 9 we notice that, at $r = 1.46$,

$$\frac{d\rho}{dr} = 0,$$

$$\frac{dp_r}{dr} = 0,$$

$$\frac{d^2\rho}{dr^2} < 0,$$

$$\frac{d^2p_r}{dr^2} < 0. \tag{32}$$

This points out that the energy density and radial pressure have maximum value at $r = 1.46$ of the quintessence strange star model in $f(T)$ gravity.

Now the anisotropic stress ($\Delta = p_t - p_r$) is as follows

$$\Delta = \frac{\beta e^{-Ar^3}}{8\pi}\left\{2B + 3C - 3Ar + \frac{3}{2}Br^3(2C + A) + (C - A)\left(\frac{9}{4}Cr^4 + \frac{3}{2}r\right) + \frac{1}{r^2}\right\} - \frac{\beta}{8\pi r^2}$$
$$+ \frac{(2B\beta + 3C\beta r + 3A\beta r) + \sqrt{256\zeta\pi^2 e^{2Ar^3}(1+\xi) + (2B\beta + 3C\beta r + 3A\beta r)^2}}{16\pi e^{Ar^3}(1+\xi)}$$
$$- \frac{\xi(2B\beta + 3C\beta r + 3A\beta r) + \xi\sqrt{256\zeta\pi^2 e^{2Ar^3}(1+\xi) + (2B\beta + 3C\beta r + 3A\beta r)^2}}{16\pi e^{Ar^3}(1+\xi)}$$
$$- \frac{\zeta}{\left((2B\beta + 3C\beta r + 3A\beta r) + \sqrt{256\zeta\pi^2 e^{2Ar^3}(1+\xi) + (2B\beta + 3C\beta r + 3A\beta r)^2}\right)/16\pi e^{Ar^3}(1+\xi)} \quad (33)$$

From Figure 10, we notice that $\Delta > 0$ for $\beta = 1, -6$ which imply that the anisotropic stress is outwardly directed and there exists repulsive gravitational force for the strange star and for $\beta = -15$, $\Delta < 0$ in somewhere implying the existence of attractive gravitational force and $\Delta > 0$ in the remaining part implying the existence of repulsive gravitational force of the strange star.

Figure 11 shows that E^2 is decreasing with the increment of the radial coordinate.

4.2. Energy Conditions. Energy conditions are very useful tools to discuss cosmological geometry in general relativity and modified gravity [10, 48, 51]. These include null energy condition (NEC), weak energy condition (WEC), and strong energy condition (SEC), given as

$$\begin{aligned} \textbf{NEC}: \rho + \frac{E^2}{8\pi} &\geq 0, \\ \textbf{WEC}: \rho + p_r &\geq 0, \\ \rho + p_t + \frac{E^2}{4\pi} &\geq 0, \\ \textbf{SEC}: \rho + p_r + 2p_t + \frac{E^2}{4\pi} &\geq 0. \end{aligned} \quad (34)$$

Figures 6, 7, 8, and 9 represent the plots of $d\rho/dr$, dp_r/dr, $d^2\rho/dr^2$, and d^2p_r/dr^2 with respect to r to show the maximality of density and radial pressure at $r = 1.46$ of the strange star taking $B = 1$, $C = 7$, $A = 0.1$, $\xi = 9$, and $\zeta = 10$.

Figure 10 represents the plots of Δ with respect to r to show the presence of repulsive and attractive force of the strange star and Figure 11 represents the plot of E^2 with respect to r taking $B = 10$, $C = 1$, $A = 10$, $\xi = 2$, and $\zeta = 1$.

From Figures 12, 4, 13, and 14, we observe that the interior of our proposed strange star model satisfies all energy conditions.

4.3. Matching Conditions. Many authors have worked on the matching condition to compare the exterior solution with the interior solution [10, 47, 49, 51]. We correspond the exterior geometry with our interior solution, evoked by the Schwarzschild solution which is given by the line element

$$\begin{aligned} ds^2 = -\left(1 - \frac{2M}{r}\right)dt^2 + \left(1 - \frac{2M}{r}\right)^{-1}dr^2 \\ + r^2\left(d\theta^2 + \sin^2\theta d\phi^2\right). \end{aligned} \quad (35)$$

The continuity of the metric components g_{tt}, g_{rr}, and $\partial g_{tt}/\partial r$ at the boundary surface $r = R$ yields

$$\begin{aligned} g_{tt}^- &= g_{tt}^+, \\ g_{rr}^- &= g_{rr}^+, \\ \frac{\partial g_{tt}^-}{\partial r} &= \frac{\partial g_{tt}^+}{\partial r}, \end{aligned} \quad (36)$$

where − and + indicate interior and exterior solutions. Now, using (36) and the metrics (9) and (35), we have

$$\begin{aligned} A &= -\frac{1}{R^3}\ln\left(1 - \frac{2M}{R}\right), \\ B &= \frac{3}{R^2}\ln\left(1 - \frac{2M}{R}\right) - \frac{2M}{R^3}\left(1 - \frac{2M}{R}\right)^{-1}, \\ C &= \frac{2M}{R^4}\left(1 - \frac{2M}{R}\right)^{-1} - \frac{2}{R^3}\ln\left(1 - \frac{2M}{R}\right). \end{aligned} \quad (37)$$

For the values of M and R for a strange stars, we compute the constants A, B, and C, specified as in Table 1.

4.4. Stability. Now we calculate the two sound speed squares v_{sr}^2, v_{st}^2 for the radial and transverse coordinate, respectively. Herrera [55] introduced cracking concept and developed a new technique to examine potential stability for the matter. If we investigate the sign of the difference $v_{st}^2 - v_{sr}^2$ then we can conclude whether our strange star is potential stable or not; i.e., if the radial speed sound is greater than the transverse speed sound, then there exists potentially stable region; otherwise, the region will be potentially unstable [10, 47, 49, 51]. **It is clear from Figures 15 and 16 that** $0 < v_{sr}^2 \leq 1$ **and** $0 < v_{st}^2 \leq 1$ **always within the stellar objects.** From Figure 17, we see that the corresponding difference is

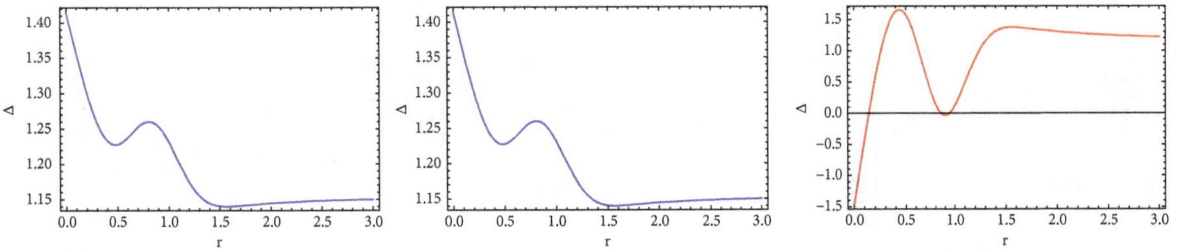

FIGURE 10: This figure represents the variation of Δ versus r (km) for the strange star taking $\beta = 1$, $\beta = -6$, and $\beta = -15$.

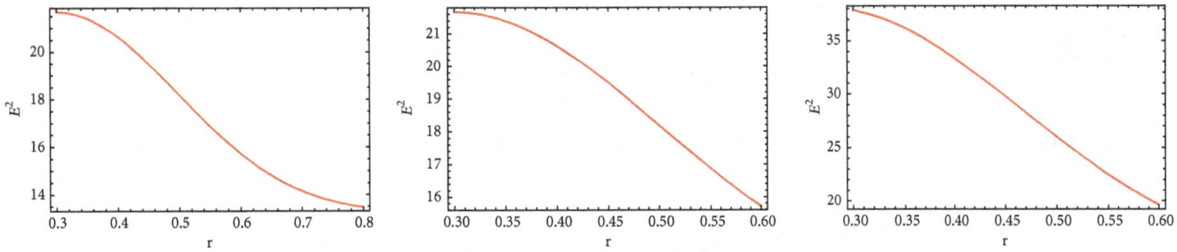

FIGURE 11: This figure represents the variation of E^2 versus r (km) for the strange star taking $\beta = 1$, $\beta = 2$, and $\beta = 3$.

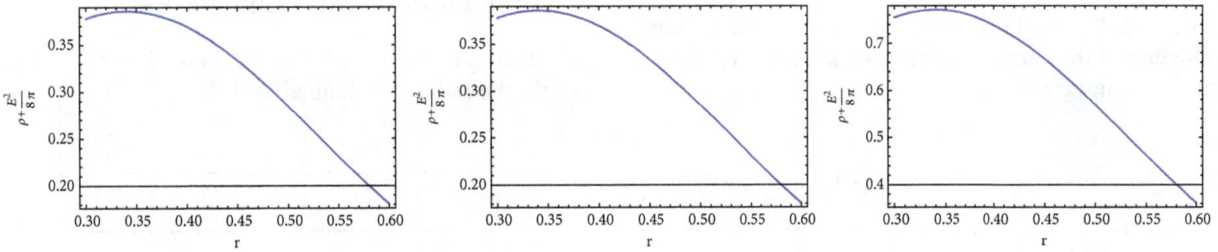

FIGURE 12: This figure represents the variation of $\rho + E^2/8\pi$ versus r (km) for the strange star taking $\beta = 1$, $\beta = 2$, and $\beta = 3$.

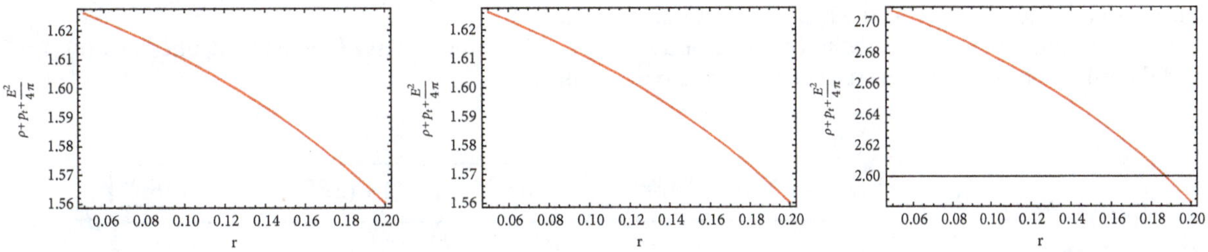

FIGURE 13: This figure represents the variation of $\rho + p_t + E^2/4\pi$ versus r (km) for the strange star taking $\beta = 1$, $\beta = 2$, and $\beta = 3$.

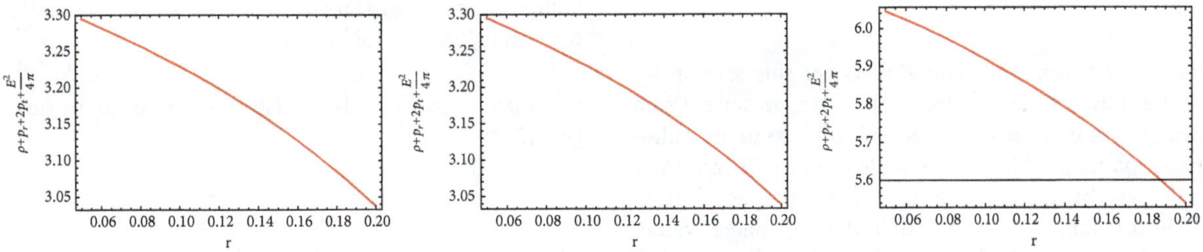

FIGURE 14: This figure represents the variation of $\rho + p_r + 2p_t + E^2/4\pi$ versus r (km) for the strange star taking $\beta = 1$, $\beta = 2$, and $\beta = 3$.

TABLE 1: The values of A, B, and C have been obtained using (37).

Compact Stars	$M(M_\odot)$	$R(Km)$	$A(Km^{-2})$	$B(Km^{-2})$	$C(Km^{-2})$
$SAX\ J$ 1808.4 − 3658(SS1)	1.435	7.07	0.001473644346	-0.044926791	0.004880923098
$4U1820-30$	2.25	10	0.0005978370008	-0.026116928	0.00201385582
$Vela\ X-12$	1.77	9.99	0.0004388196046	-0.018169038	0.001428126229
$PSR\ J$ 1614 − 2230	1.97	10.3	0.000441203995	-0.019472555	0.00144933537

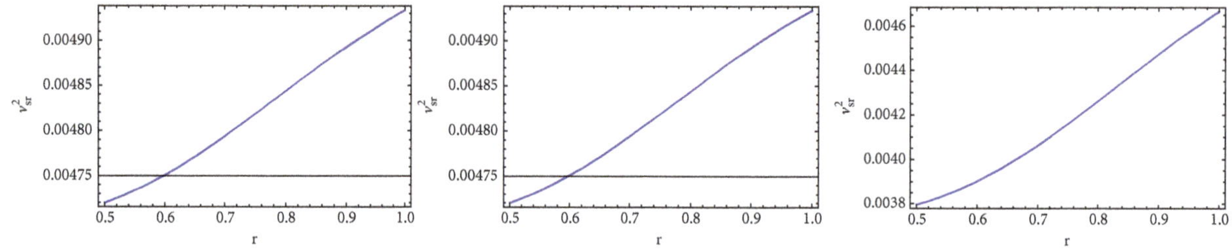

FIGURE 15: This figure represents the variation of v^2_{sr} versus r (km) for the strange star taking $\beta = 1$, $\beta = 2$, and $\beta = 3$.

negative which means the radial speed sound is greater than the transverse speed sound which implies that our proposed strange star model is potentially stable in the framework of $f(T)$ gravity. **Again Figure 18 shows that $|v^2_{st} - v^2_{sr}| \leq 1$ is satisfied [56].**

Figures 12, 13, and 14 represent the plots to understand the validation of the energy conditions taking $B = 10$, $C = 1$, $A = 10$, $\xi = 2$, and $\zeta = 1$.

Figures 15, 16, 17, and 18 represent the plots to show the stability of our proposed model taking $B = 5$, $C = 1$, $A = 2$, $\xi = 2$, and $\zeta = 1$.

5. Some Fundamental Calculations

5.1. Mass Function and Compactness. The mass function within the radius r is defined as [10]

$$m(r) = \int_0^r 4\pi r^2 \rho\, dr = 2\pi \int_0^r \frac{r^2 \left\{ (2B\beta + 3C\beta r + 3A\beta r) + \sqrt{256\zeta\pi^2 e^{2Ar^3}(1+\xi) + (2B\beta + 3C\beta r + 3A\beta r)^2} \right\}}{8\pi e^{Ar^3}(1+\xi)} dr \quad (38)$$

From Figure 19, we have seen that at origin the mass function is regular (i.e., $m(r) \longrightarrow 0$ when $r \longrightarrow 0$) and monotonic increasing with respect to radius (r). We have also evaluated the values of mass for a few strange stars from our model to compare these values with observational data (see Table 2).

The compactness of the star is defined by $u(r)$ [10] in the form of

$$u(r) = \frac{m(r)}{r} = \frac{2\pi}{r} \int_0^r \frac{r^2 \left\{ (2B\beta + 3C\beta r + 3A\beta r) + \sqrt{256\zeta\pi^2 e^{2Ar^3}(1+\xi) + (2B\beta + 3C\beta r + 3A\beta r)^2} \right\}}{8\pi e^{Ar^3}(1+\xi)} dr \quad (39)$$

We have plotted the corresponding function given by Figure 20.

5.2. Relation between Mass and Radius. In this section we discuss the mass radius relation of the strange stars. From [57], twice the maximum allowable ratio of mass to the radius for an astrophysical object is always less than 8/9 ($2M/R < 8/9$) whereas the factor M/R is called "compactification factor". From Table 3, we find that the calculated values corresponding to our model lie in the expected range [34]. Compactification factor for strange star always lies between 1/4 and 1/2. The calculated values of the compactification factor of the strange stars from our model are compatible with the condition (see Table 3).

5.3. Surface Redshift. The redshift function can be defined as [10, 47, 49, 51]

$$z_s = \frac{1}{\sqrt{1 - 2m(r)/r}} - 1, \quad (40)$$

where $m(r)$ has been obtained from (38). According to Bohmer and Harko, the surface redshift should be ≤ 5 for

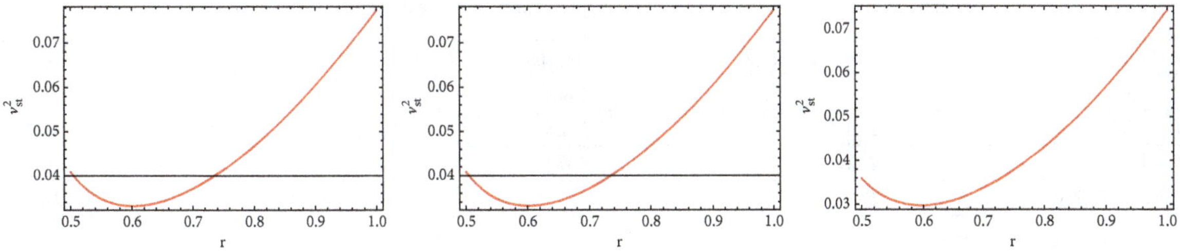

FIGURE 16: This figure represents the variation of v_{st}^2 versus r (km) for the strange star taking $\beta = 1$, $\beta = 2$, and $\beta = 3$.

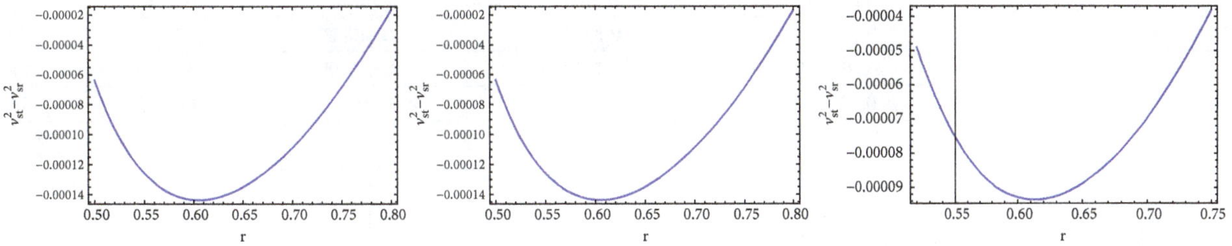

FIGURE 17: This figure represents the variation of $v_{st}^2 - v_{sr}^2$ versus r (km) for the strange star taking $\beta = 1$, $\beta = 2$, and $\beta = 3$.

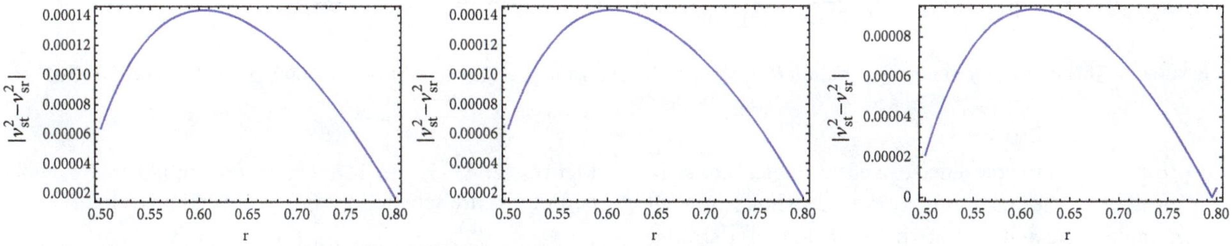

FIGURE 18: This figure represents the variation of $|v_{st}^2 - v_{sr}^2|$ versus r (km) for the strange star taking $\beta = 1$, $\beta = 2$, and $\beta = 3$.

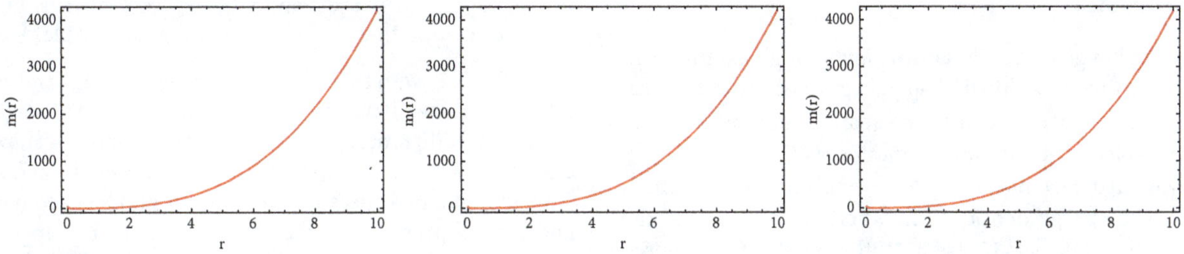

FIGURE 19: This figure represents the variation of $m(r)$ versus r (km) for the strange star taking $\beta = 1$, $\beta = 2$, and $\beta = 3$.

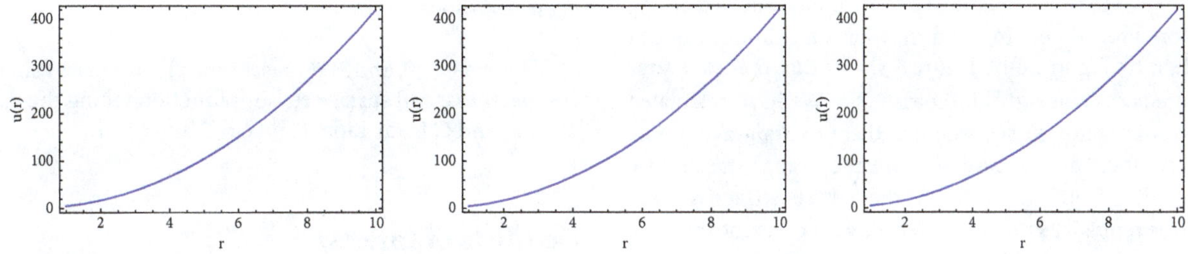

FIGURE 20: This figure represents the variation of $u(r)$ versus r (km) for the strange star taking $\beta = 1$, $\beta = 2$, and $\beta = 3$.

TABLE 2: Calculated values of mass, energy density, and pressure from our model.

Compact Stars	*Mass standard data (in km)*	*Mass from model (in km)*	$\rho_0(gm/cc)$	$\rho_R(gm/cc)$	$p_0(dyne/cm^2)$
$SAX\ J\ 1808.4-3658(SS1)$	2.116625	2.0868	1.996428×10^{-12}	1.000531×10^{-12}	-1.001789×10^{12}
$4U1820-30$	3.31875	3.34265	1.997923×10^{-12}	1.000286×10^{-12}	-1.001040×10^{12}
$Vela\ X-12$	2.61075	2.61043	1.998555×10^{-12}	1.000253×10^{-12}	-1.000723×10^{12}
$PSR\ J\ 1614-2230$	2.90575	2.91837	1.998451×10^{-12}	1.000239×10^{-12}	-1.000775×10^{12}

TABLE 3: Calculated values of the desired parameters of our model.

Compact Stars	*M/R (standard data)*	*M/R from model*	$2M/R < 8/9$	ρ_0/ρ_R	z_s
$SAX\ J\ 1808.4-3658(SS1)$	0.299381	0.295163	0.590325	1.995368	0.562358
$4U1820-30$	0.331875	0.334265	0.66853	1.997352	0.736912
$Vela\ X-12$	0.266134	0.261304	0.522609	1.998050	0.447314
$PSR\ J\ 1614-2230$	0.282112	0.283337	0.566674	1.997973	0.519122

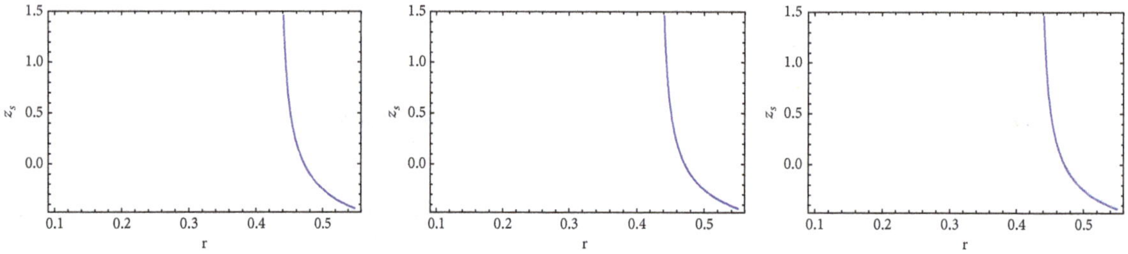

FIGURE 21: This figure represents the variation of z_s versus r (km) for the strange star taking $\beta = 0.009$, $\beta = 0.010$, and $\beta = 0.011$.

an anisotropic star in the presence of a cosmological constant [58]. We calculate the maximum surface redshift from our model in Table 3. Now, it is clear that our model for strange stars obeys the relation $z_S \leq 5$ though the cosmological constant is absent in our model which is quite reasonable.

6. Discussions

This paper has given out the anisotropic strange star model in $f(T)$ gravity with modified Chaplygin gas. Using the diagonal tetrad field we have obtained the equations of motion where we have solved the unknown function $f(T)$ as $\beta T + \beta_1$, β and β_1 being constants. Then we have solved the differential equation of energy density from where we have found the value of energy density (22) of it. With the help of this energy density, we have found out radial pressure ensuring this model as a quintessence dark energy candidate from Figures 1, 2, 3, and 4. We have also noticed that both the energy density (ρ) and radial pressure (p_r) are monotonic decreasing function with respect to r and they have maximum value at $r = 1.46$ by Figures 6–9. Figure 5 shows that the transverse pressure is decreasing with the rise of r. We have calculated anisotropic factor to see whether there exists gravitational attractive force or repulsive force for the strange star and we have studied from Figure 10 that there exists attractive force as well as repulsive gravitational force with different values of β. Here, the square of energy is monotonic decreasing with the increment of radial coordinate given by Figure 11. From Figures 12, 4, 13, and 14 we have concluded that all energy conditions are satisfied for our proposed model.

Using matching condition, the unknown parameters A, B, and C have been calculated for the different strange stars from our model which is given by Table 1. By stability analysis given on the basic of Figures 15–18, we have observed that $0 < v_{sr}^2, v_{st}^2 \leq 1$, $v_{sr}^2 > v_{st}^2$, and $|v_{st}^2 - v_{sr}^2| \leq 1$ always. Finally, we have ensured that our model is potentially stable.

In Table 2, with the help of energy density (22) and radial pressure (23) we have calculated the numerical values of the mass of the different strange stars from our model to show the closeness of these values with the observational data. Also, we have obtained the values of central and surface density and central pressure for the above-mentioned strange stars from our model which have been calculated in Table 2. From Table 3, we have observed that twice the compactification factor are always less than $< 8/9$ and maximum values of the surface redshift function are always less than 5. So, our proposed model is completely rational.

Figures 19, 20, and 21 represent the plots of mass function, compactness, and surface redshift function taking the values of A, B, and C from Table 1 and $\xi = 2$ and $\zeta = 1$.

Conflicts of Interest

The authors declare that they have no conflicts of interest.

References

[1] S. Perlmutter, G. Aldering, G. Goldhaber et al., "Measurements of Ω and Λ from 42 high-redshift supernovae," *The Astrophysical Journal*, vol. 517, no. 2, article no. 565, 1999.

[2] D. N. Spergel, R. Bean, O. Doré et al., "Three-Year Wilkinson Microwave Anisotropy Probe (WMAP) Observations: Implications for Cosmology," *The Astrophysical Journal Supplement Series*, vol. 170, no. 2, article no. 377, 2007.

[3] E. Hawkins, S. Maddox, S. Cole, O. Lahav et al., "The 2dF Galaxy Redshift Survey: correlation functions, peculiar velocities and the matter density of the Universe," *Monthly Notices of the Royal Astronomical Society*, vol. 346, no. 1, pp. 78–96, 2003.

[4] D. J. Eisenstein, I. Zehavi, D. W. Hogg et al., "Detection of the Baryon Acoustic Peak in the Large-Scale Correlation Function of SDSS Luminous Red Galaxies," *The Astrophysical Journal*, vol. 633, no. 2, article no. 560, 2005.

[5] P. Mazur and E. Mottola, "Gravitational Condensate Stars: An Alternative to Black Holes," https://arxiv.org/abs/gr-qc/0109035, 2011.

[6] P. Mazur and E. Mottola, "Gravitational Vacuum Condensate Stars," *Proceedings of the National Acadamy of Sciences of the United States of America*, vol. 101, no. 26, pp. 9545–9550, 2004.

[7] A. Usmani, F. Rahaman, S. Ray et al., "Charged gravastars admitting conformal motion," *Physics Letters B*, vol. 701, no. 4, pp. 388–392, 2011.

[8] F. Rahaman, S. Ray, A. A. Usmani, and S. Islam, "The(2+1)-dimensional gravastars," *Physics Letters B*, vol. 707, no. 3-4, pp. 319–322, 2012.

[9] F. Rahaman, A. A. Usmani, S. Ray, and S. Islam, "The (2+1) - dimensional charged gravastars," *Physics Letters B*, vol. 717, no. 1-3, pp. 1–5, 2012.

[10] P. Bhar, "Higher dimensional charged gravastar admitting conformal motion," *Astrophysics and Space Science*, vol. 354, no. 2, pp. 457–462, 2014.

[11] A. de Felice and S. Tsujikawa, "*f(R)* Theories," *Living Reviews in Relativity*, vol. 13, no. 1, article no. 3, 2010.

[12] R. Durrer and R. Maartens, *Dark energy: Observational and Theoretical Approaches*, P. Ruiz-Lapuente, Ed., Cambridge University Press, 2010.

[13] S. Nojiri and S. D. Odintsov, "Unified cosmic history in modified gravity: From F(R) theory to Lorentz non-invariant models," *Physics Reports*, vol. 505, no. 2-4, pp. 59–144, 2011.

[14] T. Harko, F. S. N. Lobo, S. Nojiri, and S. D. Odintsov, "*f(R,T)* gravity," *Physical Review D: Particles, Fields, Gravitation and Cosmology*, vol. 84, no. 2, Article ID 024020, 2011.

[15] G. Cognola, E. Elizalde, S. Nojiri, S. D. Odintsov, and S. Zerbini, "Dark energy in modified Gauss-Bonnet gravity: Late-time acceleration and the hierarchy problem," *Physical Review D: Particles, Fields, Gravitation and Cosmology*, vol. 73, no. 8, Article ID 084007, 2006.

[16] G. R. Bengochea and R. Ferraro, "Dark torsion as the cosmic speed-up," *Physical Review D: Particles, Fields, Gravitation and Cosmology*, vol. 79, no. 12, Article ID 124019, 2009.

[17] E. V. Linder, "Einstein's other gravity and the acceleration of the Universe," *Physical Review D: Particles, Fields, Gravitation and Cosmology*, vol. 81, no. 12, Article ID 127301, 2010.

[18] R. Myrzakulov, "Accelerating universe from *F(T)* gravity," *The European Physical Journal C*, vol. 71, article no. 1752, 2011.

[19] K. Bamba, C.-Q. Geng, C.-C. Lee, and L.-W. Luo, "Equation of state for dark energy in f(t) gravity," *Journal of Cosmology and Astroparticle Physics*, vol. 2011, no. 1, article no. 021, 2011.

[20] P. Wu and H. Yu, "*f(T)* models with phantom divide line crossing," *The European Physical Journal C*, vol. 71, no. 2, article no. 1552, 2011.

[21] S. Camera, V. F. Cardone, and N. Radicella, "Detectability of torsion gravity via galaxy clustering and cosmic shear measurements," *Physical Review D: Particles, Fields, Gravitation and Cosmology*, vol. 89, no. 8, Article ID 083520, 2014.

[22] C. M. Will, "The Confrontation between General Relativity and Experiment," *Living Reviews in Relativity*, vol. 9, no. 1, article no. 3, 2006.

[23] J. B. Dent, S. Dutta, and E. N. Saridakis, "f(T) gravity mimicking dynamical dark energy. Background and perturbation analysis," *Journal of Cosmology and Astroparticle Physics*, vol. 2011, no. 01, article no. 009, 2011.

[24] R.-X. Miao, M. Li, and Y.-G. Miao, "Violation of the first law of black hole thermodynamics in f(T) gravity," *Journal of Cosmology and Astroparticle Physics*, vol. 2011, no. 11, article no. 033, 2011.

[25] T. Wang, "Static solutions with spherical symmetry in *f(T)* theories," *Physical Review D*, vol. 84, no. 2, Article ID 024042, 2011.

[26] M. H. Daouda, M. E. Rodrigues, and M. J. S. Houndjo, "New static solutions in f(T) theory," *The European Physical Journal C*, vol. 71, no. 11, article no. 1817, 2011.

[27] C. G. Böhmer, A. Mussa, and N. Tamanini, "Existence of relativistic stars in *f(T)* gravity," *Classical and Quantum Gravity*, vol. 28, no. 24, Article ID 245020, 2011.

[28] S. Capozziello, P. A. González, E. N. Saridakisd, and Y. Va'squez, "Exact charged black-hole solutions in D-dimensional *f (T)* gravity: torsion vs curvature analysis," *Journal of High Energy Physics*, vol. 2013, no. 2, article no. 039, 2013.

[29] M. Sharif and S. Rani, "Wormhole solutions in *f(T)* gravity with noncommutative geometry," *Physical Review D*, vol. 88, no. 12, Article ID 123501, 2013.

[30] M. Sharif and S. Rani, "f(T) gravity and static wormhole solutions," *Modern Physics Letters A*, vol. 29, no. 27, Article ID 1450137, 2014.

[31] M. Sharif and S. Rani, "Charged noncommutative wormhole solutions in *f(T)* gravity," *The European Physical Journal Plus*, vol. 129, no. 10, article no. 237, 2014.

[32] M. Sharif and S. Rani, "Dynamical Instability of Expansion-Free Collapse in *f(T)* Gravity," *International Journal of Theoretical Physics*, vol. 54, no. 8, pp. 2524–2542, 2015.

[33] E. Witten, "Cosmic separation of phases," *Physical Review D: Particles, Fields, Gravitation and Cosmology*, vol. 30, no. 2, article no. 272, 1984.

[34] K. Jotania and R. Tikekar, "Paraboloidal space-times and relativistic models of strange stars," *International Journal of Modern Physics D*, vol. 15, no. 8, pp. 1175–1182, 2006.

[35] L. Herrera and N. O. Santos, "Local anisotropy in self-gravitating systems," *Physics Reports*, vol. 286, no. 2, pp. 53–130, 1997.

[36] G. Abbas, "Effects of electromagnetic field on the collapse and expansion of anisotropic gravitating source," *Astrophysics and Space Science*, vol. 352, no. 2, pp. 955–961, 2014.

[37] G. Abbas, "Collapse and expansion of anisotropic plane symmetric source," *Astrophysics and Space Science*, vol. 350, no. 1, pp. 307–311, 2014.

[38] G. Abbas, "Phantom energy accretion onto a black hole in Hořava-Lifshitz gravity," *Science China Physics, Mechanics and Astronomy*, vol. 57, no. 4, pp. 604–607, 2014.

[39] G. Abbas, "Cardy-Verlinde Formula of Noncommutative Schwarzschild Black Hole," *Advances in High Energy Physics*, vol. 2014, Article ID 306256, 4 pages, 2014.

[40] G. Abbas and U. Sabiullah, "Geodesic study of regular Hayward black hole," *Astrophysics and Space Science*, vol. 352, no. 2, pp. 769–774, 2014.

[41] M. K. Mak and T. Harko, "Quark stars admitting a one-parameter group of conformal motions," *International Journal of Modern Physics D*, vol. 13, no. 1, pp. 149–156, 2004.

[42] K. D. Krori and J. Barua, "A singularity-free solution for a charged fluid sphere in general relativity," *Journal of Physics A: Mathematical and General*, vol. 8, no. 4, article no. 508, 1975.

[43] M. Kalam, F. Rahaman, S. Ray, Sk. M. Hossein, I. Karar, and J. Naskar, "Anisotropic strange star with de Sitter spacetime," *The European Physical Journal C*, vol. 72, no. 12, article no. 2248, 2012.

[44] M. Kalam, F. Rahaman, Sk. M. Hossein, and S. Ray, "Central density dependent anisotropic compact stars," *The European Physical Journal C*, vol. 73, article no. 2638, 2013.

[45] C. Deliduman and B. Yapışkan, "Absence of Relativistic Stars in *f(T)* Gravity," https://arxiv.org/abs/1103.2225, 2015.

[46] G. Abbas, S. Nazeer, and M. A. Meraj, "Cylindrically symmetric models of anisotropic compact stars," *Astrophysics and Space Science*, vol. 354, no. 2, pp. 449–455, 2014.

[47] G. Abbas, A. Kanwal, and M. Zubair, "Anisotropic compact stars in *f(T)* gravity," *Astrophysics and Space Science*, vol. 357, no. 2, article no. 109, 2015.

[48] G. Abbas, D. Momeni, M. Aamir Ali, R. Myrzakulov, and S. Qaisar, "Anisotropic compact stars in *f(G)* gravity," *Astrophysics and Space Science*, vol. 357, no. 2, article no. 158, 2015.

[49] G. Abbas, S. Qaisar, and M. A. Meraj, "Anisotropic strange quintessence stars in *f(T)* gravity," *Astrophysics and Space Science*, vol. 357, no. 2, article no. 156, 2015.

[50] G. Abbas, M. Zubair, and G. Mustafa, "Anisotropic strange quintessence stars in *f(R)* gravity," *Astrophysics and Space Science*, vol. 358, no. 2, article no. 26, 2015.

[51] G. Abbas, S. Qaisar, and A. Jawad, "Strange stars in *f(T)* gravity with MIT bag model," *Astrophysics and Space Science*, vol. 359, no. 2, article no. 57, 2015.

[52] T. P. Sotiriou and V. Faraoni, "*f (R)* theories of gravity," *Reviews of Modern Physics*, vol. 82, no. 1, article no. 451, 2010.

[53] K. Bamba, S. Capozziello, S. Nojiri, and S. D. Odintsov, "Dark energy cosmology: the equivalent description via different theoretical models and cosmography tests," *Astrophysics and Space Science*, vol. 342, no. 1, pp. 155–228, 2012.

[54] H. B. Benaoum, "Modified Chaplygin Gas Cosmology," *Advances in High Energy Physics*, vol. 2012, Article ID 357802, 12 pages, 2012.

[55] L. Herrera, "Cracking of self-gravitating compact objects," *Physics Letters A*, vol. 165, no. 3, pp. 206–210, 1992.

[56] H. Andréasson, "Sharp Bounds on the Critical Stability Radius for Relativistic Charged Spheres," *Communications in Mathematical Physics*, vol. 288, no. 2, pp. 715–730, 2009.

[57] H. A. Buchdahl, "General relativistic fluid spheres," *Physical Review Journals Archieve*, vol. 116, no. 4, article no. 1027, 1959.

[58] C. G. Bohmer and T. Harko, "Bounds on the basic physical parameters for anisotropic compact general relativistic objects," *Classical and Quantum Gravity*, vol. 23, no. 22, article no. 6479, 2006.

7

Nonlocal Black Hole Evaporation and Quantum Metric Fluctuations via Inhomogeneous Vacuum Density

Alexander Y. Yosifov[1] and Lachezar G. Filipov[2]

[1]*Department of Physics and Astronomy, Shumen University, Bulgaria*
[2]*Space Research and Technology Institute, Bulgarian Academy of Sciences, Bulgaria*

Correspondence should be addressed to Alexander Y. Yosifov; alexanderyyosifov@gmail.com

Guest Editor: Farook Rahaman

Inhomogeneity of the *actual* value of the vacuum energy density is considered in a black hole background. We examine the back-reaction of a Schwarzschild black hole to the highly inhomogeneous vacuum density and argue the fluctuations lead to deviations from general relativity in the near-horizon region. In particular, we found that vacuum fluctuations *onto* the horizon trigger adiabatic release of quantum information, while vacuum fluctuations in the vicinity of the horizon produce potentially observable metric fluctuations of order of the Schwarzschild radius. Consequently, we propose a form of strong nonviolent nonlocality in which we simultaneously get nonlocal release of quantum information and observable metric fluctuations.

1. Introduction

Recently, Unruh et al. argued (see [1]) the observed small nonnegative cosmological constant can be achieved without introducing new degrees of freedom, *e.g.*, negative-pressure scalar field. Instead, they address the cosmological constant problem by embracing the diverging value of the vacuum energy density as predicted by quantum field theory, without applying any renormalization procedures. Interestingly, by studying the gravitational effects of the constantly fluctuating vacuum, they found its local energy density to be highly inhomogeneous. As a result of this inhomogeneity, the spatial distance between a pair of neighboring points undergoes constant phase transitions in the form of rapid changes between expansion and contraction. Also, the singular expectation value of the local energy density of the vacuum was argued to be harmless at the energy levels of effective field theory since the fluctuations lead to huge cancellations on cosmological scales and ultimately to the observed accelerating expansion of the universe.

In a separate work [2], we studied the back-reaction of a Schwarzschild black hole geometry to quantum vacuum fluctuations. We examined how the black hole metric back-reacts to vacuum fluctuations in two regions, *onto* the horizon, and in the vicinity of the black hole. We considered vacuum fluctuations with local energy density *below* and *above* a certain threshold ζ. We found that *"strong"* quantum fluctuations (considered onto the horizon) lead to brief nonviolent departures from classicality, which allow for adiabatic leakage of low-temperature Hawking quanta at the necessary rate. In addition, we argued that quantum information can begin leaking out of the black hole as early as $r_S \log r_S$ after the initial collapse. The *"weak"* quantum fluctuations (considered in the vicinity of the black hole) were argued to be the microscopic source, which on scales of order of the Schwarzschild radius $\mathcal{O}(r_S)$ accumulates and produces metric fluctuations [2, 3].

The current model questions the classical black hole picture of local quantum field theory on a semiclassical geometry. In this work we apply the Unruh et al. model to Schwarzschild geometry; namely, we rewrite the equations of [1] in that background and study the gravitational effects of the inhomogeneous vacuum density. As a result, we propose a form of strong nonviolent nonlocality which yields significant modifications to the well-known general relativistic picture of black holes. The conjectured deviations from classicality lead simultaneously to nonlocal release of Hawking particles and quantum metric fluctuations. In fact, the present work may be thought of as the microscopic origin of the initially proposed by Giddings nonviolent nonlocality model [4–6].

The paper is organized as follows. In Section 2 we summarize the relevant work. In Section 3 we rewrite the equations of [1] in Schwarzschild black hole background and show that we can derive nontrivial modifications to general relativity in the near-horizon region.

2. Summary of Related Work

To make the paper self-contained, we begin by making a brief review of the main results of [1, 2]. Note the Unruh et al. model will serve as a basis for the scenario we will present later in Section 3. In Section 2.1 we demonstrate the physical interpretation of the conjectured in [1] inhomogeneous microscopically diverging vacuum energy density, and its effects on cosmological scales. Then, in Section 2.2, we review [2] to show how we initially derived the "soft" but nontrivial modifications to general relativity in the near-horizon region.

2.1. Accelerating Expansion via Inhomogeneous Quantum Vacuum Density. In the generic ΛCDM model of the universe the vacuum energy density is considered to be constant and homogeneous throughout space. As it was pointed out in [1], however, this basic assumption is only true for the expectation value of the vacuum energy density. Its *actual* value, *i.e.*, the one obtained by performing repeated measurements at a particular spatial point, constantly fluctuates. Consequently, we get a picture, where although the expectation value is effectively constant and homogeneous on cosmological scales, the actual value is rapidly changing in both time, as well as from point to point. Physically, this inhomogeneity of the vacuum density implies the spatial distance between any pair of nearby points constantly changes between phases of expansion and contraction. As we show in greater detail later, this leads to very important results in a black hole background.

Taking the constantly fluctuating inhomogeneous vacuum density as a starting point, we now briefly demonstrate its effects in a general spacetime [1]. Note that in Section 3 we apply those results to Schwarzschild black hole geometry.

Suppose the local energy density $\rho_{x,x'}$ between a pair of neighboring spatial points x and x' in some general metric $g_{\mu\nu}$ is [1]

$$\rho_{x,x'} = \frac{\operatorname{cov}(T_{00}(x), T_{00}(x'))}{\sigma_x \sigma_{x'}} \tag{1}$$

where

$$\sigma_x = \sqrt{\left\langle \left(T_{00}(x) - \langle T_{00}(x)\rangle\right)^2 \right\rangle} \tag{2}$$

Here, $T_{00}(x)$ and $T_{00}(x')$ are the local vacuum densities defined at the spatial points x and x', respectively.

Evidently from (1), the value of $\rho_{x,x'}$ is determined by the covariant vacuum densities, defined at the neighboring spatial coordinates, namely, $T_{00}(x)$ and $T_{00}(x')$. Effectively, one can think of $\rho_{x,x'}$ as a 2-point correlation function. That is, in order for $\rho_{x,x'}$ to have a nontrivial value, it must always be evaluated between close spacetime points. Otherwise, $\rho_{x,x'} \longrightarrow 0$ as the separation between x and x' becomes large, in which case $T_{00}(x)$ and $T_{00}(x')$ are no longer correlated and thus evolve independently. The requirement that x and x' are close comes from the limited domain of dependence of individual vacuum fluctuations, that is, their high momentum/short wavelength.

Assuming $\rho_{x,x'}$ is nonvanishing, then in order for it to be positive/negative, both $T_{00}(x)$ and $T_{00}(x')$ need to be, respectively, above/below the zero threshold of $\langle T_{00}\rangle$ [1]

$$\rho_{x,x'} = \begin{cases} > 0 & if\ T_{00}(x), T_{00}(x') > 0 \\ < 0 & if\ T_{00}(x), T_{00}(x') < 0 \end{cases} \tag{3}$$

Therefore, the coefficient $\rho_{x,x'}$ shows the correlation between vacuum densities defined between a pair of nearby points x and x'. One should note that since in this model we do not apply any renormalization procedures, we cannot use the generic stress-energy tensor as a source in the Einstein field equations. Instead, we must slightly modify the stress tensor in order to account for the diverging expectation value of the vacuum fluctuations.

Studying the gravitational effects of the inhomogeneous vacuum density requires inhomogeneity of the underlying metric as well. So in this scenario, the scale factor has to have an extra stochastic component which would allow it to account for that inhomogeneity. Therefore, following (1), the generic scale factor of the standard Friedmann-Robertson-Walker metric is modified as

$$ds^2 = -dt^2 + a^2(t,x)\left(dx^2 + dy^2 + dz^2\right) \tag{4}$$

As a result of the vacuum inhomogeneity, the scale factor $a(t,x)$ now has additional degrees of freedom in the form of a space-dependent coupling term. That is, when the local scale factor is evaluated at a given spacetime point, its dynamics is dictated (sourced) by the stochastically varying vacuum fluctuations at that point. This richer structure of the scale factor allows for a pair of nearby points to be expanding or contracting, depending on the sign of $\rho_{x,x'}$, (3). In particular, for $\rho_{x,x'} > 0$ the spatial separation between the pair of points increases, while for $\rho_{x,x'} < 0$, the spatial separation decreases.

It was then shown in [1] that one could consider a local Hubble rate term and evaluate it between the neighboring x and x'

$$\nabla H = -4\pi G J \tag{5}$$

where $H \equiv \dot{a}/a$ is the Hubble parameter and J denotes the energy flux of the vacuum, which accumulates over a given region of space. Hence J can be thought of as a functional of the local energy density in that neighborhood $\langle J(x,x')\rangle \sim \int_x^{x'} \rho_{x,x'}$.

Interestingly enough, following the extra degrees of freedom of $a(t,x)$, ∇H was found to be *constantly* fluctuating with energy, sourced by the accumulation of the vacuum density.

The general solution of (5) reads

$$H(t,x) = H(t,x_0) - 4\pi G \int_x^{x'} J(t,x')\left(dx', dy', dz'\right) \tag{6}$$

As (6) shows, the local Hubble rate depends on the spatial accumulation of J in the region between x and x'. To be more precise, we expect there may be some dissipation of the accumulated energy to nearby coordinates. For simplicity, however, throughout the paper we ignore all such effects and focus solely on individual 2-point functions.

Using this more complex spacetime dynamics induced by the inhomogeneous vacuum density, it was proposed that the equation of motion of ∇H goes as

$$\ddot{a} + \Omega^2 (t, x) a = 0 \tag{7}$$

where

$$\Omega^2 (t, x) = \frac{4\pi G}{3} \left(\rho + \sum_{i=1}^{3} P_i \right), \quad \rho = T_{00}, \; P_i = \frac{T_{ii}}{a^2} \tag{8}$$

In this case, (7) simply states that ∇H has the behavior of a harmonic oscillator. That is, it constantly fluctuates around its equilibrium point. Of course, every crossing of the equilibrium point is associated with a change of sign.

Let us briefly summarize the spacetime back-reaction to the conjectured in [1] constantly fluctuating inhomogeneous vacuum density. Physically, (5) and (7) describe a picture where the separation between a pair of neighboring points, $\Delta x \equiv |x - x'|$, is constantly fluctuating between phases of expansion and contraction. In fact, when Δx (herein defined as the proper distance) is expanding in some region, a neighboring region must be contracting, and vice versa. When a vacuum fluctuation with local energy density *above* its equilibrium point is considered between x and x', Δx (i.e., the proper distance) grows. On the other hand, if the local energy density is *below* the equilibrium point, the proper distance Δx decreases. This enhanced spacetime dynamics is characterized by constant local phase transitions between expansion and contraction which happen as the local energy density of the vacuum goes through its equilibrium point and changes sign as it does so. Those constant phase changes accumulate and lead to massive cancellations on cosmological scales. However, assuming a slight *positive* excess, we get the observed cosmological expansion. Although the microscopic values of ρ may be huge, their infrared effects are small, *i.e.*, the wild fluctuations do not lead to $\mathcal{O}(1)$ corrections in weak gravitational regimes. Besides, (8) tell us that the scale factor $a(t, x)$, in a given neighborhood, depends on the time-dependent frequency $\Omega(t, x)$ which exhibits quasiperiodic dynamics.

2.2. Unitary Black Hole Evolution and Horizon Fluctuations via Quantum Vacuum Fluctuations. In [2] we argued that treating gravity in the near-horizon region of Schwarzschild black hole as a field theory and considering its coupling to the matter fields in this region $\int \phi^{\mu\nu} T_{\mu\nu}$ produce fluctuations which modify the general relativistic description. We considered fluctuations with local energy density *below* and *above* ζ, where ζ is an arbitrary threshold. We then studied how the black hole geometry back-reacts to those fluctuations in two distinct regions, *i.e.*, *onto* the horizon and just outside the black hole.

More precisely, we studied how the horizon geometry back-reacts to "strong" quantum fluctuations and how the near-horizon region back-reacts to "weak" fluctuations. The analysis was carried out under the assumptions that (i) black holes are fast scramblers [7], (ii) quantum information is found in the emitted Hawking particles, and (iii) the scrambled infallen information *need not* be embedded uniformly across the horizon.

Let us now precisely define what we mean by "strong" and "weak" quantum fluctuations.

(A) In a broader sense, we take a fluctuation to be *strong* if, when considered at asymptotic infinity, its local energy density leads to a localized particle production

$$a_i^\dagger |0\rangle = |x\rangle \tag{9}$$

Therefore, if we had a measuring apparatus counting the strong fluctuations in a given spacetime region Σ at asymptotic infinity, its results would be consistent with the expectation value of the number operator $\langle N \rangle$ in that region

$$\langle N \rangle = \sum_{i}^{\mathcal{N}} \int_{\Sigma} \varphi_i^{strong} \tag{10}$$

More specifically, in a black hole background, we argue a "strong" quantum fluctuation *onto* the horizon yields brief departure from local quantum field theory.

(B) A quantum fluctuation is taken as *weak* if its local energy density is below the threshold ζ. Because of the small local energy density, we assume the back-reaction of the background metric would be negligible if we considered weak fluctuations in a relatively small part of the near-horizon region. That is why we are interested in how the near-horizon metric back-reacts when the weak fluctuations are taken on scale $\mathcal{O}(r_S)$.

We argue that weak fluctuations on scale $\mathcal{O}(r_S)$ lead to nonperturbative effects which manifest in potentially observable metric fluctuations that can play an important role in observer complementarity and in gravitational wave astronomy in the form of detectable "echoes" and deviations from general relativity close to the horizon. We assume that away from a black hole the weak fluctuations do not lead to perturbations and leave the geodesic equation invariant. One should keep in mind that, although effectively negligible at infinity, when examined locally, the weak fluctuations still cause a pair of points to rapidly change phases between expansion and contraction.

To demonstrate how we define the weak fluctuations in the vicinity of the horizon we adopted the following analogy. Suppose we interpret individual fluctuations as harmonic oscillators, denoted by χ_i. Imagine we place the harmonic oscillators on a string in the vicinity of the horizon, Figure 1. Using an arbitrary normalized spacing ϵ, we can generally describe the string as

$$\sum_{i=1}^{N} \int_{S2} d\varphi \left(n_i \varphi_i \right) \tag{11}$$

where φ_i is the oscillation frequency of the different harmonic oscillators.

In this picture we get an ensemble of fluctuations which, as we will argue later, yield coherent Schwarzschild-scale metric fluctuations. We assume separate harmonic oscillators *need not* have the same frequency. In fact, due to their limited domain of dependence (of order of the wavelength of the fluctuation), even neighboring harmonic oscillators have different frequencies. Thus an evolution equation for a particular spacetime region can be given in terms of a linear combination of the harmonic oscillators in that region. Like we mentioned earlier, despite the arbitrarily high (diverging) oscillating frequency a single harmonic oscillator may have microscopically, its effect in an infrared cut-off is negligible, and we do not expect $\mathcal{O}(1)$ corrections to the background metric.

Our analysis in [2] lead us to the following two main results.

First, we found that "strong" fluctuations, considered onto the horizon, lead to nonlocal release of quantum information via substantial deviations from classicality (for similar results [4–6]). We also demonstrated the model achieves Page-like evaporation spectrum [8, 9] and predicts black holes begin emitting Hawking particles in time, logarithmic in the entropy, *i.e.*, $t_* \sim \mathcal{O}(r_S \log r_S)$, where $t_* \ll M^3$.

Second, we found that "weak" quantum fluctuations, considered in the vicinity of the horizon, produce metric fluctuations which, although locally negligible, were shown to accumulate and on scale of order of the Schwarzschild radius lead to significant modifications to general relativity. Physically, the horizon periodically shifts (with frequency ω) radially outward with an amplitude δ, where $l_P \ll \delta < M$. In Schwarzschild coordinates the metric fluctuations translate to shift of the horizon from $r = 2M$ to $r = 2M+\delta$. One can think of the weak fluctuations as exerting a drag-like force on the black hole which produces observable macroscopic quantum gravity effects. Schematically

$$\omega \sim M_{BH}^{-1} \tag{12}$$

where ω is the frequency of the metric fluctuations.

The inverse proportionality between the frequency of the metric fluctuations and the mass of the black hole was argued to lead to thermodynamic instability at late times. Imagine a freely evaporating Schwarzschild black hole with no perturbations being introduced to it. Given $M_{BH} = 1/T$, we expect T to monotonically increase as the black hole evaporates. As we can see, (12) dictates that, as the black hole loses mass, ω steadily increases. Tracing that evolution to the final stages of the evaporation we assume the black hole becomes thermodynamically unstable as $M_{BH} \longrightarrow m_P$, where m_P is the Planck mass. At that point, the black hole was conjectured to explode [10]. Here, late-time black hole evolution is in agreement with the thermodynamic instability first proposed in [10].

Recently, the authors of [11] numerically solved the Einstein equations modified with metric fluctuations. Similar to the current work, they showed that the proposed metric fluctuations lead to major deviations from the traditional general relativistic black hole description. In agreement with our results (in particular (12)), their analysis shows there is an inverse proportionality between the frequency of the fluctuations, and black hole's mass, where for a black hole binary merger, the overlap between the classical and the modified pictures of the waveform of the emitted gravitational waves decreases as the mass of the binary gets smaller. That is, as the frequency of the metric fluctuations increases, the near-horizon region deviates from general relativity even more significantly.

3. The Quantum Vacuum Origin of Metric Fluctuations and Nonlocal Evaporation

In the current section we adopt the Unruh et al. model and study its effects in Schwarzschild background. Consequently, we propose a dynamical mechanism for a form of strong nonviolent nonlocality which significantly modifies the traditional field theory picture in the near-horizon region. Specifically, we demonstrate how the conjectured in [1] inhomogeneous vacuum density leads simultaneously to quantum metric fluctuations [2, 4–6, 11–13] and nonlocal release of quantum information. One should note we do not attempt to quantize gravity in the current paper. Although the considered vacuum fluctuations may not be normalized, which modifies the stress-energy tensor and makes it nongeneric, they still contribute to its expectation value.

We aim to make the transition from cosmological scales to a black hole case more consistent. That is why we would like to first expand more on the implications of (1) for black hole backgrounds.

So let us now focus in more detail on how a pair of neighboring spacetime points x and x' is affected by the suggested extremely inhomogeneous vacuum density. Consider the following [1]:

$$\Delta\rho^2(\Delta x) = \frac{\left\langle \left\{ T_{00}(t,x) - T_{00}(t,x') \right)^2 \right\} \right\rangle}{(4/3)\left\langle T_{00}(t,x) \right\rangle^2} \tag{13}$$

where $T_{00}(x)$ is the vacuum density at x and $\Delta x \equiv |x - x'|$ denotes the separation between x and x'.

Recall our earlier discussion about the evolution of the local energy density. Specifically, $\rho \longrightarrow 0$ as Δx gets large, in which case $T_{00}(x)$ and $T_{00}(x')$ evolve independently.

Keeping that in mind, suppose a classical Schwarzschild solution

$$ds^2 = -\left(1 - \frac{2M}{r}\right) dt^2 + \left(1 - \frac{2M}{r}\right)^{-1} dr^2 + r^2 d\Omega^2 \tag{14}$$

where $d\Omega^2 = (d\theta^2 + \sin^2\theta d\phi^2)$.

Particularly, we are interested in rewriting (13) in Schwarzschild geometry. Working in a black hole background requires us to take into account the large gradient with respect to the horizon

$$\rho = \partial_r (\partial_r \phi)^2 \tag{15}$$

where ρ is the local energy density and ϕ is the gravitational potential. Thus near a matter source (15) is very sensitive to changes in the radial coordinate.

Therefore, by taking into account (15), we can rewrite (13) in Schwarzschild background as

$$\int_R \int_x^{x'} J^2 = \int_R \int_x^{x'} \frac{\left\langle \left\{ \partial_r J(t,x) - \partial_r J(t,x')\right)^2 \right\} \right\rangle}{(4/3)\,\langle \partial_r J(t,x)\rangle^2} \quad (16)$$

where R denotes the near-horizon region and J is the vacuum accumulation term.

Notice that since we no longer work on cosmological scales, (15) becomes relevant, and so we substituted the vacuum energy density terms, defined on particular spacetime points, with J. Both terms $J(x)$ and $J(x')$ are calculated with respect to the horizon. Here, J is very sensitive to radial changes because of (15). In case x and x' both lie on the same $r = const$ surface, the expectation value of J between them varies stochastically.

Because J is of particular importance in the near-horizon neighborhood, let us now focus on its general properties.

Since the local energy density in a spacetime region which includes a black hole exhibits Gaussian-like distribution (15), we assume that as $r \longrightarrow \infty$, $\langle J \rangle \longrightarrow 0$. Although this may be true for the expectation value of J, its *actual* value must still constantly fluctuate due to the extremely inhomogeneous vacuum density. Clearly, we can see that, in the near-horizon region, J has a nonzero expectation value $\int_R \langle J \rangle \gg 0$. Therefore, the strongly radial-dependent behavior of J in the near-horizon region is trivially given as

$$\int_R \frac{\partial \langle J \rangle}{\partial r} \quad (17)$$

Where depending on r with respect to the horizon, the general solutions to (17) are

$$\int_R \frac{\partial \langle J \rangle}{\partial r} = \begin{cases} 0 & for\ r > r_S \\ > 0 & for\ 2M < r < r_S \end{cases} \quad (18)$$

Lastly, we can easily extend (17) to include an arbitrary number of gauge fields as

$$\int_R \langle T_{\mu\nu} \rangle + \frac{\partial \lambda}{\partial r} \quad (19)$$

where λ is the vacuum and $T_{\mu\nu}$ is the stress-energy tensor.

Note that, at constant r from the horizon, the energy density of the vacuum fluctuations depends on the internal degrees of freedom of the black hole and varies stochastically.

3.1. Metric Fluctuations: Weak Quantum Fluctuations = Local Phase Transitions. In this subsection we demonstrate how the proposed inhomogeneous vacuum density can modify the classical near-horizon physics and thus yield metric fluctuations. Specifically, by embracing the harmonic-oscillator-like constant phase transitions of ∇H, we rewrite (5) in Schwarzschild background and study how the metric back-reacts. Hence, we show that in this scenario one can obtain the conjectured quantum corrections to general relativity in the region just outside the black hole. Like we saw earlier, the

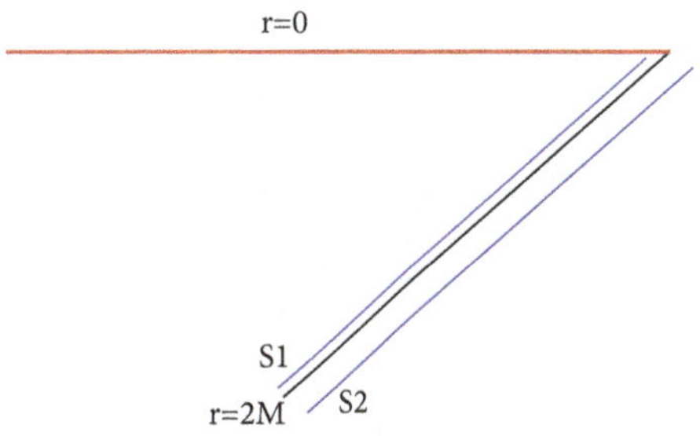

FIGURE 1: The bold red line is the singularity, and the bold black line, $r = 2M$, is the horizon. Imagine that $S1$ coincides with the horizon, and thus modes located on $S1$ are spacelike separated from an observer outside the black hole. Further, imagine that $S2$ is placed in the vicinity of the horizon.

microscopically extremely inhomogeneous vacuum density gives freedom to the scale factor to be locally expanding or contracting. Recall that in (5) we defined a local Hubble rate term ∇H and argued that, due to the inhomogeneous vacuum density, it has the dynamics of a harmonic oscillator (7). Namely, it constantly changes between phases of expansion and contraction.

For the sake of completeness we will examine coarse-grained and fine-grained version of the argument.

3.1.1. Coarse-Grained. In the particular case we neglect contributions coming from individual degrees of freedom and instead only focus on the effective (macroscopic) back-reaction on scale of order of the Schwarzschild radius.

Keeping in mind the large field-strength radial dependence in this region (15), we can rewrite (5) as

$$\int_{S1}^{S2} \nabla H = -4\pi G \int_{S1}^{S2} \partial_t \langle J \rangle \quad (20)$$

Unlike earlier, where we considered ∇H between the neighboring points x and x', we now evaluate it in the region between $S2$ and $S1$ (the horizon), Figure 1.

To better demonstrate the back-reaction of the near-horizon geometry (20), consider the following *gedanken* experiment. Imagine $S2$ is a timelike hypersurface the size of the Schwarzschild radius just outside the black hole, coupled to the horizon degrees of freedom. Suppose we wish to evaluate the constant ∇H phase transitions in the region between the hypersurface and the horizon. We assume this region constantly undergoes uniform phase transitions with characteristic oscillation cycle time T, where by "uniform" we mean that, once every T, a phase change of ∇H in that region takes place. Unlike the cosmological case [1], here we assume that on every T the phase transitions average out to a small *negative* value. Here, T is given as

$$T = \frac{2\pi}{\Omega} \quad (21)$$

where Ω is time-dependent frequency which depends on $\langle T_{00} \rangle$.

That constant change between phases of expansion and contraction just outside the black hole (assuming a slight

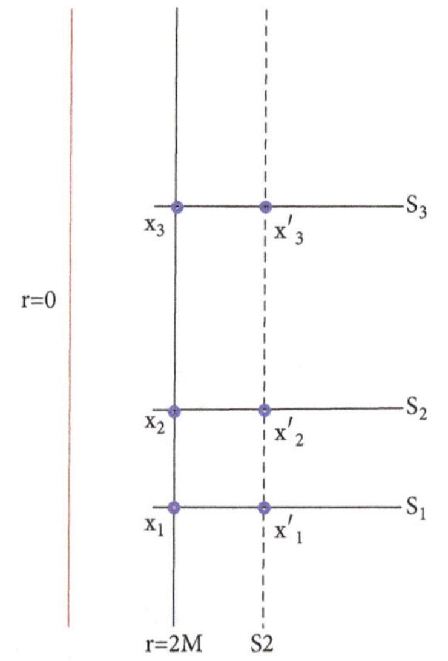

FIGURE 2: Similar to Figure 1, imagine S1 coincides with the horizon. The dotted line, S2, is an artificial (nonphysical) timelike hypersurface in the vicinity of the horizon. Note that the blue dots x_i and x_i' are pairs of corresponding points, where $x_i \in S1$, $x_i' \in S2$, and $x_i, x_i' \in S_i$. Also, $\sum_i x_i \equiv S1$ and $\sum_i x_i' \equiv S2$.

negative excess every T) yields the dragging-like effect on the horizon. For simplicity, one can imagine the conjectured dynamics outside the black hole as the back-reaction of the horizon to a vector field coupled to matter fields in Rindler space, where we consider only contributions from modes very close to the horizon.

Physically, the back-reaction of the near-horizon region manifests as a slight outward shift of the horizon from $r = 2M$ to $r = 2M + \delta$ in Schwarzschild coordinates.

Imagine a far-away observer who cannot probe the region near the horizon. As far as she is concerned, she may effectively interpret those metric fluctuations as a Planck scale structure just outside the horizon (similar to the stretched horizon in observer complementarity or gravitational wave "echoes"). Different proposals have recently been made regarding the possibility of observing such quantum gravity effects (see [11–13]).

In conclusion, we can see that by rewriting (5) in a Schwarzschild black hole background we can effectively derive the proposed in [2–6, 11–13] metric fluctuations.

3.1.2. Fine-Grained. In this case we focus on contributions coming from individual degrees of freedom, *i.e.*, phase transitions of ∇H between a pair of corresponding points (say, x and x'), where both lie on the same spacelike hypersurface S_i in the near-horizon region. Namely, $x \in S1$, $x' \in S2$, and $x, x' \in S_i$, Figure 2.

Suppose we foliate the black hole spacetime into $t = const$ spacelike hypersurfaces. The particular foliation $\mathcal{F}$ forms a family of extrinsically flat slicing. The slicing is generic and compatible with the Killing symmetries of the spherically symmetric spacetime. Thus $\mathcal{F}$ preserves the spherical symmetry of the horizon geometry. The foliation can be intuitively identified with a family of observers that only have time and radial components. We have chosen this more trivial geometry-preserving foliation as we aim to make the study of the back-reaction of the metric to individual 2-point functions easier.

Considering the particular foliation, we can straightforwardly rewrite (20) as

$$\sum_{i=1}^{\mathcal{N}} \nabla H_i = -4\pi G \sum_{i=1}^{\mathcal{N}} \left\langle J\left(x_i, x_i'\right)\right\rangle \tag{22}$$

where x and x' lie on the same spacelike hypersurface S_i and also $\sum_i x_i \equiv S1$ and $\sum_i x_i' \equiv S2$. We think of $S2$ as an artificial (nonphysical) $r = const$ timelike hypersurface, and we assume it respects the same foliation as the horizon, Figure 2.

Evidently from (22), and in agreement with what we have argued earlier (5), the phase transitions of ∇H between a pair of neighboring points depend on the vacuum accumulation term J between them. Note that, due to the stochastic nature of the inhomogeneous vacuum density, the r.h.s. of (22) fluctuates constantly with field strength of order of the expectation value of the vacuum density in the region $\langle J(x, x')\rangle \approx \int_x^{x'} \langle T_{00}\rangle$.

However, (22) does not suffice for a complete description of the near-horizon dynamics. In addition, we also need to examine the evolution of the local scale factor. The equation of motion of $a(t, x)$ away from matter source is given as [1]

$$\tan\Theta_x = \frac{\Omega(0, x)}{\Omega(0, x')}\tan\Theta_x + \frac{4\pi G}{\Omega(0, x')}\int_x^{x'} J\left(0, x'\right) dl' \tag{23}$$

where Θ_x is the initial phase of $a(t, x)$ at some arbitrary x.

Similar to what we did with (22), we now wish to express the equation of motion of $a(t, x)$ in the vicinity of the horizon. We do that in terms of the vacuum accumulation between a pair of corresponding points both of which lie on the same spacelike hypersurface. Neglecting the initial phase of $a(t, x)$ because of its constant fluctuations, we get

$$\tan\Theta_{x,x'} = \int_x^{x'} \frac{\Omega(x)}{\Omega(x')}\tan\Theta_{x,x'} + \frac{4\pi G}{\Omega(x')}\left\langle J\left(x, x'\right)\right\rangle \tag{24}$$

where $\langle J(x, x')\rangle \approx \int_x^{x'} \langle T_{00}\rangle$. One should recall we focus on the small *negative* excess of the fluctuations every T.

The back-reaction of the near black hole geometry to the fluctuations between a pair of points on a single slice (22) is negligible, regardless of their microscopically singular expectation value. However, the spatial separation between the pair, Δx, is constantly fluctuating in a harmonic-oscillator-like manner between phases of expansion and contraction due to the inhomogeneous vacuum density. Consequently, by considering the constant fluctuations of Δx on all hypersurfaces of order of the Schwarzschild radius and assuming a subtle *negative* excess on each slice every $T = 2\pi/\Omega$, the back-reaction accumulates and leads to deviations from general relativity which manifest in metric fluctuations.

In case $M_{BH} = const$, then $\langle J(x, x')\rangle$ depends on the dimensionality of the internal Hilbert space of the black hole and thus fluctuates stochastically

$$\left\langle J\left(x, x'\right)\right\rangle \sim \int_{x}^{x'} \langle T_{00}\rangle \sim \dim\left(\mathcal{H}_{int}\right) \quad (25)$$

Therefore, following (15) and considering the microscopic back-reaction of the metric to the constantly fluctuating inhomogeneous vacuum density, the action for the near-horizon region reads

$$I_{near} = \int_{R} \int dx \int dt \left[\partial_r \lambda + T_{00}\left(x, x'\right) \partial_r \left(x, x'\right) \partial_t \lambda\right] \quad (26)$$

where the radial-dependent term of the vacuum $\partial_r \lambda$ is given by (15). The time-dependent vacuum term $\partial_t \lambda$ is taken on a $r = const$ surface with respect to the horizon and varies with the change in the mass of the black hole. The action is given in terms of couplings between the inhomogeneous constantly fluctuating and radially dependent vacuum term λ and the local energy density.

In summary, we demonstrated that when we consider the accumulated back-reaction of the background metric to the weak quantum vacuum fluctuations (*i.e.*, constant phase transitions of ∇H on different hypersurfaces) across the whole horizon area, $\mathcal{O}(r_S)$, we effectively get the proposed quantum metric fluctuations. The current derivation of these "soft" quantum modifications to the effective black hole geometry may be thought of as the microscopic origin of the metric fluctuations proposed by Giddings [4–6].

3.2. Nonlocal Black Hole Evaporation via Strong Quantum Vacuum Fluctuations. In this subsection we continue the study of the back-reaction of a Schwarzschild black hole to the constantly fluctuating inhomogeneous vacuum density. More precisely, we examine what effects the conjectured in [2] strong quantum vacuum fluctuations have, when they are considered *onto* the horizon. Specifically, we focus on how the locality constraint of local quantum field theory is modified due to the back-reaction of the metric to those fluctuations. As a result, we argue the horizon geometry *need not* always respect locality.

Given the fundamental degrees of freedom are not continuously distributed, we assume the semiclassical geometry to be just an effective field theory. Thus we expect the local quantum field theory evolution to only be approximately correct. As a result, deviations from it should be present in high energy regimes. For instance, black holes are one such place where we expect nonlocal corrections to manifest. That is the case since due to the large field-strength radial dependence in the near-horizon region, $\langle T_{00}(x)\rangle \gg \langle T_{00}(y)\rangle$, where $2M \le x \le R$ and $y > R$.

The particular strong vacuum fluctuations are assumed to have high enough local energy density as to yield "soft" (*i.e.*, brief and highly localized) nonlocal corrections to local quantum field theory. Similar to [1], we assume these soft corrections manifest in the form of local singularity points.

Because in [1] the local scale factor $a(t, x)$, defined at each spacetime point, was argued to have a harmonic-oscillator-like behavior, as it changes phases, $a(t, x)$ must inevitably go through zero, *i.e.*, yield a local singularity point. However, this was shown to not be physically problematic. Since we interpret the local scale factor as a harmonic oscillator, $a(t, x) = 0$ is just a generic part of the oscillation cycle which takes place every T. Namely, a harmonic oscillator cannot change sign (phase) without passing through zero. Therefore, in a generic oscillation cycle a singularity point disappears almost immediately.

Similarly here, because we interpret individual strong vacuum fluctuations *onto* the horizon as harmonic oscillators, the microscopically singular expectation value (as predicted by quantum field theory) of a single strong fluctuation at a given point on the horizon should not be problematic. Likewise, we assume the local singularity point disappears almost immediately.

Physically, a strong vacuum fluctuation on the horizon briefly violates the generic locality constraint of local quantum field theory in that small region. As a result, this deviation from classicality allows quantum information to escape to asymptotic infinity at the necessary rate. Due to the small domain of dependence and brief lifespan of the local singularity points, their effects on a freely falling observer are negligible. These nonlocal corrections to the semiclassical geometry, although "soft," have significant effect on the black hole over periods compared to its lifetime. That being said, we assume that, since the inhomogeneous vacuum density is constantly fluctuating, it will carry out quantum information at a rate of $\mathcal{O}(1)$ per light-crossing time as to restore unitary quantum mechanics. Such local quantum field theory modifications allow evaporation of information-carrying Hawking particles to begin as early as the scrambling time.

Let us clarify what we mean when we characterize the nonlocal corrections as "soft." The local quantum field theory deviations (*i.e.*, local singularity points) are "brief" in the sense that they have very short lifespan of order of the lifetime of the fluctuation. Furthermore, the deviations are considered to be "highly localized" (*i.e.*, short wavelength/high momentum) since they manifest on scales of order of the wavelength of the fluctuation. Thus a strong vacuum fluctuation has a limited domain of dependence, and cannot lead to $\mathcal{O}(1)$ corrections to the background metric of a solar mass black hole.

Let us now point out an important distinction between the local singularity points on cosmological scales (away from a black hole) [1] and onto the horizon. We claim there is an intrinsic difference in the stage of the oscillation cycle during which a singularity point is produced. More precisely, a strong vacuum fluctuation (onto the horizon) produces a local singularity when, during an oscillation cycle T, it reaches the maximum of its local energy density, which also happens to be above a certain threshold ζ. On the other hand, the local scale factor $a(t, x)$ (away from a black hole) produces a singularity point when it reaches zero during its oscillation cycle.

Moreover, the local singularity points, although similar, should not be mistaken with spacetime defects (see [14, 15]). For instance, (i) at a local singularity point/spacetime defect the curvature is divergent, and (ii) particle passing through (near) a spacetime defect/local singularity will be scattered off, *i.e.*, experience a local Lorentz boost.

4. Casimir Stress Interpretation

In [1] the Casimir effect was used as a tool for illustrating the effects of vacuum fluctuations. In the current Section we present a toy model in which we restate the conjectured quantum metric fluctuations in terms of the Casimir effect in the near-horizon region. Specifically, we present the metric fluctuations in the language of the Casimir effect just outside the black hole (between *S1* and *S2*, Figure 1).

Likewise, we will examine two distinct cases: coarse-grained and fine-grained.

In its general form, the Casimir stress equation is given as [1]

$$S(t, x, y) = T_{zz}^{inside} - T_{zz}^{outside} \tag{27}$$

In this picture, imagine the role of the pair of conducting plates is played by *S1* and *S2*, Figure 1.

4.1. Coarse-Grained. Similar to Section 3, we are only interested in the effective (macroscopic) dynamics and thus neglect contributions from individual degrees of freedom.

We can straightforwardly expand (27) in the near-horizon region as

$$\int_{S1}^{S2} S(t, x, y) = \int_{S1}^{S2} \langle \partial_r \rho_{in} \rangle - \int_{S2}^{\infty} \langle \partial_r \rho_{out} \rangle \tag{28}$$

where, as we showed earlier, the expectation value of the radially dependent local energy density $\langle \partial_r \rho_{in} \rangle$ is of order of the vacuum accumulation term in that region

$$\int_{S1}^{S2} \langle \partial_r \rho_{in} \rangle \sim \int_{S1}^{S2} \partial_r \langle J \rangle \tag{29}$$

Following (18), we expect $\int_R \langle J \rangle \gg 0$ and $\int_\infty \langle J \rangle = 0$. That is, the second term on the r.h.s. of (28) vanishes at the asymptotic limit; namely $\langle \rho_{out} \rangle \longrightarrow 0$ as $r \longrightarrow \infty$.

Therefore, we generally get

$$\left\langle T_{zz}^{inside} \right\rangle \gg \left\langle T_{zz}^{outside} \right\rangle \tag{30}$$

One should keep in mind that, regardless of whether $\langle J \rangle$ vanishes or not, it constantly fluctuates due to the inhomogeneous vacuum density.

Notice that (30) is in agreement with the general construction of the Casimir effect (27).

Considering the microscopic back-reaction of the spherically symmetric background metric that we discussed earlier and the greater energy density in the vicinity of the horizon (30), one can easily see how the proposed quantum metric fluctuations of order of the Schwarzschild radius emerge in this setting.

4.2. Fine-Grained. Similar to our approach in Section 3.1.2, we begin by foliating the horizon region into individual spacelike hypersurfaces. As a result, we can now express the l.h.s. of (28) as

$$\int_R S(t, x, y) = \sum_{i=1}^{\mathcal{N}} \left\langle J\left(x_i, x_i'\right) \right\rangle \tag{31}$$

Evidently, the nonvanishing l.h.s. of (28) can straightforwardly be rewritten as a linear combination of vacuum accumulation terms, defined on individual slices.

Due to the inhomogeneous vacuum density, Δx on any given slice constantly fluctuates between phases of expansion and contraction. Moreover, because of the strong field gradient (15) and considering (30), we assume that in the large $\mathcal{N}$ limit (*i.e.*, of order of the Schwarzschild radius) the accumulated back-reaction is positive. Thus, an observer at asymptotic infinity sees oscillations of the horizon as a back-reaction of the underlying black hole metric.

In summary, we see that with minimal assumptions one can restate our earlier argument about quantum metric fluctuations in terms of the Casimir effect in a near black hole region.

5. Conclusions

In the current work we studied how a Schwarzschild black hole back-reacts to the constantly fluctuating inhomogeneous vacuum density proposed in [1]. More precisely, embracing the microscopically singular expectation value of the local energy density of the inhomogeneous vacuum fluctuations (predicted by quantum field theory), we examined how a black hole metric back-reacts in two distinct regions: the vicinity of the black hole and onto the horizon. As a result, we demonstrated that vacuum fluctuations above a given threshold, considered onto the horizon, cause deviations from local quantum field theory. Meanwhile, fluctuations below that threshold, considered in the vicinity of the horizon, lead to potentially observable metric fluctuations of order of the Schwarzschild radius.

Physically, the conjectured modifications of local quantum field theory, induced by the strong vacuum fluctuations onto the horizon, were argued to lead to nonlocal release of information-carrying Hawking particles. In fact, we argued that in this scenario a black hole can begin radiating quantum information to infinity as early as the scrambling time [2].

On the other hand, we argued that weak fluctuations in the near-horizon region yield observable macroscopic quantum gravity effects in the form of metric fluctuations of order of the Schwarzschild radius, that is, constant oscillations of the horizon between $r = 2M$ and $r = 2M + \delta$. As far as a distant observer is concerned, we assume she may interpret the conjectured metric fluctuations as a physical membrane just outside the horizon. Thus the proposed metric fluctuations may serve as the microscopic origin of the stretched horizon in observer complementarity [16]. Also, we assume the proposed metric fluctuations play a significant role in binary black hole mergers. In particular, they may produce observable postmerger gravitational wave "echoes" similar to [17].

The recent advances in gravitational wave astronomy have opened new possibilities for experimentally testing models, similar to this one, which predict deviations from general relativity in the near-horizon region. In addition, the current scenario could also be approached from an accretion disk perspective as we believe the metric fluctuations may have measurable effects on accretion disk flows around a black hole.

Conflicts of Interest

The authors declare that they have no conflicts of interest.

Acknowledgments

This research is supported in part by project RD-08-112/2018 of Shumen University.

References

[1] Q. Wang, Z. Zhu, and W. G. Unruh, "How the huge energy of quantum vacuum gravitates to drive the slow accelerating expansion of the Universe," *Physical Review D: Particles, Fields, Gravitation and Cosmology*, vol. 95, no. 10, Article ID 103504, 2017.

[2] A. Y. Yosifov and L. G. Filipov, "Entropic entanglement: information prison break," *Advances in High Energy Physics*, vol. 2017, Article ID 8621513, 7 pages, 2017.

[3] A. Y. Yosifov and L. G. Filipov, "Oscillations for Equivalence Preservation and Information Retrieval from Young Black Holes," *Electronic Journal of Theoretical Physics*, vol. 36, pp. 183–198, 2016.

[4] S. B. Giddings, "Models for unitary black hole disintegration," *Physical Review D: Particles, Fields, Gravitation and Cosmology*, vol. 85, no. 4, Article ID 044038, 2012.

[5] S. B. Giddings, "Nonviolent information transfer from black holes: a field theory parametrization," *Physical Review D*, vol. 88, no. 2, Article ID 024018, 2013.

[6] S. B. Giddings, "Modulated Hawking radiation and a nonviolent channel for information release," *Physics Letters B*, vol. 738, pp. 92–96, 2014.

[7] P. Hayden and J. Preskill, "Black holes as mirrors: quantum information in random subsystems," *Journal of High Energy Physics*, vol. 2007, no. 9, article 120, 2007.

[8] D. N. Page, "Information in black hole radiation," *Physical Review Letters*, vol. 71, no. 23, pp. 3743–3746, 1993.

[9] D. N. Page, "Black hole information," in *Proceedings of the 5th Canadian Conference on General Relativity and Relativistic Astrophysics*, R. B. Mann and R. G. McLenaghan, Eds., pp. 13–15, University of Waterloo, Singapore, 1993.

[10] S. W. Hawking, "Black hole explosions?" *Nature*, vol. 248, no. 5443, pp. 30-31, 1974.

[11] S. L. Liebling, M. Lippert, and M. Kavic, "Probing near-horizon fluctuations with black hole binary mergers," *Journal of High Energy Physics*, vol. 2018, no. 3, 2018.

[12] S. B. Giddings, "Possible observational windows for quantum effects from black holes," *Physical Review D: Particles, Fields, Gravitation and Cosmology*, vol. 90, no. 12, 2014.

[13] S. B. Giddings, "Gravitational wave tests of quantum modifications to black hole structure - with post-GW150914 update," *Classical and Quantum Gravity*, vol. 33, p. 235010, 2016.

[14] S. Hossenfelder, "Phenomenology of Space-time Imperfection I: Nonlocal Defects," *Physical Review*, vol. 88, p. 124030, 2013.

[15] S. Hossenfelder and R. G. Torromé, "General relativity with space-time defects," *Classical and Quantum Gravity*, vol. 35, 2018.

[16] L. Susskind, L. Thorlacius, and J. Uglum, "The stretched horizon and black hole complementarity," *Physical Review D: Particles, Fields, Gravitation and Cosmology*, vol. 48, no. 8, pp. 3743–3761, 1993.

[17] J. Abedi and N. Afshordi, "Echoes from the Abyss: A highly spinning black hole remnant for the binary neutron star merger GW170817," https://arxiv.org/abs/1803.10454.

8

Revisiting the Black Hole Entropy and the Information Paradox

Ovidiu Cristinel Stoica

Department of Theoretical Physics, National Institute of Physics and Nuclear Engineering – Horia Hulubei, Bucharest, Romania

Correspondence should be addressed to Ovidiu Cristinel Stoica; holotronix@gmail.com

Academic Editor: Theophanes Grammenos

The black hole information paradox and the black hole entropy are currently extensively researched. The consensus about the solution of the information paradox is not yet reached, and it is not yet clear what can we learn about quantum gravity from these and the related research. It seems that the apparently irreducible paradoxes force us to give up on at least one well-established principle or another. Since we are talking about a choice between the principle of equivalence from general relativity and some essential principles from quantum theory, both being the most reliable theories we have, it is recommended to proceed with caution and search more conservative solutions. These paradoxes are revisited here, as well as the black hole complementarity and the firewall proposals, with an emphasis on the less obvious assumptions. Some arguments from the literature are reviewed, and new counterarguments are presented. Some less considered less radical possibilities are discussed, and a conservative solution, which is more consistent with both the principle of equivalence from general relativity and the unitarity from quantum theory, is discussed.

1. Introduction

By applying general relativity and quantum field theory on curved spacetime, Hawking arrives at the conclusion that the information is lost in the black holes, and this breaks the predictability [1]. Apparently, no matter how was formed and what information was contained in the matter falling in a black hole, the only degrees of freedom characterizing it are its mass, angular momentum, and electric charge, so black holes are "hairless" [2–5]. This means that the information describing the matter crossing the event horizon is lost, because nothing outside the black hole reminds us of it. In general relativity, this information loss is irreversible, not only because we cannot extract it from beyond the event horizon, but also because in a finite time the infalling matter reaches the singularity of the black hole. And the occurrence of singularities is unavoidable, according to the singularity theorems [6–8]. This already seemed to be a problem, but it would not be so severe if we at least know that the information is still there, censored behind the horizon [9, 10]. But we are not even left with this possibility, since Hawking proved that quantum effects make the black holes evaporate [11]. It was already expected that black holes should radiate, after the realization that they have entropy and temperature [11, 12], and these should be part of an extension of thermodynamics which includes matter as well. This evaporation is thermal, and after the black hole reaches a planckian size, it explodes and reveals to the exterior world that the information is indeed lost. In addition, if the quantum state prior to the formation of the black hole was pure, the final state is mixed, increasing the drama even more. Moreover, a problem seems to occur long before the complete evaporation, since the black hole entropy seems to increase during evaporation, until the Page time is reached [13]. Some consider this to be the real black hole information paradox [14].

Mainly for general relativists the information loss seemed to be definitive and yet not a big problem [15], position initially endorsed by Hawking too. On the other hand, for high energy physicists, loss of unitarity was considered a problem, and various proposals to fix it appeared (see, e.g., [16–18] and references therein). For example, *remnants* were proposed, containing the information remaining in the black hole after evaporation. The remnant is in a mixed state but together with the Hawking radiation forms a pure state. A possible cause for remnants is the yet unknown quantum corrections expected to occur when the black hole becomes too small, comparable to the Plank scale, and the usual analysis of Hawking radiation no longer applies [19–21]. There are other possibilities, some being discussed in the above-mentioned reviews. For example, it was proposed that the information leaks out of the black hole through evaporation, including by quantum tunneling, that it escapes at the final explosion or that it leaks out of the universe in a baby universe [22, 23]. Another possibility is that the information escapes as Hawking radiation by *quantum teleportation* [24], which actually happens as if the particle zig-zags forward and backward in time to escape without exceeding the speed of light. This is not so unnatural, if we assume that the final boundary condition at the future singularity of the black hole forces the maximally entangled particles to be in a singlet state. There are also *bounce scenarios* [25] or by using local scale invariance to avoid singularities [26]. Some bounce scenarios are based on *loop quantum gravity*, like [27, 28], as well as *black hole to white hole tunneling scenarios* in which quantum tunneling is supposed to break the Einstein equation, and the apparent horizon is prevented to evolve into an event horizon [29, 30]. It would take a long review to do justice to the various proposals, and this is beyond the scope of this article.

The dominating proposed solution was, for two decades, *black hole complementarity* [31–33]. This was later challenged by the *firewall paradox* [34]. The debate is not settled down yet, but the dominant opinion seems to be that we have to give up at least one principle considered fundamental so far, and the unlucky one is most likely the principle of equivalence from general relativity. One of the objectives of the present article is to show that we can avoid this radical solution while keeping unitarity.

The problems related to the black hole information loss are considered important, being seen as a benchmark for the candidate theories of quantum gravity, which are expected to solve these problems.

The main purpose of this discussion is to identify the main assumptions and see if it is possible to solve the problem in a less radical way. I argue that some of the usually made assumptions are unnecessary, that there are less radical possibilities, and that the black hole information problem is not a decisive test for candidate theories of quantum gravity. New counterarguments to some popular models proposed in relation to the black hole information problem are the following. Black hole complementarity is discussed in Section 3, in particular the fact that an argument by Susskind, aiming to prove that no-cloning is satisfied by the black hole complementarity, does not apply to most black holes (Section 3.1), the fact that its main argument, the "no-omniscience" proposal, does not really hold for black holes in general (Section 3.3), and the fact that black hole complementarity is also at odds with the principle of equivalence (Section 3.2). As for the firewall proposal, in Section 4.1, I explain why the tacit assumption that unitarity should apply only to the exterior of the black hole, and that we should ignore the interior, is not justified, and anyway if taken as true, it imposes boundary conditions to the field, which is why the firewall seems to emerge. Section 5 is dedicated to black hole entropy. In Section 5.1, I present an argument based on time symmetry that the true entropy is not necessarily proportional to the area of the event horizon and at best in the usual cases is bounded. This has negative implications to the various proposals that the event horizon would contain some bits representing the microstates of the black hole, discussed in Section 5.4. This may also explain the so-called "real black hole information paradox," discussed in Section 6. Section 5.2 contains an explanation of the fact that if the laws of black hole mechanics should be connected with those of thermodynamics, this happens already at the classical level, so they are not necessarily indications of quantum gravity or tests of such approaches. Section 5.3 contains arguments that one should not read too much in the so-called no-hair theorems; in particular they do not constrain, contrary to a widespread belief, neither the horizon nor the interior of a black hole. A major motivation invoked for the theoretical research of the black hole information and entropy is that these may provide a benchmark to test approaches to quantum gravity, but in Section 5.5 I argue that these features appear merely by considering quantum fields on spacetime. Consequently, any approach to quantum gravity which includes both quantum field theory and the curved spacetime of general relativity, as a minimal requirement, will also satisfy the consequences derived from them.

To my knowledge, the above-mentioned arguments, presented in more detail in the following, are new, and in the cases when I was aware of other results seeming to point in the same direction, I gave the relevant references. While most part of the article may look like a review of the literature, it is a critical review, aiming to point out some assumptions which, in my opinion, drove us too far from the starting point, which is just the most straightforward and conservative combination of quantum field theory with the curved background of general relativity. The entire structure of arguments converges therefore towards a more conservative picture than that suggested by the more popular proposals. The counterarguments are meant to build up the willingness to consider the less radical proposal that I made, which follows naturally from my work on singularities in standard general relativity ([35] and references therein) and is discussed in Section 7. The background theory is presented in Section 7.1, and a new, enhanced version of the proposal is made in Section 7.2.

2. Black Hole Evaporation

Hawking's derivation of the black hole evaporation [1, 11] has been disputed and checked many times and redone in different settings, and it turned valid, at most allowing some improvements of the unavoidable approximations, as well as mild generalizations. But the result is correct; the radiation is as predicted and thermal in the Kubo-Martin-Schwinger sense [36, 37]. Moreover, it is corroborated via the principle of equivalence with the Unruh radiation, which takes place in the Minkowski spacetime for accelerated observers [38]. Hawking's derivation is obtained in the framework of quantum field theory on curved spacetime, but since the black hole is considered large and the time scale is also large, the spacetime curvature induced by the radiation is ignored.

The derivation, as well as the discussion surrounding black hole information, requires the framework of quantum field theory on curved spacetime [39–41]. Quantum field theory on curved spacetime is a good effective limit of the true but yet unknown theory of quantum gravity. On curved background, there is no Poincaré symmetry to select a preferred vacuum, so there is no canonical Fock space construction of the Hilbert space. The stress-energy expectation value of the quantum fields, $\langle \hat{T}_{ab}(x) \rangle$, is connected with the spacetime geometry via Einstein's equation,

$$R_{ab} - \frac{1}{2} R g_{ab} + \Lambda g_{ab} = \frac{8\pi G}{c^4} \left\langle \hat{T}_{ab}(x) \right\rangle, \tag{1}$$

where R_{ab} is the *Ricci tensor*, R is the *scalar curvature*, g_{ab} is the *metric tensor*, Λ is the *cosmological constant*, G is Newton's gravitational constant, and c is the speed of light constant.

But in the calculations of the Hawking radiation, the gravitational backreaction is ignored, being very small. To have well behaved solutions, the spacetime slicing is such that the intrinsic and extrinsic curvatures of the spacelike slices are considered small compared to the Plank length; the curvature in a neighborhood of the spacelike surface is also taken to be small. The wavelengths of particles are considered large compared to the Plank length. The energy and momentum densities are assumed small compared to the Plank density. The stress-energy tensor satisfies the positive energy conditions. The solution evolves smoothly into future slices that also satisfy these conditions.

The canonical (anti)commutation relations at distinct points of the slice are imposed. A decomposition into positive and negative frequency solutions is assumed to which the Fock construction is applied to obtain the Hilbert space. The renormalizability of the stress-energy expectation value $\langle \hat{T}_{ab}(x) \rangle$ and the uniqueness of the n-point function $\langle \hat{\phi}(x_1) \dots \hat{\phi}(x_n) \rangle$ are ensured by imposing the *Hadamard condition* to the quantum states [41]. This condition is needed because when two of the n-points coincide, there is no invariant way to define the n-point function on curved spacetime. The Hadamard condition is imposed on the *Wightman functionG*$(x, y) = \langle \hat{\phi}(x) \hat{\phi}(y) \rangle$, and it is preserved under time evolution. This condition is naturally satisfied in the usual quantum field theory in Minkowski spacetime. It ensures the possibility to renormalize the stress-energy tensor and to prevent it from diverging.

The Fock space construction of the Hilbert space can be made in many different ways in curved spacetime, since the decomposition into positive and negative frequency solutions depends on the choice of the slicing of spacetime into spacelike hypersurfaces.

Suppose that a basis of annihilation operators is $(\hat{a}_\nu)$, and they satisfy the canonical commutation relations if they are bosons and the canonical anticommutation relations if they are fermions. Another observer has a different basis of annihilation operators $(\hat{b}_\omega)$, assuming that the spacetime is curved or that one observer accelerates with respect to the other. The two bases are related by the *Bogoliubov transformations*,

$$\hat{b}_\omega = \frac{1}{2\pi} \int_0^\infty \left(\alpha_{\omega\nu} \hat{a}_\nu + \beta_{\omega\nu} \hat{a}_\nu^\dagger \right) d\nu, \tag{2}$$

where $\alpha_{\omega\nu}$ and $\beta_{\omega\nu}$ are the *Bogoliubov coefficients*.

The Bogoliubov transformation preserves the canonical (anti)commutation relations and expresses the change of basis of the Fock space, allowing us to move from one construction to another. The Bogoliubov transformations are linear but not unitary. They are symplectic for bosons and orthogonal for fermions though. The number of particles is not preserved, so there is no invariant notion of particles.

This is in fact the reason for both the Unruh effect near a Rindler horizon and the Hawking evaporation near a black hole event horizon. Because of the nonunitarity of the Bogoliubov transformation relating the Fock space representations of two distinct observers, particles can be produced [38–40], including for black holes [11]. This means that what is a vacuum state for an inertial observer is a state with many particles for an accelerated one. This is true in the Minkowski spacetime, if one observer is accelerated with respect to the other, but also for two inertial observers, if the curvature is relevant, as in the case of infalling and escaping observers near a black hole. Moreover, the many-particle state in which the vacuum of one observer appears to the other is thermal. The particle and the antiparticle created in pair during the evaporation are maximally entangled.

3. Black Hole Complementarity

While Hawking's derivation of the black hole evaporation is rigorous and the result is correct, the implication that the information is definitively lost can be challenged. In fact, most of the literature on this problem is trying to find a workaround to restore the lost information and the unitarity. The most popular proposals like black hole complementarity and firewalls do not actually dispute the calculations, but rather they add the requirement that the Hawking radiation should contain the complete information.

Additional motivation for unitarity comes from the AdS/CFT correspondence [43]. The AdS/CFT is not yet rigorously proven, and it is in fact against the current cosmological observations that the cosmological constant is positive [44, 45], but it is widely considered true or standing for a correct gauge-gravity duality, and it is likely that it convinced Hawking to change his mind about information loss [46].

The favorite scenario among high-energy physicists was, for two decades, the idea of *black hole complementarity* [31–33], which supposedly resolves the conflict between unitarity, essential for quantum theory, and the principle of equivalence from general relativity. Susskind and collaborators framed the black hole information paradox as implying a contradiction between unitarity and the principle of equivalence. They proposed a radical solution of this apparent conflict by admitting two distinct Hilbert space descriptions for the infalling matter and the escaping radiation [31].

Assuming that unitarity is to be restored by evaporation alone, the infalling information should be found in the Hawking radiation or should somehow remain above the black hole event horizon, forming the *stretched horizon* [31], similar to the *membrane paradigm* [47]. But since this information falls in the black hole, it would violate the *no-cloning theorem* of quantum mechanics [48–50]. If the cloning does not happen, either the information is not recovered (and unitarity is violated) or no information can cross the horizon, which would violate the principle of equivalence from general relativity, which implies that nothing dramatic should happen at the event horizon, assuming that the black hole is large enough. The black hole complementarity assumes that both unitarity and the principle of equivalence hold true, by allowing cloning, but the cloning cannot be observed, because each observer sees only one copy. The infalling copy of the information is accessible to an infalling observer only (usually named Alice) and the escaping one to an escaping observer (Bob). Susskind and collaborators conjectured that Alice and Bob can never meet to confirm that the infalling quantum information was cloned and the copy escaped the black hole.

At first sight, it may seem that the black hole complementarity solves the contradiction by allowing it to exist, as long as no experiment is able to prove it. Alice and Bob's lightcones intersect, but none of them is included in the other, and they cannot be made so. This means that whatever slicing of spacetime they choose in their reference frames, the Hilbert space constructions they make will be different. So it would be impossible to compare quantum information from the interior of the black hole with the copy of quantum information escaping it. And it is impossible to conceive an observer able to see both copies of information—this would be the so-called *omniscience condition*, which is rejected by Susskind and collaborator to save both unitarity and the principle of equivalence.

3.1. No-Cloning and Timelike Singularities. An early objection to the proposal that Alice and Bob can never compare the two copies of quantum information was that the escaping observer Bob can collect the escaping copy of the information and jump into the black hole to collect the infalling copy. This objection was rejected because, in order to collect a single bit of infalling information from the Hawking radiation, Bob should wait until the black hole loses half of its initial mass by evaporation—the time needed for this to happen is called the *Page time* [13]. So if Bob decides to jump in the black hole to compare the escaping information with the infalling one, it would be too late, because the infalling information will have just enough time to reach the singularity.

The argument based on the Page time works well, but it applies only to black holes of the Schwarzschild type (more precisely this is an Oppenheimer-Snyder black hole [51]), whose singularity is a spacelike hypersurface. For rotating or electrically charged black holes, the singularity is a timelike curve or cylinder. In this case, Alice can carry the infalling information around the singularity for an indefinitely long time, without reaching the singularity. So Bob will be able to reach Alice and confirm that the quantum information was cloned.

This objection is relevant, because for the black hole to be of Schwarzschild type, two of the three parameters defining the black hole, the angular momentum and the electric charge, have to vanish, which is very unlikely. The things are even more complicated if we take into account the fact that, during evaporation or any additional particle falling in the black hole, the type of the black hole changes. Usually particles have nonvanishing electric charges and spin, and even if an infalling particle is electrically neutral and has the spin equal to 0, most likely it will not collide with the black hole radially. This continuous change of the type of the black hole may result in changes of type of the singularity, rendering the argument based on the Page time invalid.

In Section 3.3, we will see that even if the black hole somehow manages to remain of Schwarzschild type, the cloning can be made manifest to a single observer.

3.2. No-Cloning and the Principle of Equivalence. Because of the principle of equivalence, Susskind's argument should also hold for Rindler horizons in Minkowski spacetime. The equivalence implies that Bob is an accelerated observer, and Alice is an inertial observer, who crosses Bob's Rindler horizon. Because of the Unruh effect, Bob will perceive the vacuum state as thermal radiation, while for Alice it would be just vacuum. Bob can see Alice being burned at the Rindler horizon by the thermal radiation, but Alice will experience nothing of this sort. But since they are now in the Minkowski spacetime, Bob can stop and go back to check the situation with Alice, and he will find that she did not experience the thermal bath he saw her experiencing. While we can just say that the complementarity should be applied only to black holes, to rule it out for the Rindler horizon and still maintain the idea of stretched horizon only for black holes, this would be at odds with the principle of equivalence which black hole complementarity is supposed to rescue.

3.3. The "No-Omniscience" Proposal. The resolution proposed by black hole complementarity appeals to the fact that the Hilbert spaces constructed by Alice and Bob are distinct, which would allow quantum cloning, as long as the two copies belong to distinct Hilbert spaces and there is no observer to see the violation of the no-cloning theorem. This means that the patches of spacetime covered by Alice and Bob are distinct, such that apparently no observer can cover both of them. If there was such an "omniscient" observer, he or she would see the cloning of quantum information and see that the laws of quantum theory are violated.

Yet, there is such an observer, albeit moving backwards in time (see Figure 1). Remember that the whole point of trying to restore the loss information and unitarity is because quantum theory should be unitary. This means not only deterministic, but also that the time evolution laws have to be time symmetric, as quantum theory normally is, so that we can recover the lost information. So everything quantum evolution does forward in time should be accessible by backwards in time evolution. An observer going backwards in time, Charlie, can then in principle be able to perceive both copies of the information carried by Alice and Bob, so he is "omniscient."

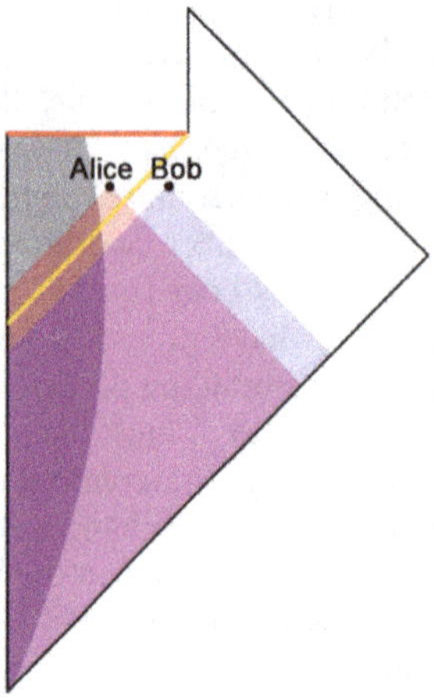

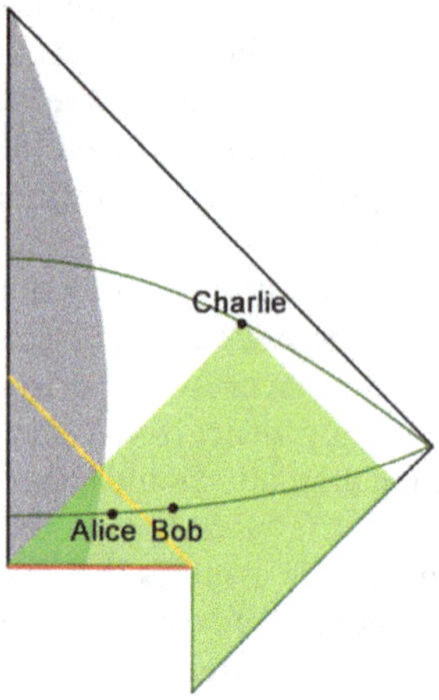

FIGURE 1: (a) The Penrose diagram of black hole evaporation, depicting Alice and Bob and their past lightcones. (b) The Penrose diagram of a backwards in time observer Charlie, depicting how he observes Alice and Bob, and the quantum information each of them caries, even if this information is cloned, therefore disclosing a violation of quantum theory.

One can try to rule Charlie out, on the grounds that he violates causality or more precisely the second law of thermodynamics [52]. But from the point of view of quantum theory, the von Neumann entropy is preserved by unitary evolution, and the quantum evolution is reversible anyway, so it is irrelevant that if in our real universe there is a thermodynamic arrow of time, this does not invalidate a principial thought experiment like this one.

4. The Firewall Paradox

After two decades since the proposal of black hole complementarity, this solution was disputed by the *firewall paradox* [34], which suggested that the equivalence principle should be violated at the event horizon, where a highly energetic curtain or a singularity should form to prevent the information falling inside the black hole.

The firewall argument takes place in the same settings as the black hole complementarity proposal, but this time it involves the *monogamy of entanglement*. More precisely, it is shown that the late radiation has to be maximally entangled with both the early radiation and the infalling counterpart of the late radiation. Since the monogamy of entanglement forbids this, it is proposed that one of the assumptions has to go, most likely the principle of equivalence. The immediate reaction varied from quick acceptance to arguments that the paradox is solved too by the black hole complementarity [53, 54]. After all, we can think of the late radiation as being entangled with the early one in Bob's Hilbert space and with the infalling radiation in Alice's Hilbert space. But it turned out that, unlike the case of the violation of the no-cloning theorem, the violation of monogamy cannot be resolved by Alice and Bob having different Hilbert spaces [55].

One can argue that if the firewall experiment is performed, it creates the firewall, and if it is not performed, Alice sees no firewall, so black hole complementarity is not completely lost. Susskind and Maldacena proposed the *ER=EPR* solution, which states that if entangled particles are thrown in different black holes, then they become connected by a wormhole [56]; also see [57]. The firewall idea also stimulated various discussions about the relevance of complexity of quantum computation and error correction codes in the black hole evaporation and decoding the information from the Hawking radiation using unitary operations (see [54, 58, 59] and references therein).

Various proposals to rescue both the principle of equivalence and unitarity were made, for example, based on the entropy of entanglement across the event horizon in [60, 61]. Hawking proposed that the black hole horizons are only apparent horizons and never actual event horizons [62]. Later, Hawking proposed that supertranslations allow the preservation of information and further expanded the idea with Perry and Strominger [63–65].

Having to give up the principle of equivalence or unitarity is a serious dilemma, so it is worth revisiting the arguments to find a way to save both.

4.1. The Meaning of "Unitarity". In the literature about black hole complementarity and firewalls, by the assumption or requirement of "unitarity," we should understand "unitarity of the Hawking radiation" or, more precisely, "unitarity of the quantum state exterior to the black hole." Let us call this *exterior unitarity* to emphasize that it ignores the interior of the black hole. It is essential to clarify this, because when we feel that we are forced to choose between unitarity and the principle of equivalence, we are in fact forced to choose between exterior unitarity and the principle of equivalence. This assumption is also at the origin of the firewall proposal. So no choice between unitarity and the principle of equivalence is enforced to us, unless by "unitarity" we understand "exterior unitarity."

The idea that unitarity should be restored from the Hawking radiation alone, ignoring the interior of the black hole, was reinforced by the holographic principle and the idea of stretched horizon [31, 32, 66], a place just above the event horizon which presumably stores the infalling information until it is restored through evaporation, and it was later reinforced even more by the AdS/CFT conjecture [43]. But it is not excluded to solve the problem by taking into consideration both the exterior and interior of the black hole and the corresponding quantum states. A proposal accounting for the interior in the AdS/CFT correspondence, based on the impossibility to localize the quantum operators in quantum gravity in a background-independent manner, was made in [67]. A variation of the AdS/CFT leading to a regularization was made in [68].

In fact, considering both the exterior and the interior of the black hole is behind proposals like remnants and baby universes. But we will see later that there is a less radical option.

Exterior unitarity, or the proposal that the full information and purity are restored from Hawking radiation alone, simply removes the interior of the black hole from the reference frame of an escaping observer, consequently from his Hilbert space. This type of unitarity imposes a boundary condition to the quantum fields, which is simply the fact that there is no relevant information inside the black hole. So it is natural that, at the boundary of the support of the quantum fields, which is the black hole event horizon, quantum fields behave as if there is a firewall. This is what the various estimates revealing the existence of a highly energetic firewall or horizon singularity confirm. Note that since the boundary condition which aims to rescue the purity of the Hawking radiation is a condition about the final state, sometimes its consequences give the impression of a conspiracy, as sometimes Bousso and Hayden put it [69].

While I have no reason to doubt the validity of the firewall argument [34], I have reservations about assuming unitarity as referring only to quantum fields living only to the exterior of the black hole, while ignoring those from its interior.

4.2. Firewalls versus Complementarity. The initial Hilbert spaces of Alice and Bob are not necessarily distinct. Even if they and their Fock constructions are distinct, each state from one of the spaces may correspond to a state from the other. The reason is that a basis of annihilation operators in Alice's frame, say $(\widehat{a}_\gamma)$, is related to a basis of annihilation operators in Bob's frame, $(\widehat{b}_\omega)$, by a Bogoliubov transformation (2). The Bogoliubov transformation is linear, although not unitary.

Thus, one may hope that the Hilbert spaces of Alice and Bob may be identified, even though through a very scrambled vector space isomorphism, so that black hole complementarity saves the day. However, exterior unitarity imposes that the evolved quantum fields from the Hilbert spaces have different supporting regions in spacetime. While before the creation of the black hole they may have the same support in the spacelike slice, they evolve differently because of the exterior unitarity condition. Bob's system evolves so that his quantum fields are constrained to the exterior of the black hole, while Alice's quantum fields include the interior too. Bob's Hilbert space is different, because when the condition of exterior unitarity was imposed, it excluded the interior of the black hole. So even if the initial underlying vector space is the same for both the Hilbert space constructed by Alice and that constructed by Bob, their coordinate systems diverged in time, so the way they slice spacetime became different. While normally Alice's vacuum is perceived by Bob as loaded with particles in a thermal state, this time in Bob's frame Alice's vacuum energy becomes singular at the horizon. This makes the firewall paradox a problem for black hole complementarity. A cleaner argument based on purity rather than monogamy is made by Bousso [70].

An interesting issue is that Bob can infer that if the modes he detects passed very close to the event horizon, they were redshifted. So, evolving the modes backwards in time, it must be that the particle passes close to the horizon at a very high frequency, maybe even higher than the Plank frequency. Does this mean that Alice should feel dramatically this radiation? There is the possibility that, for Alice, Bob's high frequency modes are hidden in her vacuum state. This is also confirmed by acoustic black holes [71]. Only if these modes are somehow disclosed, for example, if Bob, being accelerated, performs some temperature detection nearby Alice, these modes may become manifest due to the projection postulate; otherwise, they remain implicit in Alice's vacuum.

It seems that the strength of the firewall proposal comes from rendering black hole complementarity unable to solve the firewall paradox. They are two competing proposals, both aiming to solve the same problem. While one can logically think that proposals that take into account the interior of black holes to restore unitarity are good candidates as well and that they may have the advantage of rescuing the principle of equivalence, sometimes they are dismissed as not addressing the "real" black hole information paradox. I will say more about this in Section 6.

5. Black Hole Entropy

The purposes of this section are to prepare for Section 6 and to discuss the implications of black hole entropy for the black hole information paradox and for quantum gravity.

The entropy bound of a black hole is proportional to the area of the event horizon [12, 72, 73],

$$S_{BH} = \frac{k_B A}{4\ell_P^2}, \tag{3}$$

where k_B is the Boltzmann constant, A is the area of the event horizon, and ℓ_P is the Plank length.

The black hole entropy bound (3) was suggested by Hawking's result that the black hole horizon area never decreases [74], as well as the development of this result into the four laws of black hole mechanics [72].

5.1. The Area of the Event Horizon and the Entropy. It is tempting to think that the true entropy of quantum fields in spacetime should also include the areas of the event horizons.

In fact, there are computational indications that the black hole evaporation leaks the right entropy to compensate the decrease of the area of the black hole event horizon.

But there is a big difference between the entropy of quantum fields and the areas of horizons. First, entropy is associated with the state of the matter (including radiation, of course). If we look at the phase space, we see that the entropy is a property of the state alone, so it is irrelevant if the system evolves in one direction of time or the opposite; the entropy corresponding to the state at a time t is the same. The same is true for quantum entropy, associated with the quantum states, which in fact is preserved by unitary evolution and is the same in either time direction.

On the other hand, the very notion of event horizon in general relativity depends on the direction of time. By looking again at Figure 1(b), this time without being interested in black hole complementarity, we can see that for Charlie there is no event horizon. But the entropy corresponding to matter is the same independently of his time direction. So even if we are able to put the area on the event horizon in the same formula with the entropy of the fields and still have the second law of thermodynamics, the two terms behave completely differently. So if the area of the event horizon is required to compensate for the disappearance of entropy beyond the horizon and for its reemergence as Hawking radiation, for Charlie the things are quite different, because he has full clearance to the interior of the black hole, which for him is white. In other words, he is so omniscient that he knows the true entropy of the matter inside the black hole and not a mere bound given by the event horizon.

This is consistent with the usual understanding of entropy as hidden information; indeed, the true information about the microstates is not accessible (only the macrostate), and this is what entropy stands for. But it is striking, nevertheless, to see that black holes do the same, yet in a completely time-asymmetric manner. This is because the horizon entropy is just a bound for the entropy beyond the horizon; the true entropy is a property of the state.

5.2. Black Hole Mechanics and Thermodynamics: Matter or Geometry? The four laws of black hole mechanics are the following [72, 75]:

(i) **0th law:** the surface gravity κ is constant over the event horizon

(ii) **1st law:** for nearby solutions, the differences in mass are equal to differences in area times the surface gravity, plus some additional terms similar to work

(iii) **2st law:** in any physical process, the area of the event horizon never decreases (assuming positive energy of matter and regularity of spacetime)

(iv) **3rd law:** there is no procedure, consisting of a finite number of steps, to reduce the surface gravity to zero.

The analogy between the laws of black hole mechanics and thermodynamics is quite impressive [75]. In particular, enthalpy, temperature, entropy, and pressure correspond, respectively, to the mass of the black hole, its surface gravity, its horizon area, and the cosmological constant.

These laws of black hole mechanics are obtained in purely classical general relativity but were interpreted as laws of black hole thermodynamics [11, 76, 77]. Their thermodynamical interpretation occurs when considering quantum field theory on curved spacetime, and it is expected to follow more precisely from the yet to be found quantum gravity.

Interestingly, despite their analogy with the laws of thermodynamics, the laws of black hole mechanics hold in purely classical general relativity. While we expect general relativity to be at least a limit theory of a more complete, quantized one, it is a standalone and perfectly selfconsistent theory. This suggests that it is possible that the laws of black hole mechanics already have thermodynamic interpretation in the geometry of spacetime. And this turns out to be true, since black hole entropy can be shown to be the Noether charge of the diffeomorphism symmetry [78]. This works exactly for general relativity, and it is different for gravity modified so that the action is of higher order in terms of curvature. In addition, we already know that Einstein's equation can be understood from an entropic perspective, which has a geometric interpretation [79, 80].

This is not to say that the interpretations of the laws of black hole mechanics in terms of thermodynamics of quantum fields do not hold, because there are strong indications that they do. My point is rather that there are thermodynamics of the spacetime geometry, which are tied somehow with the thermodynamics of quantum matter and radiation. This connection is probably made via Einstein's equation or whatever equation whose classical limit is Einstein's equation.

5.3. Do Black Holes Have No Hair? Classically, black holes are considered to be completely described by their mass, angular momentum, and electric charge. This idea is based on the *no-hair theorems*. These results were obtained for the Einstein-Maxwell equations, assuming that the solutions are asymptotically flat and stationary. While it is often believed that these results hold universally, they are in fact similar to Birkhoff's theorem [81], which states that any spherically symmetric solution of the vacuum field equations must be static and asymptotically flat; hence the exterior solution must be given by the Schwarzschild metric. Werner Israel establishes that the Schwarzschild solution is the unique asymptotically flat static nonrotating solution of Einstein's equation in vacuum, under certain conditions [2]. This was generalized to the Einstein-Maxwell equations (electrovac) [3–5], the result being the characterization of static asymptotically flat solutions only by mass, electric charge, and angular momentum. It is conjectured that this result is general, but counterexamples are known [82, 83].

In classical general relativity, the black holes radiate gravitational waves and are expected to converge to a no-hair solution very fast. If this is true, it happens asymptotically, and the gravitational waves carry the missing information about the initial shape of the black hole horizon, because classical general relativity is deterministic on regular globally hyperbolic regions of spacetime.

Moreover, it is not known what happens when quantum theory is applied. If the gravitational waves are quantized (resulting in gravitons), it is plausible to consider the possibility that quantum effects prevent such a radiation, like in the case of the electron in the atom. Therefore, it is not clear that the information about the infalling matter is completely lost in the black hole, even in the absence of Hawking evaporation. So we should expect at most that black holes converge asymptotically to the simple static solutions, but if they would reach them in finite time, there would be no time reversibility in GR.

Nevertheless, this alone is unable to provide a solution to the information loss paradox, especially since spacetime curvature does not contain the complete information about matter fields. But we see that we have to be careful when we use the no-hair conjecture as an assumption in other proofs.

5.4. Counting Bits. While black hole mechanics suggest that the entropy of a black hole is limited by the Bekenstein bound (3), it is known that the usual classical entropy of a system can be expressed in terms of its microstates:

$$S_Q = -k_B \sum_i p_i \ln p_i, \tag{4}$$

where p_i denotes the number of microstates which cannot be distinguished because of the coarse graining, macroscopically appearing as the i-th macrostate. A similar formula gives the quantum von Neumann entropy, in terms of the density matrix ρ:

$$S = -k_B \mathrm{tr}\left(\rho \ln \rho\right). \tag{5}$$

Because of the *no-hair theorem* (see Section 5.3), it is considered that classical black holes can be completely characterized by the mass, angular momentum, and electric charge, at least from the outside. This is usually understood as suggesting that quantum black holes have to contain somewhere, most likely on their horizons, some additional degrees of freedom corresponding to their microstates, so that (3) can be interpreted in terms of (4).

It is often suggested that there are some horizon microstates, either floating above the horizon but not falling because of a *brick wall* [84–86] or being horizon gravitational states [87].

Other counting proposals are based on counting string excited microstates [88–90]. There are also proposals of counting microstates in LQG, for example, by using a Chern-Simons field theory on the horizon, as well as choosing a particular Immirzi parameter [91].

Another interesting possible origin of entropy comes from *entropy of entanglement* resulting by the reduced density matrix of an external observer [92, 93]. This is proportional but for short distances requires renormalization.

But, following the arguments in Section 5.1, I think that the most natural explanation of black hole entropy seems to be to consider the internal states of matter and gravity [94]. A model of the internal state of the black hole similar to the atomic model was proposed in [95–97]. Models based on Bose-Einstein condensates can be found in [98–100] and references therein.

Since in Section 5.1 it was explained that the horizons just hide matter, and hence entropy, and are not in fact the carriers of the entropy, it seems more plausible to me that the structure of the matter inside the black hole is just bounded by the Bekenstein bound, and does not point to an unknown microstructure.

5.5. A Benchmark to Test Quantum Gravity Proposals? The interest in the black hole information paradox and black hole entropy is not only due to the necessity of restoring unitarity. This research is also motivated by testing various competing candidate theories of quantum gravity. Quantum gravity seems to be far from our experimental possibilities, because it is believed to become relevant at very small scales. On the other hand, black hole information loss and black hole entropy pose interesting problems, and the competing proposals of quantum gravity are racing to solve them. The motivation is that it is considered that black hole entropy and information loss can be explained by one of these quantum gravity approaches.

On the other hand, it is essential to remember how black hole evaporation and black hole entropy were derived. The mathematical proofs are done within the framework of quantum field theory on curved spacetime, which is considered a good effective limit of the true but yet to be discovered theory of quantum gravity. The calculations are made near the horizon; they do not involve extreme conditions like singularities or planckian scales, where quantum gravity is expected to take the lead. The main assumptions are

(1) quantum field theory on curved spacetime

(2) the Einstein equation, with the stress-energy tensor replaced by the stress-energy expectation value $\langle \widehat{T}_{ab}(x)\rangle$ (see (1))

For example, when we calculate the Bekenstein entropy bound, we do this by throwing matter in a black hole and see how much the event horizon area increases.

These conditions are expected to hold in the effective limit of any theory of quantum gravity.

But since both the black hole entropy and the Hawking evaporation are obtained from the two conditions mentioned above, this means that any theory in which these conditions are true, at least in the low energy limit, is also able to imply both the black hole entropy and the Hawking evaporation. In other words, if a theory of quantum gravity becomes in some limit the familiar quantum field theory and also describes Einstein's gravity, it should also reproduce the black hole entropy and the Hawking evaporation.

Nevertheless, some candidate theories to quantum gravity do not actually work in a dynamically curved spacetime, being, for example, defined on flat or AdS spacetime, yet they still are able to reproduce a microstructure of black hole entropy. This should not be very surprising, given that, even in nonrelativistic quantum mechanics, quantum systems bounded in a compact region of space have discrete spectrum. So it may be very well possible that these results are due to the fact that even in nonrelativistic quantum mechanics entropy bounds hold [101]. In flat spacetime, we can think that the number of states in the spectrum is proportional with the volume. However, when we plug in the masses of the particles in the formula for the Schwarzschild radius (which incidentally is the same as Michell's formula in Newtonian gravity [102]), we should obtain a relation similar to (3).

The entropy bound (3) connects the fundamental constants usually considered to be characteristic for general relativity, quantum theory, and thermodynamics. This does not necessarily mean that the entropy of the black hole witnesses about quantum gravity. This should be clear already from the fact that the black hole entropy bound was not derived by assuming quantum gravity but simply from the assumptions mentioned above. It is natural that if we plug the information and the masses of the particles in the formula for the Schwarzschild radius, we obtain a relation between the constants involved in general relativity, quantum theory, and thermodynamics. It is simply a property of the system itself, not a witness of a deeper theory. But, of course, if a candidate theory of quantum gravity fails to pass even this test, this may be a bad sign for it.

6. The Real Black Hole Information Paradox

Sometimes it is said that the true black hole information paradox is the one following from Don Page's article [13]. For example, Marolf considers that here lies the true paradoxical nature of the black hole information, while he calls the mere information loss and loss of purity, "the straw man information problem" [14]. Apparently, the black hole von Neumann entropy should increase with one bit for each emitted photon. At the same time, its area decreases by losing energy, so the black hole entropy should also decrease by the usual Bekenstein-Hawking kind of calculation. So what happens with the entropy of the black hole? Does it increase or decrease? This problem occurs much earlier in the evolution of the black hole, when the black hole area is reduced to half of its initial value (the *Page time*), so we do not have to wait for the complete evaporation to notice this problem. Marolf put it as follows[14]:

> This is now a real problem. Evaporation causes the black hole to shrink and thus to reduce its surface area. So S_{BH} decreases at a steady rate. On the other hand, the actual von Neumann entropy of the black hole must increase at a steady rate. But the first must be larger than the second. So some contradiction is reached at a finite time.

I think there are some assumptions hidden in this argument. We compare the von Neumann entropy of the black hole calculated during evaporation with the black hole entropy calculated by Bekenstein and Hawking by throwing particles in the black hole. While the proportionality of the black hole entropy with the area of the event horizon has been confirmed by various calculations for numerous cases, the two types of processes are different, so it is natural that they lead to different states of the black hole and hence to different values for the entropy. This is not a paradox; it is just an evidence that the entropy contained in the black hole depends on the way it is created, despite the bound given by the horizon. So it seems more natural not to consider that the entropy of the matter inside the black hole reached the maximum bound at the beginning but rather that it reaches its maximum at the Page time, due to the entanglement entropy with the Hawking radiation. Alternatively, we may still want to consider the possibility of having more entropy in the black hole than the Bekenstein bound allows. In fact, Rovelli made another argument pointing in the same direction that the Bekenstein-Bound is violated, by counting the number of states that can be distinguished by local observers (as opposed to external observers) using local algebras of observables [103]. This argument provided grounds for a proposal of a white hole remnant scenario discussed in [104].

7. A More Conservative Solution

We have seen in the previous sections that some important approaches to the black hole information paradox and the related topics assume that the interior of the black hole is irrelevant or does not exist, and the event horizon plays the important role. I also presented arguments that if it is to recover unitarity without losing the principle of equivalence, then the interior of the black hole should be considered as well, and the event horizon should not be endowed with special properties. More precisely, given that the original culprit of the information loss is its supposed disappearance at singularities, then singularities should be closely investigated. The least radical approach is usually considered the avoidance of singularity, by modifying gravity (*i.e.*, the relation between the stress-energy tensor and the spacetime curvature as expressed by the Einstein equation), so that one or more of the three assumptions of the singularity theorems [6–8] no longer hold. In particular, it is hoped that this may be achieved by the quantum effects in a theory of quantum gravity. However, it would be even less radical if the problem could be solved without modifying general relativity, and such an approach is the subject of this section.

But singularities are accompanied by divergences in the very quantities involved in the Einstein equation, in particular the curvature and the stress-energy tensor. So even if it is possible to reformulate the Einstein equation in terms of variables that do not diverge, remaining instead finite at the singularity, the question remains whether the physical fields diverge or break down. In other words, what are in fact the true, fundamental physical fields, the diverging variables, or those that remain finite? This question will be addressed soon.

An earlier mention of the possibility of changing the variables in the Einstein equation was made by Ashtekar, for example, in [105] and references therein, where it is also proposed that the new variables could remain finite at singularities even in the classical theory. However, it turned out that one of his two new variables diverges at singularities (see, *e.g.*, [106]). Eventually this formulation led to loop quantum gravity, where the avoidance is instead achieved on some toy bounce models (see *e.g.*, [28, 29]). But the problem whether standard general relativity can admit a formulation free of infinities at singularities remained open for a while.

7.1. Singular General Relativity. In [107, 108], the author introduced a mathematical formulation of semi-Riemannian geometry which allows a description of a class of singularities free of infinities. The fields that allowed this are invariant, and in the regions without singularities they are equivalent to the standard formulation. To understand what the problem is and how it is solved, recall that in geometry the metric tensor is assumed to be smooth and regular, that is, without infinite components and nondegenerate, which means that its determinant is nonvanishing. If the metric tensor has infinite components or if it is degenerate, the metric is called singular. If the determinant is vanishing, one cannot define the Levi-Civita connection, because the definition relies on the Christoffel symbols of the second kind,

$$\Gamma^i_{jk} := \frac{1}{2} g^{is} \left(g_{sj,k} + g_{sk,j} - g_{jk,s} \right), \tag{6}$$

which involve the contraction with g^{is}, which is the inverse of the metric tensor g_{ij}; hence it assumes it to be nondegenerate. This makes it impossible to define the covariant derivative and the Riemann curvature (hence the Ricci and scalar curvatures as well) at the points where the metric is degenerate. These quantities blow up while approaching the singularities. Therefore, Einstein's equation as well breaks down at singularities.

However, it turns out that, on the space obtained by factoring out the subspace of isotropic vectors, an inverse can be defined in a canonical and invariant way and that there is a simple condition that leads to a finite Riemann tensor, which is defined smoothly over the entire space, including at singularities. This allows the contraction of a certain class of tensors and the definition of all quantities of interest to describe the singularities without running into infinities and is equivalent to the usual, nondegenerate semi-Riemannian geometry outside the singularities [107]. Moreover, it works well for warped products [108], allowing the application for big bang models [109, 110]. This approach also works for black hole singularities [42, 111, 112], allowing the spacetime to be globally hyperbolic even in the presence of singularities [113]. More details can be found in [35, 114] and the references therein. Here I will first describe some of the already published results and continue with new and more general arguments.

An essential difficulty related to singularities is given by the fact that, despite the Riemann tensor being smooth and finite at such singularities, the Ricci tensor $R_{ij} := R^s_{isj}$ usually continues to blow up. The Ricci tensor and its trace, the scalar curvature $R = R^s_s$, are necessary to define the Einstein tensor, $G_{ij} = R_{ij} - (1/2)Rg_{ij}$. Now here is the part where the physical interpretation becomes essential. In the Einstein equation, the Einstein tensor is equated to the stres-energy tensor. So they both seem to blow up, and indeed they do. Physically, the stress-energy tensor represents the density of energy and momentum at a point. However, what is physically measurable is never such a density at a point but its integral over a volume. The energy or momentum in a finite measure volume is obtained by integrating with respect to the volume element. And the quantity to be integrated, for example, the energy density $T_{00}\mathrm{d}_{vol}$, where $T_{00} = T(u,u)$ for a timelike vector u and $\mathrm{d}_{vol} := \sqrt{-\det g}\,\mathrm{d}x^0 \wedge \mathrm{d}x^1 \wedge \mathrm{d}x^2 \wedge \mathrm{d}x^3$, is finite, even if $T_{00} \longrightarrow \infty$, since $\mathrm{d}_{vol} \longrightarrow 0$ in the proper way. The mathematical theory of integration on manifolds makes it clear that what we integrate are differential forms, like $T_{00}\mathrm{d}_{vol}$, and not scalar functions like T_{00}. So I suggest that we should do in physics the same as in geometry, because it makes more sense to consider the physical quantities to be the differential forms rather than the scalar components of the fields [109]. This is also endorsed by two other mathematical reasons. On one hand, when we define the stress-energy T_{ij}, we do it by functional derivative of the Lagrangian with respect to the metric tensor, and the result contains the volume element, which we then divide out to get T_{ij}. Should we keep it, we would get instead $T_{ij}\mathrm{d}_{vol}$. Also, when we derive the Einstein equation from the Lagrangian density R, we in fact vary the integral of the differential form $R\mathrm{d}_{vol}$ and not of the scalar R. And the resulting Einstein equation has again a factor d_{vol}, which we leave out of the equation on the grounds that it is never vanishing. Well, at singularities it vanishes, so we should keep it, because otherwise we divide by 0 and we get infinities. The resulting densitized form of the Einstein equation,

$$G_{ij}\mathrm{d}_{vol} + \Lambda g_{ij}\mathrm{d}_{vol} = \frac{8\pi G}{c^4} T_{ij}\mathrm{d}_{vol}, \tag{7}$$

is equivalent to Einstein's outside singularities, but, as already explained, I submit that it better represents the physical quantities and not only because these quantities remain finite at singularities. I call this *densitized Einstein equation*, but they are in fact tensorial as well; the fields involved are tensors, being the tensor products between other tensors and the volume form, which itself is a completely antisymmetric tensor. Note that Ashtekar's variables are also densities, and they are more different from the usual tensor fields involved in the semi-Riemannian geometry and Einstein's equation, yet they were proposed to be the real variables both for quantization and for eliminating the infinities in the singularities [105]. But the formulation I proposed remains finite even at singularities, and it is closer as interpretation to the original fields.

Another difficulty this approach had to solve was that it applies to a class of degenerate metrics, but the black holes are nastier, since the metric has components that blow up at the singularities. For example, the metric tensor of the Schwarzschild black hole solution, expressed in the Schwarzschild coordinates, is

$$\mathrm{d}s^2 = -\left(1 - \frac{2m}{r}\right)\mathrm{d}t^2 + \left(1 - \frac{2m}{r}\right)^{-1}\mathrm{d}r^2 + r^2\mathrm{d}\sigma^2, \tag{8}$$

where m is the mass of the body, the units were chosen so that $c = 1$ and $G = 1$, and

$$\mathrm{d}\sigma^2 = \mathrm{d}\theta^2 + \sin^2\theta\mathrm{d}\phi^2 \tag{9}$$

is the metric of the unit sphere S^2.

For the horizon $r = 2m$, the singularity of the metric can be removed by a singular coordinate transformation; see, for example, [115, 116]. Nothing of this sort could be done for the $r = 0$ singularity, since no coordinate transformation can make the Kretschmann scalar $R^{ijkl}R_{ijkl}$ finite. However, it turns out that it is possible to make the metric at the singularity $r = 0$ into a degenerate and analytic metric by coordinate transformations. In [111], it was shown that this is possible, and an infinite number of solutions were found, which lead to an analytic metric degenerate at $r = 0$. Among these solutions, there is a unique one that satisfies the condition of semiregularity from [107], which ensures the smoothness and analyticity of the solution for the interior of the black hole. This transformation is

$$r = \tau^2$$
$$t = \xi\tau^4 \quad (10)$$

and the resulting metric describing the interior of the Schwarzschild black hole is

$$ds^2 = -\frac{4\tau^4}{2m - \tau^2}d\tau^2 + \left(2m - \tau^2\right)\tau^4 (4\xi d\tau + \tau d\xi)^2 + \tau^4 d\sigma^2. \quad (11)$$

This is not to say that physics depend on the coordinates. It is similar to the case of switching from polar to Cartesian coordinates in plane or like the Eddington-Finkelstein coordinates. In all these cases, the transformation is singular at the singularity, so it is not a diffeomorphism. The atlas, the differential structure, is changed, and in the new atlas, with its new differential structure, the diffeomorphisms preserve, of course, the semiregularity of the metric. And just like in the case of the polar or spherical coordinates and the Eddington-Finkelstein coordinates, it is assumed that the atlas in which the singularity is regularized is the real one, and the problems were an artifact of the Schwarzschild coordinates, which themselves were in fact singular.

Similar transformations were found for the other types of black holes (Reissner-Nordström, Kerr, and Kerr-Newman) and for the electrically charged ones the electromagnetic field also no longer blows up [42, 112].

7.2. Beyond the Singularity. Returning to the Schwarzschild black hole in the new coordinates (11), the solution extends analytically through the singularity. If we plug this solution in the Oppenheimer-Snyder black hole solution, we get an analytic extension depicting a black hole which forms and then evaporates, whose Penrose-Carter diagram is represented in Figure 2.

The resulting spacetime does not have Cauchy horizons, being hyperbolic, which allows the partial differential equations describing the fields on spacetime to be well posed and continued through the singularity. Of course, there is still the problem that the differential operators in the field equations of the matter and gauge fields going through the singularity should be replaced with the new ones. Such formulations are introduced in [117], and sufficient conditions are to be satisfied by the fields at the singularities so that their evolution equations work was given, in the case of Maxwell and Yang-Mills equations.

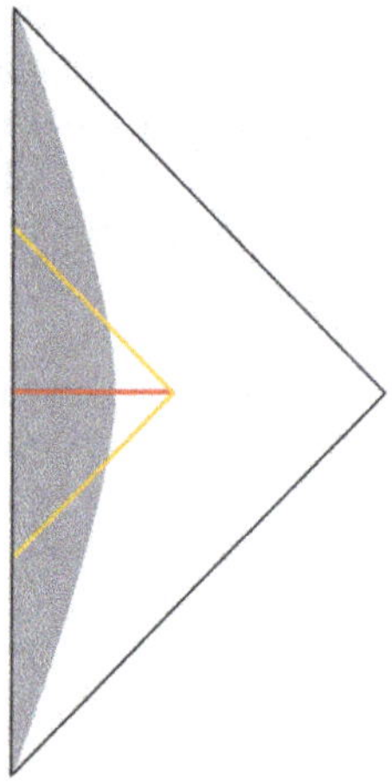

Figure 2: An analytic extension of the black hole solution beyond the singularity.

It is an open problem whether the backreaction will make the spacetime to curve automatically so that these conditions are satisfied for all possible initial conditions of the field. This should be researched in the future, including for quantum fields. It is to be expected that the problem is difficult, and what is given here is not the general solution but rather a toy model. Anyway, no one should expect very soon an exact treatment of real case situations, so the whole discussion here is in principle to establish whether this conservative approach is plausible enough.

However, I would like to propose here a different, more general argument, which avoids the difficulties given by the necessity that the field equations should satisfy at the singularities special conditions like the sufficient conditions found in [117] and also the open problem of which are the conditions to be satisfied by the fermionic fields at singularities.

First consider Fermat's principle in optics. A ray of light in geometric optics is straight, but if it passes from one medium to another having a different refraction index, the ray changes its direction and appears to be broken. It is still continuous, but the velocity vector is discontinuous, and it appears that the acceleration blows up at the surface separating the two media. But Fermat's principle still allows us to know exactly what happens with the light ray in geometric optics.

On a similar vein, I think that, in the absence of a proof that the fields satisfy the exact conditions [117] when crossing a singularity, we can argue that the singularities are not a threat to the information contained in the field by using the least action principle instead.

The least action principle involves the integration of the Lagrangian densities of the fields. While the conditions the fields have to satisfy at the singularity in order to behave well are quite restrictive, the Lagrangian formulation is much more general. The reason is that integration can be done over fields with singularities, also on distributions, and the result can still be finite.

Consider first classical, point-like particles falling in the black hole, crossing the singularity, and exiting through the

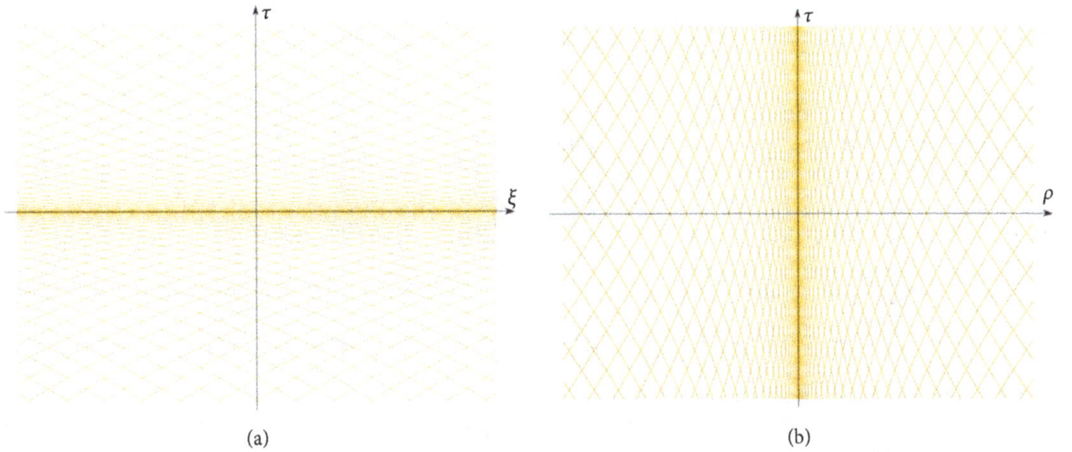

FIGURE 3: (a) The causal structure of the Schwarzschild black hole in coordinates (τ, ξ) from (10). (b) The causal structure of the Reissner-Nordström black hole, in coordinates (τ, ρ) playing a similar role (see [42]).

white hole which appears after the singularity disappears. The history of such a test particle is a geodesic, and to understand the behavior of geodesics, we need to understand first the causal structure. In Figure 3, the causal structures of (a) a Schwarzschild black hole and (b) a Reissner-Nordström black hole are represented in the coordinates which smoothen the singularity (see [118]).

If the test particle is massless, its path is a null geodesic. In [118], I showed that, for the standard black holes, the causal structure at singularities is not destroyed. The lightcones will be squashed, but they will remain lightcones. Therefore, the history of a massless particle like a photon is, if we apply the least action principle, just a null geodesic crossing the singularity and getting out.

If the test particle is massive, its history is a timelike geodesic. In this case, a difficulty arises, because in the new coordinates the lightcones are squashed. This allows for distinct geodesics to intersect the singularity at the same point and to have the same spacetime tangent direction. In the Schwarzschild case, this does not happen for timelike geodesics, but in the Reissner-Nordström case [42] all of the timelike geodesics crossing the singularity at the same point become tangent. Apparently, this seems to imply that a geodesic crossing a timelike singularity can get out of it in any possible direction in a completely undetermined way. To fix this, one may want to also consider the second derivative or to use the local cylindrical symmetry around the timelike singularity.

But the least action principle allows this to be solved regardless of the specific local solution of the problem at the singularity. The timelike geodesics are tangent only at the singularity, which is a zero-measure subset of spacetime. So we can apply the least action principle to obtain the history of a massive particle and obtain a unique solution. The least action principle can be applied for classical test particles because a particle falling in the black hole reaches the singularity in finite proper time, and similarly a finite proper time is needed for it to get out. Moreover, the path integral quantization will consider anyway all possible paths, so even if there would be an indeterminacy at the classical level, it will be removed by integrating them all.

For classical fields, the same holds as for point-like classical particles; only the paths are much more difficult to visualize. The least action principle is applied in the configuration space even for point-like particles, and the same holds for fields, the only difference being the dimension of the configuration space and the Lagrangian. The points from the singularity form again a zero-measure subset compared to the full configuration space, so finding the least action path is similar to the case of point-like particles. The Lagrangian density is finite at least at the points of the configuration space outside the singularities, which means almost everywhere. But the volume element vanishes at singularities, which improves the situation. So its integral can very well be finite, even if the Lagrangian density would be divergent at the singularities. It may be the case that the fields have singular Lagrangian density at the singularity and that when we integrate them it is not excluded that even the integral may diverge, but in this case the least action principle will force us anyway to choose the paths that have a finite action density at the singularities, and such paths exist, for example, those satisfying the conditions found in [117].

So far we have seen that the principle of least action allows determining the history of classical, point-like particles or fields, from the initial and final conditions, even if they cross the singularity. This is done so far on fixed background, so no backreaction via Einstein's equation is considered, only particles or fields. But the Lagrangian approach extends easily to include the backreaction; we simply add the Hilbert-Einstein Lagrangian to that of the fields or point-like particles. So now we vary not only the path of point-like particles or fields in the configuration space but also the geometry of spacetime in order to find the least action history. This additional variation gives even more freedom to choose the least action path, so even if on fixed background the initial condition of a particular field will not evolve to become, at the singularity, a field satisfying the conditions from [117], because the spacetime geometry is varied as well to include backreaction; the spacetime adjusts itself to minimize the action, and it is not too wild to conjecture that it adjusts itself to satisfy such conditions.

Now let us consider quantum fields. When moving to quantum fields on curved background, since the proper time of all classical test particles is finite, we can apply the path integral formulation of quantum field theory [119, 120]. Since the proper time is finite along each path φ joining two points, including for the paths crossing a singularity, and since the action $S(\varphi, t)$ is well defined for almost all times t, then $e^{(i/\hbar)S(\varphi,t)}$ is also well defined. So at least on fixed curved background, even with singularities, it seems to exist little difference from special relativistic quantum field theory via path integrals.

Of course the background geometry should also depend on the quantum fields. Can we account for this, in the absence of a theory of quantum gravity? We know that at least the framework of path integrals works on curved classical spacetime (see, *e.g.*, [121]), where the Einstein equation becomes (1). To also include quantized gravity is more difficult because of its nonrenormalizability by perturbative methods. Add to this the fact that at least for the Standard Model we know that in flat background renormalization helps and even on curved background without singularities. But what about singularities? Is not it possible that they make renormalization impossible? In fact, quite the contrary may be true: in [122], it is shown that singularities improve the behavior of the quantum fields, including for gravity, at UV scales. These results are applied to already existing results obtained by various researchers who use various types of dimensional reduction to improve this behavior for quantum fields, including gravity. In fact, some of these approaches improve the renormalizability of quantum fields so well that even the Landau poles disappear even for nonrenoramlizable theories [123, 124]. But the various types of dimensional reduction are, in these approaches, postulated somehow *ad hoc*, for no other reason than to improve perturbative renormalizability. On the contrary, if the perturbative expansion is made in terms of point-like particles, these behave like black holes with singularities, and some of the already postulated types of dimensional reduction emerge automatically, with no additional assumption, from the properties of singularities [122]. Thus, the very properties of the singularities lead automatically to improved behavior at the UV scale, even for theories thought to be perturbatively nonrenormalizable.

The proposal I described in this section is still at the beginning, compared to the difficulty of the remaining open problems to be addressed. First, there is obviously no experimental confirmation, and it is hard to imagine that the close future can provide one. The plausibility rests mainly upon making as few new assumptions as possible, in addition to those coming from general relativity and quantum theory, theories well established and confirmed, but not in the regimes where both become relevant. For some simple examples, there are mathematical results, but a truly general proof, with fully developed mathematical steps and no gaps, does not exist yet. And, considering the difficulty of the problem, it is hard to believe that it is easy to have very soon a completely satisfying proof in this or other approaches. Nevertheless, I think that promising avenues of research are opened by this proposal.

Conflicts of Interest

The author declares that there are no conflicts of interest.

References

[1] S. W. Hawking, "Breakdown of predictability in gravitational collapse," *Physical Review D: Particles, Fields, Gravitation and Cosmology*, vol. 14, no. 10, pp. 2460–2473, 1976.

[2] W. Israel, "Event horizons in static vacuum space-times," *Physical Review A: Atomic, Molecular and Optical Physics*, vol. 164, no. 5, pp. 1776–1779, 1967.

[3] W. Israel, "Event horizons in static electrovac space-times," *Communications in Mathematical Physics*, vol. 8, no. 3, pp. 245–260, 1968.

[4] B. Carter, "Axisymmetric black hole has only two degrees of freedom," *Physical Review Letters*, vol. 26, no. 6, pp. 331–333, 1971.

[5] W. K. Misner, S. Thorne, and J. A . Wheeler, *Gravitation*, W. H. Freeman and Company, 1973.

[6] R. Penrose, "Gravitational collapse and space-time singularities," *Physical Review Letters*, vol. 14, pp. 57–59, 1965.

[7] S. W. Hawking and R. Penrose, "The singularities of gravitational collapse and cosmology," *Proceedings of the Royal Society of London*, vol. 314, no. 1519, pp. 529–548, 1970.

[8] S. W. Hawking and G. F. R. Ellis, *The Large Scale Structure of Space-Time*, Cambridge University Press, 1995.

[9] R. Penrose, "Gravitational Collapse: the Role of General Relativity," *Revista del Nuovo Cimento; Numero speciale 1*, pp. 252–276, 1969.

[10] R. Penrose, "The Question of Cosmic Censorship," in *Black Holes and Relativistic Stars*, R. M. Wald, Ed., pp. 233–248, niversity of Chicago Press, Chicago, IL, USA, 1998.

[11] S. W. Hawking, "Particle creation by black holes," *Communications in Mathematical Physics*, vol. 43, no. 3, pp. 199–220, 1975.

[12] J. D. Bekenstein, "Black holes and entropy," *Physical Review D: Particles, Fields, Gravitation and Cosmology*, vol. 7, pp. 2333–2346, 1973.

[13] D. N. Page, "Average entropy of a subsystem," *Physical Review Letters*, vol. 71, no. 9, pp. 1291–1294, 1993.

[14] D. Marolf, "The black hole information problem: Past, present, and future," *Reports on Progress in Physics*, vol. 80, no. 9, 2017.

[15] W. G. Unruh and R. M. Wald, "Information loss," *Reports on Progress in Physics*, vol. 80, no. 9, p. 092002, 2017.

[16] J. Preskill, "Do black holes destroy information?" in *Black Holes, Membranes, Wormholes and Superstrings*, vol. 1, p. 22, World Scientific, River Edge, NJ, USA, 1993.

[17] S. B. Giddings, "The black hole information paradox," 1995, https://arxiv.org/abs/hep-th/9508151.

[18] S. Hossenfelder and L. Smolin, "Conservative solutions to the black hole information problem," *Physical Review D: Particles, Fields, Gravitation and Cosmology*, vol. 81, no. 6, Article ID 064009, 13 pages, 2010.

[19] S. W. Hawking, "The unpredictability of quantum gravity," *Communications in Mathematical Physics*, vol. 87, no. 3, pp. 395–415, 1982/83.

[20] S. B. Giddings, "Constraints on black hole remnants," *Physical Review D: Particles, Fields, Gravitation and Cosmology*, vol. 49, no. 2, pp. 947–957, 1994.

[21] S. B. Giddings, "Why aren't black holes infinitely produced?" *Physical Review D: Particles, Fields, Gravitation and Cosmology*, vol. 51, no. 12, pp. 6860–6869, 1995.

[22] M. A. Markov, "Problems of a perpetually oscillating universe," *Annals of Physics*, vol. 155, no. 2, pp. 333–357, 1984.

[23] M. K. Parikh and F. Wilczek, "Hawking radiation as tunneling," *Physical Review Letters*, vol. 85, no. 24, pp. 5042–5045, 2000.

[24] S. Lloyd, "Almost certain escape from black holes in final state projection models," *Physical Review Letters*, vol. 96, no. 6, 061302, 4 pages, 2006.

[25] V. P. Frolov, "Information loss problem and a 'black hole' model with a closed apparent horizon," *Journal of High Energy Physics*, vol. 2014, no. 5, 2014.

[26] D. P. Prester, "Curing Black Hole Singularities with Local Scale Invariance," *Advances in Mathematical Physics*, vol. 2016, Article ID 6095236, 9 pages, 2016.

[27] A. Ashtekar, V. Taveras, and M. Varadarajan, "Information is not lost in the evaporation of 2D black holes," *Physical Review Letters*, vol. 100, no. 21, 211302, 4 pages, 2008.

[28] A. Ashtekar, F. Pretorius, and F. M. Ramazanoglu, "Evaporation of two-dimensional black holes," *Physical Review D*, vol. 83, no. 4, Article ID 044040, 2011.

[29] C. Rovelli and F. Vidotto, "Planck stars," *International Journal of Modern Physics D*, vol. 23, no. 12, Article ID 1442026, 2014.

[30] H. M. Haggard and C. Rovelli, "Quantum-gravity effects outside the horizon spark black to white hole tunneling," *Physical Review D: Particles, Fields, Gravitation and Cosmology*, vol. 92, no. 10, 104020, 11 pages, 2015.

[31] L. Susskind, L. Thorlacius, and J. Uglum, "The stretched horizon and black hole complementarity," *Physical Review D: Particles, Fields, Gravitation and Cosmology*, vol. 48, no. 8, pp. 3743–3761, 1993.

[32] C. R. Stephens, G. 't Hooft, and B. F. Whiting, "Black hole evaporation without information loss," *Classical and Quantum Gravity*, vol. 11, no. 3, pp. 621–647, 1994.

[33] S. Leonard and L. James, *The holographic universe - An introduction to black holes, information and the string theory revolution*, World Scientific, 2004.

[34] A. Almheiri, D. Marolf, J. Polchinski, and J. Sully, "Black holes: Complementarity or firewalls?" *Journal of High Energy Physics*, vol. 2013, no. 2, pp. 1–19, 2013.

[35] O. C. Stoica, *Singular General Relativity [Ph.D. Thesis]*, Minkowski Institute Press, 2013.

[36] R. Kubo, "Statistical-mechanical theory of irreversible processes. I. general theory and simple applications to magnetic and conduction problems," *Journal of the Physical Society of Japan*, vol. 12, no. 6, pp. 570–586, 1957.

[37] P. C. Martin and J. Schwinger, "Theory of many-particle systems. I," *Physical Review A: Atomic, Molecular and Optical Physics*, vol. 115, no. 6, pp. 1342–1373, 1959.

[38] W. G. Unruh, "Notes on black-hole evaporation," *Physical Review D: Particles, Fields, Gravitation and Cosmology*, vol. 14, no. 4, pp. 870–892, 1976.

[39] S. A. Fulling, "Nonuniqueness of canonical field quantization in riemannian space-time," *Physical Review D: Particles, Fields, Gravitation and Cosmology*, vol. 7, no. 10, pp. 2850–2862, 1973.

[40] P. C. Davies, "Scalar production in Schwarzschild and Rindler metrics," *Journal of Physics A: Mathematical and General*, vol. 8, no. 4, pp. 609–616, 1975.

[41] R. M. Wald, *Quantum Field Theory in Curved Space-Time and Black Hole Thermodynamics*, University of Chicago Press, 1994.

[42] O. Stoica, "Analytic Reissner-Nordström singularity," *Physica Scripta*, vol. 85, no. 5, p. 055004, 2012.

[43] M. Maldacena, "The large-N limit of superconformal field theories and supergravity," *International Journal of Theoretical Physics*, vol. 38, no. 4, pp. 1113–1133, 1999.

[44] A. G. Riess, A. V. Filippenko, P. Challis et al., "Observational evidence from supernovae for an accelerating universe and a cosmological constant," *The Astronomical Journal*, vol. 116, no. 3, pp. 1009–1038, 1998.

[45] S. Perlmutter, G. Aldering, and G. Goldhaber, "Measurements of Ω and Λ from 42 High-Redshift Supernovae," *The Astrophysical Journal*, vol. 517, no. 2, pp. 565–586, 1999.

[46] S. W. Hawking, "Information loss in black holes," *Physical Review D: Particles, Fields, Gravitation and Cosmology*, vol. 72, Article ID 084013, 2005.

[47] R. H. Price and K. S. Thorne, "Membrane viewpoint on black holes: properties and evolution of the stretched horizon," *Physical Review D: Particles, Fields, Gravitation and Cosmology*, vol. 33, no. 4, pp. 915–941, 1986.

[48] J. L. Park, "The concept of transition in quantum mechanics," *Foundations of Physics*, vol. 1, no. 1, pp. 23–33, 1970.

[49] W. K. Wootters and W. H. Zurek, "A single quantum cannot be cloned," *Nature*, vol. 299, no. 5886, pp. 802-803, 1982.

[50] D. Dieks, "Communication by EPR devices," *Physics Letters A*, vol. 92, no. 6, pp. 271-272, 1982.

[51] J. R. Oppenheimer and H. Snyder, "On continued gravitational contraction," *Physical Review A: Atomic, Molecular and Optical Physics*, vol. 56, no. 5, pp. 455–459, 1939.

[52] L. S. Schulman, *Time's arrows and quantum measurement*, Cambridge University Press, 1997.

[53] R. Bousso, "Observer complementarity upholds the equivalence principle," 2012, https://arxiv.org/abs/1207.5192.

[54] D. Harlow and P. Hayden, "Quantum computation vs. firewalls," *Journal of High Energy Physics*, vol. 6, no. 85, 2013.

[55] R. Bousso, "Complementarity is not enough," *Physical Review D: Particles, Fields, Gravitation and Cosmology*, vol. 87, no. 12, 2013.

[56] J. Maldacena and L. Susskind, "Cool horizons for entangled black holes," *Fortschritte der Physik/Progress of Physics*, vol. 61, no. 9, pp. 781–811, 2013.

[57] K. L. H. Bryan and A. J. M. Medved, "Black holes and information: a new take on an old paradox," *Advances in High Energy Physics*, vol. 2017, Article ID 7578462, 8 pages, 2017.

[58] D. Stanford and L. Susskind, "Complexity and shock wave geometries," *Physical Review D: Particles, Fields, Gravitation and Cosmology*, vol. 90, no. 12, 2014.

[59] S. Aaronson, "The complexity of quantum states and transformations: from quantum money to black holes," 2016, https://arxiv.org/abs/1607.05256.

[60] S. L. Braunstein, S. Pirandola, and K. Zyczkowski, "Better late than never: Information retrieval from black holes," *Physical Review Letters*, vol. 110, no. 10, Article ID 101301, 2013.

[61] A. Y. Yosifov and L. G. Filipov, "Entropic Entanglement: Information Prison Break," *Advances in High Energy Physics*, vol. 2017, Article ID 8621513, 7 pages, 2017.

[62] S. W. Hawking, "Information preservation and weather forecasting for black holes," 2014, https://arxiv.org/abs/1401.5761.

[63] SW. Hawking, "The information paradox for black holes," Tech. Rep. DAMTP-2015-49, 2015.

[64] S. W. Hawking, M. J. Perry, and A. Strominger, "Soft Hair on Black Holes," *Physical Review Letters*, vol. 116, no. 23, Article ID 231301, 2016.

[65] S. W. Hawking, M. J. Perry, and A. Strominger, "Superrotation charge and supertranslation hair on black holes," *Journal of High Energy Physics*, vol. 5, p. 161, 2017.

[66] L. Susskind, "The world as a hologram," *Journal of Mathematical Physics*, vol. 36, no. 11, pp. 6377–6396, 1995.

[67] K. Papadodimas and S. Raju, "Black Hole Interior in the Holographic Correspondence and the Information Paradox," *Physical Review Letters*, vol. 112, no. 5, 2014.

[68] Z.-L. Wang and Y. Yan, "Bulk Local Operators, Conformal Descendants, and Radial Quantization," *Advances in High Energy Physics*, vol. 2017, Article ID 8185690, 11 pages, 2017.

[69] A. Gefter, "Complexity on the horizon," *Nature*, 2014.

[70] R. Bousso, "Firewalls from double purity," *Physical Review D: Particles, Fields, Gravitation and Cosmology*, vol. 88, no. 8, 2013.

[71] S. Weinfurtner, E. W. Tedford, M. C. Penrice, W. G. Unruh, and G. A. Lawrence, "Measurement of Stimulated Hawking Emission in an Analogue System," *Physical Review Letters*, vol. 106, no. 2, 2011.

[72] J. M. Bardeen, B. Carter, and S. W. Hawking, "The four laws of black hole mechanics," *Communications in Mathematical Physics*, vol. 31, pp. 161–170, 1973.

[73] R. Bousso, "The holographic principle," *Reviews of Modern Physics*, vol. 74, no. 3, pp. 825–874, 2002.

[74] S. W. Hawking, "Gravitational radiation from colliding black holes," *Physical Review Letters*, vol. 26, no. 21, pp. 1344–1346, 1971.

[75] R. B. Mann, *Black Holes: Thermodynamics, Information, And Firewalls*, Springer, New York, NY, USA, 2015.

[76] L. Parker, "Quantized fields and particle creation in expanding universes. I," *Physical Review A: Atomic, Molecular and Optical Physics*, vol. 183, no. 5, pp. 1057–1068, 1969.

[77] B. P. Dolan, *Where is the pdv term in the first law of black hole thermodynamics?*, 2014.

[78] R. M. Wald, "Black hole entropy is the Noether charge," *Physical Review D: Particles, Fields, Gravitation and Cosmology*, vol. 48, no. 8, pp. R3427–R3431, 1993.

[79] T. Jacobson, "Thermodynamics of spacetime: the Einstein equation of state," *Physical Review Letters*, vol. 75, p. 1260, 1995.

[80] E. Verlinde, "On the origin of gravity and the laws of Newton," *Journal of High Energy Physics*, vol. 4, p. 29, 2011.

[81] G. D. Birkhoff and R. E. Langer, *Relativity and Modern Physics*, vol. 1, Harvard University Press Cambridge, 1923.

[82] M. Heusler, "No-hair theorems and black holes with hair," *Helvetica Physica Acta. Physica Theoretica. Societatis Physicae Helveticae Commentaria Publica*, vol. 69, no. 4, pp. 501–528, 1996.

[83] N. E. Mavromatos, "Eluding the no-hair conjecture for black holes," 1996, https://arxiv.org/abs/gr-qc/9606008.

[84] W. H. Zurek and K. S. Thorne, "Statistical mechanical origin of the entropy of a rotating, charged black hole," *Physical Review Letters*, vol. 54, no. 20, pp. 2171–2175, 1985.

[85] G. 't Hooft, "On the quantum structure of a black hole," *Nuclear Physics B*, vol. 256, no. 4, pp. 727–745, 1985.

[86] R. B. Mann, L. Tarasov, and A. Zelnikov, "Brick walls for black holes," *Classical and Quantum Gravity*, vol. 9, no. 6, pp. 1487–1494, 1992.

[87] S. Carlip, "Entropy from conformal field theory at Killing horizons," *Classical and Quantum Gravity*, vol. 16, no. 10, pp. 3327–3348, 1999.

[88] A. Strominger and C. Vafa, "Microscopic origin of the Bekenstein-Hawking entropy," *Physics Letters B*, vol. 379, no. 1–4, pp. 99–104, 1996.

[89] G. T. Horowitz and A. Strominger, "Counting States of Near-Extremal Black Holes," *Physical Review Letters*, vol. 77, no. 12, pp. 2368–2371, 1996.

[90] A. Dabholkar, "Exact counting of supersymmetric black hole microstates," *Physical Review Letters*, vol. 94, no. 24, 241301, 4 pages, 2005.

[91] A. Ashtekar, J. Baez, A. Corichi, and K. Krasnov, "Quantum geometry and black hole entropy," *Physical Review Letters*, vol. 80, no. 5, pp. 904–907, 1998.

[92] L. Bombelli, R. K. Koul, J. Lee, and R. D. Sorkin, "Quantum source of entropy for black holes," *Physical Review D: Particles, Fields, Gravitation and Cosmology*, vol. 34, no. 2, pp. 373–383, 1986.

[93] M. Srednicki, "Entropy and area," *Physical Review Letters*, vol. 71, no. 5, pp. 666–669, 1993.

[94] V. Frolov and I. Novikov, "Dynamical origin of the entropy of a black hole," *Physical Review D: Particles, Fields, Gravitation and Cosmology*, vol. 48, no. 10, pp. 4545–4551, 1993.

[95] C. Corda, "Effective temperature, hawking radiation and quasinormal modes," *International Journal of Modern Physics D*, vol. 21, no. 11, Article ID 1242023, 2012.

[96] C. Corda, "Black hole quantum spectrum," *The European Physical Journal C*, vol. 73, p. 2665, 2013.

[97] C. Corda, "Bohr-like model for black-holes," *Classical and Quantum Gravity*, vol. 32, no. 19, article 5007, 2015.

[98] G. Dvali and C. Gomez, "Quantum compositeness of gravity: black holes, AdS and inflation," *Journal of Cosmology and Astroparticle Physics*, no. 1, 023, front matter+46 pages, 2014.

[99] R. Casadio, A. Giugno, O. Micu, and A. Orlandi, "Black holes as self-sustained quantum states and Hawking radiation," *Physical Review D: Particles, Fields, Gravitation and Cosmology*, vol. 90, no. 8, 2014.

[100] R. Casadio, A. Giugno, O. Micu, and A. Orlandi, "Thermal BEC black holes," *Entropy*, vol. 17, no. 10, pp. 6893–6924, 2015.

[101] J. D. Bekenstein, "How does the entropy/information bound work?" *Foundations of Physics. An International Journal Devoted to the Conceptual Bases and Fundamental Theories of Modern Physics*, vol. 35, no. 11, pp. 1805–1823, 2005.

[102] S. Schaffer, "John michell and black holes," *Journal for the History of Astronomy*, vol. 10, no. 1, pp. 42-43, 1979.

[103] C. Rovelli, "Black holes have more states than those giving the Bekenstein-Hawking entropy: a simple argument," 2017, https://arxiv.org/abs/1710.00218.

[104] E. Bianchi, M. Christodoulou, F. D'Ambrosio, H. M. Haggard, and C. Rovelli, "White holes as remnants: A surprising scenario for the end of a black hole," 2018, https://arxiv.org/abs/1802.04264.

[105] A. Ashtekar, *Lectures on Non-Perturbative Canonical Gravity*, World Scientific, Singapore, 1991.

[106] G. Yoneda, H.-a. Shinkai, and A. Nakamichi, "Trick for passing degenerate points in the Ashtekar formulation," *Physical Review D: Particles, Fields, Gravitation and Cosmology*, vol. 56, no. 4, pp. 2086–2093, 1997.

[107] O. C. Stoica, "On singular semi-Riemannian manifolds," *International Journal of Geometric Methods in Modern Physics*, vol. 11, no. 5, 1450041, 40 pages, 2014.

[108] O. C. Stoica, "The geometry of warped product singularities," *International Journal of Geometric Methods in Modern Physics*, vol. 14, no. 2, 1750024, 16 pages, 2017.

[109] O. C. Stoica, "The Friedmann-Lemaître-Robertson-Walker Big Bang Singularities are Well Behaved," *International Journal of Theoretical Physics*, vol. 55, no. 1, pp. 71–80, 2016.

[110] O. C. Stoica, "Beyond the Friedmann-Lemaître-Robertson-Walker Big Bang singularity," *Communications in Theoretical Physics*, vol. 58, pp. 613–616, 2012.

[111] O. C. Stoica, "Schwarzschild singularity is semi-regularizable," *The European Physical Journal Plus*, vol. 127, no. 83, pp. 1–8, 2012.

[112] O. C. Stoica, "Kerr-Newman solutions with analytic singularity and no closed timelike curves," *"Politehnica" University of Bucharest. Scientific Bulletin. Series A. Applied Mathematics and Physics*, vol. 77, no. 1, pp. 129–138, 2015.

[113] O. C. Stoica, "Spacetimes with singularities," *Analele stiintifice ale Universitatii Ovidius Constanta*, vol. 20, no. 2, pp. 213–238, 2012.

[114] O. C. Stoica, "The geometry of singularities and the black hole information paradox," *Journal of Physics: Conference Series*, vol. 626, Article ID 012028, 2015.

[115] A. S. Eddington, "A Comparison of Whitehead's and Einstein's Formulæ," *Nature*, vol. 113, no. 2832, p. 192, 1924.

[116] D. Finkelstein, "Past-future asymmetry of the gravitational field of a point particle," *Physical Review Journals Archive*, vol. 110, p. 965, 1958.

[117] O. C. Stoica, "Gauge theory at singularities," 2014, https://arxiv.org/abs/1408.3812.

[118] O. C. Stoica, "Causal structure and spacetime singularities," 2015, https://arxiv.org/abs/1504.07110.

[119] PAM Dirac, "The Lagrangian in quantum mechanics," *Physikalische Zeitschrift der Sowjetunion*, vol. 1, no. 3, 1933.

[120] R. P. Feynman and A. R. Hibbs, *Quantum Mechanics and Path Integrals: Emended Edition*, Dover Publications, Incorporated, 2012.

[121] H. Kleinert, *Path integrals in quantum mechanics, statistics, polymer physics, and financial markets*, World Scientific, Singapore, 2009.

[122] O. C. Stoica, "Metric dimensional reduction at singularities with implications to quantum gravity," *Annals of Physics*, vol. 347, pp. 74–91, 2014.

[123] P. P. Fiziev and D. V. Shirkov, "Solutions of the Klein-Gordon equation on manifolds with variable geometry including dimensional reduction," *Theoretical and Mathematical Physics*, vol. 167, no. 2, pp. 680–691, 2011.

[124] D. V. Shirkov, "Dream-land with Classic Higgs field, Dimensional Reduction and all that," in *Proceedings of the Steklov Institute of Mathematics*, vol. 272, pp. 216–222, 2011.

9

Thermodynamics of the FRW Universe at the Event Horizon in Palatini $f(R)$ Gravity

A. S. Sefiedgar and M. Mirzazadeh

Department of Physics, Faculty of Basic Sciences, University of Mazandaran, Babolsar 47416-95447, Iran

Correspondence should be addressed to A. S. Sefiedgar; a.sefiedgar@umz.ac.ir

Academic Editor: Elias C. Vagenas

In an accelerated expanding universe, one can expect the existence of an event horizon. It may be interesting to study the thermodynamics of the Friedmann-Robertson-Walker (FRW) universe at the event horizon. Considering the usual Hawking temperature, the first law of thermodynamics does not hold on the event horizon. To satisfy the first law of thermodynamics, it is necessary to redefine Hawking temperature. In this paper, using the redefinition of Hawking temperature and applying the first law of thermodynamics on the event horizon, the Friedmann equations are obtained in $f(R)$ gravity from the viewpoint of Palatini formalism. In addition, the generalized second law (GSL) of thermodynamics, as a measure of the validity of the theory, is investigated.

1. Introduction

The existence of a deep connection between gravity and thermodynamics is one of the greatest discoveries in theoretical physics [1–5]. Based on the Hawking proposal, a Schwarzschild black hole emits a thermal radiation. For a black hole with mass M, the Hawking temperature is given by $T = 1/8\pi M$ [4, 5]. Moreover, the black hole entropy is given by $S = A/4l_p^2$, which is introduced by Bekenstein. The parameter A is the area of the black hole horizon and l_p is the Planck length. The black hole mass, temperature, and the horizon entropy obey the first law of thermodynamics. Hence, one can conclude that a black hole can be assumed as a thermodynamical object [6–8]. In 1995, Jacobson has considered the relation between the horizon area and the entropy and derived the Einstein field equations from the first law of thermodynamics [9]. Following Jacobson, the relation between the thermodynamics and the field equations has been investigated in modified theories of gravity [10–15]. The thermodynamical interpretation of gravity is also important in modern cosmology. It is possible to derive the Friedmann equations from the first law of thermodynamics and vice versa [16–18]. The deep connection between the first law of thermodynamics and Friedmann equations has been investigated in gravity with Gauss-Bonnet term [19], Lovelock gravity theory [19], the braneworld scenarios [20], rainbow gravity [21], and scalar-tensor gravity and $f(R)$ gravity [22].

The thermodynamics of the FRW universe bounded by the apparent horizon has been studied by many authors [16–18]. However, astronomical observations show that the universe is currently in an accelerated expanding phase [23]. In an accelerated expanding universe, there may exist an event horizon [24, 25]. Certainly, it is necessary to investigate the thermodynamics on the event horizon. Through the researches about the thermodynamics of the universe, Wang et al. have found that the laws of thermodynamics do not hold on the event horizon. Therefore, one may conclude that the event horizon is unphysical [24]. Chacraborty has proposed a redefinition of Hawking temperature to obtain the universe bounded by the event horizon as a Bekenstein system [26]. Using the Chakraborty redefinition of Hawking temperature, Tu and Chen have considered a universe dominated by tachyon fluid and obtained a good thermodynamical description on the event horizon [27]. However, the temperature redefinition proposed by Chakraborty can be applied only in a flat universe. Tu and Chen have introduced a new redefinition of the temperature in a universe with an arbitrary spatial curvature [25]. They have investigated the thermodynamics on the event horizon in metric $f(R)$ gravity. Utilizing the redefinition of the temperature, it is possible to study the thermodynamics of the event horizon in Palatini $f(R)$ gravity.

In this paper, we are going to apply Hawking temperature redefinition introduced by Tu and Chen in [25] to investigate the event horizon thermodynamics from the viewpoint of Palatini $f(R)$ gravity. Redefining the gravitational energy density and the gravitational pressure density, the continuity equation has been satisfied. Then, we have considered the new gravitational source accompanying with the ordinary matter as a modified energy-momentum source in the context of general relativity. Applying the first law of thermodynamics on the event horizon and using the usual entropy-area relation, we have derived the Friedmann equations the same as the ones obtained via other approaches. Since the first law of thermodynamics is reasonably applicable on the event horizon, one can consider the event horizon as a physical system with an equilibrium thermodynamics in Palatini $f(R)$ gravity. Moreover, the equivalency between the first law of thermodynamics and Friedmann equations has been appeared. Finally, the generalized second law of thermodynamics (GSL) has been studied. The generalized second law of thermodynamics can be satisfied by choosing suitable $f(R)$ functions.

2. Hawking Temperature Redefinition on the Event Horizon

Considering a homogeneous and isotropic FRW universe, one can write the metric as follows:

$$ds^2 = h_{ij}dx^i dx^j + R^2 d\Omega_2^2, \tag{1}$$

where h_{ij} is the two-dimensional metric

$$h_{ij} = \text{diag}\left(-1, \frac{a^2(t)}{1-Kr^2}\right). \tag{2}$$

The parameters i and j take the values 0 and 1. The parameter K is the spatial curvature constant, the parameter $R = a(t)r$ is the area radius, and $a(t)$ is the scale factor. Using R as a scalar field in the normal 2-dimensional space, one can define a scalar quantity

$$\chi = h^{ij}\partial_i R \partial_j R. \tag{3}$$

The apparent horizon can be found via $\chi = 0$ as

$$R_A = \frac{1}{\sqrt{H^2 + K/a^2}}, \tag{4}$$

where $H = (1/a)(da/dt) = \dot{a}/a$ is the Hubble parameter.

The surface gravity on the apparent horizon is given by

$$\kappa_A = -\frac{1}{2}\frac{\partial\chi}{\partial R}\bigg|_{R=R_A} = \frac{1}{R_A}. \tag{5}$$

Then, one can derive Hawking temperature on the apparent horizon as

$$T_A = \frac{|\kappa_A|}{2\pi} = \frac{1}{2\pi R_A}. \tag{6}$$

On the other hand, one can consider the definition of the event horizon

$$R_E = a(t)\int_t^{\infty}\frac{dt}{a(t)}. \tag{7}$$

In an accelerated expanding universe, one can find that the infinite integral converges. In other words, it is possible to have a cosmological event horizon in an accelerated expanding universe. Then, the thermodynamics on the event horizon is important to be studied.

Through the researches about the thermodynamics on the event horizon, Wang et al. found that the thermodynamical description based on the standard definitions of boundary entropy and temperature breaks down in a universe bounded by the cosmological event horizon [24]. They concluded that the cosmological event horizon is unphysical from the viewpoint of thermodynamic laws. However, a redefinition of Hawking temperature has been introduced by Chakraborty in a flat universe to preserve the laws of thermodynamics on the event horizon [26]. The temperature redefinition has been upgraded by Tu et al. to be applicable in a universe with an arbitrary curvature constant [25]. According to [25], one can write the surface gravity on the event horizon as

$$\kappa_E = -\frac{1}{2}\frac{\partial\chi}{\partial R}\bigg|_{R=R_E}\frac{\dot{R}_A}{R_A}\frac{R_E}{\dot{R}E}. \tag{8}$$

Using the relation between the surface gravity and Hawking temperature on spacetime horizons, one can find

$$T_E = \frac{|\kappa_E|}{2\pi} = \frac{H}{2\pi}\left(\frac{K}{a^2} - \dot{H}\right)\frac{R_E^2}{\dot{R}_E}. \tag{9}$$

To investigate the universality of the redefinition of Hawking temperature on the event horizon, one can investigate the validity of the first law of thermodynamics. During an infinitesimal time interval, one can write the energy flux across the event horizon as [24–26, 28, 29]

$$\delta Q = AT_{\mu\nu}k^\mu k^\nu dt\Big|_{r=R_E}, \tag{10}$$

where k^μ is a null vector, $T_{\mu\nu} = (\rho + p)u_\mu u_\nu + pg_{\mu\nu}$ is the perfect fluid energy-momentum tensor, and u^μ is the timelike 4-velocity vector. Thus, the energy flux can be obtained as

$$\delta Q = 4\pi R_E^3 H(\rho + p)\,dt. \tag{11}$$

It can also be written as

$$\delta Q = \frac{HR_E^3}{G}\left(\frac{K}{a^2} - \dot{H}\right)dt, \tag{12}$$

in which the Friedmann equation in Einstein gravity; $\dot{H} - K/a^2 = -4\pi G(\rho + p)$ has been used. On the other hand, the

Bekenstein entropy-area relation in Einstein gravity is given by

$$S_E = \frac{A}{4G}. \tag{13}$$

Using (9) and (13), one can find

$$T_E dS_E = \frac{HR_E^3}{G}\left(\frac{K}{a^2} - \dot{H}\right) dt. \tag{14}$$

Comparing (12) and (14), the validity of the first law of thermodynamics, $\delta Q = T_E dS_E$, can be concluded on the event horizon [25]. It seems that the temperature redefinition can lead to a good thermodynamical description on the event horizon.

3. From the First Law of Thermodynamics into Friedmann Equations in Palatini $f(R)$ Gravity

The action of $f(R)$ gravity in Palatini formalism is written as

$$I = \int d^4x\sqrt{-g}\left[\frac{f(\mathcal{R})}{2k^2} + \mathcal{L}_{matter}\right], \tag{15}$$

where g is the determinant of the metric tensor, $\mathcal{L}_{matter}$ is the matter Lagrangian, and $k^2 = 8\pi G$. The connection and the metric tensor are considered as independent variables in Palatini formalism. The Ricci tensor and the Ricci scalar constructed with the independent connection are denoted, respectively, by $\mathcal{R}_{\mu\nu}$ and $\mathcal{R} = g^{\mu\nu}\mathcal{R}_{\mu\nu}$. Certainly, $\mathcal{R}_{\mu\nu}$ is different from the Ricci tensor $R_{\mu\nu}$ which is constructed with the Levi-Civita connection of the metric. Variation of the action with respect to the metric leads to

$$F(\mathcal{R})\mathcal{R}_{\mu\nu} - \frac{1}{2}f(\mathcal{R})g_{\mu\nu} = k^2 T^m_{\mu\nu}, \tag{16}$$

in which $F(\mathcal{R}) \equiv df(\mathcal{R})/d\mathcal{R}$ and $T^m_{\mu\nu}$ is the energy-momentum tensor related to the ordinary matter. Varying the action with respect to the connection and using (16) yield

$$\begin{aligned} FG_{\mu\nu} &= k^2 T^m_{\mu\nu} - \frac{1}{2}g_{\mu\nu}(F\mathcal{R} - f) + \nabla_\mu\nabla_\nu F \\ &\quad - g_{\mu\nu}g^{\alpha\beta}\nabla_\alpha\nabla_\beta F \\ &\quad - \frac{3}{2F}\left[\nabla_\mu F\nabla_\nu F - \frac{1}{2}g_{\mu\nu}g^{\alpha\beta}\nabla_\alpha F\nabla_\beta F\right]. \end{aligned} \tag{17}$$

Here $G_{\mu\nu} = R_{\mu\nu} - (1/2)Rg_{\mu\nu}$ is the Einstein tensor and $T^m_{\mu\nu} = (\rho_m + p_m)u_\mu u_\nu + p_m g_{\mu\nu}$ is the energy-momentum tensor of the ordinary matter. One can rewrite the field equations as

$$G_{\mu\nu} = \frac{k^2}{F}\left(T^m_{\mu\nu} + T^g_{\mu\nu}\right), \tag{18}$$

where

$$\begin{aligned} T^g_{\mu\nu} &= \frac{1}{k^2}\left(\frac{f - F\mathcal{R}}{2}g_{\mu\nu} + \nabla_\mu\nabla_\nu F - g_{\mu\nu}g^{\alpha\beta}\nabla_\alpha\nabla_\beta F\right) \\ &\quad - \frac{3}{2Fk^2}\left[\nabla_\mu F\nabla_\nu F - \frac{1}{2}g_{\mu\nu}g^{\alpha\beta}\nabla_\alpha F\nabla_\beta F\right] \end{aligned} \tag{19}$$

is the gravitational energy-momentum tensor arising from $f(R)$ gravity. Now, the gravitational energy density and the gravitational pressure can be written respectively as

$$\rho_g = \frac{1}{k^2}\left[\frac{1}{2}(F\mathcal{R} - f) - 3H\dot{F} - \frac{3}{4}\frac{\dot{F}^2}{F}\right], \tag{20}$$

and

$$p_g = \frac{1}{k^2}\left[-\frac{1}{2}(F\mathcal{R} - f) + \ddot{F} + 2H\dot{F} - \frac{3}{4}\frac{\dot{F}^2}{F}\right]. \tag{21}$$

The perfect fluid energy-momentum tensor satisfies the continuity equation

$$\dot{\rho}_m + 3H(\rho_m + p_m) = 0. \tag{22}$$

It is easy to investigate the continuity equation for the gravitational sector of the energy-momentum tensor arising from $f(R)$ gravity. Clearly, one can find

$$\dot{\rho}_g + 3H(\rho_g + p_g) = \frac{3}{k^2}\left(H^2 + \frac{K}{a^2}\right)\dot{F}. \tag{23}$$

To satisfy the continuity equation, it is possible to redefine the gravitational energy density and the gravitational pressure density as

$$\begin{aligned} \widehat{\rho}_g &= \frac{1}{k^2}\left[\frac{1}{2}(F\mathcal{R} - f) - 3H\dot{F} - \frac{3}{4}\frac{\dot{F}^2}{F}\right. \\ &\quad \left. + 3(1 - F)\left(H^2 + \frac{K}{a^2}\right)\right], \end{aligned} \tag{24}$$

and

$$\begin{aligned} \widehat{p}_g &= \frac{1}{k^2}\left[-\frac{1}{2}(F\mathcal{R} - f) + \ddot{F} + 2H\dot{F} - \frac{3}{4}\frac{\dot{F}^2}{F}\right. \\ &\quad \left. - (1 - F)\left(2\dot{H} + 3H^2 + \frac{K}{a^2}\right)\right]. \end{aligned} \tag{25}$$

Now, one can rewrite the continuity equation for the gravitational energy density and gravitational pressure density as

$$\dot{\widehat{\rho}}_g + 3H(\widehat{\rho}_g + \widehat{p}_g) = 0. \tag{26}$$

From (22) and (26), one can write the total continuity equation as

$$\dot{\rho}_T + 3H(\rho_T + p_T) = 0, \tag{27}$$

in which $\rho_T = \rho_m + \widehat{\rho}_g$ and $p_T = p_m + \widehat{p}_g$. It is clear that the redefinition of the energy density in (24) and the pressure density in (25) lead to the standard continuity equation. It is now possible to consider ρ_T and p_T as the energy and the pressure density of a modified source in the context of general relativity. Hence, one can apply the usual entropy-area relation in (13).

To derive the Friedmann equation, one can start with the first law of thermodynamics. Using (10), the energy flux can be written as

$$\delta Q = 4\pi R_E^3 H \left(\rho_T + p_T\right) dt. \tag{28}$$

Utilizing (14) and (28), the first law of thermodynamics yields

$$\left(\frac{K}{a^2} - \dot{H}\right) = 4\pi G \left(\rho_T + p_T\right). \tag{29}$$

Now, one can substitute (24) and (25) into (29) to find the Friedmann equation

$$-2\left(\dot{H} - \frac{K}{a^2}\right) F = k^2 \left(\rho_m + p_m\right) + \ddot{F} - H\dot{F} - \frac{3}{2}\frac{\dot{F}^2}{F}. \tag{30}$$

The other Friedmann equation can be obtained via (27) and (30)

$$3\left(H^2 + \frac{K}{a^2}\right) F = k^2 \rho_m + \frac{1}{2}\left(F\mathcal{R} - f\right) - 3H\dot{F} - \frac{3}{4}\frac{\dot{F}^2}{F}. \tag{31}$$

The modified Friedmann equations in (30) and (31) are the same as the ones obtained in [30]. In other words, the first law of thermodynamics on the event horizon leads to the Friedmann equations which are consistent with the ones obtained via the other approach in [30]. Hence, one can rely on the validity of the first law of thermodynamics on the event horizon and consider the event horizon as a thermodynamical system in Palatini $f(R)$ gravity. It is also clear that the first law of thermodynamics is equivalent with Friedmann equations.

Equation (31) can also be rewritten as

$$\left(H^2 + \frac{K}{a^2}\right) = \frac{8\pi G}{3}\rho_{eff}, \tag{32}$$

where

$$\rho_{eff} = \frac{1}{F}\left[\rho_m + \frac{1}{8\pi G}\left(\frac{1}{2}\left(F\mathcal{R} - f\right) - 3H\dot{F} - \frac{3}{4}\frac{\dot{F}^2}{F}\right)\right] \tag{33}$$

can be considered as an effective energy density. Such an effective energy density may provide an accelerated expanding universe to solve the problem of dark energy. The modified Friedmann equations can be reduced to the Friedmann equations in Einstein gravity by putting $f(\mathcal{R}) = \mathcal{R}$.

4. The Generalized Second Law of Thermodynamics

Applying the usual Hawking temperature, the first law of thermodynamics does not hold on the event horizon. However, using the redefinition of Hawking temperature, it has been shown that the first law of thermodynamics may hold on the event horizon in Einstein gravity and metric $f(R)$ gravity [25]. Hence, the FRW universe bounded by the event horizon may be described by equilibrium thermodynamics. In previous section, we have considered Hawking temperature redefinition on the event horizon in Palatini $f(R)$ gravity. We have shown that the first law of thermodynamics can yield the Friedmann equations just the same as the ones obtained via the other approach in [30]. It means that he FRW universe bounded by the event horizon can be described by an equilibrium thermodynamics in Palatini approach too.

It is now possible to investigate the generalized second law of thermodynamics in the FRW universe bounded by the event horizon in Palatini approach. Considering a universe with the dust energy-momentum tensor in $f(R)$ gravity, one can write the continuity equations as

$$\dot{\rho}_d + 3H\rho_d = 0, \tag{34}$$

and

$$\dot{\widehat{\rho}}_g + 3H\left(\widehat{\rho}_g + \widehat{p}_g\right) = 0. \tag{35}$$

The parameter ρ_d is the energy density of dust and the pressure of dust is $p_d = 0$. The energy flux can be obtained from (10) as

$$\delta Q = 4\pi R_E^3 H \left(\rho_d + \widehat{\rho}_g + \widehat{p}_g\right) dt. \tag{36}$$

Based on the generalized second law of thermodynamics, the sum of the entropy on the event horizon and the one inside the bulk must never decrease. To find the total entropy changes, one can start with Gibb's relation [31–33]

$$T_E dS_{in} = dE_{in} + \widehat{p}_g dV, \tag{37}$$

where $V = 4\pi R_E^3/3$. The parameters $E_{in} = (4\pi R_E^3/3)(\rho_d + \widehat{\rho}_g)$ and S_{in} are the energy and the entropy inside the boundary respectively. From (7), one can find

$$dR_E = \left(HR_E - 1\right) dt. \tag{38}$$

Now, (37) leads to

$$\frac{dS_{in}}{dt} = -\frac{4\pi R_E^2}{T_E}\left(\rho_d + \widehat{\rho}_g + \widehat{p}_g\right), \tag{39}$$

in which we have used relation (38). Since the universe is assumed to be in thermal equilibrium, the temperature inside the universe and the temperature on the event horizon are considered the same. Using the definition $S_{tot} = S_{in} + S_E$ and applying (14), one can find

$$\frac{dS_{tot}}{dt} = \frac{4\pi R_E^2}{T_E}\left(\rho_d + \widehat{\rho}_g + \widehat{p}_g\right)\left(HR_E - 1\right). \tag{40}$$

It is possible to substitute (24) and (25) into (40) to find

$$\frac{dS_{tot}}{dt} = \frac{R_E^2}{2GT_E}\left(HR_E - 1\right)\left(\rho_d + \ddot{F} - H\dot{F} - \frac{3}{2}\frac{\dot{F}^2}{F}\right) - \frac{R_E^2}{GT_E}\left(1 - F\right)\left(HR_E - 1\right)\left(\dot{H} - \frac{K}{a^2}\right). \tag{41}$$

Based on the redefinition of Hawking temperature, we have found the event horizon as an equilibrium thermodynamical system. Hence, the first law of thermodynamics and Gibb's relation have been applied to obtain the change of the total entropy. To satisfy GSL, it is necessary to have $\dot{S}_{tot} \geq 0$. Of course, the validity of the generalized second law of thermodynamics depends on the $f(R)$ functions. In other words, GSL can provide some constraints on the choice of $f(R)$ gravity models.

To investigate GSL, one can substitute T_E from (9) into (40) to find

$$\frac{dS_{tot}}{dt} = \frac{8\pi^2}{H\left(K/a^2 - \dot{H}\right)}\left(\rho_d + \widehat{\rho}_g + \widehat{p}_g\right)\dot{R}_E^2. \quad (42)$$

Based on (29), one can conclude that $(K/a^2 - \dot{H}) \geq 0$ in the weak energy condition. Hence, GSL can be satisfied, when

$$\left(\rho_d + \widehat{\rho}_g + \widehat{p}_g\right) \geq 0. \quad (43)$$

It means that the function f should satisfy the condition

$$\rho_d + \ddot{F} - H\dot{F} - \frac{3}{2}\frac{\dot{F}^2}{F^2} + 2(1-F)\left(\frac{K}{a^2} - \dot{H}\right) \geq 0. \quad (44)$$

In the case $f(R) = R$, one can substitute $F = 1$ in (44) to investigate the validity of GSL in general relativity. Since $\rho_d \geq 0$, one can find $\dot{S}_{tot} \geq 0$ in the dust dominated universe bounded by the event horizon in general relativity. The Hawking temperature redefinition has provided the validity of the GSL in the universe bounded by the event horizon in general relativity.

5. Conclusions

Although the apparent horizon thermodynamics has been studied by many authors in FRW universe, the event horizon thermodynamics is not investigated enough. Since the universe is undergoing an accelerated expansion phase, one may expect the existence of an event horizon. Certainly, the thermodynamics of the event horizon is important to be studied. It has been shown that applying the usual Hawking temperature leads to a nonequilibrium thermodynamics on the event horizon [24]. Hence, the event horizon may be considered unphysical from the view point of the thermodynamic laws. However, redefining Hawking temperature can solve this problem [25, 26]. The Hawking temperature redefinition introduced by Tu and Chen [25] leads to an equilibrium thermodynamics on the event horizon in general relativity and metric $f(R)$ gravity. In this paper, the thermodynamics of the event horizon in FRW universe is investigated in Palatini $f(R)$ gravity. Apparently, the continuity equation does not hold in Palatini formalism. To satisfy the continuity equation, redefinitions of the energy density and the pressure density have been introduced in the gravitational sector of the energy-momentum tensor. The redefined gravitational source accompanying with the ordinary matter may play the role of a new modified source in the context of Einstein gravity. Considering this modified energy-momentum source, using the first law of thermodynamics on the event horizon and applying Bekenstein-Hawking entropy-area relation, the Friedmann equations have been derived. The corrections to the Friedmann equations are the same as the ones obtained via other approaches in [30]. It means that considering the first law of thermodynamics on the event horizon is reliable. In addition, the equivalency of the first law of thermodynamics and the Friedmann equations has been shown. The generalized second law of thermodynamics as a measure of the validity of the gravitational models is investigated in Palatini $f(R)$ gravity. Of course, GSL can provide some constraints on choosing $f(R)$ functions.

Conflicts of Interest

The authors declare that they have no conflicts of interest.

References

[1] J. D. Bekenstein, "Black holes and the second law," *Lettere al Nuovo Cimento (1971-1985)*, vol. 4, no. 15, pp. 737–740, 1972.

[2] J. D. Bekenstein, "Generalized second law of thermodynamics in black-hole physics," *Physical Review D: Particles, Fields, Gravitation and Cosmology*, vol. 9, article 3292, 1974.

[3] J. D. Bekenstein, "Black holes and entropy," *Physical Review D: Particles, Fields, Gravitation and Cosmology*, vol. 7, pp. 2333–2346, 1973.

[4] S. W. Hawking, "Black hole explosions?" *Nature*, vol. 248, no. 5443, pp. 30-31, 1974.

[5] S. W. Hawking, "Particle creation by black holes," *Communications in Mathematical Physics*, vol. 43, no. 3, pp. 199–220, 1975.

[6] P. C. W. Davies, "Scalar production in Schwarzschild and Rindler metrics," *Journal of Physics A: Mathematical and General*, vol. 8, no. 4, p. 609, 1975.

[7] W. G. Unruh, "Notes on black-hole evaporation," *Physical Review D: Particles, Fields, Gravitation and Cosmology*, vol. 14, no. 4, p. 870, 1976.

[8] L. Susskind, "A predictive Yukawa unified SO(10) model: higgs and sparticle masses," *Journal of Mathematical Physics*, vol. 36, no. 7, article 139, pp. 6377–6396, 1995.

[9] T. Jacobson, "Thermodynamics of spacetime: the Einstein equation of state," *Physical Review Letters*, vol. 75, p. 1260, 1995.

[10] C. Eling, R. Guedens, and T. Jacobson, "Nonequilibrium thermodynamics of spacetime," *Physical Review Letters*, vol. 96, no. 12, Article ID 121301, 2006.

[11] A. Sheykhi, B. Wang, and R.-G. Cai, "Thermodynamical properties of apparent horizon in warped DGP braneworld," *Nuclear Physics B*, vol. 779, no. 1-2, pp. 1–12, 2007.

[12] R. G. Cai and L. M. Cao, "Thermodynamics of apparent horizon in brane world scenario," *Nuclear Physics B*, vol. 785, no. 1-2, pp. 135–148, 2007.

[13] A. Sheykhi, B. Wang, and R.-G. Cai, "Deep connection between thermodynamics and gravity in Gauss-Bonnet braneworlds," *Physical Review D: Particles, Fields, Gravitation and Cosmology*, vol. 76, no. 2, Article ID 023515, 5 pages, 2007.

[14] A. Sheykhi and B. Wang, "Generalized second law of thermodynamics in Gauss-Bonnet braneworld," *Physics Letters B*, vol. 678, no. 5, pp. 434–437, 2009.

[15] T. Padmanabhan, "Classical and quantum thermodynamics of horizons in spherically symmetric spacetimes," *Classical and Quantum Gravity*, vol. 19, no. 21, pp. 5387–5408, 2002.

[16] R.-G. Cai and L. M. Cao, "Unified first law and the thermodynamics of the apparent horizon in the FRW universe," *Physical Review D*, vol. 75, Article ID 064008, 2007.

[17] S. Mitra, S. Saha, and S. Chakraborty, "Universal thermodynamics in different gravity theories: modified entropy on the horizons," *Physics Letters B*, vol. 734, pp. 173–177, 2014.

[18] M. Akbar and R.-G. Cai, "Thermodynamic behavior of the Friedmann equation at the apparent horizon of the FRW universe," *Physical Review D*, vol. 75, Article ID 084003, 2007.

[19] R.-G. Cai and S. P. Kim, "First Law of Thermodynamics and Friedmann Equations of Friedmann-Robertson-Walker Universe," *Journal of High Energy Physics*, vol. 2005, artilce 02, 2005, arXiv.

[20] X.-H. Ge, "First law of thermodynamics and Friedmann-like equations in braneworld cosmology," *Physics Letters B*, vol. 651, no. 1, pp. 49–53, 2007.

[21] A. S. Sefiedgar and M. Daghigh, "Thermodynamics of the FRW universe in rainbow gravity," *International Journal of Modern Physics D*, vol. 26, no. 13, Article ID 1750139, 2017.

[22] M. Akbar and R. G. Cai, "Friedmann equations of FRW universe in scalar–tensor gravity, f(R) gravity and first law of thermodynamics," *Physics Letters B*, vol. 635, no. 1, pp. 7–10, 2006.

[23] R. A. Knop, G. Aldering, R. Amanullah et al., "New Constraints on ΩM, ΩΛ, and w from an Independent Set of 11 High-Redshift Supernovae Observed with the Hubble Space Telescope," *The Astrophysical Journal*, vol. 598, no. 1, pp. 102–137, 2003, The Astrophysical Journal.

[24] B. Wang, Y. G. Gong, and E. Abdalla, "Thermodynamics of an accelerated expanding universe," *Physical Review D: Particles, Fields, Gravitation and Cosmology*, vol. 74, Article ID 083520, 2006.

[25] F.-Q. Tu and Y.-X. Chen, "A general thermodynamical description of the event horizon in the FRW universe," *The European Physical Journal C*, vol. 76, article 28, 2016.

[26] S. Chakraborty, "Is thermodynamics of the universe bounded by event horizon a Bekenstein system?" *Physics Letters B*, vol. 718, no. 2, pp. 276–278, 2012.

[27] F.-Q. Tu and Y.-X. Chen, "Thermodynamics of the universe bounded by the cosmological event horizon and dominated by the tachyon fluid," https://arxiv.org/abs/1310.7295.

[28] Q.-J. Cao, Y.-X. Chen, and K.-N. Shao, "Clausius relation and Friedmann equation in FRW universe model," *Journal of Cosmology and Astroparticle Physics*, vol. 2010, article 030, 2010.

[29] R. S. Bousso, "Cosmology and the S matrix," *Physical Review D: Particles, Fields, Gravitation and Cosmology*, vol. 71, no. 6, Article ID 064024, 2005.

[30] K. Bamba and C.-Q. Geng, "Thermodynamics in f(R) gravity in the Palatini formalism," *Journal of Cosmology and Astroparticle Physics*, vol. 2010, article 014, 2010.

[31] S. H. Pereira and J. A. S. Lima, "On phantom thermodynamics," *Physics Letters B*, vol. 669, no. 5, pp. 266–270, 2008.

[32] E. N. Saridakis, P. F. González-Díaz, and C. L. Sigüenza, "Unified dark energy thermodynamics: varying w and the −1-crossing," *Classical and Quantum Gravity*, vol. 26, Article ID 165003, 8 pages, 2009.

[33] G. Izquierdo and D. Pavon, "Dark energy and the generalized second law," *Physics Letters B*, vol. 633, no. 4-5, pp. 420–426, 2006.

10

Phase Transition of RN-AdS Black Hole with Fixed Electric Charge and Topological Charge

Shan-Quan Lan

Department of Physics, Lingnan Normal University, Zhanjiang, 524048, Guangdong, China

Correspondence should be addressed to Shan-Quan Lan; shanquanlan@126.com

Academic Editor: Elias C. Vagenas

Phase transition of RN-AdS black hole is investigated from a new perspective. Not only is the cosmological constant treated as pressure but also the spatial curvature of black hole is treated as topological charge ϵ. We obtain the extended thermodynamic first law from which the mass is naturally viewed as enthalpy rather than internal energy. In canonical ensemble with fixed topological charge and electric charge Q, interesting van der Waals like oscillatory behavior in $T-S$ and $P-V$ graphs and swallow tail behavior in $G-T$ and $G-P$ graphs is observed. By applying the Maxwell equal area law and analysing the Gibbs free energy, we obtain analytical phase transition coexistence curves which are consistent with each other. The phase diagram is four dimensional with T, P, Q, ϵ.

1. Introduction

Black hole is a simple object which can be described by only a few physical quantities, such as mass, charge, and angular momentum, while it is also a complicate thermodynamic system. Since the discovery of black hole's entropy [1], the four-thermodynamic law [2], and the Hawking radiation [3] in 1970s, thermodynamic of black hole has become an interesting and challenging topic. Especially, in the anti-de Sitter (AdS) space, there exists Hawking-Page phase transition between stable large black hole and thermal gas [4]. Due to the AdS/CFT correspondence [5–7], the Hawking-Page phase transition is explained as the confinement/deconfinement phase transition of a gauge field [8].

When the AdS black hole is electrically charged, its thermodynamic properties become more rich. In the canonical ensemble with fixed electric charge, there is a first-order phase transition between small and large black holes [9–12]. Increasing the temperature, the phase transition coexistence curve ends at the critical point, where the first-order phase transition becomes a second-order one. In the grand canonical ensemble with fixed temperature, there is also a critical temperature. Below the critical temperature, $\Phi(Q)$ is a single-valued function, where Q is electric charge and Φ is the conjugate potential. Above the critical temperature, $\Phi(Q)$ is a multivalued function with phase transitions [12]. The phase transition behavior of AdS black hole is reminiscent to the liquid-gas phase transition in a van der Waals system.

Viewing the cosmological constant as a dynamical pressure and the black hole volume as its conjugate quantity [13], the analogy of charged AdS black hole as a van der Waals system has been further enhanced in Ref. [14]. Both the systems share the same oscillatory behavior in pressure-volume graph and swallow tail behavior in Gibbs free energy-temperature (pressure) graph. What's more, they have very similar phase diagrams and have exactly the same critical exponents. The phase transition property is also investigated in temperature-entropy graph [15]. Later, this analogy has been generalized to different AdS black holes, such as rotating black holes, higher dimensional black holes, Gauss-Bonnet black holes, f(R) black holes, black holes with scalar hair, etc [15–57], where more interesting phenomena are found.

Recently, the spatial curvature of electrically charged AdS black hole is viewed as variable and treated as topological charge [58, 59] in Einstein-Maxwell's gravity and Lovelock-Maxwell theory. The authors found that the topological charge naturally arose in holography. What is more, together with all other known charges (electric charge, mass, and entropy), they satisfy an extended first law and the Gibbs-Duhem-like relation as a completeness. In our last paper [60], when the cosmological constant is not viewed as variable, we find a van der Waals type but new phase transition relating to the topological charge, while in this paper, we will treat both the cosmological constant and the spatial curvature as variables, then following one of their methods to derive the extended first law, from which one can see the cosmological constant is naturally viewed as pressure and the mass is viewed as enthalpy. Based on the extended first law, the black hole's phase transition property will be investigated in canonical ensemble with fixed electric charge and topological charge.

This paper is organized as follows. In Section 2, following the method in Ref. [59], we will derive the extended first law in d dimensional space-time. In Section 3, by analysing the specific heat, the phase transition of AdS black hole in 4-dimensional space-time is studied and the critical point is determined. In Section 4, the van der Waals like oscillatory behavior is observed in both $T-S$ and $P-V$ graphs. Then we use the Maxwell equal area law to obtain the phase transition coexistence curve. In Section 5, the van der Waals like swallow tail behavior is observed in $G-T$ and $G-P$ graphs, then we will obtain the phase transition coexistence curve by analysing the gibbs free energy. Finally, we summarize and discuss our results in Section 6.

2. The Extended Thermodynamic First Law

The d dimensional space-time AdS black hole solutions with maximal symmetry in the Einstein-Maxwell theory are

$$ds^2 = \frac{dr^2}{f(r)} - f(r)\,dt^2 + r^2 d\Omega_{d-2}^{(k)2}, \tag{1}$$

where

$$\begin{aligned} f(r) &= k + \frac{r^2}{l^2} - \frac{m}{r^{d-3}} + \frac{q^2}{r^{2d-6}}, \\ d\Omega_{d-2}^{(k)2} &= \hat{g}_{ij}^{(k)}(x)\,dx^i dx^j, \\ A &= -\frac{\sqrt{d-2}q}{\sqrt{2(d-3)}r^{d-3}}dt. \end{aligned} \tag{2}$$

m, q, l are related to the ADM mass M, electric charge Q, and cosmological constant Λ by

$$\begin{aligned} M &= \frac{(d-2)\,\Omega_{d-2}^{(k)}}{16\pi}m, \\ Q &= \sqrt{2(d-2)(d-3)}\left(\frac{\Omega_{d-2}^{(k)}}{8\pi}\right)q, \\ \Lambda &= -\frac{(d-1)(d-2)}{2l^2}, \end{aligned} \tag{3}$$

and $\Omega_{d-2}^{(k)}$ is the volume of the "unit" sphere, plane, or hyperbola, and k stands for the spatial curvature of the black hole. Under suitable compactifications for $k \le 0$, we assume that the volume of the unit space is a constant $\Omega_{d-2} = \Omega_{d-2}^{(k=1)}$ hereafter [58, 59].

Following [59], the first law of thermodynamics can be derived. As the first law of thermodynamics is about the differential relation of every physical quantity, one can first find an equation containing these physical quantities and then differentiate it to obtain the first law of thermodynamics. Considering an equipotential surface $f(r) = c$ with fixed c (here set $c = 0$), we variate both sides of the equation and obtain

$$\begin{aligned} df(r_+, k, m, q) &= \frac{\partial f}{\partial r_+}dr_+ + \frac{\partial f}{\partial k}dk + \frac{\partial f}{\partial (1/l^2)}d\frac{1}{l^2} \\ &\quad + \frac{\partial f}{\partial m}dm + \frac{\partial f}{\partial q}dq = 0, \end{aligned} \tag{4}$$

where r_+ is the radius of event horizon. Noting

$$\begin{aligned} \partial_{r_+} f &= 4\pi T, \\ \partial_k f &= 1, \\ \partial_{1/l^2} f &= r_+^2, \\ \partial_m f &= -\frac{1}{r_+^{d-3}}, \\ \partial_q f &= \frac{2q}{r_+^{2d-6}}, \end{aligned} \tag{5}$$

we obtain

$$dm = \frac{4\pi T}{d-2}dr_+^{d-2} + r_+^{d-3}dk + r_+^{d-1}d\frac{1}{l^2} + \frac{2q}{r_+^{d-3}}dq. \tag{6}$$

Multiplying both sides with an constant factor $(d-2)\Omega_{d-2}/16\pi$, the above equation becomes

$$\begin{aligned} dM &= TdS + \frac{(d-2)\,\Omega_{d-2}}{16\pi}r_+^{d-3}dk \\ &\quad + \frac{(d-2)\,\Omega_{d-2}}{16\pi}r_+^{d-1}d\frac{1}{l^2} + \Phi dQ, \end{aligned} \tag{7}$$

where $T = \partial_{r_+} f/4\pi$ is the temperature, $S = (\Omega_{d-2}/4)r_+^{d-2}$ is the entropy, and $\Phi = \sqrt{(d-2)/2(d-3)}(q/r_+^{d-3})$ is the electric potential. If we introduce a new "charge" as in [58, 59]

$$\epsilon = \Omega_{d-2}k^{(d-2)/2}, \tag{8}$$

then its conjugate potential is obtained as $\omega = (1/8\pi)k^{(4-d)/2}r_+^{d-3}$. If we define the black hole volume as $V = (\Omega_{d-2}/(d-1))r_+^{d-1}$, then its conjugate pressure is naturally arisen as $P = (d-1)(d-2)/16\pi l^2$, and the black hole mass is naturally viewed as enthalpy instead of energy. Finally, the extended first law is obtained as

$$dM = TdS + \omega d\epsilon + VdP + \Phi dQ. \quad (9)$$

3. The Specific Heat and Phase Transition

Hereafter, the investigation will be limited in $d = 4$ dimensional space-time and in canonical ensemble with fixed electric charge and topological charge, leaving other situations for further study. First of all, we would like to analyse the behavior of the specific heat and the related possible phase transition phenomena. The first law can be rewritten in terms of energy $E = M - PV$,

$$dE = TdS - PdV. \quad (10)$$

So the isobaric specific heat can be written as

$$\begin{aligned} C_{P,Q,\epsilon} &= T\left(\frac{\partial S}{\partial T}\right)_{P,Q,\epsilon} \\ &= \frac{2\pi r_+^2\left(32\pi^2 P r_+^4 + \epsilon r_+^2 - 4\pi Q^2\right)}{32\pi^2 P r_+^4 - \epsilon r_+^2 + 12\pi Q^2}. \end{aligned} \quad (11)$$

Since we are in canonical ensemble, $C_{P,Q,\epsilon}$ can be abbreviated as C_P. From the denominator, we can conclude the following:

(1) When $P < \epsilon^2/1536\pi^3 Q^2$, C_P has two diverge points at

$$r_{+(1,2)} = \frac{1}{8\pi}\sqrt{\frac{\epsilon \pm \sqrt{\epsilon^2 - 1536(\pi)^3 Q^2 P}}{P}}, \quad (12)$$

which signals a phase transition.

(2) When $P = P_c = \epsilon^2/1536\pi^3 Q^2$, the two diverge points of C_P merge into one at

$$r_+ = r_c = 2\sqrt{\frac{6\pi}{\epsilon}}Q, \quad (13)$$

which is the phase transition critical point. The critical temperature $T_c = \epsilon^{3/2}/24\sqrt{6}(\pi)^{5/2}Q$.

(3) When $P > \epsilon^2/1536\pi^3 Q^2$, C_P is always larger than zero, so there is no phase transition.

Comparing with the van der Waals equation, the specific volume is defined as [14]

$$v = 2r_+. \quad (14)$$

At the critical point, we obtain an interesting relation

$$\frac{P_c v_c}{T_c} = \frac{3}{8}, \quad (15)$$

which is exactly the same as for the van der Waals fluid and RN-AdS black holes. Note that this number which seems to be universal does not depend on the topological charge or electric charge.

All the physical quantities can be rescaled by those at the critical point. Defining

$$\begin{aligned} r_+ &= \tilde{r} r_c, \\ P &= \tilde{P} P_c, \end{aligned} \quad (16)$$

the isobaric specific heat becomes

$$C_P = \frac{16\pi^2 Q^2}{\epsilon}\frac{\tilde{r}^2\left(3\tilde{P}\tilde{r}^4 + 6\tilde{r}^2 - 1\right)}{\tilde{P}\tilde{r}^4 - 2\tilde{r}^2 + 1} \equiv \frac{16\pi^2 Q^2}{\epsilon}\widetilde{C}_{\tilde{P}} \quad (17)$$

The behaviors of the rescaled specific heat $\widetilde{C}_{\tilde{P}}$ for the cases $P < P_c$, $P = P_c$, and $P > P_c$ are shown in Figure 1. The curve of specific heat for $P < P_c$ has two divergent points which divide the region into three parts. Both the large radius region and the small radius region are thermodynamically stable with positive specific heat, while the medium radius region is unstable with negative specific heat. So there is a phase transition which takes place between small black hole and large black hole. The curve of specific heat for $P = P_c$ has only one divergent point and always positive denoting that ϵ_c is exactly the critical point. The curve of specific heat for $P > P_c$ has no divergent point and is always positive, implying that the black holes are stable and no phase transition will take place. This behavior of specific heat is very similar to that of the liquid-gas var der Waals system.

4. Oscillatory Behavior in T–S and P–V Graphs: Phase Transition Coexistence Curve

In the last section, we have determined the critical point and found a phase transition when $P \leq P_c$. In this section and in the next section, we will derive the analytical phase transition coexistence curve by using different methods.

4.1. Maxwell Equal Area Law in $T - S$ Graph and Phase Transition Coexistence Curve. The temperature and entropy are

$$\begin{aligned} T &= \frac{f'(r_+)}{4\pi} = \frac{1}{4\pi r_+}\left(k - \frac{q^2}{r_+^2} + \frac{3r_+^2}{l^2}\right) \\ &= \frac{\epsilon}{16\pi^2 r_+} - \frac{Q^2}{4\pi r_+^3} + 2Pr_+, \\ S &= \pi r_+^2, \end{aligned} \quad (18)$$

which can be rescaled by (16) to be

$$\begin{aligned} T &= \frac{3\tilde{P}\tilde{r}^4 + 6\tilde{r}^2 - 1}{8\tilde{r}^3}\frac{\epsilon^{3/2}}{24\sqrt{6}\pi^{5/2}Q} = \tilde{T}T_c, \\ S &= \tilde{r}^2\frac{24\pi^2 Q^2}{\epsilon} = \tilde{S}S_c, \end{aligned} \quad (19)$$

here $S_c \equiv \pi r_c^2 = 24\pi^2 Q^2/\epsilon$. Thus we obtain

$$\tilde{T} = \frac{3\tilde{P}\tilde{S}^2 + 6\tilde{S} - 1}{8\tilde{S}^{3/2}}. \quad (20)$$

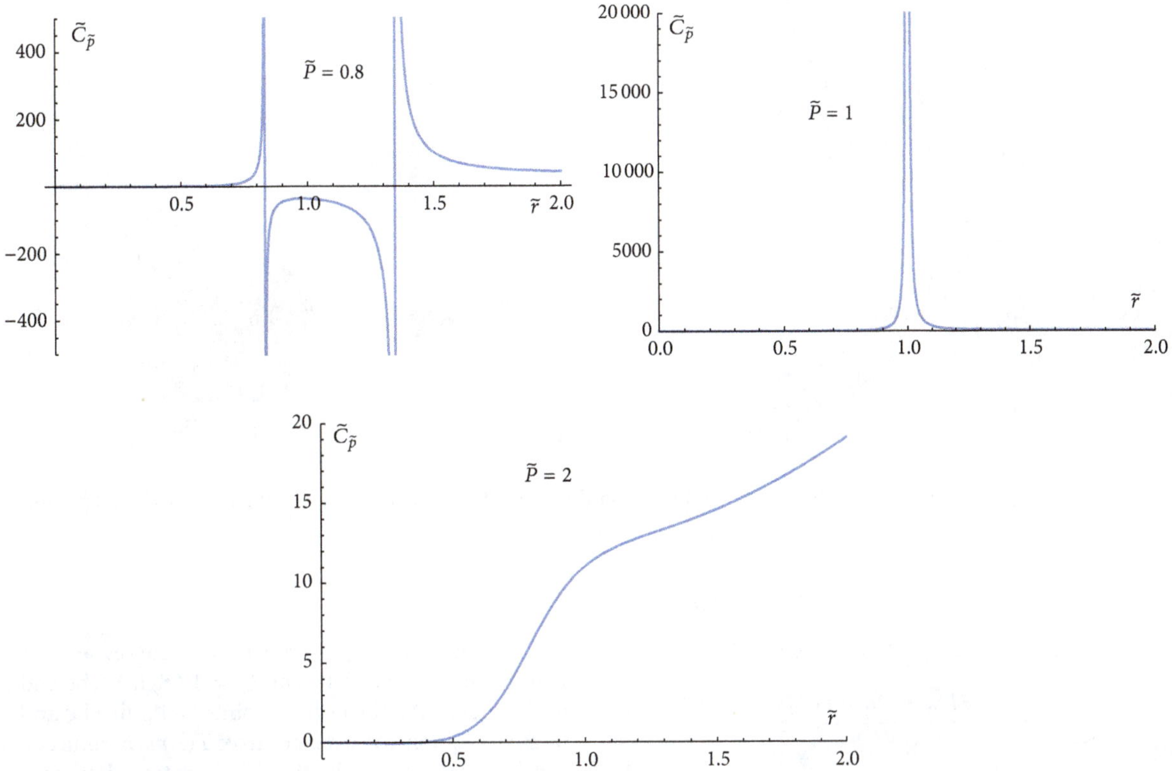

FIGURE 1: The specific heat $\widetilde{C}_{\widetilde{P}}$ vs. $\tilde{r}$ for $\widetilde{P} = 0.8 < \widetilde{P}_c$ which has two divergent points,$\widetilde{P} = 1 = \widetilde{P}_c$ which has only one divergent point and $\widetilde{P} = 2 > \widetilde{P}_c$ which has no divergent point.

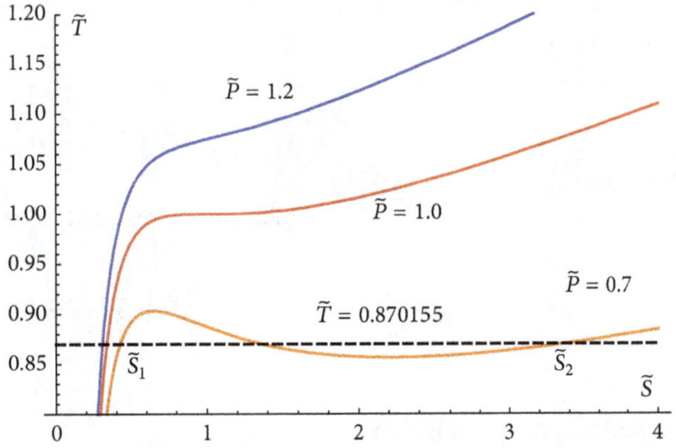

FIGURE 2: $\widetilde{T}$ vs. $\widetilde{S}$ for $\widetilde{P} = 0.7, 1.0, 1.2$. The phase transition take place for $\widetilde{P} \le 1.0$. The dashed black line $\widetilde{T} = 0.870155$ equally separates the oscillatory part. According to Maxwell's equal area law, the phase transition point is ($\widetilde{T} = 0.870155, \widetilde{P} = 0.7$).

From the above equation, we can plot the curve $\widetilde{T}(\widetilde{S})$ for different $\widetilde{P}$ in Figure 2. One can see that, for pressure $\widetilde{P} \le 1.0$, temperature $\widetilde{T}(\widetilde{S})$ curves show interesting var der Waals system's oscillatory behavior which denote the existence of phase transition. The oscillatory part needs to be replaced by an isobar (denote as $\widetilde{T}^*$) such that the areas above and below it are equal to each other. This treatment obeys Maxwell's equal area law. In what follows, we will analytically determine this isobar $\widetilde{T}^*$ for different $\widetilde{P}$.

Maxwell's equal area law is manifest as

$$\begin{aligned}\widetilde{T}^*\left(\widetilde{S}_2 - \widetilde{S}_1\right) &= \int_{\widetilde{S}_1}^{\widetilde{S}_2} \widetilde{T}\left(\widetilde{S}, \widetilde{P}\right) d\widetilde{S} \\ &= \frac{\widetilde{P}}{4}\left(\widetilde{S}_2^{3/2} - \widetilde{S}_1^{3/2}\right) + \frac{3}{2}\left(\widetilde{S}_2^{1/2} - \widetilde{S}_1^{1/2}\right) \\ &\quad + \frac{1}{4}\left(\widetilde{S}_2^{-1/2} - \widetilde{S}_1^{-1/2}\right).\end{aligned} \tag{21}$$

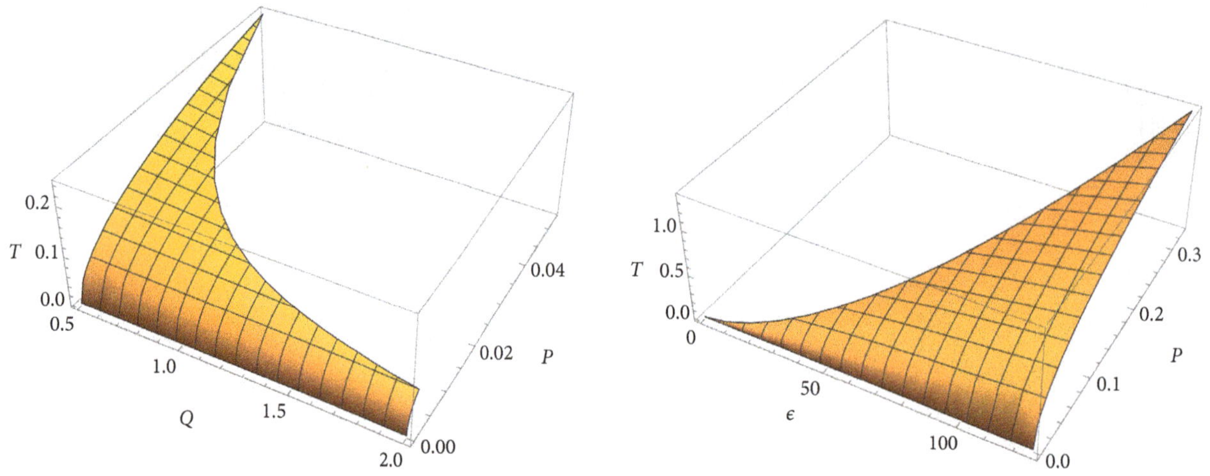

FIGURE 3: The phase transition coexistence curves for $\epsilon = 8\pi$ (left) and $Q = 1$ (right). The end points are the critical points.

At points $(\widetilde{S}_1, \widetilde{T}^*)$ and $(\widetilde{S}_2, \widetilde{T}^*)$, we have two equations

$$\widetilde{T}^* = \frac{3\widetilde{P}\widetilde{S}_1^2 + 6\widetilde{S}_1 - 1}{8\widetilde{S}_1^{3/2}},$$
$$\widetilde{T}^* = \frac{3\widetilde{P}\widetilde{S}_2^2 + 6\widetilde{S}_2 - 1}{8\widetilde{S}_2^{3/2}}. \tag{22}$$

The above three equations can be solved as

$$\widetilde{S}_1 = \frac{\left(\sqrt{3 - \sqrt{\widetilde{P}}} - \sqrt{3 - 3\sqrt{\widetilde{P}}}\right)^2}{2\widetilde{P}},$$
$$\widetilde{S}_2 = \frac{\left(\sqrt{3 - \sqrt{\widetilde{P}}} + \sqrt{3 - 3\sqrt{\widetilde{P}}}\right)^2}{2\widetilde{P}}, \tag{23}$$
$$\widetilde{T}^* = \sqrt{\frac{\widetilde{P}\left(3 - \sqrt{\widetilde{P}}\right)}{2}}.$$

The last equation $\widetilde{T}^*(\widetilde{P})$ is the rescaled phase transition coexistence curve. Then we can make a backward rescale to obtain the phase transition coexistence curve,

$$T = \frac{\sqrt{2P\left(3\epsilon - 16\sqrt{6}\pi^{3/2}Q\sqrt{P}\right)}}{3\pi}. \tag{24}$$

Note that the phase diagram is four dimensional (T, P, Q, ϵ). The condition for the phase transition is that $\epsilon > 16\sqrt{6}\pi^{3/2}Q\sqrt{P}/3$. When the topological charge $\epsilon = 0$, there will be no phase transition. While when the electric charge $Q \longrightarrow 0$ $(Q > 0)$, there will be phase transition with the critical temperature $T_c \longrightarrow \infty$ and pressure $P_c \longrightarrow \infty$. The detailed dependence of the phase transition on the topological charge can be seen in (24) and in right graph of Figure 3.

The phase transition coexistence curves are plotted in Figure 3 for $\epsilon = 8\pi$ (left) and $Q = 1$ (right). The end points in the graphs are the critical points. With fixed ϵ and Q, the phase transition coexistence curve $T(P)$ is reminiscent of the var der Waals system's liquid-gas phase transition coexistence curve.

4.2. Maxwell Equal Area Law in P − V Graph and Phase Transition Coexistence Curve. The pressure and volume are

$$P = \frac{T}{2r_+} - \frac{\epsilon}{32\pi^2 r_+^2} + \frac{Q^2}{8\pi r_+^4},$$
$$V = \frac{4}{3}\pi r_+^3, \tag{25}$$

which can be rescaled to be

$$P = \left(\frac{8\widetilde{T}}{3\widetilde{r}} - \frac{2}{\widetilde{r}^2} + \frac{1}{3\widetilde{r}^4}\right)\frac{\epsilon^2}{1536\pi^3 Q^2} = \widetilde{P}P_c,$$
$$V = \widetilde{r}^3 64\sqrt{6}\pi^{5/2}\left(\frac{Q^2}{\epsilon}\right)^{3/2} = \widetilde{V}V_c. \tag{26}$$

Thus we obtain

$$\widetilde{P} = \frac{8\widetilde{T}}{3}\widetilde{V}^{-1/3} - 2\widetilde{V}^{-2/3} + \frac{1}{3}\widetilde{V}^{-4/3}. \tag{27}$$

From the above equation, we can plot the curve $\widetilde{P}(\widetilde{V})$ for different $\widetilde{T}$ in Figure 4. One can see that for temperature $\widetilde{T} \le 1.0$, pressure $\widetilde{P}(\widetilde{V})$ curves show interesting var der Waals system's oscillatory behavior which corresponds to the phase transition. Similarly, the oscillatory part needs to be replaced by an isobar (denote as $\widetilde{P}^*$) such that the areas above and below it are equal to each other. This treatment follows Maxwell's equal area law. The analytical phase transition curve is derived as follows.

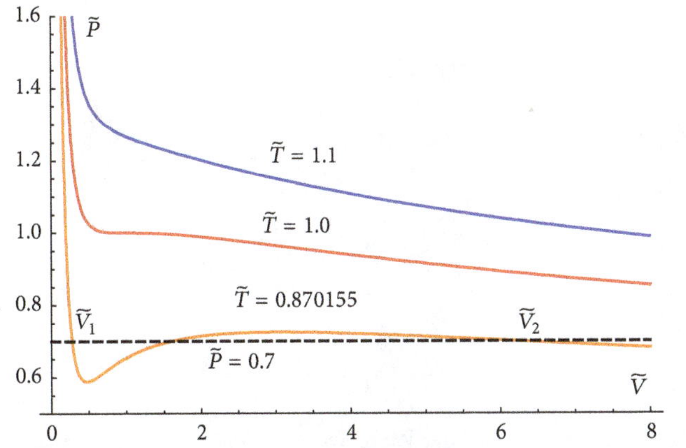

FIGURE 4: $\widetilde{P}$ vs. $\widetilde{V}$ for $\widetilde{T} = 0.870155, 1.0, 1.1$. The phase transition takes place for $\widetilde{T} \leq 1.0$. The dashed black line $\widetilde{P} = 0.7$ equally separates the oscillatory part. According to Maxwell's equal area law, the phase transition point is $(\widetilde{T} = 0.870155, \widetilde{P} = 0.7)$.

Maxwell's equal area law is manifest as

$$\begin{aligned}\widetilde{P}^*\left(\widetilde{V}_2 - \widetilde{V}_1\right) &= \int_{\widetilde{V}_1}^{\widetilde{V}_2} \widetilde{P}\left(\widetilde{V}, \widetilde{T}\right) d\widetilde{V} \\ &= 4\widetilde{T}\left(\widetilde{V}_2^{2/3} - \widetilde{V}_1^{2/3}\right) - 6\left(\widetilde{V}_2^{1/3} - \widetilde{V}_1^{1/3}\right) \\ &\quad - \frac{1}{3}\left(\widetilde{V}_2^{-1/3} - \widetilde{V}_1^{-1/3}\right).\end{aligned} \tag{28}$$

At points $(\widetilde{V}_1, \widetilde{P}^*)$ and $(\widetilde{V}_2, \widetilde{P}^*)$, we have two equations

$$\begin{aligned}\widetilde{P}^* &= \frac{8\widetilde{T}}{3}\widetilde{V}_1^{-1/3} - 2\widetilde{V}_1^{-2/3} + \frac{1}{3}\widetilde{V}_1^{-4/3}, \\ \widetilde{P}^* &= \frac{8\widetilde{T}}{3}\widetilde{V}_2^{-1/3} - 2\widetilde{V}_2^{-2/3} + \frac{1}{3}\widetilde{V}_2^{-4/3}.\end{aligned} \tag{29}$$

The above three equations can be solved as

$$\begin{aligned}\widetilde{V}_1 &= \left(\frac{2\cos^2\varphi}{\widetilde{T}} - \sqrt{\frac{4\cos^4\varphi}{\widetilde{T}^2} - \frac{\sqrt{2}\cos\varphi}{\widetilde{T}}}\right)^3, \\ \widetilde{V}_2 &= \left(\frac{2\cos^2\varphi}{\widetilde{T}} + \sqrt{\frac{4\cos^4\varphi}{\widetilde{T}^2} - \frac{\sqrt{2}\cos\varphi}{\widetilde{T}}}\right)^3, \\ \varphi &= \frac{\pi - \theta}{3}, \\ \cos\theta &= \frac{\sqrt{2}}{2}\widetilde{T} \\ \widetilde{P}^* &= \left(4\cos\frac{\theta}{3}\cos\frac{\pi+\theta}{3}\right)^2 \\ &= \left(1 - 2\cos\frac{\arccos\left(1 - \widetilde{T}^2\right) + \pi}{3}\right)^2.\end{aligned} \tag{30}$$

The last equation $\widetilde{P}^*(\widetilde{T})$ is the rescaled phase transition coexistence curve, and it can be rewritten as

$$\widetilde{T} = \sqrt{\frac{\widetilde{P}^*\left(3 - \sqrt{\widetilde{P}^*}\right)}{2}}, \tag{31}$$

which is exactly the same as (23). So the phase transition coexistence curves obtained by applying the Maxwell equal area law in $T-S$ graph and in $P-V$ graph are consistent with each other.

5. Swallow Tail Behavior in G–T and G–P Graphs: Phase Transition Coexistence Curve

In Section 2, we see that the black hole mass can be interpreted as enthalpy. Thus the Gibbs free energy is

$$G = M - TS, \tag{32}$$

and its differential form in canonical ensemble can be obtained from (9)

$$dG = -SdT + \omega d\epsilon + VdP + \Phi dQ = -SdT + VdP, \tag{33}$$

which denotes that the Gibbs free energy is a function of temperature and pressure.

Substituting black hole mass, temperature, and entropy into (32), then making a rescaling by the quantities at the critical point, we will obtain

$$\begin{aligned}G &= \frac{\epsilon}{16\pi}r_+ - \frac{2\pi}{3}Pr_+^3 + \frac{3Q^2}{4r_+} = \frac{3 + 6\widetilde{r}^2 - \widetilde{P}\widetilde{r}^4}{8\widetilde{r}}\sqrt{\frac{\epsilon Q^2}{6\pi}} \\ &= \widetilde{G}G_c, \\ T &= \frac{\epsilon}{16\pi^2 r_+} - \frac{Q^2}{4\pi r_+^3} + 2Pr_+ \\ &= \frac{3\widetilde{P}\widetilde{r}^4 + 6\widetilde{r}^2 - 1}{8\widetilde{r}^3}\frac{\epsilon^{3/2}}{24\sqrt{6}\pi^{5/2}Q} = \widetilde{T}T_c\end{aligned} \tag{34}$$

From the above equation, we can plot $\widetilde{G}-\widetilde{T}$ curves for different $\widetilde{P}$ and plot $\widetilde{G}-\widetilde{P}$ curves for different $\widetilde{T}$ in Figure 5. One can see that both panels display var der Waals system's swallow tail behavior when $\widetilde{P} \leq 1.0$ and $\widetilde{T} \leq 1.0$. Note that $\widetilde{G}(\widetilde{T})$ does not depend on the topological charge or the electric charge. But by applying a backward rescale, one can find that the swallow tail curves of $G(T)$ in (34) for different ϵ and Q have conformal symmetry. Since phase transition takes place where the system's two phases have equal Gibbs free energy, temperature, and pressure, the swallow tail's intersection point is exactly the phase transition point. As a result, left panel and right panel have a duality relation and they are equal to each other. Then we will only analyse $\widetilde{G}-\widetilde{T}$ graph to derive the phase transition coexistence curve as follows.

In $\widetilde{G}-\widetilde{T}$ graph, at the phase transition point$(\widetilde{T}^*, \widetilde{G}^*)$ for fixed $\widetilde{P}$, we assume that the black hole radius is $\widetilde{r}_1$ for one

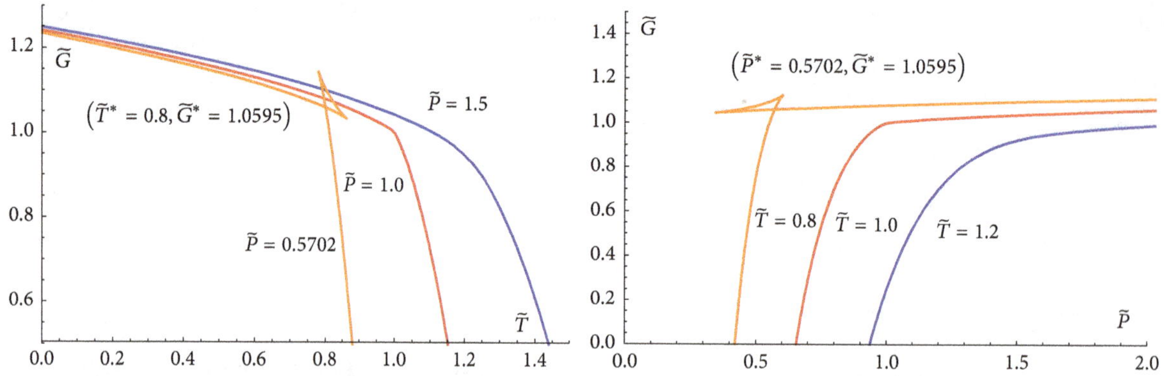

FIGURE 5: Left panel shows $\tilde{G}$ vs. $\tilde{T}$ for different $\tilde{P} = 0.5702, 1.0, 1.5$. When $\tilde{P} \leq 1.0$, there is var der Waals system's swallow tail behavior, and the cross point is the phase transition point. Right panel shows $\tilde{G}$ vs. $\tilde{P}$ for different $\tilde{T} = 0.8, 1.0, 1.2$. When $\tilde{T} \leq 1.0$, there is also var der Waals system's swallow tail behavior, and the cross point is the phase transition point.

phase and $\tilde{r}_2 > \tilde{r}_1$ for the other phase. Thus we will have the following equations from (34):

$$\tilde{G}^* = \frac{3 + 6\tilde{r}_1^2 - \tilde{P}\tilde{r}_1^4}{8\tilde{r}_1} = \frac{3 + 6\tilde{r}_2^2 - \tilde{P}\tilde{r}_2^4}{8\tilde{r}_2},$$
$$\tilde{T}^* = \frac{3\tilde{P}\tilde{r}_1^4 + 6\tilde{r}_1^2 - 1}{8\tilde{r}_1^3} = \frac{3\tilde{P}\tilde{r}_2^4 + 6\tilde{r}_2^2 - 1}{8\tilde{r}_2^3}. \quad (35)$$

The above equations can be solved as

$$\tilde{r}_1 = \frac{\sqrt{3 - \sqrt{\tilde{P}}} - \sqrt{3 - 3\sqrt{\tilde{P}}}}{\sqrt{2\tilde{P}}},$$
$$\tilde{r}_2 = \frac{\sqrt{3 - \sqrt{\tilde{P}}} + \sqrt{3 - 3\sqrt{\tilde{P}}}}{\sqrt{2\tilde{P}}},$$
$$\tilde{G}^* = \frac{\sqrt{6 - 2\sqrt{\tilde{P}}}}{2}, \quad (36)$$
$$\tilde{T}^* = \sqrt{\frac{\tilde{P}\left(3 - \sqrt{\tilde{P}}\right)}{2}}.$$

The last equation is exactly the same as (23). So the phase transition coexistence curves obtained by analysing the Gibbs free energy in $G - T$ graph and $G - P$ graph, or by applying the Maxwell equal area law in $T - S$ graph and $P - V$ graph, are consistent with each other.

6. Conclusion and Discussion

Treating the cosmological constant as a variable [13, 14] and the spatial curvature as topological charge [58, 59], thermodynamics of electrically charged Reissner-Nordström AdS black holes are investigated. Firstly, by variation of the equipotential equation on horizon, the extended thermodynamic first law is obtained. From the extended first law, a conjugate potential correspondent to the topological charge is arisen. Meanwhile, if the black hole volume is defined as $V = (\Omega_{d-2}/(d-1))r_+^{d-1}$, then its conjugate pressure is naturally assigned as the cosmological constant and the black hole mass as enthalpy.

Secondly, in four dimensional space-time and canonical ensemble with fixed electric charge and topological charge, the isobaric specific heat C_P is calculated and the corresponding divergent solutions are derived. The two solutions merge into one at the critical point with $P_c = \epsilon^2/(1536\pi^3 Q^2), r_c = 2\sqrt{6\pi/\epsilon}Q$. When $P < P_c$, the curve of specific heat has two divergent points and is divided into three regions. The specific heat is positive for both the large radius region and the small radius region which are thermodynamically stable, while it is negative for the medium radius region which is unstable. When $P > P_c$, the specific heat is always positive implying the black holes are stable and no phase transition will take place.

Thirdly, rescaling the quantities by those at the critical point, the behavior of temperature in $\tilde{T} - \tilde{S}$ graph and the behavior of pressure in $\tilde{P} - \tilde{V}$ graph are studied. They exhibit the interesting van de Waals gas-liquid system's behavior. When $\tilde{P} > 1, \tilde{T} > 1$, the curves vary monotonically and no phase transition will take place. When $\tilde{P} < 1, \tilde{T} < 1$, the curves display an oscillatory behavior which signals phase transition. The oscillatory part is replaced by an isobar according to the Maxwell's equal area law and the analytical phase transition coexistence curves (rescaled) are obtained which are consistent with each other. Then by making a backward rescale, the explicit phase transition coexistence curve is derive in (24) and the phase diagrams are shown in Figure 3.

Fourthly, van der Waals system's swallow tail behavior is observed in the $\tilde{G}-\tilde{T}$ graph and $\tilde{G}-\tilde{P}$ graph when $\tilde{P} < 1, \tilde{T} < 1$. The swallow tail's intersection point is the phase transition point. By analytically solving the constraint equations, the rescaled phase transition coexistence curve is obtained which is consistent with those derived in $\tilde{T}-\tilde{S}$ graph and $\tilde{P}-\tilde{V}$ graph.

From the above detailed study in canonical ensemble, the analogy of RN-AdS black hole as van der Waals system has been examined when the spatial curvature is treated as topological charge and the cosmological constant is treated as pressure. Both the systems share the same oscillatory behavior and swallow tail behavior. Comparing with the case when the spatial curvature is fixed [14, 15], our phase transition diagram in Figure 3 is four dimensional (T, P, Q, ϵ), which is more rich with an extra parameter, the topological charge.

A further investigation in grand canonical ensemble is outside the scope of this paper, but it is surely a very interesting direction for future research. The influence of this topological charge on black hole thermodynamics in other gravity theories (such as the Lovelock, Gauss-Bonnet theory, and f(R) theory) and different dimensional space-time also deserves to be disclosed in the future research. Another interesting future research line is the comparison of the influence between the electric charge and the topological charge.

In the end, we would like to point out that the black hole thermodynamics discussed in this paper is based on the first law derived from the equipotential surface $f(r) = c$ with $c = 0$. A more conventional way to compute thermodynamics quantities is the Euclidean formalism, where the free energy is computed firstly and then the other quantities follow from it. The difference between these two formalisms remains unknown which deserves to be investigated in future.

Conflicts of Interest

The author declares that they have no conflicts of interest.

Acknowledgments

This research is supported by Department of Education of Guangdong Province, China (Grant no. 2017KQNCX124).

References

[1] J. D. Bekenstein, "Black holes and entropy," *Physical Review D: Particles, Fields, Gravitation and Cosmology*, vol. 7, pp. 2333–2346, 1973.

[2] J. M. Bardeen, B. Carter, and S. W. Hawking, "The four laws of black hole mechanics," *Communications in Mathematical Physics*, vol. 31, pp. 161–170, 1973.

[3] S. W. Hawking, "Particle creation by black holes," *Communications in Mathematical Physics*, vol. 43, no. 3, pp. 199–220, 1975.

[4] S. W. Hawking and D. N. Page, "Thermodynamics of black holes in anti-de Sitter space," *Communications in Mathematical Physics*, vol. 87, no. 4, pp. 577–588, 1982/83.

[5] J. Maldacena, "The large N limit of superconformal field theories and supergravity," *Advances in Theoretical and Mathematical Physics*, vol. 2, no. 2, pp. 231–252, 1998.

[6] S. S. Gubser, I. R. Klebanov, and A. M. Polyakov, "Gauge theory correlators from non-critical string theory," *Physics Letters B*, vol. 428, no. 1-2, pp. 105–114, 1998.

[7] E. Witten, "Anti de Sitter space and holography," *Advances in Theoretical and Mathematical Physics*, vol. 2, no. 2, pp. 253–291, 1998.

[8] E. Witten, "Anti-de Sitter space, thermal phase transition, and confinement in gauge theories," *Advances in Theoretical and Mathematical Physics*, vol. 2, no. 3, pp. 505–532, 1998.

[9] A. Chamblin, R. Emparan, C. V. Johnson, and R. C. Myers, "Charged AdS black holes and catastrophic holography," *Physical Review D: Particles, Fields, Gravitation and Cosmology*, vol. 60, no. 6, 1999.

[10] R. Banerjee and D. Roychowdhury, "Thermodynamics of phase transition in higher dimensional AdS black holes," *Journal of High Energy Physics*, vol. 2011, article 4, 2011.

[11] R. Banerjee, S. K. Modak, and D. Roychowdhury, "A unified picture of phase transition: From liquid-vapour systems to AdS black holes," *Journal of High Energy Physics*, vol. 2012, no. 10, 2012.

[12] A. Chamblin, R. Emparan, C. V. Johnson, and R. C. Myers, "Holography, thermodynamics, and fluctuations of charged AdS black holes," *Physical Review D: Particles, Fields, Gravitation and Cosmology*, vol. 60, no. 10, Article ID 104026, 1999.

[13] D. Kastor, S. Ray, and J. Traschen, "Enthalpy and the mechanics of AdS black holes," *Classical and Quantum Gravity*, vol. 26, no. 19, Article ID 195011, 2009.

[14] D. Kubiznak and R. B. Mann, "P-V criticality of charged AdS black holes," *Journal of High Energy Physics*, vol. 2012, no. 7, article 033, 2012.

[15] E. Spallucci and A. Smailagic, "Maxwell's equal-area law for charged Anti-de Sitter black holes," *Physics Letters B*, vol. 723, no. 4-5, pp. 436–441, 2013.

[16] S. Gunasekaran, D. Kubiznak, and R. B. Mann, "Extended phase space thermodynamics for charged and rotating black holes and Born-Infeld vacuum polarization," *Journal of High Energy Physics*, vol. 2012, no. 11, article no. 110, 2012.

[17] S. H. Hendi and M. H. Vahidinia, "Extended phase space thermodynamics and $P-V$ criticality of black holes with a nonlinear source," *Physical Review D: Particles, Fields, Gravitation and Cosmology*, vol. 88, no. 8, Article ID 084045, pp. 1–11, 2013.

[18] S. Chen, X. Liu, C. Liu, and J. Jing, "*P-V* criticality of AdS black hole in $f(R)$ gravity," *Chinese Physics Letters*, vol. 30, no. 6, Article ID 060401, 2013.

[19] R. Zhao, H. H. Zhao, M. S. Ma, and L. C. Zhang, "On the critical phenomena and thermodynamics of charged topological dilaton AdS black holes," *The European Physical Journal C*, vol. 73, no. 12, pp. 1–10, 2013.

[20] N. Altamirano, D. Kubiznak, and R. Mann, "Reentrant phase transitions in rotating AdS black holes," *Physical Review D: Particles, Fields, Gravitation and Cosmology*, vol. 88, no. 10, Article ID 101502, 2013.

[21] R.-G. Cai, L.-M. Cao, L. Li, and R.-Q. Yang, "P-V criticality in the extended phase space of Gauss-Bonnet black holes in AdS space," *Journal of High Energy Physics*, vol. 2013, no. 9, article no. 5, 2013.

[22] N. Altamirano, D. Kubizňák, R. B. Mann, and Z. Sherkatghanad, "Kerr-AdS analogue of triple point and solid/liquid/gas phase transition," *Classical and Quantum Gravity*, vol. 31, no. 4, Article ID 042001, 2014.

[23] W. Xu, H. Xu, and L. Zhao, "Gauss-bonnet coupling constant as a free thermodynamical variable and the associated criticality," *The European Physical Journal C*, vol. 74, no. 7, article no. 2970, 2014.

[24] J.-X. Mo and W.-B. Liu, "Ehrenfest scheme for *P-V* criticality in the extended phase space of black holes," *Physics Letters B*, vol. 727, no. 1–3, pp. 336–339, 2013.

[25] D.-C. Zou, S.-J. Zhang, and B. Wang, "Critical behavior of Born-Infeld AdS black holes in the extended phase space thermodynamics," *Physical Review D: Particles, Fields, Gravitation and Cosmology*, vol. 89, Article ID 044002, 2014.

[26] J.-X. Mo and W.-B. Liu, "*P-V* criticality of topological black holes in Lovelock-Born-Infeld gravity," *The European Physical Journal C*, vol. 74, article 2836, 2014.

[27] N. Altamirano, D. Kubizňák, R. Mann, and Z. Sherkatghanad, "Thermodynamics of rotating black holes and black rings: phase transitions and thermodynamic volume," *Galaxies*, vol. 2, no. 1, pp. 89–159, 2014.

[28] S.-W. Wei and Y.-X. Liu, "Critical phenomena and thermodynamic geometry of charged Gauss-Bonnet AdS black holes," *Physical Review D: Particles, Fields, Gravitation and Cosmology*, vol. 87, Article ID 044014, 2013.

[29] S. W. Wei and Y. X. Liu, "Triple points and phase diagrams in the extended phase space of charged Gauss-Bonnet black holes in AdS space," *Physical Review D*, vol. 90, Article ID 044057, 2014.

[30] L.-C. Zhang, M.-S. Ma, H.-H. Zhao, and R. Zhao, "Thermodynamics of phase transition in higher-dimensional Reissner-Nordström-de Sitter black hole," *The European Physical Journal C*, vol. 74, article no. 3052, 2014.

[31] J.-X. Mo and W.-B. Liu, "Ehrenfest scheme for P–V criticality of higher dimensional charged black holes, rotating black holes, and Gauss-Bonnet AdS black holes," *Physical Review D: Particles, Fields, Gravitation and Cosmology*, vol. 89, no. 8, Article ID 084057, 2014.

[32] D. C. Zou, Y. Liu, and B. Wang, "Critical behavior of charged Gauss-Bonnet-AdS black holes in the grand canonical ensemble," *Physical Review D: Particles, Fields, Gravitation and Cosmology*, vol. 90, no. 4, Article ID 044063, 2014.

[33] Z. Zhao and J. Jing, "Ehrenfest scheme for complex thermodynamic systems in full phase space," *Journal of High Energy Physics*, vol. 2014, no. 11, article no. 37, 2014.

[34] R. Zhao, M. Ma, H. Zhao, and L. Zhang, "The critical phenomena and thermodynamics of the reissner-nordstrom-de sitter black hole," *Advances in High Energy Physics*, vol. 2014, Article ID 124854, 6 pages, 2014.

[35] H. Xu, W. Xu, and L. Zhao, "Extended phase space thermodynamics for third-order Lovelock black holes in diverse dimensions," *The European Physical Journal C*, vol. 74, article 3074, 2014.

[36] A. M. Frassino, D. Kubizňák, R. B. Mann, and F. Simovic, "Multiple reentrant phase transitions and triple points in Lovelock thermodynamics," *Journal of High Energy Physics*, vol. 2014, no. 9, article no. 80, 2014.

[37] J. L. Zhang, R. G. Cai, and H. Yu, "Phase transition and thermodynamical geometry for Schwarzschild AdS black hole in $AdS_5 \times S_5$ spacetime," *Journal of High Energy Physics*, vol. 2015, no. 2, article no. 143, 2015.

[38] B. Mirza and Z. Sherkatghanad, "Phase transitions of hairy black holes in massive gravity and thermodynamic behavior of charged AdS black holes in an extended phase space," *Physical Review D: Particles, Fields, Gravitation and Cosmology*, vol. 90, no. 8, article no. 084006, 2014.

[39] B. P. Dolan, A. Kostouki, D. Kubizňák, and R. B. Mann, "Isolated critical point from Lovelock gravity," *Classical and Quantum Gravity*, vol. 31, no. 24, Article ID 242001, 2014.

[40] A. Rajagopal, D. Kubiznak, and R. B. Mann, "Van der Waals black hole," *Physics Letters B*, vol. 737, pp. 277–279, 2014.

[41] Y. Liu, D.-C. Zou, and B. Wang, "Signature of the Van der Waals like small-large charged AdS black hole phase transition in quasinormal modes," *Journal of High Energy Physics*, vol. 2014, no. 09, article no. 179, 2014.

[42] S. H. Hendi, R. B. Mann, S. Panahiyan, and B. Eslam Panah, "Van der Waals like behavior of topological AdS black holes in massive gravity," *Physical Review D: Particles, Fields, Gravitation and Cosmology*, vol. 95, no. 2, Article ID 021501, 2017.

[43] S. H. Hendi, B. Eslam Panah, and S. Panahiyan, "Topological charged black holes in massive gravity's rainbow and their thermodynamical analysis through various approaches," *Physics Letters B*, vol. 769, pp. 191–201, 2017.

[44] S.-Q. Lan, J.-X. Mo, and W.-B. Liu, "A note on Maxwell's equal area law for black hole phase transition," *The European Physical Journal C*, vol. 75, no. 9, article no. 419, 2015.

[45] M.-S. Ma and R.-H. Wang, "Peculiar P-V criticality of topological Horava-Lifshitz black holes," *Physical Review D: Particles, Fields, Gravitation and Cosmology*, vol. 96, no. 2, article no. 024052, 2017.

[46] S. W. Wei, B. Liang, and Y. Liu, "Critical phenomena and chemical potential of a charged AdS black hole," *Physical Review D: Particles, Fields, Gravitation and Cosmology*, vol. 96, no. 12, article no. 124018, 2017.

[47] K. Bhattacharya and B. R. Majhi, "Thermogeometric description of the van der Waals like phase transition in AdS black holes," *Physical Review D: Particles, Fields, Gravitation and Cosmology*, vol. 95, no. 10, article no. 104024, 2017.

[48] S. H. Hendi, B. E. Panah, S. Panahiyan, and M. S. Talezadeh, "Geometrical thermodynamics and P–V criticality of the black holes with power-law Maxwell field," *The European Physical Journal C*, vol. 77, no. 2, article no. 133, 2017.

[49] X. M. Kuang and O. Miskovic, "Thermal phase transitions of dimensionally continued AdS black holes," *Physical Review D: Particles, Fields, Gravitation and Cosmology*, vol. 95, no. 4, Article ID 046009, 2017.

[50] S. Fernando, "P-V criticality in AdS black holes of massive gravity," *Physical Review D: Particles, Fields, Gravitation and Cosmology*, vol. 94, no. 12, Article ID 124049, 2016.

[51] B. R. Majhi and S. Samanta, "P-V criticality of AdS black holes in a general framework," *Physics Letters B*, vol. 773, pp. 203–207, 2017.

[52] S. He, L. F. Li, and X. X. Zeng, "Holographic Van der Waals-like phase transition in the Gauss-Bonnet gravity," *Nuclear Physics B*, vol. 915, p. 243, 2017.

[53] J. Sadeghi, B. Pourhassan, and M. Rostami, "P-V criticality of logarithm-corrected dyonic charged AdS black holes," *Physical Review D: Particles, Fields, Gravitation and Cosmology*, vol. 94, no. 6, Article ID 064006, 2016.

[54] X. X. Zeng and L. F. Li, "Van der waals phase transition in the framework of holography," *Physics Letters B*, vol. 764, article no. 100, 2017.

[55] P. H. Nguyen, "An equal area law for holographic entanglement entropy of the AdS-RN black hole," *Journal of High Energy Physics*, vol. 2015, no. 12, article no. 139, 2015.

[56] J. F. Xu, L.-M. Cao, and Y.-P. Hu, "P-V criticality in the extended phase space of black holes in massive gravity," *Physical Review D: Particles, Fields, Gravitation and Cosmology*, vol. 91, no. 12, Article ID 124033, 2015.

[57] Z.-Y. Nie and H. Zeng, "P-T phase diagram of a holographic s+p model from Gauss-Bonnet gravity," *Journal of High Energy Physics*, no. 10, article no. 047, 2015.

[58] Y. Tian, X.-N. Wu, and H. Zhang, "Holographic entropy production," *Journal of High Energy Physics*, no. 10, article no. 170, 2014.

[59] Y. Tian, "The last (lost) charge of a black hole," https://arxiv.org/abs/1804.00249, 2018.

[60] S. Q. Lan, G. Q. Li, J. X. Mo, and X. B. Xu, "New phase transition related to the the black hole's topological charge," https://arxiv.org/abs/1804.06652, 2018.

11

The DKP Oscillator in Spinning Cosmic String Background

Mansoureh Hosseinpour and Hassan Hassanabadi

Faculty of Physics, Shahrood University of Technology, Shahrood, P.O. Box 3619995161-316, Iran

Correspondence should be addressed to Mansoureh Hosseinpour; hosseinpour.mansoureh@gmail.com

Academic Editor: Diego Saez-Chillon Gomez

In this article, we investigate the behaviour of relativistic spin-zero bosons in the space-time generated by a spinning cosmic string. We obtain the generalized beta-matrices in terms of the flat space-time ones and rewrite the covariant form of Duffin-Kemmer-Petiau (DKP) equation in spinning cosmic string space-time. We find the solution of DKP oscillator and determine the energy levels. We also discuss the influence of the topology of the cosmic string on the energy levels and the DKP spinors.

1. Introduction

The Duffin-Kemmer-Petiau (DKP) equation has been used to describe relativistic spin-0 and spin-1 bosons [1–4]. The DKP equation has five- and ten-dimensional representation, respectively, for spin-0 and spin-1 bosons [5]. This equation is compared to the Dirac equation for fermions [6]. The DKP equation has been widely investigated in many areas of physics. The DKP equation has been investigated in the momentum space with the presence of minimal length [7, 8] and for spins 0 and 1 in a noncommutative space [9–12]. Also, the DKP oscillator has been studied in the presence of topological defects [13]. Recently, there has been growing interest in the so-called DKP oscillator [14–23] in particular in the background of a magnetic cosmic string [13]. The cosmic strings and other topological defects can form at a cosmological phase transition [24, 25]. The conical nature of the space-time around the string causes a number of interesting physical effects. Until now, some problems have been investigated in the gravitational fields of topological defects including the one-electron atom problem [26–28]. Spinning cosmic strings are similar usual cosmic string, characterized by an angular parameter α that depends on their linear mass density μ. The DKP oscillator is described by performing the nonminimal coupling with a linear potential. The name distinguishes it from the system called a DKP oscillator with Lorentz tensor couplings of [7–12, 14–16]. The DKP oscillator for spin-0 bosons has been investigated by Guo et al. in [10] in noncommutative phase space. The DKP oscillator with spin-0 has been studied by Yang et al. [11]. Exact solution of DKP oscillator in the momentum space with the presence of minimal length has been analysed in [8]. De Melo et al. construct the Galilean DKP equation for the harmonic oscillator in a noncommutative phase space [29]. Falek and Merad investigated the DKP oscillator of spins 0 and 1 bosons in noncommutative space [9]. Recently, there has been an increasing interest on the DKP oscillator [13–16, 22, 29–31]. The nonrelativistic limit of particle dynamics in curved space-time is considered in [32–36]. Also, the dynamics of relativistic bosons and fermions in curved space-time is considered in [17, 20, 31].

The influence of topological defect in the dynamics of bosons via DKP formalism has not been established for spinning cosmic strings. In this way, we consider the quantum dynamics of scalar bosons via DKP formalism embedded in the background of a spinning cosmic string. We solve DKP equation in presence of the spinning cosmic string space-time whose metric has off diagonal terms which involves time and space. The influence of this topological defect in the energy spectrum and DKP spinor presented graphically.

The structure of this paper is as follows: Section 2 describes the covariant form of DKP equation in a spinning cosmic string background. In Section 3, we introduce the DKP oscillator by performing the nonminimal coupling in this space-time, and we obtain the radial equations that are solved. We plotted the DKP spinor, density of probability, and the energy spectrum for different conditions involving the deficit angle and the oscillator frequency. In the Section 4 we present our conclusions.

2. Covariant Form of the DKP Equation in the Spinning Cosmic String Background

We choose the cosmic string space-time background, where the line element is given by

$$ds^2 = -dT^2 + dX^2 + dY^2 + dZ^2 \quad (1)$$

The space-time generated by a spinning cosmic string without internal structure, which is termed ideal spinning cosmic string, can be obtain by coordinate transformation as

$$\begin{aligned} T &= t + a\alpha^{-1}\varphi \\ X &= r\cos(\varphi) \\ Y &= r\sin(\varphi) \\ \varphi &= \alpha\varphi' \end{aligned} \quad (2)$$

With this transformation, the line element ((1)) becomes [37–43]

$$\begin{aligned} ds^2 &= -(dt + ad\varphi)^2 + dr^2 + \alpha^2 r^2 d\varphi^2 + dz^2 \\ &= -dt^2 + dr^2 - 2adtd\varphi + (\alpha^2 r^2 - a^2)\, d\varphi^2 + dz^2 \end{aligned} \quad (3)$$

with $-\infty < z < \infty$, $\rho \geq 0$, and $0 \leq \varphi \leq 2\pi$. From this point on, we will take $c = 1$. The angular parameter α which runs in the interval (0, 1] is related to the linear mass density μ of the string as $\alpha = 1 - 4\mu$ and corresponds to a deficit angle $\gamma = 2\pi(1-\alpha)$. We take $a = 4Gj$ where G is the universal gravitation constant and j is the angular momentum of the spinning string; thus a is a length that represents the rotation of the cosmic string. Note that, in this case, the source of the gravitational field relative to a spinning cosmic string possesses angular momentum and the metric (1) has an off diagonal term involving time and space.

The DKP equation in the cosmic string space-time (1) reads [13, 17, 31]

$$\left(i\beta^\mu(x)\nabla_\mu - M\right)\Psi(x) = 0. \quad (4)$$

The covariant derivative in (4) is

$$\nabla_\mu = \partial_\mu + \Gamma_\mu(x) \quad (5)$$

where Γ_μ are the spinorial affine connections given by

$$\Gamma_\mu = \frac{1}{2}\omega_{\mu ab}\left[\beta^a, \beta^b\right]. \quad (6)$$

The matrices β^a are the standard Kemmer matrices in Minkowski space-time.

$$\beta^\mu = e^\mu_a \beta^a \quad (7)$$

The Kemmer matrices are an analogous to Dirac matrices in Dirac equation. There has been an increasing interest Dirac equation for spin half particles [44–47]. The matrices β^a satisfies the DKP algebra,

$$\beta^\mu\beta^\nu\beta^\lambda + \beta^\lambda\beta^\nu\beta^\mu = g^{\mu\nu}\beta^\mu + g^{\lambda\nu}\beta^\mu. \quad (8)$$

The conserved four-current is given by

$$J^\mu = \frac{1}{2}\overline{\Psi}\beta^\mu\Psi, \quad (9)$$

and the conservation law for J^μ takes the form

$$\nabla_\mu J^\mu + \frac{i}{2}\overline{\Psi}\left(U - \eta^0 U^\dagger \eta^0\right)\Psi = \frac{1}{2}\overline{\Psi}\left(\nabla_\mu \beta^\mu\right)\Psi \quad (10)$$

The adjoint spinor $\overline{\Psi}$ is defined as $\overline{\Psi} = \Psi^\dagger\eta^0$ with $\eta^0 = 2\beta^0\beta^0 - 1$, in such a way that $(\eta^0\beta^\mu)^\dagger = \eta^0\beta^\mu$. The factor 1/2 which multiplies $\overline{\Psi}\beta^\mu\Psi$ is of no importance for the conservation law and ensures the charge density is compatible with the one used in the Klein-Gordon theory and its nonrelativistic limit. Thus, if U is Hermitian with respect to η^0 and the curved-space beta-matrices are covariantly constant, then the four-current will be conserved if [30]

$$\nabla_\mu \beta^\mu = 0. \quad (11)$$

The algebra expressed by these matrices generates a set of 126 independent matrices whose irreducible representations comprise a trivial representation, a five-dimensional representation describing the spin-zero particles, and a ten-dimensional representation associated with spin-one particles. We choose the 5×5 beta-matrices as follows [31]:

$$\begin{aligned} \beta^0 &= \begin{pmatrix} \theta & 0_{2\times3} \\ 0_{3\times2} & 0_{3\times3} \end{pmatrix}, \\ \vec{\beta} &= \begin{pmatrix} 0_{2\times2} & \vec{\tau} \\ -\vec{\tau}^T & 0_{3\times3} \end{pmatrix}, \\ \theta &= \begin{pmatrix} 0 & 1 \\ 1 & 0 \end{pmatrix}, \\ \tau^1 &= \begin{pmatrix} -1 & 0 & 0 \\ 0 & 0 & 0 \end{pmatrix}, \\ \tau^2 &= \begin{pmatrix} 0 & -1 & 0 \\ 0 & 0 & 0 \end{pmatrix}, \\ \tau^3 &= \begin{pmatrix} 0 & 0 & -1 \\ 0 & 0 & 0 \end{pmatrix}. \end{aligned} \quad (12)$$

In (7), e^μ_a denote the tetrad basis that we can choose as

$$e^\mu_a = \begin{pmatrix} 1 & \dfrac{a\sin(\varphi)}{r\alpha} & \dfrac{-a\cos(\varphi)}{r\alpha} & 0 \\ 0 & \cos(\varphi) & \sin(\varphi) & 0 \\ 0 & \dfrac{-\sin(\varphi)}{r\alpha} & \dfrac{\cos(\varphi)}{r\alpha} & 0 \\ 0 & 0 & 0 & 1 \end{pmatrix}. \quad (13)$$

For the specific tetrad basis given by (13), we find from (7) that the curved-space beta-matrices read

$$\begin{aligned}
\beta^{(0)} &= e^t_a\beta^a = \beta^0 - \frac{a}{r\alpha}\beta^\varphi \\
\beta^{(1)} &= e^r_a\beta^a = \beta^r \\
\beta^r &= \cos\varphi\beta^1 + \sin\varphi\beta^2 \\
\beta^{(2)} &= e^2_a\beta^a = \frac{\beta^\varphi}{r\alpha} \\
\beta^\varphi &= -\sin\varphi\beta^1 + \cos\varphi\beta^2 \\
\beta^{(3)} &= e^z_a\beta^a = \beta^3 = \beta^z
\end{aligned} \tag{14a}$$

and the spin connections are given by

$$\Gamma_\varphi = (1-\alpha)\left[\beta^1,\beta^2\right] \tag{15}$$

We consider only the radial component in the nonminimal substitution. Since the interaction is time-independent, one can write $\Psi(r,t) \propto e^{im\varphi}e^{ik_z z}e^{-iEt}\Phi(r)$, where E is the energy of the scalar boson, m is the magnetic quantum number, and k_z is the wave number. The five-component DKP spinor can be written as $\Phi^T = (\Phi_1,\Phi_2,\Phi_3,\Phi_4,\Phi_5)$, and the DKP equation (4) leads to the five

$$\begin{aligned}
&(r\alpha(-\mathrm{M}\Phi_1(r) + \mathrm{E}\Phi_2(r) + kz\Phi_5(r)) + \cos\varphi\big((a\mathrm{E} \\
&\quad + m)\Phi_4(r) - \mathrm{i}\big((-1+\alpha)\Phi_3(r) + r\alpha\Phi'_3(r)\big)\big) \\
&\quad - \sin\varphi\big((a\mathrm{E}+m)\Phi_3(r) \\
&\quad + \mathrm{i}\big((-1+\alpha)\Phi_4(r) + r\alpha\Phi'_4(r)\big)\big)\big) = 0 \\
&(\mathrm{E}\Phi_1(r) - \mathrm{M}\Phi_2(r)) = 0 \\
&\big((a\mathrm{E}+m)\sin\varphi\Phi_1(r) + r\alpha\big(-\mathrm{M}\Phi_3(r) + \mathrm{i} \\
&\quad \cdot\cos\phi\Phi'_1(r)\big)\big) = 0 \\
&\big(-(a\mathrm{E}+m)\cos\varphi\Phi_1(r) + r\alpha\big(-\mathrm{M}\Phi_4(r) + \mathrm{i} \\
&\quad \cdot\sin\varphi\Phi'_1(r)\big)\big) = 0 \\
&(kz + \mathrm{M}\psi\Phi_5(r)) = 0
\end{aligned} \tag{16}$$

Then we obtain the following equation of motion for the first component Φ_1 of the DKP spinor:

$$\begin{aligned}
&\left(-\frac{(a\mathrm{E}+m)^2}{\alpha^2 r^2} + \mathrm{E}^2 - \mathrm{kz}^2 - \mathrm{M}^2\right)\Phi_1(r) \\
&\quad + \frac{(\alpha-1)\Phi'_1(r)}{\alpha r} + \Phi''_1(r) = 0
\end{aligned} \tag{17}$$

$$\Phi_1 = r^{1/2\alpha}R_{n,\ell}(r). \tag{18}$$

Then (17) changes to

$$\begin{aligned}
&R''_{n,m}(r) + \frac{R'_{n,m}(r)}{r} \\
&\quad + \left(E^2 - k_z{}^2 - M^2 - \frac{1+4(aE+m)^2}{4r^2\alpha^2}\right)R(r) \\
&= 0
\end{aligned} \tag{19}$$

By the change of variable $r = x\eta$, we can write (19) in the form

$$R''_{n,m}(x) + \frac{R'_{n,m}(x)}{x} + \left(1 - \frac{\lambda^2}{x^2}\right)R(x) = 0 \tag{20}$$

where $\lambda = ((1+4(aE+m)^2)/4\alpha^2)^{1/2}$ and $\eta = (E^2 - k_z{}^2 - M^2)^{-1/2}$. The physical solution of (20) is *Bessel J* function. Therefore the general solution to (20) is given by

$$R(r) = A_{\lambda\eta}J_\lambda\left(\frac{r}{\eta}\right) + B_{\lambda\eta}Y_\lambda\left(\frac{r}{\eta}\right) \tag{21}$$

where $Y_\lambda(r/\eta)$ is the Bessel function of the second kind. Sometimes this family of functions is also called Neumann functions or Weber functions. $J_\lambda(r/\eta)$ is the Bessel function of the first kind, given by

$$J_\nu(x) = \sum_{k=0}^{\infty}\frac{(-1)^k}{k!\Gamma(k+\nu+1)}\left(\frac{x}{2}\right)^{\nu+2k} \tag{22}$$

and $Y_\lambda(r/\eta)$ is the Bessel function of the second kind, given by

$$Y_\nu(x) = \frac{J_\nu(x)\cos(\nu\pi) - J_{-\nu}(x)}{\sin(\nu\pi)} \tag{23}$$

By considering the boundary condition for (21) such that $B_{\lambda\eta} = 0$, we find

$$R(r) = A_{\lambda\eta}J_\lambda\left(\frac{r}{\eta}\right) \tag{24}$$

3. The DKP Oscillator in Spinning Cosmic String Background

The DKP oscillator is introduced via the nonminimal substitution [17, 30, 31]

$$\frac{1}{i}\vec{\nabla}_\alpha \longrightarrow \frac{1}{i}\vec{\nabla}_\alpha - iM\omega\eta_0\vec{r} \tag{25}$$

where ω is the oscillator frequency, M is the mass of the boson already found in (4), and $\vec{\nabla}$ is defined in (5). We consider only the radial component in the nonminimal substitution. The DKP equation (4) leads to the five equations:

$$\begin{aligned}&(r\alpha(-M\Phi_1(r)+E\Phi_2(r)+kz\Phi_5(r))+(\cos\varphi(aE\\&+m)\Phi_4(r))+(\mathrm{i}\cos\varphi(\Phi_3(r)\\&+\alpha(-1+r^2M\omega)\Phi_3(r)-r\alpha\Phi'_3(r)))\\&-\sin\varphi((aE+m)\Phi_3(r)\\&-\mathrm{i}(1-\alpha+r^2\alpha M\omega)\Phi_4(r)+\mathrm{i}r\alpha\Phi'_4(r)))=0\\&((aE+m)\sin\varphi\Phi_1(r)-Mr\alpha\Phi_3(r)+\mathrm{i}r\alpha\\&\cdot\cos\varphi(rM\omega\Phi_1(r)+\Phi'_1(r)))=0\\&(-(aE+m)\cos\varphi\Phi_1(r)-Mr\alpha\Phi_4(r)+\mathrm{i}r\alpha\\&\cdot\sin\varphi(rM\omega\Phi_1(r)+\Phi'_1(r)))=0\\&(E\Phi_1(r)-M\Phi_2(r))=0\\&(kz\Phi_1(r)+M\Phi_4(r))=0\end{aligned} \tag{26}$$

By solving the above system of (26) in favour of Φ_1 we get

$$\begin{aligned}\Phi_2(r)&=\frac{E}{M}\Phi_1(r)\\\Phi_5(r)&=-\frac{k_z}{M}\Phi_1(r)\\\Phi_4(r)&=\frac{-aE\cos\varphi\Phi_1(r)-m\sin\varphi\Phi_1(r)+\mathrm{i}(r^2\alpha M\omega\sin\varphi\Phi_1(r)+r\alpha\sin\varphi\Phi'_1(r))}{Mr\alpha}\\\Phi_3(r)&=\frac{aE\sin\varphi\Phi_1(r)+m\sin\varphi\Phi_1(r)+\mathrm{i}(r^2\alpha M\omega\cos\varphi\Phi_1(r)+r\alpha\cos\varphi\Phi'_1(r))}{Mr\alpha}\end{aligned} \tag{27}$$

Combining these results we obtain (23) of motion for the first component of the DKP spinor:

$$\begin{aligned}&\Phi''_1(r)+\frac{(-1+\alpha)\Phi'_1(r)}{r\alpha}+\Bigg(E^2-kz^2-M^2\\&+2M\omega-\frac{M\omega}{\alpha}-\frac{(aE+m)^2}{r^2\alpha^2}-r^2M^2\omega^2\Bigg)\Phi_1(r)\\&=0\end{aligned} \tag{28}$$

Let us take Φ_1 as

$$\Phi_1=r^{1/2\alpha}R_{n,\ell}(r). \tag{29}$$

Then (28) changes to

$$\begin{aligned}&R''_{n,m}(r)+\frac{R'_{n,m}(r)}{r}+\Bigg(E^2-{k_Z}^2-M^2+2M\omega\\&-\frac{M\omega}{\alpha}-\frac{1+4(aE+m)^2}{4r^2\alpha^2}-r^2M^2\omega^2\Bigg)R_{n,m}(r)=0\end{aligned} \tag{30}$$

In order to solve the above equation, we employ the change of variable: $s=r^2$; thus we rewrite the radial equation (34) in the form

$$\begin{aligned}&R_{n,m}''(s)+\frac{1}{s}R_{n,m}'(s)+\frac{1}{s^2}(-\xi_1s^2+\xi_2s-\xi_3)R_{n,m}(r)\\&=0\end{aligned} \tag{31}$$

If we compare with this second-order differential equation with the Nikiforov-Uvarov (NU) form, given in (A.1) of Appendix, we see that

$$\begin{aligned}\xi_1&=\frac{M^2\omega^2}{4}\\\xi_2&=\frac{1}{4}\left(E^2-kz^2-M^2+2M\omega-\frac{M\omega}{\alpha}\right),\\\xi_3&=\frac{1-\alpha^2+4(aE+m)^2}{16\alpha^2}\end{aligned} \tag{32}$$

which gives the energy levels of the relativistic DKP equation from

$$(2n+1)\sqrt{\xi_1}-\xi_2+2\sqrt{\xi_3\xi_1}=0, \tag{33}$$

where

$$\begin{aligned}\alpha_1&=1,\\\alpha_2&=\alpha_3=\alpha_4=\alpha_5=0,\\\alpha_6&=\xi_1,\\\alpha_7&=-\xi_2,\\\alpha_8&=\xi_3,\\\alpha_9&=\xi_1,\\\alpha_{10}&=1+2\sqrt{\xi_3},\end{aligned}$$

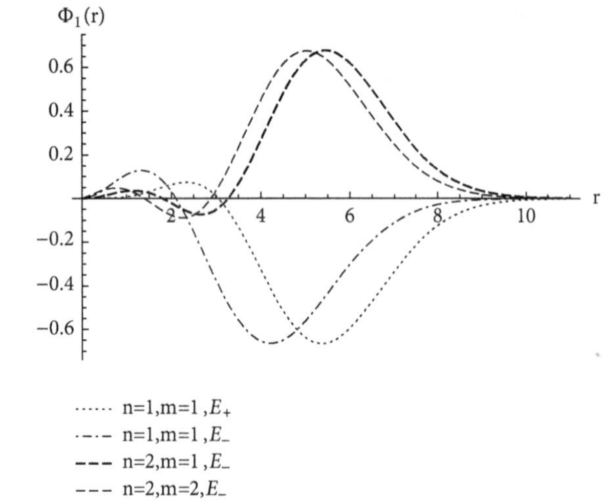

FIGURE 1: The wave function Φ_1 for $n = 1, 2$ and $0.0 \leq r \leq 10.0$ GeV $^{-1}$, with the parameters $M = 1$ GeV, $\alpha = 0.9$, $\omega = 0.25$, and $m = k_z = a = 1$.

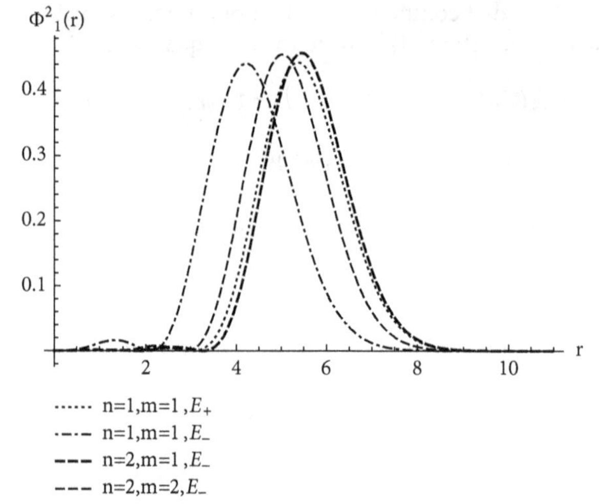

FIGURE 2: Density of probability $|\Phi_1|^2$ for $n = 1, 2$ and $0.0 \leq r \leq 10.0$ GeV $^{-1}$, with the parameters $M = 1$ GeV, $\alpha = 0.9$, $\omega = 0.25$, and $m = k_z = a = 1$.

$$\alpha_{11} = 2\sqrt{\xi_1}$$

$$\alpha_{12} = \sqrt{\xi_3},$$

$$\alpha_{13} = -\sqrt{\xi_1}. \tag{34}$$

As the final step, it should be mentioned that the corresponding wave function is

$$R_{n,m}(r) = N r^{2\alpha_{12}} e^{\alpha_{13} r^2} L_n^{\alpha_{10}-1}\left(\alpha_{11} r^2\right). \tag{35}$$

where N is the normalization constant. In limit a$\longrightarrow 0$ we have the usual metric in cylindrical coordinates where described by the line element

$$ds^2 = -dt^2 + dr^2 + \alpha^2 r^2 d\varphi^2 + dz^2 \tag{36}$$

as pointed out by authors in [17] dynamic of DKP oscillator in the presence of this metric describe by

$$\begin{aligned} \varphi''_1(r) + \frac{\alpha - 1}{\alpha r}\varphi'_1(r) + \Bigg(E^2 - M^2 - kz^2 \\ + \frac{(2\alpha - 1) M\omega}{\alpha} - \frac{m^2}{\alpha^2 r^2} - M^2\omega^2 r^2\Bigg)\varphi_1(r) = 0 \end{aligned} \tag{37}$$

and the corresponding wave function is

$$\varphi(r) = N r^{2A} e^{Br^2} L_n^{C-1}\left(Dr^2\right) \tag{38}$$

where A,B,C, and D are constant and L_n^{C-1} denotes the generalized Laguerre polynomial. In Figure 1, $\Phi_1(r)$ is plotted versus r for different quantum number with the parameters listed under it. The density of probability $|\Phi_1|^2$ is shown in

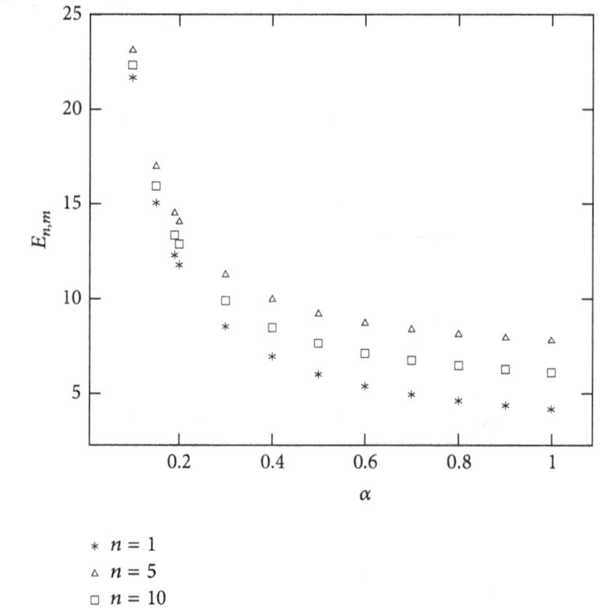

FIGURE 3: The energy as a function of α with $M = m = a = \omega = k_z = 1$.

Figure 2. The negative and positive solution of energy versus α is shown in Figures 3 and 4 for $n = 1, 5$ and 10. As in Figures 3 and 4, we observe that the absolute value of energy decreases with α. Also in Figure 5 energy is plotted versus ω for quantum numbers. We see that absolute value of energy increases with ω. The negative and positive solution of energy versus n is shown in Figure 6 for different parameter α. We obtained the energy levels of the DKP oscillator in that background and observed that the energy increases with the level number. In Figure 7, energy is plotted versus a for different quantum numbers. We see energy increases with parameter a. Also we observed that the energy levels of the

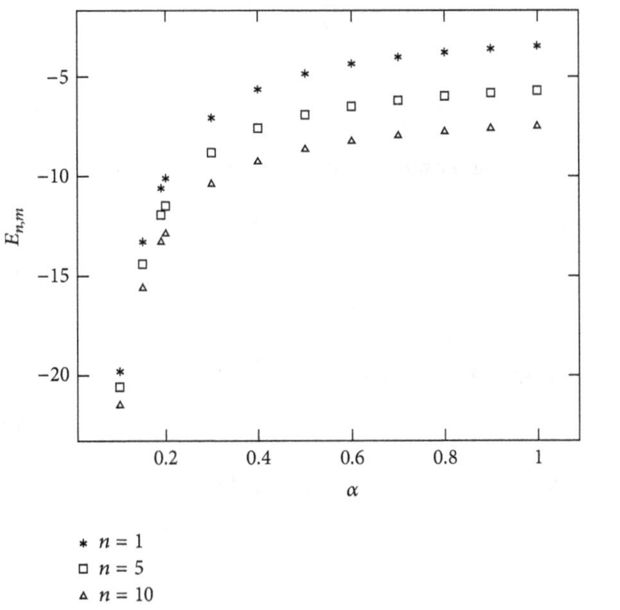

FIGURE 4: The energy as a function of α with $M = m = a = \omega = k_z = 1$.

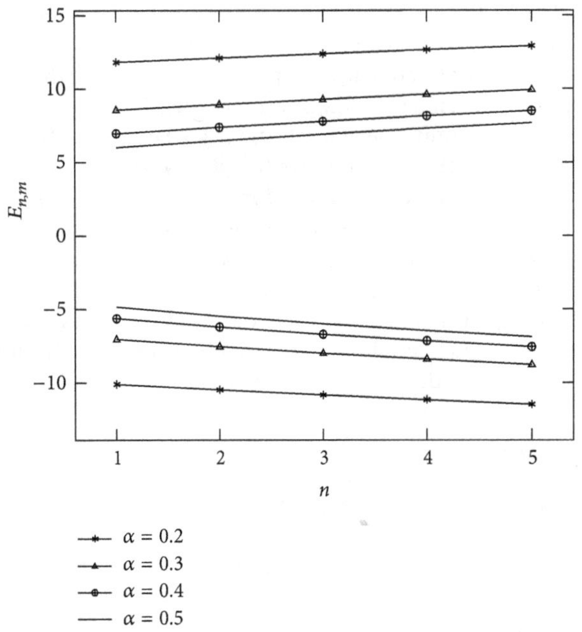

FIGURE 6: The energy as a function of n with $M = k_z = m = 1 = \omega = 1$, and $\alpha = 0.5$.

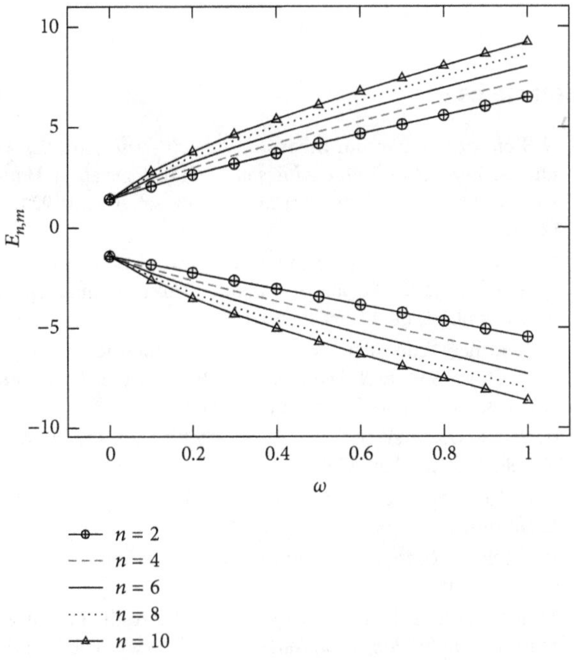

FIGURE 5: The energy as a function of ω with $M = 1$, $k_z = 1$, $m = 1$, $a = 1$, and $\alpha = 0.5$.

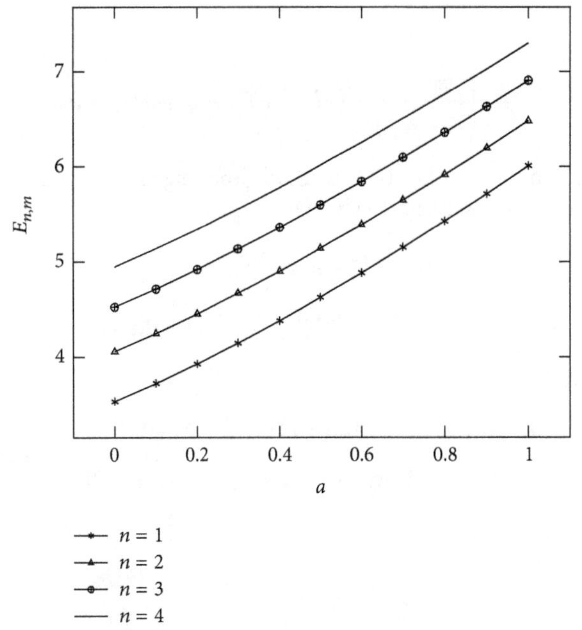

FIGURE 7: The energy as a function of a with $M = m = \omega = k_z = 1$, and $\alpha = 0.5$.

DKP oscillator in that background increases with the level number.

4. Conclusion

The overall objective of this paper is the study of the relativistic quantum dynamics of a DKP oscillator field for spin-0 particle in the spinning cosmic string space-time. The line element in this background is obtained by coordinate transformation of Cartesian coordinate. The metric has off diagonal terms which involves time and space. We considered the covariant form of DKP equation in the spinning cosmic string background and obtained the solutions of DKP equation for spin-0 bosons. Second we introduced DKP oscillator via the nonminimal substitution and considered DKP oscillator in that background. From the corresponding DKP equation, we obtained a system of five equations. By combining the results of this system we obtained a second-order differential equation for first component of DKP spinor that the solutions are Laguerre polynomials. We see that the results are dependent on the linear mass density of the cosmic string. In the limit case of $a = 0$ and $\alpha = 1$, i.e., in the absence of a topological defect, recover the general solution for flat space- time. We plotted $\Phi_1(r)$, for $n = 1, 2$. We examined the behaviour of the density of probability $|\Phi_1|^2$. We observed that $|\Phi_1|^2$ for any parameter by increasing r have a very small peak at beginning and then have a taller peak and then by increasing r it tends to zero. We obtained the behaviour of energy spectrum as a function of α. We see that the absolute value of energy decreases as α increasing.

Appendix

Nikiforov-Uvarov (NU) Method

The Nikiforov-Uvarov method is helpful in order to find eigenvalues and eigenfunctions of the Schrödinger equation, as well as other second-order differential equations of physical interest. More details can be found in [48, 49]. According to this method, the eigenfunctions and eigenvalues of a second- order differential equation with potential are

$$\Phi''(s) + \frac{\alpha_1 - \alpha_2 s}{s(1-\alpha_3 s)}\Phi'(s) + \frac{1}{(s(1-\alpha_3 s))^2}\left(-\xi_1 s^2 + \xi_2 s - \xi_3\right)\Phi(s) = 0 \quad (A.1)$$

According to the NU method, the eigenfunctions and eigenenergies, respectively, are

$$\Phi(s) = s^{\alpha_{12}}(1-\alpha_3 s)^{-\alpha_{12}-(\alpha_{13}/\alpha_3)} \cdot P_n^{(\alpha_{10}-1,(\alpha_{11}/\alpha_3)-\alpha_{10}-1)}(1-2\alpha_3 s) \quad (A.2)$$

and

$$\alpha_2 n - (2n+1)\alpha_5 + (2n+1)\left(\sqrt{\alpha_9} + \alpha_3\sqrt{\alpha_8}\right) + n(n-1)\alpha_3 + \alpha_7 + 2\alpha_3\alpha_8 + 2\sqrt{\alpha_8\alpha_9} = 0 \quad (A.3)$$

where

$$\alpha_4 = \frac{1}{2}(1-\alpha_1),$$
$$\alpha_5 = \frac{1}{2}(\alpha_2 - 2\alpha_3),$$
$$\alpha_6 = \alpha_5^2 + \xi_1,$$
$$\alpha_7 = 2\alpha_4\alpha_5 - \xi_2,$$
$$\alpha_8 = \alpha_4^2 + \xi_3,$$
$$\alpha_9 = \alpha_3\alpha_7 + \alpha_3^2\alpha_8 + \alpha_6,$$
$$\alpha_{10} = \alpha_1 + 2\alpha_4 + 2\sqrt{\alpha_8},$$
$$\alpha_{11} = \alpha_2 - 2\alpha_5 + 2\left(\sqrt{\alpha_9} + \alpha_3\sqrt{\alpha_8}\right),$$
$$\alpha_{12} = \alpha_4 + \sqrt{\alpha_8},$$
$$\alpha_{13} = \alpha_5 - \left(\sqrt{\alpha_9} + \alpha_3\sqrt{\alpha_8}\right) \quad (A.4)$$

In the rather more special case of $\alpha = 0$,

$$\lim_{\alpha_3 \to 0} P_n^{(\alpha_{10}-1,(\alpha_{11}/\alpha_3)-\alpha_{10}-1)}(1-2\alpha_3 s) = L_n^{\alpha_{10}-1}(\alpha_{11}s)$$
$$\lim_{\alpha_3 \to 0}(1-\alpha_3 s)^{-\alpha_{12}-(\alpha_{13}/\alpha_3)} = e^{\alpha_{13}s} \quad (A.5)$$

and, from (14a), we find for the wave function

$$\Phi(s) = s^{\alpha_{12}}e^{\alpha_{13}s}L_n^{\alpha_{10}-1}(\alpha_{11}s) \quad (A.6)$$

where $L_n^{\alpha_{10}-1}$ denotes the generalized Laguerre polynomial.

Conflicts of Interest

The authors declare that they have no conflicts of interest.

References

[1] N. Kemmer, "Quantum theory of einstein-bose particles and nuclear interaction," *Proceedings of the Royal Society A Mathematical, Physical and Engineering Sciences*, vol. 166, no. 924, pp. 127–153, 1938.

[2] R. J. Duffin, "On the characteristic matrices of covariant systems," *Physical Review A: Atomic, Molecular and Optical Physics*, vol. 54, no. 12, article no. 1114, 1938.

[3] N. Kemmer, "The particle aspect of meson theory," *Proceedings of the Royal Society A Mathematical, Physical and Engineering Sciences*, vol. 173, no. 952, pp. 91–116, 1939.

[4] G. Petiau, University of Paris, Thesis (1936). Published in Acad. R. Belg. Cl. Sci. Mem. Collect. 8 16 (2) (1936).

[5] E. M. Corson, *Introduction to Tensors, Spinors Relativistic Wave Equations*, Chelsea Publishing, 1953.

[6] W. Greiner, *Relativistic Quantum Mechanics*, Springer, Berlin, Germany, 2000.

[7] M. Falek and M. Merad, "Bosonic oscillator in the presence of minimal length," *Journal of Mathematical Physics*, vol. 50, no. 2, Article ID 023508, 2009.

[8] M. Falek and M. Merad, "A generalized bosonic oscillator in the presence of a minimal length," *Journal of Mathematical Physics*, vol. 51, no. 3, Article ID 033516, 2010.

[9] M. Falek and M. Merad, "DKP oscillator in a noncommutative space," *Communications in Theoretical Physics*, vol. 50, no. 3, pp. 587–592, 2008.

[10] G. Guo, C. Long, Z. Yang, and S. Qin, "DKP oscillator in noncommutative phase space," *Canadian Journal of Physics*, vol. 87, no. 9, pp. 989–993, 2009.

[11] Z.-H. Yang, C.-Y. Long, S.-J. Qin, and Z.-W. Long, "DKP oscillator with spin-0 in three-dimensional noncommutative phase space," *International Journal of Theoretical Physics*, vol. 49, no. 3, pp. 644–651, 2010.

[12] H. Hassanabadi, Z. Molaee, and S. Zarrinkamar, "DKP oscillator in the presence of magnetic field in (1+2)-dimensions for spin-zero and spin-one particles in noncommutative phase space," *The European Physical Journal C*, vol. 72, no. 11, article no. 2217, 2012.

[13] L. B. Castro, "Quantum dynamics of scalar bosons in a cosmic string background," *The European Physical Journal C*, vol. 75, no. 6, article no. 287, 2015.

[14] N. Debergh, J. Ndimubandi, and D. Strivay, "On relativistic scalar and vector mesons with harmonic oscillator - like interactions," *Zeitschrift für Physik C Particles and Fields*, vol. 56, pp. 421–425, 1992.

[15] Y. Nedjadi and R. C. Barrett, "The Duffin-Kemmer-Petiau oscillator," *Journal of Physics A: Mathematical and General*, vol. 27, no. 12, pp. 4301–4315, 1994.

[16] Y. Nedjadi, S. Ait-Tahar, and R. C. Barrett, "An extended relativistic quantum oscillator for $S = 1$ particles," *Journal of Physics A: Mathematical and General*, vol. 31, no. 16, pp. 3867–3874, 1998.

[17] M. Hosseinpour, H. Hassanabadi, and F. M. Andrade, "The DKP oscillator with a linear interaction in the cosmic string space-time," *The European Physical Journal C*, vol. 78, no. 2, p. 93, 2018.

[18] A. Boumali and L. Chetouani, "Exact solutions of the Kemmer equation for a Dirac oscillator," *Physics Letters A*, vol. 346, no. 4, pp. 261–268, 2005.

[19] I. Boztosun, M. Karakoc, F. Yasuk, and A. Durmus, "Asymptotic iteration method solutions to the relativistic Duffin-Kemmer-Petiau equation," *Journal of Mathematical Physics*, vol. 47, no. 6, Article ID 062301, 2006.

[20] M. de Montigny, M. Hosseinpour, and H. Hassanabadi, "The spin-zero Duffin-Kemmer-Petiau equation in a cosmic-string space-time with the Cornell interaction," *International Journal of Modern Physics A*, vol. 31, no. 36, Article ID 1650191, 2016.

[21] F. Yasuk, M. Karakoc, and I. Boztosun, "The relativistic Duffin–Kemmer–Petiau sextic oscillator," *Physica Scripta*, vol. 78, no. 4, Article ID 045010, 2008.

[22] A. Boumali, "On the eigensolutions of the one-dimensional Duffin-Kemmer-Petiau oscillator," *Journal of Mathematical Physics*, vol. 49, no. 2, Article ID 022302, 2008.

[23] Y. Kasri and L. Chetouani, "Energy spectrum of the relativistic Duffin-Kemmer-Petiau equation," *International Journal of Theoretical Physics*, vol. 47, no. 9, pp. 2249–2258, 2008.

[24] A. Vilenkin and E. P. Shellard, *Cosmic Strings and Other Topological Defects*, Cambridge University Press, Cambridge, UK, 1994.

[25] A. Vilenkin, "Cosmic strings and domain walls," *Physics Reports*, vol. 121, no. 5, pp. 263–315, 1985.

[26] N. G. Marchuknmarchuk, "Dirac equation in Riemannian space without tetrads," *Il Nuovo Cimento B*, vol. 115, no. 11, 2000.

[27] L. D. Landau and E. M. Lifshitz, *Quantum Mechanics, Non-Relativistic Theory*, Pergamon, New York, USA, 1977.

[28] G. de A. Marques and V. B. Bezerra, "Hydrogen atom in the gravitational fields of topological defects," *Physical Review D: Covering Particles, Fields, Gravitation and Cosmology*, vol. 66, no. 10, Article ID 105011, 2002.

[29] G. R. de Melo, M. de Montigny, and E. S. Santos, "Spinless Duffin-Kemmer-Petiau oscillator in a Galilean non-commutative phase space," *Journal of Physics: Conference Series*, vol. 343, Article ID 012028, 2012.

[30] L. B. Castro, "Noninertial effects on the quantum dynamics of scalar bosons," *The European Physical Journal C*, vol. 76, no. 2, article no. 61, 2016.

[31] H. Hassanabadi, M. Hosseinpour, and M. de Montigny, "Duffin-Kemmer-Petiau equation in curved space-time with scalar linear interaction," *The European Physical Journal Plus*, vol. 132, no. 12, p. 541, 2017.

[32] R. Bausch, R. Schmitz, and L. A. Turski, "Single-Particle Quantum States in a Crystal with Topological Defects," *Physical Review Letters*, vol. 80, no. 11, pp. 2257–2260, 1998.

[33] E. Aurell, "Torsion and electron motion in quantum dots with crystal lattice dislocations," *Journal of Physics A: Mathematical and General*, vol. 32, no. 4, article no. 571, 1999.

[34] C. R. Muniz, V. B. Bezerra, and M. S. Cunha, "Landau quantization in the spinning cosmic string spacetime," *Annals of Physics*, vol. 350, pp. 105–111, 2014.

[35] V. B. Bezerra, "Global effects due to a chiral cone," *Journal of Mathematical Physics*, vol. 38, no. 5, pp. 2553–2564, 1997.

[36] C. Furtado, V. B. Bezerra, and F. Moraes, "Quantum scattering by a magnetic flux screw dislocation," *Physics Letters A*, vol. 289, no. 3, pp. 160–166, 2001.

[37] M. S. Cunha, C. R. Muniz, H. R. Christiansen, and V. B. Bezerra, "Relativistic Landau levels in the rotating cosmic string spacetime," *The European Physical Journal C*, vol. 76, no. 9, p. 512, 2016.

[38] G. Clément, "Rotating string sources in three-dimensional gravity," *Annals of Physics*, vol. 201, no. 2, pp. 241–257, 1990.

[39] C. Furtado, F. Moraes, and V. B. Bezerra, "Global effects due to cosmic defects in Kaluza-Klein theory," *Physical Review D: Particles, Fields, Gravitation and Cosmology*, vol. 59, no. 10, Article ID 107504, 1999.

[40] R. A. Puntigam and H. H. Soleng, "Volterra distortions, spinning strings, and cosmic defects," *Classical and Quantum Gravity*, vol. 14, no. 5, article no. 1129, 1997.

[41] P. S. Letelier, "Spinning strings as torsion line spacetime defects," *Classical and Quantum Gravity*, vol. 12, no. 2, 1995.

[42] P. O. Mazur, "Spinning cosmic strings and quantization of energy," *Physical Review Letters*, vol. 57, no. 8, pp. 929–932, 1986.

[43] J. R. Gott and M. Alpert, "General relativity in a (2+1)-dimensional space-time," *General Relativity and Gravitation*, vol. 16, no. 3, pp. 243–247, 1984.

[44] K. Bakke and C. Furtado, "On the interaction of the Dirac oscillator with the Aharonov–Casher system in topological defect backgrounds," *Annals of Physics*, vol. 336, pp. 489–504, 2013.

[45] G. Q. Garcia, J. R. de S. Oliveira, K. Bakke, and C. Furtado, "Fermions in Gödel-type background space-times with torsion and the Landau quantization," *The European Physical Journal Plus*, vol. 132, no. 3, article no. 123, 2017.

[46] J. Carvalho, C. Furtado, and F. Moraes, "Dirac oscillator interacting with a topological defect," *Physical Review A: Atomic, Molecular, and Optical Physics and Quantum Information*, vol. 84, no. 3, Article ID 032109, 2011.

[47] E. R. F. Medeiros and E. R. B. de Mello, "Relativistic quantum dynamics of a charged particle in cosmic string spacetime in the presence of magnetic field and scalar potential," *The European Physical Journal C*, vol. 72, no. 6, article no. 2051, 2012.

[48] A. F. Nikiforov and V. B. Uvarov, *Special Functions of Mathematical Physics*, Birkhäauser, Basel, Switzerland, 1988.

[49] C. Tezcan and R. Sever, "A general approach for the exact solution of the Schrödinger equation," *International Journal of Theoretical Physics*, vol. 48, no. 2, pp. 337–350, 2009.

12

Generalization of the Randall-Sundrum Model Using Gravitational Model $F(T,\Theta)$

S. Davood Sadatian and S. M. Hosseini

Department of Physics, Faculty of Basic Sciences, University of Neyshabur, P. O. Box 9318713331, Neyshabur, Iran

Correspondence should be addressed to S. Davood Sadatian; sd-sadatian@um.ac.ir

Academic Editor: George Siopsis

In this letter, we explore a generalized model based on two scenarios including the Randall-Sundrum model and gravity model $F(T,\Theta)$. We first study the standard Randall-Sundrum gravitational model and then add a function containing two parameters as torsion and trace energy-momentum tensor to the main action of the model. Next, we derive the equations of the generalized model and obtain a new critical value for the energy density of the brane. The results showed that inflation and the dark energy-dominated stage can be realized in this model. We pointed out one significant category of dark energy models that had greatly developed the knowledge about dark energy. To be specific, dark energy could either be quintessence-like, phantom-like, or the so-called "quintom"-like. The models of quintom type suggest that the equation of state parameter of dark energy can cross the cosmological constant boundary $\omega = -1$. Interestingly, this quintom scenario exactly appeared in this paper.

1. Introduction

According to the standard model of cosmology, the Planck Era refers to a time starting from the creation of the universe (about 14 billion years ago) until 10^{-43} seconds. In this time lapse, quantum effects are important parameters that can be formulated using the quantum gravity theory [1]. One of the ideas for describing quantum gravity is the string theory [2]. Some features of this theory include the replacement of one-dimensional quantum operators with pseudopoint operators in Hilbert space and the need for the presence of supersymmetry instead of symmetry [3]. The idea of the cosmology of the string emerged in the early 1990s by extending the string idea into the cosmological model [4]. According to this theory, from the mathematical point of view, we need a space with higher dimensions. The outcome of this idea is that the fundamental constants of the physics have a variable form with respect to time. The image of these constants in a (3 + 1)-dimensional world of our space-time universe is constant with respect to time [5].

Kaluza (1921) and Klein (1926) proposed the unified field theory of gravitation and electromagnetism, which needs an extra spatial dimension that is a function of the Planck length [6, 7].

Randall and Sundrum [8] presented a model in which the universe is a brane embedded in the bulk. In other words, the universe is located in a less spatial dimension than the main dimension of the bulk with five-dimensional space-time [9]. The main feature of the bulk is the gravity which can freely propagate in it. Because only a graviton particle can exist inside the bulk in this model, the other particles of the standard model are in the braneworld. Based on the number of branes available in this model, two Randall-Sundrum (RS) models (I) and (II) have been presented [8, 9]. The five-dimensional space-time determines the type of model. In type (I) Randall model, the distance L is limited while in type (II) Randall model L tends to infinity. The adjustable parameters of this model include cosmic bulk constant and energy density (tension) of the brane [8–10].

In this model, the torsional metric in the five-dimensional space-time is defined according to the following:

$$ds^2 = e^{-2A|y|}\eta_{\mu\nu}dx^{\mu}dx^{\nu} + dy^2, \quad (1)$$

where $0 \le y \le \pi r_c$ is the fifth dimension of space and r_c is essentially a compactification "radius" [8].

Given the abovementioned explanations, we first study the Randall-Sundrum standard model. Then, we obtain the equations of the model by generalizing this model by adding a function $F(T, \Theta)$ to the main action of the model. Finally, we determine some cosmological parameters by our generalized model equations.

2. Teleparallel Gravity Model

The equations of the gravitational field in general relativity are based on the interaction between the curvature of space-time and the energy density content of the universe. Other models have been proposed for gravity [1–5] by generalizing a gravity model based on scalar and Ricci tensor (curvature). One of these models is based on the replacement of the torsion tensor rather than the curvature of space-time in the gravity model [11]. This modified model proves the inflation at the beginning of the universe.

In the following, a brief overview of the abovementioned model is presented. We assume Planck constant, Boltzmann constant, and the speed of light as $\hbar = K_B = c = 1$ and define the gravitational constant $8\pi G$ by the relation $K^2 = 8\pi/M_{pl}^2$ (where M_{pl} is Planck mass). In this model, the torsion tensor is denoted by the relation $T^{\rho}{}_{\mu\nu} = r^{\rho}{}_{\nu\mu} - r^{\rho}{}_{\mu\nu}$, where $r^{\rho}{}_{\nu\mu} = e^{\rho}_{A}\partial_{\mu}e^{A}_{\nu}$ is the connection without the Weitzenbock curvature [9]. The contortion tensor is also defined by the relation $K^{\mu\nu}{}_{\rho} = (1/2)(T^{\mu\nu}{}_{\rho} - T^{\nu\mu}{}_{\rho} - T_{\rho}{}^{\mu\nu})$. In addition, the torsion scalar is presented as $T = S_{\rho}{}^{\mu\nu}K_{\mu\nu}{}^{\rho}$ using the super potential $S_{\rho}{}^{\mu\nu}$ and the contortion tensor $K^{\rho}{}_{\mu\nu}$. The action of this modified model with the matter in terms of $F(T)$ is shown by the following [11]:

$$S = \int d^4x \, |e| \left(\frac{F(T)}{2K^2} + L_M \right), \tag{2}$$

where $|e| = \det(e^A_\mu) = \sqrt{-g}$ and L_M is the Lagrangian of matter. Regardless of material Lagrangian, gravity action in five dimensions is defined by the following relations:

$$\left({}^{(5)}S\right) = \int d^5x \left|{}^{(5)}e\right| \left(\frac{F\left({}^{(5)}T\right)}{2K_5^2} \right) \tag{3}$$

$$\left({}^{(5)}T\right) = \frac{1}{4}T^{abc}T_{abc} + \frac{1}{2}T^{abc}T_{cba} - T_{ab}{}^{a}T^{cb}{}_{c}, \tag{4}$$

where ${}^{(5)}e = \sqrt{{}^{(5)}g}$ and the gravitational constant and Planck mass in five dimensions are related to each other by the relation $K_5^2 = 8\pi G_5 = 8\pi/M_{pl}^2$ [11].

2.1. Compression of Kaluza-Klein. In this subsection, we need to explain the compression mechanism of the Kaluza-Klein [6, 7]. In these five dimensions, the metric is represented by the following diagonal matrix:

$${}^{(5)}g_{ab} = \begin{bmatrix} g_{\mu\nu} & 0 \\ 0 & -\varphi^2 \end{bmatrix}, \tag{5}$$

where the scalar field is uniform and dependent on time according to the relation $\varphi^2 = \mathfrak{R}^2\theta^2$. Here, φ is related to $\mathfrak{R}$ (compressed space radius) and the quadratic orthogonal components are represented in a compact space dimension with a dimensionless characteristic θ [11]. According to the relationship $g_{\mu\nu} = \eta_{AB}e^{A}{}_{\nu}e^{B}{}_{\nu}$, the second-order metric tensors in five dimensions and tangent spaces in five dimensions are defined according to the relations $\eta_{AB} = \mathrm{diag}(1, -1, -1, -1, -1)$ and $e^{A}{}_{\nu} = \mathrm{diag}(1, 1, 1, 1, \varphi)$, respectively. Therefore, using these relations in (3) and (4), the effective action is given by the compression mechanism of the Kaluza-Klein as follows:

$$S = \int d^4x \frac{1}{2K^2} \varphi \left|{}^{(4)}e\right| F\left(T + \varphi^{-2}\partial_\mu\varphi\partial^\mu\varphi\right), \tag{6}$$

where the shape of the gravitation function and $|{}^{(5)}e| = |e|\varphi$ does not change in comparison with relation (3). This function contains a torsion scalar component and a scalar field.

In the simplest case, assume a gravitational function in the form of $F(T) = T - 2\Lambda_4$, where Λ_4 is a positive cosmological constant. If we define a scalar field σ in the form of $\varphi = \sigma^2\xi$, where $\xi = 1/4$, by rewriting (6), the effective action is obtained as follows:

$$S_{KK}^{eff} = \int d^4x \frac{1}{K^2} \left|{}^{(4)}e\right| \left[\frac{1}{8}\sigma^2 T + \frac{1}{2}\partial_\mu\sigma\partial^\mu\sigma - \Lambda_4 \right], \tag{7}$$

where the metric in the flat universe of FLRW is defined as $ds^2 = dt^2 - a^2(t)\Sigma_{i=1,2,3}(dx^i)^2$, a is the scale factor, and $H = \dot{a}/a$ is the Hubble parameter. Moreover, the vector of the space tangent and metric are defined with the relations $e^A_\mu = \mathrm{diag}(1, a, a, a)$ and $g_{\mu\nu} = \mathrm{diag}(1, -a^2, -a^2, -a^2)$, respectively. If $T = -6H^2$, the equations of the gravity field are calculated by the following [11]:

$$\frac{1}{2}\dot{\sigma}^2 - \frac{3}{4}H^2\sigma^2 + \Lambda_4 = 0 \tag{8}$$

$$\dot{\sigma} + H\sigma\dot{\sigma} + \frac{1}{2}\dot{H}\sigma^2 = 0, \tag{9}$$

And the equation of motion for the scalar field is obtained by

$$\ddot{\sigma} + 3H\dot{\sigma} + \frac{3}{2}H^2\sigma = 0. \tag{10}$$

By combining the abovementioned gravity field equations, the following is obtained:

$$\frac{3}{2}H^2\sigma^2 - 2\Lambda_4 + H\sigma\dot{\sigma} + \frac{1}{2}\dot{H}\sigma^2 = 0. \tag{11}$$

With solving this equation in terms of H, the obtained value (H) is achieved corresponding to the Hubble parameter in the inflation time [11].

2.2. Effective Gravity in Randall-Sundrum Model. Considering the solution of vacuum equations in a five-dimensional space-time and (RS) type (II) model, which has only one mass within a five-dimensional bulk with positive energy density, it is concluded that the five-dimensional space-time is an anti-desitter (AdS) space. In the following, according to Randall model type II [12, 13] in the teleparallel gravity theory, Friedmann equations for the brane in the FLRW background metric for effective gravity $F(T)$ are determined as follows:

$$H^2 \frac{dF(T)}{dT} = -\frac{1}{12}\left[F(T) - 4\Lambda - 2K^2\rho_M - \left(\frac{K_5^2}{2}\right)Q\rho_m^2\right] \quad (12)$$

where $Q = (11 - 60\omega_M + 93\omega_M^2)/4$. In this relation, $\omega_M = P_M/\rho_M$ corresponds to the perfect fluid equation of state for the pressure and density of the confined matter in the brane. $\Lambda = \Lambda_5 + (K_5^2/2)^2\lambda^2$ with $0 < \lambda$ brane tension, and $G = [1/(3\pi)](K_5^2)^2\lambda$. With substituting $F(T) = T - 2\Lambda_5$ in (11), the approximate solution of desitter for brane is obtained as $H = H_{DE} = \sqrt{(\Lambda_5 + K_5^4\lambda^2/6)}$, where $a(t) = a_{DE}e^{(H_{DE}t)}$ with $a_{DE} > 0$, and $T = -6H^2$ is assumed. Therefore, this relation can describe the accelerating expansion of the universe [11].

3. Teleparallel Gravity Model with a Trace of Momentum-Energy Tensor

Another example of the generalized model based on the $F(T)$ model that we discuss here is the $F(T, \Theta)$ model. There are two main reasons for implementing the $F(T, \Theta)$ model. First, it is a novel gravitational theory. Here, the only restriction imposed on F is that it must be an analytic function, meaning that $F(T, \Theta)$ is a real function locally given by a convergent power series and is infinitely differentiable. Second, due to the extra freedom in the Lagrangian imposed, the $F(T, \Theta)$ cosmology allows for a vast class of scenarios and behaviors. For example, the scalar perturbations at the linear level reveal that $F(T, \Theta)$ cosmology can be free of ghosts and instabilities for a wide class of ansatzes and model parameters [13, 14].

An important feature of this model is the effect of geometry and the universe content through space-time torsion and trace of momentum-energy tensors, where T is torsion scalar and Θ is energy-momentum tensor trace. Therefore, in order to study this model, we consider the proposed gravity function $F(T, \Theta)$ in a five-dimensional space-time compressed by the Kaluza-Klein theory [11, 14]:

$$F(T, \Theta) = \alpha(-T)^n(\Theta)^m \tanh\left(\frac{T_0}{T}\right), \quad (13)$$

where $\alpha = 1, m = 1, n = 2, T_0 = -6H_0^2$. The physical motivation, in this case, is a possible expression that leads to an accelerated expansion phase. It is particularly interesting to look at models that are able to give an effective equation of state with crossing the phantom divide. Also, interpreting gravitational interactions in terms of the torsion rather than the scalar curvature results in the equivalent teleparallel formulation of the general relativity. Now, let us use the abovementioned gravity function in a type II Randall-Sundrum model.

First, we transfer the five-dimensional space-time in which the gravity model is related to the scalar field and coupled with the additional dimension of space, into a four-dimensional space-time in a brane with a flat metric FLRW using the compression mechanism of the Kaluza-Klein. We have already investigated that in the gravity model $F(T)$ in the four-dimensional space we get the Friedmann equations according to (12) [11]. Therefore, due to the characteristics of the gravity model $F(T, \Theta)$ mentioned above, we rewrite Friedmann equations for this gravitational model. In this model, the torsion is based on the relation $T = -6H^2$ and Θ is the momentum-energy tensor trace. The total momentum-energy tensor on brane is written according to (14) [13]:

$$T_B^A = S_B^A\delta(y), \quad (14)$$

where $S_B^A = \mathrm{diag}(-\rho_b, P_b, P_b, P_b, 0)$ and P_b, ρ_b are pressure and density of the total brane energy, respectively. By the conjugation condition in $y = 0$ in the five-dimensional space and simplifying the relations, we have

$$H^2 = \left(\varepsilon\frac{\rho_b^2}{36}\chi^4\right) + \frac{\Lambda}{6} - \frac{k}{a^2} + \frac{C}{a^4}, \quad (15)$$

where C is constant of integration. It is of note that this relation is established in the brane and the energy of conservation law will be conserved according to the following relation:

$$\dot{\rho}_b + 3H(\rho_b + P_b) = 0. \quad (16)$$

Assuming that $\rho_b = \rho + \lambda$, where λ is a brane tension and $\Lambda = \varepsilon\lambda^2(\chi^4/6)(\varepsilon = 1$, if the extra dimension is space-like, while $\varepsilon = -1$, if it is time-like), we get the following relation in a flat metric ($k = 0$), where $C = 0$ is the constant of integration and called dark radiation:

$$H^2 = \left(\frac{\rho}{3}\right)\left(1 + \frac{\rho}{2\lambda}\right). \quad (17)$$

Now, by applying the gravity function $F(T, \Theta)$, (13), in the Friedman equation obtained from the compression of space KK, (12), we have

$$\tanh\left(\frac{T_0}{T}\right)\left[A(-T)^n + BT^{n-1} + C(-T)^{n+1}\right] = \frac{8\pi G}{12(3\omega - 1)} + \frac{\pi G}{16\lambda(3\omega - 1)}\left(11 - 60\omega + 93\omega^2\right)\rho, \quad (18)$$

where $A = -n/6, B = T_0/6, C = 1/72$. Assuming $n = 2$ and due to the relation of momentum-energy tensor (14), we get $\Theta = 3P - \rho$, which can be rewritten as $\Theta = (-1 + 3\omega)$ using the equation of state. Using the Maclaurin expansion for the hyperbolic tangent and using the first-order approximation, we have

$$T = \frac{T_0}{72}\left[\frac{8\pi G}{12(3\omega - 1)} + \frac{\pi G}{16\lambda(3\omega - 1)}\left(11 - 60\omega + 93\omega^2\right)\rho\right] - 12. \quad (19)$$

Here, assuming that $\rho_b = \lambda\sqrt{\rho}$ and substituting it in (14), we get (18). Moreover, the Hubble parameter is obtained according to (20) using the relation $T = -6H^2$. Therefore, we have

$$H = \pm\left(-\frac{T_0}{432}\left[\frac{8\pi G}{12(3\omega - 1)} + \frac{\pi G}{16\lambda(3\omega - 1)}\left(11 - 60\omega + 93\omega^2\right)\rho\right] + 2\right)^{1/2}. \quad (20)$$

Using (20) and assuming that $\rho_b = \lambda\sqrt{\rho}$, the critical value for the energy density of the brane is determined according to the following equation:

$$\rho_b < \sqrt{\frac{863\Theta\lambda^3}{\pi G T_0}}. \quad (21)$$

Now, we can obtain the universe scale factor from the combination of Hubble parameter and relation (20), for the inflation period using the hyperbolic tangent series and the first-order approximation. Also, using an approximate of the deSitter solution on the brane (assuming $T_0 = -6H_0^2 = -24 \times 10^{-36}$), we obtain scale factor $a(t)$, shown in Figure 1. To calculate the equation of state $\omega(t)$, it is necessary to solve simultaneously three nonlinear differential equations according to the obtained relations for the gravity model $F(T, \Theta)$ in the five-dimensional space. For this purpose, first, by combining (16) and the equation of state $p = \omega\rho c^2$, we get the differential equation between the energy density of the brane and $\omega(t)$ as follows:

$$\dot{\rho}_b + 3H\rho_b(1 + \omega(t)) = 0. \quad (22)$$

Now, according to the pressure and energy relation for the teleparallel gravity model [15] and combining it with the equation of state, we get the nonlinear differential equation for the scalar field and the Hubble parameter as follows:

$$\left(\frac{\dot{\varphi}}{2} - \frac{3}{4}\varphi^2 H^2 + \Lambda_4\right)\omega = \frac{\dot{\varphi}^2}{2} + H\varphi\dot{\varphi} + \frac{3}{4}H^2\varphi^2 + \frac{1}{2}\varphi^2\dot{H} - \Lambda_4. \quad (23)$$

By solving (22) and (23) simultaneously and also the equation of motion for the scalar field (10), the equation of state $\omega(t)$ is obtained. In the following, we used the first-order approximation to solve differential equations. Moreover, by adjusting the brane tension parameter, the energy density, and the potential of the scalar field with the cosmological constant on the brane, we get the equation of state in terms of time shown in Figure 2. As shown in this model, the phantom boundary crossing occurred [16, 17].

It is shown that inflation or the dark energy-dominated stage can be realized only by the effect of the torsion and trace energy-momentum tensor without the curvature. As a result, it can be interpreted that these models may be equivalent to the Kaluza-Klein and RS models without gravitational effects of the curvature but just due to those of the torsion and trace energy-momentum tensor in teleparallelism. Indeed, this is the new work on the concrete cosmological solutions to describe the cosmic accelerated expansion of the KK and RS models in $F(T, \Theta)$ gravity. Based on these results, it can be stated that phenomenological $F(T, \Theta)$ gravity models in the four-dimensional space-time can be derived from more fundamental theories. In this regard, the observational constraints on the derivative of $F(T)$ and similar function as $F(T, \Theta)$ until the fifth order were presented in [18] with cosmographic parameters acquired from the observational data of Supernovae Ia and the baryon acoustic oscillations. The results of the model presented in this work, as a concrete example of $F(T, \Theta)$ gravity models, are consistent with those obtained in [18].

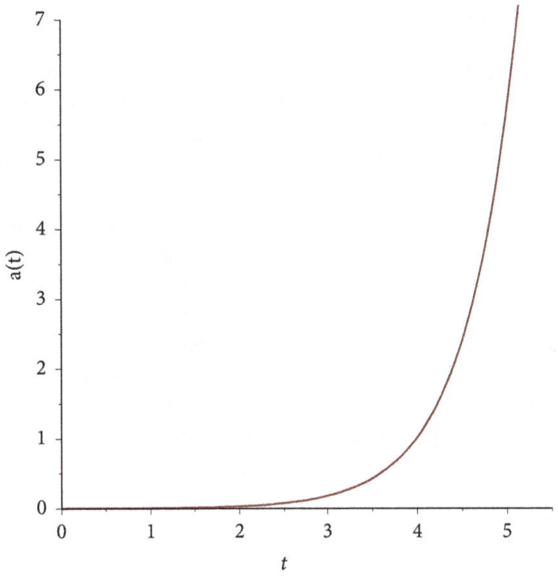

FIGURE 1: The universe scale factor for the inflation period (where $\alpha = 1, m = 1, n = 2$.)

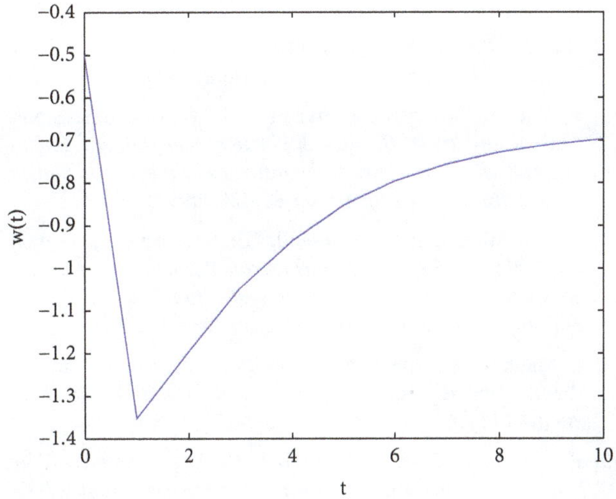

FIGURE 2: The equation of state $\omega(t)$ in terms of time for generalized RS model with $F(T, \Theta)$ for $\alpha = 1, m = 1, n = 2$.

4. Conclusion

In this paper, a generalized gravity model was proposed based on the time-space torsion and interaction with the universe content. The study of this gravity model in a five-dimensional space according to the Randall-Sundrum approach included a four-dimensional brane in a five-dimensional bulk. In this regard, the Kaluza-Klein theory was used to compress the fifth dimension of space in this gravity model. Then, in accordance with the inflation period of the standard cosmological model, the new critical value for the energy density of the brane, the Hubble parameter, and the scale factor were obtained. Finally, it has been illustrated that, in $F(T, \Theta)$ gravity, inflation in the early universe and the late-time cosmic acceleration can be realized.

Conflicts of Interest

The authors declare that they have no conflicts of interest.

References

[1] P. P. Coles and F. Lucchin, *Cosmology: The Origin and Evolution of Cosmic Structure*, John Wiley and Sons, 2003.

[2] E. Witten, "String theory dynamics in various dimensions," *Nuclear Physics. B. Theoretical, Phenomenological, and Experimental High Energy Physics. Quantum Field Theory and Statistical Systems*, vol. 443, no. 1-2, pp. 85–126, 1995.

[3] C. M. Hull and P. K. Townsend, "Unity of superstring dualities," *Nuclear Physics. B. Theoretical, Phenomenological, and Experimental High Energy Physics. Quantum Field Theory and Statistical Systems*, vol. 438, no. 1-2, pp. 109–137, 1995.

[4] J. Maldacena, "The large N limit of superconformal field theories and supergravity," *Advances in Theoretical and Mathematical Physics*, vol. 2, no. 2, pp. 231–252, 1998.

[5] P. P. Avelino, E. P. S. Shellard, J. H. P. Wu, and B. Allen, "Non-Gaussian features of linear cosmic string models," *The Astrophysical Journal*, vol. 507, no. 2, pp. L101–L104, 1998.

[6] T. Kaluza, "On the problem of unity in physics," *Sitzungsberichte der Königlich Preussischen Akademie der Wissenschaften zu Berlin*, vol. 1921, pp. 966–972, 1921.

[7] O. Klein, "Quantentheorie und fünfdimensionale relativitätstheorie," *Zeitschrift für Physik*, vol. 37, no. 12, pp. 895–906, 1926.

[8] L. Randall and R. Sundrum, "An alternative to compactification," *Physical Review Letters*, vol. 83, no. 23, pp. 4690–4693, 1999.

[9] L. Randall and R. Sundrum, "Large mass hierarchy from a small extra dimension," *Physical Review Letters*, vol. 83, no. 17, pp. 3370–3373, 1999.

[10] M. Gogberashvili, "Hierarchy problem in the shell-Universe model," *International Journal of Modern Physics D*, vol. 11, no. 10, pp. 1635–1638, 2002.

[11] K. Bamba, S. Nojiri, and S. D. Odintsov, "Effective $F(T)$ gravity from the higher-dimensional Kaluza-KLEin and Randall-Sundrum theories," *Physics Letters B*, vol. 725, no. 4-5, pp. 368–371, 2013.

[12] A. Behboodi, S. Akhshabi, and K. Nozari, "Braneworld teleparallel gravity," *Physics Letters. B. Particle Physics, Nuclear Physics and Cosmology*, vol. 723, no. 1-3, pp. 201–206, 2013.

[13] A. V. Astashenok, E. Elizalde, J. de Haro, S. D. Odintsov, and A. V. Yurov, "Brane cosmology from observational surveys and its comparison with standard FRW cosmology," *Astrophysics and Space Science*, vol. 347, no. 1, pp. 1–13, 2013.

[14] S. D. Sadatian and A. Tahajjodi, "Cosmological parameters in a generalized multi-function gravitation model $f(T,\Theta)$," *Indian Journal of Physics*, vol. 91, no. 11, pp. 1447–1450, 2017.

[15] C.-Q. Geng, C.-C. Lee, E. N. Saridakis, and Y.-P. Wu, "'Teleparallel' dark energy," *Physics Letters B*, vol. 704, no. 5, pp. 384–387, 2011.

[16] K. Nozari and S. D. Sadatian, "A Lorentz invariance violating cosmology on the DGP brane," *Journal of Cosmology and Astroparticle Physics*, vol. 2009, no. 1, article 005, 2009.

[17] S. D. Sadatian, "Warm inflation and WMAP9," *International Journal of Geometric Methods in Modern Physics*, vol. 15, no. 1, Article ID 1850010, 12 pages, 2018.

[18] S. Capozziello, V. F. Cardone, H. Farajollahi, and A. Ravanpak, "Cosmography in $f(T)$ gravity," *Physical Review D: Particles, Fields, Gravitation and Cosmology*, vol. 84, no. 4, Article ID 043527, 2011.

13

Classical Polymerization of the Schwarzschild Metric

Babak Vakili [1,2]

[1] *Research Institute for Astronomy and Astrophysics of Maragha (RIAAM), Maragha, P.O. Box 55134-441, Iran*
[2] *Department of Physics, Central Tehran Branch, Islamic Azad University, Tehran, Iran*

Correspondence should be addressed to Babak Vakili; b.vakili@iauctb.ac.ir

Guest Editor: Farook Rahaman

We study a spherically symmetric setup consisting of a Schwarzschild metric as the background geometry in the framework of classical polymerization. This process is an extension of the polymeric representation of quantum mechanics in such a way that a transformation maps classical variables to their polymeric counterpart. We show that the usual Schwarzschild metric can be extracted from a Hamiltonian function which in turn gets modifications due to the classical polymerization. Then, the polymer corrected Schwarzschild metric may be obtained by solving the polymer-Hamiltonian equations of motion. It is shown that while the conventional Schwarzschild space-time is a vacuum solution of the Einstein equations, its polymer-corrected version corresponds to an energy-momentum tensor that exhibits the features of dark energy. We also use the resulting metric to investigate some thermodynamical quantities associated with the Schwarzschild black hole, and in comparison with the standard Schwarzschild metric the similarities and differences are discussed.

1. Introduction

One of the most important arenas that show the power of general relativity in describing the gravitational phenomena is the classical theory of black hole physics. However, when we introduce the quantum considerations to study of a gravitational systems, general relativity does not provide a satisfactory description of the physics of the system. The phenomena such as black hole radiation and all kinds of cosmological singularities are among the phenomena in which the use of quantum mechanics in their description is inevitable. This means that although general relativity is a classical theory, in its most important applications, the system under consideration originally obeys the rules of quantum mechanics. Therefore, any hope in the accurate description of gravitational systems in high energies depends on the development of a complete theory of quantum gravity. That is why the quantum gravity is one of the most important challenges in theoretical physics which from its DeWitt's traditional canonical formulation [1] to the more modern viewpoints of string theory and loop quantum gravity (LQG) [2–4] has gone a long way. One of the main features of the space-time proposed in LQG is its granular structure which in turn, supports the idea of existence of a minimal measurable length. In the absence of a full theory of quantum gravity, effective theories which somehow exhibit quantum effects in gravitational systems play a significant role. These are theories which show some phenomenological aspects of quantum gravity and usually use a certain deformation in their formalism. For example, theories like generalized uncertainty principles and noncommutative geometry are in this category [5–13].

Among the effective theories that also use a minimal length scale in their formalism, we can mention the polymer quantization [14], which uses the methods very similar to the effective theories of LQG [15]. In polymer quantum approach a polymer length scale, λ, which shows the scale of the segments of the granular space, enters into the Hamiltonian of the system to deform its functional form into a so-called polymeric Hamiltonian. This means that, in a polymeric quantized system in addition of a quantum parameter $\hbar$, which is responsible to canonical quantization of the system, there is also another quantum parameter λ that labels the granular properties of the underlying space. This approach then opened new windows for the theories which are dealing with the quantum gravitational effects in physical systems such as quantum cosmology and black hole physics; see, for instance [16–31] and the references therein.

To polymerize a dynamical system one usually begins with a classical system described by Hamiltonian H. The canonical quantization of such a system transforms its Hamiltonian to an Hermitian operator, which now contains the parameter $\hbar$, in such a way that in the limit $\hbar \longrightarrow 0$, the quantum Hamiltonian $H_{\hbar}$ returns to its classical counterpart. By polymerization, the Hamiltonian gets an additional quantum parameter λ, which is rooted in the ideas of granular structure of the space-time. Therefore, by taking the classical limit of the resulting Hamiltonian $H_{\hbar,\lambda}$, we arrive at a semiclassical theory in which the parameter λ is still present. To achieve the initial classical theory, one should once again take the limit $\lambda \longrightarrow 0$ from this intermediate theory. It is believed that such effective classical theories H_{λ} have enough rich structure to exhibit some important features of the system related to the quantum effects without quantization of the system. The process by which the theory H_{λ} is obtained from the classical theory is called *classical polymerization*. A detailed explanation of this process with some of its cosmological applications can be found in [32].

In this paper, we are going to study how the metric of the Schwarzschild black hole gets modifications due to the classical polymerization. Since the thermodynamical properties of the black hole come from its geometrical structure, the corrections to the black hole's geometry yield naturally modifications to its thermodynamics. To do this, we begin with a general form of a spherically symmetric space-time and then construct a Hamiltonian in such a way that the Schwarzschild metric is resulted from the corresponding Hamiltonian equations of motion. We then follow the procedure described above and by applying it to the mentioned Hamiltonian we get the classical polymerized Hamiltonian, by means of which we expect to obtain the polymer-corrected Schwarzschild metric. The paper is organized as follows. In Section 2 we have presented a brief review of the polymer representation and classical polymerization. Section 3 is devoted to the Hamiltonian formalism of a general spherically symmetric space time. We show in this section that the resulting Hamiltonian equations of motion yield the Schwarzschild solution. In Section 4, we will apply the classical polymerization on the Hamiltonian of the spherically symmetric space-time given in Section 3 to get the polymerized Hamiltonian. We then construct the deformed Hamiltonian equations of motion and solve them to arrive the polymer corrected Schwarzschild metric. The energy-momentum tensor of the matter field corresponding to this metric as well as some of its thermodynamical properties are also presented in this section. The radial geodesics of the light and particles are obtained in Section 5 and, finally, we summarize the results in Section 6.

2. Classical Polymerization: A Brief Review

As is well known, in Schrödinger picture of quantum mechanics, the coordinates and momentum representations are equivalent and may be easily converted to each other by a Fourier transformation. However, in the presence of the quantum gravitational effects the space-time may take a discrete structure so that such well-defined representations are no longer applicable. As an alternative, polymer quantization provides a suitable framework for studying these situations [14, 15]. The Hilbert space of this representation of quantum mechanics is $\mathscr{H}_{\text{poly}} = L^2(R_d, d\mu_d)$, where $d\mu_d$ is the Haar measure and R_d denotes the real discrete line whose segments are labeled by an extra dimension-full parameter λ such that the standard Schrödinger picture will be recovered in the continuum limit $\lambda \longrightarrow 0$. This means that, by a classical limit $\hbar \longrightarrow 0$, the polymer quantum mechanics tends to an effective λ-dependent classical theory which is somehow different from the classical theory from which we have started. Such an effective theory may also be obtained directly from the standard classical theory, without referring to the polymer quantization, by using of the Weyl operator [32]. The process is known as *polymerization* with which we will deal with in the rest of this paper.

According to the mentioned above form of the Hilbert space of the polymer representation of quantum mechanics, the position space (with coordinate q) has a discrete structure with discreteness parameter λ. Therefore, the associated momentum operator $\hat{p}$, which is the generator of the displacement, does not exist [15]. However, the Weyl exponential operator (shift operator) corresponding to the discrete translation along q is well defined and effectively plays the role of momentum associated to q [14]. This allows us to utilize the Weyl operator to find an effective momentum in the semiclassical regime. So, considering a state $f(q)$, its derivative with respect to the discrete position q may be approximated by means of the Weyl operator as [32]

$$\begin{aligned} \partial_q f(q) &\approx \frac{1}{2\lambda}\left[f(q+\lambda) - f(q-\lambda)\right] \\ &= \frac{1}{2\lambda}\left(\widehat{e^{ip\lambda}} - \widehat{e^{-ip\lambda}}\right) f(q) = \frac{i}{\lambda}\widehat{\sin(\lambda p)} f(q), \end{aligned} \quad (1)$$

and similarly the second derivative approximation will be

$$\begin{aligned} \partial_q^2 f(q) &\approx \frac{1}{\lambda^2}\left[f(q+\lambda) - 2f(q) + f(q-\lambda)\right] \\ &= \frac{2}{\lambda^2}\left(\widehat{\cos(\lambda p)} - 1\right) f(q). \end{aligned} \quad (2)$$

Having the above approximations at hand, we define the polymerization process for the finite values of the parameter λ as

$$\begin{aligned} \hat{p} &\longrightarrow \frac{1}{\lambda}\widehat{\sin(\lambda p)}, \\ \hat{p}^2 &\longrightarrow \frac{2}{\lambda^2}\left(1 - \widehat{\cos(\lambda p)}\right). \end{aligned} \quad (3)$$

This replacement suggests the idea that a classical theory may be obtained via this process, but now without any attribution to the Weyl operator. This is what is dubbed usually as *classical Polymerization* in literature [14, 32]:

$$q \longrightarrow q, \quad p \longrightarrow \frac{\sin(\lambda p)}{\lambda}, \quad p^2 \longrightarrow \frac{2}{\lambda^2}\left[1-\cos(\lambda p)\right], \tag{4}$$

where now (q, p) are a pair of classical phase space variables. Hence, by applying the transformation (4) to the Hamiltonian of a classical system we get its classical polymerized counterpart. A glance at (4) shows that the momentum is periodic and varies in a bounded interval as $p \in [-\pi/\lambda, +\pi/\lambda)$. In the limit $\lambda \longrightarrow 0$, one recovers the usual range for the canonical momentum $p \in (-\infty, +\infty)$. Therefore, the polymerized momentum is compactified and topology of the momentum sector of the phase space is S^1 rather than the usual R [33]. Our set-up to explain the classical polymerization of a dynamical system is now complete. In Section 4, we will return to this issue by some more explanations and apply it to the Hamiltonian dynamics of a spherically symmetric space-time.

3. Hamiltonian Model of the Spherically Symmetric Space-Time

We start with the general spherically symmetric line element as (it can be shown that, by introducing of new radial and time coordinates as $b(r) \longrightarrow r'$ and $I(r)[a(r)dt - B(r)dr] \longrightarrow dt'$, this metric takes the standard form of static spherically symmetric line elements: $ds^2 = -A(r)dt^2 + C(r)dr^2 + r^2(d\vartheta^2 + \sin^2\vartheta d\varphi^2)$) [34, 35]

$$ds^2 = -a(r)\,dt^2 + N(r)\,dr^2 + 2B(r)\,dt\,dr + b^2(r)\left(d\vartheta^2 + \sin^2\vartheta d\varphi^2\right), \tag{5}$$

where $a(r)$, $B(r)$, $N(r)$, and $b(r)$ are some functions of r. Upon substitution this metric into the Einstein-Hilbert action

$$\mathcal{S} = \frac{1}{16\pi G}\int d^4x\sqrt{-g}\mathcal{R}, \tag{6}$$

the action taking the form

$$\mathcal{S} = \int dt \int dr L(a,b,n), \tag{7}$$

where [34, 35]

$$L = 2\sqrt{n}\left(\frac{a'b'b}{n} + \frac{ab'^2}{n} + 1\right) \tag{8}$$

is an effective Lagrangian in which the primes denote differentiation with respect to r and the Lagrange multiplier n is given by

$$n(r) = a(r)\,N(r) + B^2(r). \tag{9}$$

In metric (5) the function $N(r)$ plays the role of a lapse function with respect to the r-slicing in the ADM terminology; see [34–36]. On the other hand, according to the relation (9), the functions N and B are related to the Lagrange multiplier n which means that we can arbitrarily choose them. This is a reflection of this fact that we have freedom in the definition of the coordinates r and t in the metric (5). Hence, the only independent variables that can be determined by the Einstein field equations are the functions $a(r)$ and $b(r)$. In order to write the Hamiltonian the momenta conjugate to these variables should be evaluated, that is,

$$p_a = \frac{\partial L}{\partial a'} = \frac{2bb'}{\sqrt{n}}, \quad p_b = \frac{\partial L}{\partial b'} = 2\frac{\left(2ab' + a'b\right)}{\sqrt{n}}. \tag{10}$$

Also, the momentum associated with n vanishes which gives the primary constraint

$$p_n = \frac{\partial L}{\partial n'} = 0. \tag{11}$$

In terms of these conjugate momenta the canonical Hamiltonian is given by its standard definition $H = \sum_{q=a,b,n} q' p_q - L$, leading to

$$H = \sqrt{n}\left(\frac{p_a p_b}{2b} - \frac{a}{2b^2}p_a^2 - 2\right) + \Lambda p_n, \tag{12}$$

in which due to existence of the constraint (11) we have added the last term that is the primary constraints multiplied by an arbitrary functions $\Lambda(r)$. The Hamiltonian equation for n then reads

$$n' = \{n, H\} = \Lambda. \tag{13}$$

Now, let us restrict ourselves to a certain class of gauges, namely, $n = \text{const.}$, which is equivalent to the choice $\Lambda = 0$. With a constant n we assume $n = 1$ without losing general character of the solutions. By this choice, the Hamiltonian equations of motion for the other variables are as

$$a' = \{a, H\} = \frac{p_b}{2b} - \frac{a}{b^2}p_a, \quad p_a' = \{p_a, H\} = \frac{p_a^2}{2b^2}, \quad b' = \{b, H\} = \frac{p_a}{2b}, \quad p_b' = \{p_b, H\} = \frac{p_a p_b}{2b^2} - \frac{a}{b^3}p_a^2. \tag{14}$$

From the second and third equations of (14) we obtain

$$p_a = k_1 b, \tag{15}$$

from which one gets

$$b(r) = \frac{k_1}{2}r + k_2, \quad p_a(r) = \frac{k_1^2}{2}r + k_1 k_2, \tag{16}$$

where k_1 and k_2 are integration constants. Upon substituting these results into the first and fourth equations of the system (14), we arrive at the following system:

$$a' = \frac{1}{k_1 r + 2k_2} p_b - \frac{2k_1}{k_1 r + 2k_2} a,$$
$$p_b' = \frac{k_1}{k_1 r + 2k_2} p_b - \frac{2k_1^2}{k_1 r + 2k_2} a, \tag{17}$$

which results $p_b' = k_1 a'$ and then $p_b = k_1 a + k_3$. Therefore,

$$a'(r) = -\frac{k_1 a + k_3}{k_1 r + 2k_2}, \tag{18}$$

In which after integration we obtain

$$a(r) = \frac{k_3}{k_1} + \frac{k_1 k_4}{k_1 r + 2k_2}, \tag{19}$$

and

$$p_b(r) = 2k_3 + \frac{k_1^2 k_4}{k_1 r + 2k_2}, \tag{20}$$

with k_3 and k_4 being integration constants. Now, all of the above results should satisfy the constraint equation $H = 0$. Thus, with the help of (12) we get $k_1 k_3 = 4$, where we fix them as $k_1 = k_3 = 2$. Also, k_2 and k_4 remain arbitrary where we take their values as $k_2 = 0$ and $k_4 = -2M$ with M being a constant. Therefore, the metric functions take the form

$$a(r) = 1 - \frac{2M}{r},$$
$$b(r) = r, \tag{21}$$

and their conjugate momenta are

$$p_a(r) = 2r,$$
$$p_b(r) = 4 - \frac{4M}{r}. \tag{22}$$

Finally, with using these relations in (5) and (9), the metric is obtained as

$$ds^2 = -\left(1 - \frac{2M}{r}\right) dt^2 + N(r) dr^2$$
$$+ 2\left[1 - \left(1 - \frac{2M}{r}\right) N(r)\right]^{1/2} dt\, dr \tag{23}$$
$$+ r^2 \left(d\vartheta^2 + \sin^2 \vartheta d\varphi^2\right).$$

In the final stage we have to eliminate the function $N(r)$. This function should be interpreted as a Lagrange-multiplier and, thus, cannot be considered as a real dynamical variable. As we mentioned before, one may freely choose it. From the physical point of view the function $N(r)$ corresponds to a gauge freedom in choice of coordinates r and t in the above metric. If we choose the lapse function as $N(r) = (1-2M/r)^{-1}$ this metric takes its canonical form

$$ds^2 = -\left(1 - \frac{2M}{r}\right) dt^2 + \left(1 - \frac{2M}{r}\right)^{-1} dr^2$$
$$+ r^2 \left(d\vartheta^2 + \sin^2 \vartheta d\varphi^2\right), \tag{24}$$

which is nothing but the familiar form for the metric of the Schwarzschild black hole. However, we may identify the line element (23) with the Eddington-Finkelstein metric

$$ds^2 = -\left(1 - \frac{2M}{r}\right) dt^2 + \frac{4M}{r} dt\, dr + \left(1 + \frac{2M}{r}\right) dr^2$$
$$+ r^2 \left(d\vartheta^2 + \sin^2 \vartheta d\varphi^2\right), \tag{25}$$

for $N(r) = 1 + 2M/r$, or with some other kinds of spherically symmetric metrics for $N = 1$; see [37]. In summary, from physical viewpoint choosing different gauge functions $N(r)$ is actually looking at a space-time from a different perspective. For example, the metric (25) can be obtained from (24) by introducing a new time coordinate $\bar{t} = t + 2M \ln(r - 2M)$ in which the radial null geodesics (see Section 5) become straight lines. In this sense, the two metrics may differ from some aspects. While the Schwarzschild metric is singular at $r = 2M$ the Eddington-Finkelstein metric is regular not only at $r = 2M$ but also for the whole range $0 < r < 2M$. Indeed, the coordinate range is extended from $2M < r < \infty$ to $0 < r < \infty$.

From now on we focus on Schwarzschild black hole metric and to justify the meaning of the constant M, noting that the Newtonian gravitational potential of a point mass m situated at the origin is given by the relation $\phi = -Gm/r$. On the other hand in the weak-field limit the g_{00} component of the metric takes the form $g_{00} = -(1 + 2\phi/c^2)$ [38]. Therefore, comparing this with (24) we see that $M = Gm/c^2$. This means that we may interpret the constant M as due to the mass of the above mentioned point particle in relativistic units.

4. Polymerization of the Model

As explained in the second section the method of polymerization is based on the modification of the Hamiltonian to get a deformed Hamiltonian H_λ, where λ is the deformation parameter. Quantum polymerization of the spherically symmetric space-time is studied in [39, 40] in which the interior of the Schwarzschild black hole as described by a Kantowski-Sachs cosmological model is quantized by loop quantization method. For our system this method will be done by applying the transformation (4) on the Hamiltonian (12). However, since all of the thermodynamical properties of the black hole are encoded in the function $a(r)$, we will polymerize only the $a(r)$-sector of the Hamiltonian. So, by means of the transformation

$$p_a \longrightarrow \frac{1}{\lambda} \sin(\lambda p_a), \quad a \longrightarrow a,$$
$$p_b \longrightarrow p_b, \quad b \longrightarrow b \tag{26}$$

the Hamiltonian takes the form

$$H_\lambda = -\frac{a}{2b^2}\frac{\sin^2(\lambda p_a)}{\lambda^2} + \frac{p_b}{2b}\frac{\sin(\lambda p_a)}{\lambda} - 2. \qquad (27)$$

As we mentioned earlier, by this one-parameter λ-dependent classical theory, we expect to address the quantum features of the system without a direct reference to the quantum mechanics. Indeed, here instead of first dealing with the quantum pictures based on the quantum Hamiltonian operator, one modifies the classical Hamiltonian according to the transformation (4) and then deals with classical dynamics of the system with this deformed Hamiltonian. In the resulting classical system the discreteness parameter λ plays an essential role since its supports the idea that the λ-correction to the classical theory is a signal from quantum gravity. Under these conditions the Hamiltonian equations of motion for the above Hamiltonian are

$$\begin{aligned} a' &= \{a, H_\lambda\} \\ &= \frac{p_b}{2b}\cos(\lambda p_a) - \frac{a}{\lambda b^2}\sin(\lambda p_a)\cos(\lambda p_a), \\ p_a' &= \{p_a, H_\lambda\} = \frac{\sin^2(\lambda p_a)}{2\lambda^2 b^2}, \\ b' &= \{b, H_\lambda\} = \frac{\sin(\lambda p_a)}{2\lambda b}, \\ p_b' &= \{p_b, H_\lambda\} = \frac{p_b}{2\lambda b^2}\sin(\lambda p_a) - \frac{a}{\lambda^2 b^3}\sin^2(\lambda p_a). \end{aligned} \qquad (28)$$

The second and the third equations of this system give $dp_a/db = \sin(\lambda p_a)/\lambda b$, integration of which results in $b = C_1 \tan((1/2)\lambda p_a)$, where C_1 is an integration constant. We note that, in the limit $\lambda \longrightarrow 0$, this relation should back to $b = (1/2)p_a$, obtained in the previous section. So, taking this limit fixes the integration constant as $C_1 = 1/\lambda$. Therefore,

$$b = \frac{1}{\lambda}\tan\left(\frac{1}{2}\lambda p_a\right). \qquad (29)$$

Now, we may use this result in the second equation of (28) to arrive at

$$p_a' = 2\cos^4\left(\frac{1}{2}\lambda p_a\right), \qquad (30)$$

whose integral is

$$\frac{2}{3}\frac{\tan((1/2)\lambda p_a)}{\lambda} + \frac{1}{3}\frac{\tan((1/2)\lambda p_a)}{\lambda\cos^2((1/2)\lambda p_a)} = r. \qquad (31)$$

From (29) we get $\cos^2((1/2)\lambda p_a) = (1+\lambda^2 b^2)^{-1}$. With the help of these relations (31) takes the following algebraic form for the function $b(r)$:

$$\lambda^2 b^3 + 3b - 3r = 0, \qquad (32)$$

which admits the exact solution

$$b(r) = \frac{\left[3\lambda r + \sqrt{4+9\lambda^2 r^2}\right]^{2/3} - 2^{2/3}}{2^{1/3}\lambda\left[3\lambda r + \sqrt{4+9\lambda^2 r^2}\right]^{1/3}}. \qquad (33)$$

Up to second order of λ, we have

$$b(r) = r - \frac{1}{3}\lambda^2 r^3 + \mathcal{O}(\lambda^3). \qquad (34)$$

Now, let us go back to the first and the fourth equations of the system (28). Using (29), they take the form

$$a' = \frac{p_b}{2b}\frac{1-\lambda^2 b^2}{1+\lambda^2 b^2} - \frac{2a}{b}\frac{1-\lambda^2 b^2}{(1+\lambda^2 b^2)^2}, \qquad (35)$$

and

$$p_b' = \frac{p_b}{b}\frac{1}{1+\lambda^2 b^2} - \frac{4a}{b}\frac{1}{(1+\lambda^2 b^2)^2}, \qquad (36)$$

in which we have used the trigonometric relations: $\sin(\lambda p_a) = 2\lambda b/(1+\lambda^2 b^2)$ and $\cos^2(\lambda p_a) = (1-\lambda^2 b^2)/(1+\lambda^2 b^2)$. From these two equations we get

$$p_b' = \frac{2}{1-\lambda^2 b^2}a', \qquad (37)$$

where up to second order of λ, using (34) is

$$p_b' = 2\left[1 + \lambda^2 r^2 + \mathcal{O}(\lambda^3)\right]a', \qquad (38)$$

and thus

$$p_b = 2\int\left(1+\lambda^2 r^2\right)a'\,dr. \qquad (39)$$

We may use this relation in (35) to get a differential equation for $a(r)$. However, since the resulting equation seems to be too complicated to have an exact solution, we rely on an approximation according to which we ignore all powers of λ in the r.h.s. of (35) and so obtain

$$a'(r) = \frac{1}{r}\int\left(1+\lambda^2 r^2\right)a'\,dr - \frac{2}{r}a, \qquad (40)$$

or after differentiation of both sides

$$r a''(r) = \left(\lambda^2 r^2 - 2\right)a', \qquad (41)$$

with solution

$$a(r) = C_2 + C_3\left[-\frac{1}{r}e^{\lambda^2 r^2/2} + \lambda\sqrt{\frac{\pi}{2}}\operatorname{erfi}\left(\frac{\lambda r}{\sqrt{2}}\right)\right], \qquad (42)$$

where C_2 and C_3 are two integration constant and $\operatorname{erfi}(z)$ is the imaginary error function. Up to second order of λ this expression has the form

$$a(r) = C_2 - \frac{C_3}{r} + \frac{1}{2}C_3\lambda^2 r, \qquad (43)$$

comparison of which with (21) suggests that the integration constants should fix as $C_2 = 1$ and $C_3 = 2M$. So,

$$a(r) = 1 + 2M\left[-\frac{1}{r}e^{\lambda^2 r^2/2} + \lambda\sqrt{\frac{\pi}{2}}\text{erfi}\left(\frac{\lambda r}{\sqrt{2}}\right)\right]. \quad (44)$$

Therefore, by choosing the lapse function in the form $N(r) = a^{-1}(r)$ (see the discussion after (23)), the polymerized metric takes the form

$$\begin{aligned} ds^2 &= -a(r)\,dt^2 + a^{-1}(r)\,dr^2 \\ &\quad + b^2(r)\left(d\vartheta^2 + \sin^2\vartheta d\varphi^2\right), \end{aligned} \quad (45)$$

where $a(r)$ and $b(r)$ are given in (44) and (34), respectively. It is seen that in the limit $\lambda \longrightarrow 0$ the line element (45) returns to the usual Schwarzschild metric (24). However, its asymptotic behavior, which comes from the expansion

$$\begin{aligned} a(r) &= 1 - \frac{2M}{r} \\ &\quad + M\lambda^2 r\left[1 + \frac{1}{12}(\lambda r)^2 + \frac{1}{120}(\lambda r)^4 + \cdots\right], \end{aligned} \quad (46)$$

shows that, in spite of the Schwarzschild case, the metric is not flat for large values of r. Later in this section, we attribute such an asymptotic behavior to the matter field that created this metric. In what follows, we will deal with the physical properties, including thermodynamics, of the space-time (45). Since such properties of a black hole can be derived from its geometry, we expect that the deformed forms of these properties return to their ordinary form in the limit $\lambda \longrightarrow 0$.

At first, let us take a look at the horizon(s) radius of the metric (45) which may be deduced from the roots of equation $a(r) = 0$. Up to the leading order of parameter λ, the positive root of this equation is

$$r_H \simeq \frac{\sqrt{1 + 8M^2\lambda^2} - 1}{2M\lambda^2}. \quad (47)$$

On the other hand since $da(r)/dr = 2Me^{\lambda^2 r^2/2}/r^2 > 0$, the function $a(r)$ is monotonically increasing and thus the metric cannot have more than one horizon whose radius is approximately given in (47). To see the behavior of the above metric near the Schwarzschild essential singularity $r = 0$, we may evaluate some scalars associated with the metric such as Ricci scalar R, $R_{\mu\nu}R^{\mu\nu}$, and the Kretschmann scalar $K = R_{\mu\nu\sigma\delta}R^{\mu\nu\sigma\delta}$. A straightforward calculation shows that

$$R = \frac{2\lambda M}{r^2}e^{\lambda^2 r^2/2}\left[\lambda r + 2\sqrt{2}F\left(\frac{\lambda r}{\sqrt{2}}\right)\right], \quad (48)$$

where $F(x) = e^{-x^2}\int_0^x e^{y^2}dy$ is the Dawson function. Near $r = 0$, the above relation behaves as $6M\lambda^2/r$, so given that the value of the parameter λ is also very small we have $\lim_{r\longrightarrow 0}R \simeq \mathcal{O}(\lambda)$. Computing of the scalar $R_{\mu\nu}R^{\mu\nu}$ shows the similar behavior near $r = 0$, while the Kretschmann scalar takes the form

$$\begin{aligned} K &= \frac{4M^2e^{\lambda^2 r^2}}{r^6}\left[4\lambda^2 r^2\left(2F\left(\frac{r\lambda}{\sqrt{2}}\right)^2 - 1\right)\right. \\ &\quad \left. - 8\sqrt{2}\lambda rF\left(\frac{r\lambda}{\sqrt{2}}\right) + \lambda^4 r^4 + 12\right], \end{aligned} \quad (49)$$

which behaves as $K \simeq 48M^2/r^6 + 20M^2\lambda^4/3r^2 + \mathcal{O}(\lambda^5)$. Thus near $r = 0$ we have $K \simeq 48M^2/r^6$. This shows that the space-time described by the metric (45) has an essential singularity at $r = 0$, which cannot be removed by a coordinate transformation.

Now, let us investigate the properties of the matter corresponding to the metric (45). Considering the Einstein equations $G^\mu_\nu = R^\mu_\nu - (1/2)R\delta^\mu_\nu \sim T^\mu_\nu$, the components of the energy-momentum tensor become

$$\begin{aligned} T^\mu_\nu &= \text{diag}\left(-\rho, p_r, p_\perp, p_\perp\right) \\ &= \text{diag}\left(-\frac{\sqrt{2\pi}\lambda M\,\text{erfi}\left(\lambda r/\sqrt{2}\right)}{r^2},\right. \\ &\quad -\frac{\sqrt{2\pi}\lambda M\,\text{erfi}\left(\lambda r/\sqrt{2}\right)}{r^2}, -\frac{\lambda^2 Me^{\lambda^2 r^2/2}}{r}, \\ &\quad \left.-\frac{\lambda^2 Me^{\lambda^2 r^2/2}}{r}\right). \end{aligned} \quad (50)$$

Before going any further, a remark is in order. The usual Schwarzschild metric is often considered a vacuum solution since it solves $R_{\mu\nu} = 0$ which is equivalent to the Einstein vacuum field equations $G_{\mu\nu} = 0$. However, as (50) explicitly shows the polymer corrected metric (45) is not a vacuum solution. Then, a question arises: what mechanism made it possible starting from a vacuum solution we get a nonvacuum solution? To deal with this question note that any vacuum solution must be found in the absence of matter, strictly speaking, only the Minkowski metric can be considered as a vacuum solution. As shown in [41], in the Schwarzschild case there is a source term (energy-momentum tensor) concentrated on the origin, the origin which usually excluded from the space-time manifold. So we are faced with an unacceptable physical situation in which a curved metric is generated by a zero energy-momentum tensor. In [41] with more accurate calculations based on distributional techniques the energy-momentum tensor of the Schwarzschild geometry is obtained and it has been shown that its Ricci scalar is equal to $8\pi M\delta(r)$ which yields an energy-momentum tensor proportional to $M\delta(r)$. Now, what is happening in the effective theories such as noncommutative, see [42], and polymeric counterparts of the Schwarzschild solution is that the concentrated matter on the origin will spread throughout space by the polymer parameter λ (or noncommutative parameter θ in noncommutative theories).

The energy-momentum tensor (50) shows a fluid with radial pressure $p_r = -\rho$ and tangential pressure $p_\perp = -\rho - (r/2)\partial_r\rho$. In comparison with the conventional perfect

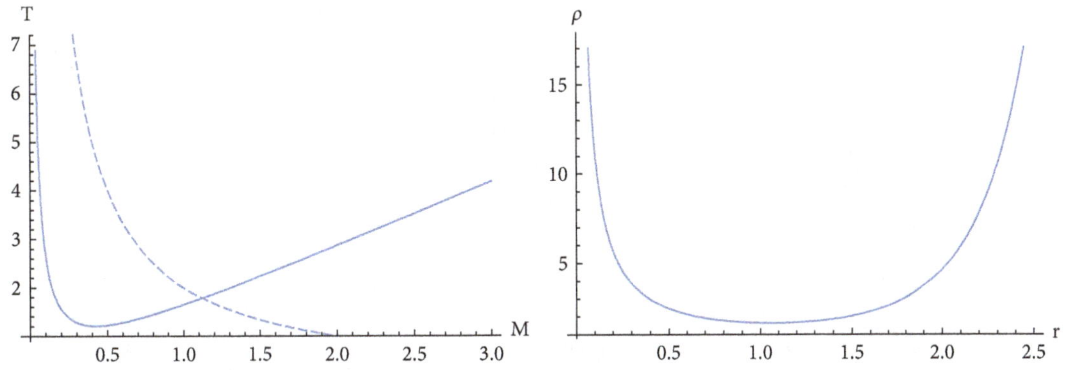

FIGURE 1: Left: Temperature versus mass. The solid line shows the qualitative behavior of the relation (53) while the dashed line refers to the conventional temperature of the Schwarzschild black hole. Right: the density of the matter distribution versus r. The figures are plotted for $\lambda = 0.1$ and $M = 1$.

fluid with isotropic pressure the above energy-momentum tensor shows an unusual behavior since its pressure exhibits an anisotropic behavior. At short distances the difference between p_r and $p_\perp$ is of order λ^2, which shows that the fluid behaves approximately like a perfect fluid. However, when r grows the anisotropy between the pressure's components increases and the behavior of the fluid is far from the perfect fluid behavior. The nonvanishing radial pressure of the above anisotropic fluid may be interpreted as a result of the quantum fluctuation of given space-time. The large amount of this pressure near the origin prevents the matter collapsing into this point. Such an unusual equation of state for fluids also appeared in the noncommutative theories of black holes [42]. In view of the validity of the energy conditions, we see that

$$\rho + p_r + 2p_\perp = -2\frac{\lambda^2 M e^{\lambda^2 r^2/2}}{r} < 0, \tag{51}$$

which shows the violation of the strong energy condition for this exotic distribution of matter. On the other hand, in view of the weak energy conditions, while the relation $\rho + p_r \geq 0$ is always satisfied, the condition $\rho + p_\perp \geq 0$ is violated for $r > \mathcal{O}(1/\lambda)$. The violation of the energy conditions shows that the classical description of this type of matter field is not credible and thus the corresponding gravity should be described by an effective quantum theory (here the polymerized theory) rather than the usual general relativity.

Finally, let us take a quick look at thermodynamics of the metric (45). According to the Hawking formulation the black hole's temperature is proportional to the surface gravity at the black hole horizon. It can be shown that for a diagonal metric such as (45) the surface gravity is [43]

$$\kappa = \sqrt{-\frac{1}{4}g^{tt}g^{rr}\left(\frac{\partial g_{tt}}{\partial r}\right)^2} = \frac{M}{r^2}e^{\lambda^2 r^2/2}. \tag{52}$$

Evaluating this expression at the horizon radius (47) gives the temperature as

$$T \propto \frac{4\lambda^4 M^3 e^{(\sqrt{8\lambda^2M^2+1}-1)^2/8\lambda^2M^2}}{\left(\sqrt{8\lambda^2M^2+1}-1\right)^2}. \tag{53}$$

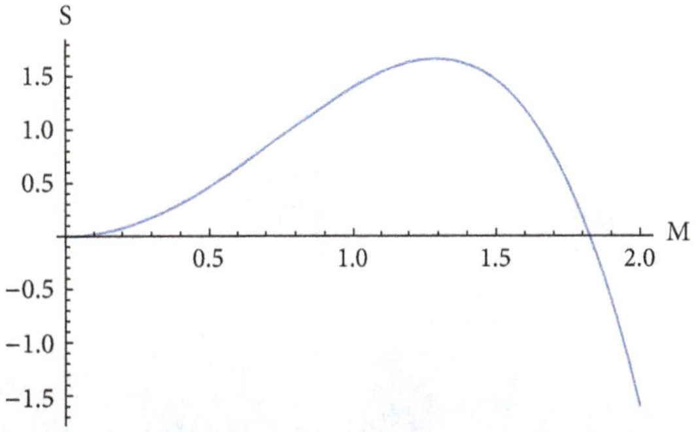

FIGURE 2: The entropy versus mass. The figure is plotted for $\lambda = 0.1$ and $M = 1$.

In Figure 1 we have plotted the qualitative behavior of the above results. As this figure shows near the origin the matter has a dense core like its conventional Schwarzschild counterpart. Thus, the temperature changes like Schwarzschild in this regime. However, in a global look, the exotic properties of the matter cause different behavior for temperature. Unlike the usual Schwarzschild case, by decreasing the mass, the radiation temperature first decreases to a minimum value and then exhibits the normal behavior; i.e., the temperature increases while the mass is decreasing. The reason for this abnormal behavior in the temperature of the radiation may be found in the nature of the dark energy-like of the matter field described by the energy-momentum tensor (50). Now, by the second law of thermodynamics $dS = dM/T$, we may compute the entropy as

$$S = \int \frac{e^{-(\sqrt{8\lambda^2M^2+1}-1)^2/8\lambda^2M^2}\left(\sqrt{8\lambda^2M^2+1}-1\right)^2}{4\lambda^4M^3}dM. \tag{54}$$

We see that this integral cannot be evaluated analytically. In Figure 2, employing numerical methods, we have shown the approximate behavior of the entropy for typical values of the parameters. As the figure shows with decreasing mass, the entropy grows from negative values up to a maximum positive value and then behaves like the Schwarzschild case and decreases to zero.

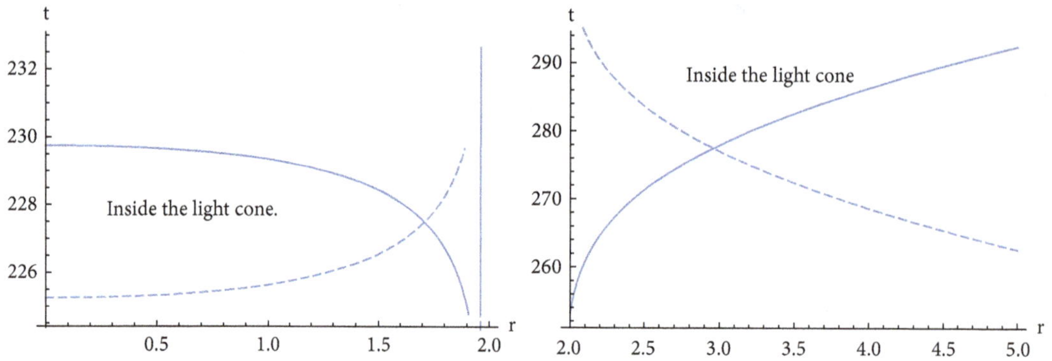

FIGURE 3: Left: the outgoing (dashed line) and incoming (solid line) geodesics for $r < r_H$. Right:tThe outgoing (solid line) and incoming (dashed line) geodesics $r > r_H$. The figures are plotted for $\lambda = 0.1$ and $M = 1$.

5. Geodesics of the Polymerized Metric

In this section we are going to study how light and particles will move in the geometrical background given by metric (45). This is important because from the classical trajectories of light or falling particles we understand that the corresponding space-time behaves really like a black hole. First, consider the radial null geodesics are defined by $ds = 0$ and $d\vartheta = d\varphi = 0$. Therefore, we have

$$-a(r)\,dt^2 + a^{-1}(r)\,dr^2 = 0 \Longrightarrow \qquad dt = \pm\frac{dr}{a(r)}. \tag{55}$$

In order to get an analytical solution we use the approximate $a(r) \sim 1 - 2M/r + M\lambda^2 r$, for which we obtain

$$t - t_0 = \pm\frac{2\left|\tanh^{-1}\left(\left(2Mr\lambda^2 + 1\right)/\sqrt{8M^2\lambda^2 + 1}\right)\right|/\sqrt{8\lambda^2M^2 + 1} + \log\left|Mr^2\lambda^2 - 2M + r\right|}{2\lambda^2 M}. \tag{56}$$

For $r > r_H$, the above expression with positive (negative) sign shows that r increases (decreases) as t increases and thus the corresponding curve is an *outgoing* (*incoming*) radial null geodesics. For $r < r_H$, the situation is reversed; i.e., the positive and negative signs correspond to the incoming and outgoing curves, respectively; see Figure 3. A glance at this figure makes it clear that none of the null geodesics can pass through the horizon which shows the black hole nature of the underlying space-time. It is clear that in comparison with region $r > r_H$ the local light cones tip over in region $r < r_H$. This is because while the coordinates r and t are space-like and time-like, respectively, in region $r > r_H$ and in region $r < r_H$ they reverse their character. The orientation of the light cones inside the horizon shows that nothing can stay at rest in this region but will be forced to move towards the black hole center.

To complete our geodesics analysis, let us now consider the radial trajectory of a falling free particle. It moves along the time-like geodesics which results the following equations of motion [38]:

$$a(r)\,\dot{t} = k, \tag{57}$$

$$a(r)\,\dot{t}^2 - a^{-1}(r)\,\dot{r}^2 = 1, \tag{58}$$

where a dot denotes differentiation with respect to the proper time τ and k is a constant that depends on the initial conditions. If we assume that the particle begins to fall with zero initial velocity from a distance r_0 for which $a(r_0) = 1$, then $k = 1$. Also, for the motion around this region we have $\dot{t} \simeq 1 \Longrightarrow t \simeq \tau$. Therefore, we may analyse the path of the particle in view of a comoving observer which uses the proper time. Then, (57) and (58) give

$$\frac{d\tau}{dr} = -\left(\frac{r}{2M - M\lambda^2 r^2}\right)^{1/2}, \tag{59}$$

in which we have used the same previous approximation for $a(r)$. Upon integration we get

$$\tau - \tau_0 = -\frac{1}{3} r\sqrt{4 - 2\lambda^2 r^2}\, {}_2F_1\left(\frac{1}{2}, \frac{3}{4}; \frac{7}{4}; \frac{r^2\lambda^2}{2}\right) \cdot \sqrt{\frac{r}{2M - \lambda^2 M r^2}}, \tag{60}$$

where τ_0 is an integration constant and ${}_2F_1(a, b; c; z)$ is a Hypergeometric function. The above equation shows that in view of the proper observer no singular behavior occurs at the horizon radius and the particle falls to the center of the black hole; see Figure 4. If instead one describes the motion in terms of the coordinate time t, the situation becomes like the null geodesics; i.e., in view of a distance observer the particle cannot pass through the horizon and it takes infinite

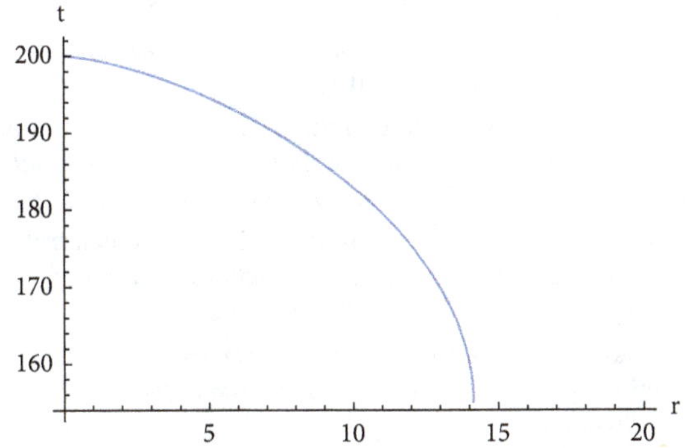

Figure 4: The trajectory of an infalling particle in terms of the proper time τ. The particle falls continuously to the singularity $r = 0$ in a finite proper time. The figure is plotted for $\lambda = 0.1$ and $M = 1$.

time for the falling particle to reach the horizon, so that r_H is approached but never passed. All of these results show that in terms of light and particles motion the space-time given by the metric (45) behaves like a black hole as its Schwarzschild counterpart.

6. Summary

In this paper we have studied the classical polymerization procedure applied on the Schwarzschild metric. This procedure is based on a classical transformation under which the momenta are transformed like their polymer quantum mechanical counterpart. After a brief review of the polymer representation of quantum mechanics, we have introduced the classical polymerization by means of which the Hamiltonian of the theory under consideration gets modification in such a way that a parameter λ, coming from polymer quantization, plays the role of a deformation parameter. In order to apply this mechanism on the Schwarzschild black hole, we first presented a Hamiltonian function for a general spherically symmetric space-time and showed that the resulting Hamiltonian equations yield the conventional Schwarzschild metric. Then, we have applied the polymerization on this minisuperspace model and solved the Hamiltonian equations once again to achieve the polymer corrected Schwarzschild metric. We saw that while the usual Schwarzschild metric is a vacuum spherical symmetric solution of the Einstein equations, this is not the case for its polymerized version obtained by the above mentioned method. Interestingly, the energy-momentum tensor of the matter field corresponding to the polymerized metric has anisotropic negative pressure sector with a dark energy-like equation of state. As expected, the unusual behavior of such a matter field resulted in an uncommon behavior for the thermodynamical quantities like temperature and entropy in comparison with the traditional Schwarzschild solution. Finally, to clarify that the polymerized metric has also the black hole nature, we have investigated the null geodesics and verified that the outgoing and incoming geodesics curves can never pass through the horizon. Also, we proved that in view of a comoving observer which uses the proper time an infalling particle continuously falls to the center $r = 0$, without experiencing something passing through the horizon. All these indicate that the underlying space-time in which the light and particles are traveling is really a black hole.

Conflicts of Interest

The author declares that they have no conflicts of interest.

Acknowledgments

This work has been supported financially by Research Institute for Astronomy and Astrophysics of Maragha (RIAAM) under research project no. 1/5237-107.

References

[1] B. S. DeWitt, "Quantum theory of gravity. I. The canonical theory," *Physical Review A: Atomic, Molecular and Optical Physics*, vol. 160, 1967.

[2] C. Rovelli and L. Smolin, "Discreteness of area and volume in quantum gravity," *Nuclear Physics B*, vol. 442, no. 3, pp. 593–619, 1995.

[3] A. Ashtekar and J. Lewandowski, "Quantum theory of geometry. I. Area operators," *Classical and Quantum Gravity*, vol. 14, no. 1A, pp. A55–A81, 1997.

[4] C. Rovelli, "Loop quantum gravity," *Living Reviews*, vol. 1, p. 1, 1998.

[5] A. Kempf, G. Mangano, and R. B. Mann, "Hilbert space representation of the minimal length uncertainty relation," *Physical Review D: Particles, Fields, Gravitation and Cosmology*, vol. 52, no. 2, pp. 1108–1118, 1995.

[6] S. Hossenfelder, "Minimal length scale scenarios for quantum gravity," *Living Reviews in Relativity*, vol. 16, no. 2, 2013.

[7] A. Kempf, "Quantum field theory with nonzero minimal uncertainties in positions and momenta," *High Energy Physics*, vol. 11-12, pp. 1041–1048, 1994.

[8] A. Kempf and G. Mangano, "Minimal length uncertainty relation and ultraviolet regularization," *Physical Review D: Particles, Fields, Gravitation and Cosmology*, vol. 55, no. 12, 1997.

[9] S. Hossenfelder, "Interpretation of quantum field theories with a minimal length scale," *Physical Review D: Particles, Fields, Gravitation and Cosmology*, vol. 73, no. 10, 2006.

[10] J. B. Achour and S. Brahma, "Covariance in self dual inhomogeneous models of effective quantum geometry: Spherical symmetry and Gowdy systems," *Physical Review D: Particles, Fields, Gravitation and Cosmology*, vol. 97, 2018.

[11] M. Bojowald, S. Brahma, and J. D. Reyes, "Covariance in models of loop quantum gravity: spherical symmetry," *Physical Review D: Covering Particles, Fields, Gravitation, and Cosmology*, vol. 92, no. 4, 2015.

[12] P. Nicolini, A. Smailagic, and E. Spallucci, "Noncommutative geometry inspired Schwarzschild black hole," *Physics Letters B*, vol. 632, no. 4, pp. 547–551, 2006.

[13] E. Spallucci, A. Smailagic, and P. Nicolini, "Trace anomaly on a quantum spacetime manifold," *Physical Review D: Particles, Fields, Gravitation and Cosmology*, vol. 73, 2006.

[14] A. Corichi, T. Vukašinac, and J. A. Zapata, "Polymer quantum mechanics and its continuum limit," *Physical Review D: Particles, Fields, Gravitation and Cosmology*, vol. 76, no. 4, 2007.

[15] A. Ashtekar, S. Fairhurst, and J. L. Willis, "Quantum gravity, shadow states and quantum mechanics," *Classical and Quantum Gravity*, vol. 20, no. 6, pp. 1031–1061, 2003.

[16] G. De Risi, R. Maartens, and P. Singh, "Graceful exit via polymerization of pre-big-bang cosmology," *Physical Review D: Particles, Fields, Gravitation and Cosmology*, vol. 76, no. 10, 2007.

[17] F. Wu and M. Zhong, "TeV scale Lee-Wick fields out of large extra dimensional gravity," *Physical Review D: Particles, Fields, Gravitation and Cosmology*, vol. 78, 2008.

[18] G. M. Hossain, V. Husain, and S. S. Seahra, "Nonsingular inflationary universe from polymer matter," *Physical Review D: Particles, Fields, Gravitation and Cosmology*, vol. 81, no. 2, 2010.

[19] A. Corichi and A. Karami, "Loop quantum cosmology of $k = 1$ FRW: A tale of two bounces," *Physical Review D*, vol. 84, 2011.

[20] S. M. Hassan and V. Husain, "Semiclassical cosmology with polymer matter," *Classical and Quantum Gravity*, vol. 34, no. 8, 2017.

[21] B. Vakili, K. Nozari, V. Hosseinzadeh, and M. A. Gorji, "Bouncing scalar field cosmology in the polymeric minisuperspace picture," *Modern Physics Letters A*, vol. 29, no. 32, 2014.

[22] J. B. Achour, F. Lamy, H. Liu, and K. Noui, "Polymer schwarzschild black hole: an effective metric," *Europhysics letters*, vol. 123, 2018.

[23] C. G. BOehmer and K. Vandersloot, "Loop quantum dynamics of the Schwarzschild interior," *Physical Review D: Particles, Fields, Gravitation and Cosmology*, vol. 76, no. 10, 2007.

[24] A. Peltola and G. Kunstatter, "Effective polymer dynamics of *D*-dimensional black hole interiors," *Physical Review D: Particles, Fields, Gravitation and Cosmology*, vol. 80, no. 4, 2009.

[25] E. Bianchi, "Black hole entropy, loop gravity, and polymer physics," *Classical and Quantum Gravity*, vol. 28, no. 11, 2011.

[26] E. R. Livine and D. R. Terno, "Entropy in the classical and quantum polymer black hole models," *Classical and Quantum Gravity*, vol. 29, no. 22, 2012.

[27] G. Chacón-Acosta, E. Manrique, L. Dagdug, and H. A. Morales-Técotl, "Statistical thermodynamics of polymer quantum systems," *Loop Quantum Gravity and Cosmology*, vol. 7, no. 110, p. 23, 2011.

[28] G. M. Hossain, V. Husain, and S. S. Seahra, "Background-independent quantization and the uncertainty principle," *Classical and Quantum Gravity*, vol. 27, no. 16, 2010.

[29] M. Gorji, K. Nozari, and B. Vakili, "Polymeric quantization and black hole thermodynamics," *Physics Letters B*, vol. 735, pp. 62–68, 2014.

[30] M. Gorji, K. Nozari, and B. Vakili, "Thermostatistics of the polymeric ideal gas," *Physical Review D: Particles, Fields, Gravitation and Cosmology*, vol. 90, no. 4, 2014.

[31] M. A. Gorji, K. Nozari, and B. Vakili, "Polymer quantization versus the Snyder noncommutative space," *Classical and Quantum Gravity*, vol. 32, no. 15, 2015.

[32] A. Corichi and T. Vukašinac, "Effective constrained polymeric theories and their continuum limit," *Physical Review D: Particles, Fields, Gravitation and Cosmology*, vol. 86, no. 6, 2012.

[33] K. Nozari, M. A. Gorji, V. Hosseinzadeh, and B. Vakili, "Natural cutoffs via compact symplectic manifolds," *Classical and Quantum Gravity*, vol. 33, no. 2, 2016.

[34] M. Cavaglia, V. De Alfaro, and A. T. Filippov, "Hamiltonian formalism for black holes and quantization," *International Journal of Modern Physics D*, vol. 4, no. 5, pp. 661–672, 1995.

[35] M. Cavaglia, V. De Alfaro, and A. T. Filippov, "Quantization of the Schwarzschild black hole," *International Journal of Modern Physics D*, vol. 5, no. 3, pp. 227–250, 1996.

[36] B. Vakili, "Quantization of the Schwarzschild black hole: a Noether symmetry approach," *International Journal of Theoretical Physics*, vol. 51, no. 1, pp. 133–145, 2012.

[37] P. Kraus and F. Wilczek, "Some applications of a simple stationary line element for the Schwarzschild geometry," *Modern Physics Letters A*, vol. 9, no. 40, pp. 3713–3719, 1994.

[38] R. d'Inverno, *Introducing Einstein's Relativity*, Oxford University Press, New York, NY, USA, 1998.

[39] A. Corichi and P. Singh, "Loop quantization of the Schwarzschild interior revisited," *Classical and Quantum Gravity*, vol. 33, no. 5, 2016.

[40] J. Olmedo, S. Saini, and P. Singh, "From black holes to white holes: a quantum gravitational, symmetric bounce," *Classical and Quantum Gravity*, vol. 34, no. 22, 2017.

[41] H. Balasin and H. Nachbagauer, "The energy-momentum tensor of a black hole, or What curves the Schwarzschild geometry?" *Classical and Quantum Gravity*, vol. 10, no. 11, pp. 2271–2278, 1993.

[42] P. Nicolini, "Noncommutative black holes, the final appeal to quantum gravity: a review," *International Journal of Modern Physics A*, vol. 24, no. 7, 2009.

[43] Ø. Grøn and S. Hervik, *Einsteins General Theory of Relativity: With Modern Applications in Cosmology*, Springer-Verlag, New York, NY, USA, 2007.

14

More on the Non-Gaussianity of Perturbations in a Nonminimal Inflationary Model

R. Shojaee,[1] **K. Nozari**,[2,3] **and F. Darabi**[1]

[1]*Department of Physics, Azarbaijan Shahid Madani University, P.O. Box 53714-161, Tabriz, Iran*
[2]*Department of Physics, Faculty of Basic Sciences, University of Mazandaran, P.O. Box 47416-95447, Babolsar, Iran*
[3]*Research Institute for Astronomy and Astrophysics of Maragha (RIAAM), P.O. Box 55134-441, Maragha, Iran*

Correspondence should be addressed to K. Nozari; knozari@umz.ac.ir

Academic Editor: Elias C. Vagenas

We study nonlinear cosmological perturbations and their possible non-Gaussian character in an extended nonminimal inflation where gravity is coupled nonminimally to both the scalar field and its derivatives. By expansion of the action up to the third order, we focus on the nonlinearity and non-Gaussianity of perturbations in comparison with recent observational data. By adopting an inflation potential of the form $V(\phi) = (1/n)\lambda\phi^n$, we show that, for $n = 4$, for instance, this extended model is consistent with observation if $0.013 < \lambda < 0.095$ in appropriate units. By restricting the equilateral amplitude of non-Gaussianity to the observationally viable values, the coupling parameter λ is constrained to the values $\lambda < 0.1$.

1. Introduction

The idea of cosmological inflation is capable of addressing some problems of the standard big bang theory, such as the horizon, flatness, and monopole problems. Also, it can provide a reliable mechanism for generation of density perturbations responsible for structure formation and therefore temperature anisotropies in Cosmic Microwave Background (CMB) spectrum [1–8]. There are a wide variety of cosmological inflation models where viability of their predictions in comparison with observations makes them acceptable or unacceptable (see, for instance, [9, 10] for this purpose). The simplest inflationary model is a single scalar field scenario in which inflation is driven by a scalar field called the inflaton that predicts adiabatic, Gaussian, and scale-invariant fluctuations [11]. But recently observational data have revealed some degrees of scale-dependence in the primordial density perturbations. Also, Planck team have obtained some constraints on the primordial non-Gaussianity [12–14]. Therefore, it seems that extended models of inflation which can explain or address this scale-dependence and non-Gaussianity of perturbations are more desirable. There are a lot of studies in this respect, some of which can be seen in [15–20] with references therein. Among various inflationary models, the nonminimal models have attracted much attention. Nonminimal coupling of the inflaton field and gravitational sector is inevitable from the renormalizability of the corresponding field theory (see, for instance, [21]). Cosmological inflation driven by a scalar field nonminimally coupled to gravity is studied, for instance, in [22–29]. There were some issues on the unitarity violation with nonminimal coupling (see, for instance, [30–32]) which have forced researchers to consider possible coupling of the derivatives of the scalar field with geometry [33]. In fact, it has been shown that a model with nonminimal coupling between the kinetic terms of the inflaton (derivatives of the scalar field) and the Einstein tensor preserves the unitary bound during inflation [34]. Also, the presence of nonminimal derivative coupling is a powerful tool to increase the friction of an inflaton rolling down its own potential [34]. Some authors have considered the model with this coupling term and have studied the early time accelerating expansion of the universe as well as the late time dynamics [35–37]. In this paper we extend the nonminimal inflation models to the case that a canonical inflaton field is coupled nonminimally to the gravitational sector and in the same time the derivatives of the field are also coupled to the background geometry (Einstein's tensor). This model provides a more realistic framework for treating cosmological inflation in essence. We study in detail the cosmological perturbations and possible non-Gaussianities in the distribution of these perturbations in this nonminimal inflation. We expand the action of the model up to the third order and compare our results with observational data from Planck2015 to see the viability of this extended model. In this manner, we are able to constrain parameter space of the model in comparison with observation.

2. Field Equations

We consider an inflationary model where both a canonical scalar field and its derivatives are coupled nonminimally to gravity. The four-dimensional action for this model is given by the following expression:

$$S = \frac{1}{2}\int d^4x \cdot \sqrt{-g}\left[M_p^2 f(\phi) R + \frac{1}{\widetilde{M}^2} G_{\mu\nu}\partial^\mu\phi\partial^\nu\phi - 2V(\phi)\right], \tag{1}$$

where M_p is a reduced Planck mass, ϕ is a canonical scalar field, $f(\phi)$ is a general function of the scalar field, and $\widetilde{M}$ is a mass parameter. The energy-momentum tensor is obtained from action (1) as follows:

$$T_{\mu\nu} = \frac{1}{2\widetilde{M}^2}\left[\nabla_\mu\nabla_\nu\left(\nabla^\alpha\phi\nabla_\alpha\phi\right) - g_{\mu\nu}\Box\left(\nabla^\alpha\phi\nabla_\alpha\phi\right) + g_{\mu\nu}g^{\alpha\rho}g^{\beta\lambda}\nabla_\rho\nabla_\lambda\left(\nabla_\alpha\phi\nabla_\beta\phi\right) + \Box\left(\nabla_\mu\phi\nabla_\nu\phi\right)\right] - \frac{g^{\alpha\beta}}{\widetilde{M}^2} \cdot \nabla_\beta\nabla_\mu\left(\nabla_\alpha\phi\nabla_\nu\phi\right) - M_p^2\nabla_\mu\nabla_\nu f(\phi) + M_p^2 g_{\mu\nu}\Box f(\phi) + g_{\mu\nu}V(\phi). \tag{2}$$

On the other hand, variation of the action (1) with respect to the scalar field gives the scalar field equation of motion as

$$\frac{1}{2}M_p^2 R f'(\phi) - \frac{1}{\widetilde{M}^2}G^{\mu\nu}\nabla_\mu\nabla_\nu\phi - V'(\phi) = 0, \tag{3}$$

where a prime denotes derivative with respect to the scalar field. We consider a spatially flat Friedmann-Robertson-Walker (FRW) line element as

$$ds^2 = -dt^2 + a^2(t)\,\delta_{ij}dx^i dx^j, \tag{4}$$

where $a(t)$ is scale factor. Now, let us assume that $f(\phi) = (1/2)\phi^2$. In this framework, $T_{\mu\nu}$ leads to the following energy density and pressure for this model, respectively,

$$\rho = \frac{9H^2}{2\widetilde{M}^2}\dot{\phi}^2 - \frac{3}{2}M_p^2 H\phi\left(2\dot{\phi} + H\phi\right) + V(\phi) \tag{5}$$

$$p = -\frac{3}{2}\frac{H^2\dot{\phi}^2}{\widetilde{M}^2} - \frac{\dot{\phi}^2\dot{H}}{\widetilde{M}^2} - \frac{2H}{\widetilde{M}^2}\dot{\phi}\ddot{\phi} + \frac{1}{2}M_p^2\left[2\dot{H}\phi^2 + 3H^2\phi^2 + 4H\phi\dot{\phi} + 2\phi\ddot{\phi} + 2\dot{\phi}\right] - V(\phi), \tag{6}$$

where a dot refers to derivative with respect to the cosmic time. The equations of motion following from action (1) are

$$H^2 = \frac{1}{3M_p^2}\left[-\frac{3}{2}M_p^2 H\phi\left(2\dot{\phi} + H\phi\right) + \frac{9H^2}{2\widetilde{M}^2}\dot{\phi}^2 + V(\phi)\right], \tag{7}$$

$$\dot{H} = -\frac{1}{2M_p^2}\left[\dot{\phi}^2\left(\frac{3H^2}{\widetilde{M}^2} - \frac{\dot{H}}{\widetilde{M}^2}\right) - \frac{2H}{\widetilde{M}^2}\dot{\phi}\ddot{\phi} - \frac{3}{2}M_p^2 H\phi\left(2\dot{\phi} + H\phi\right) + \frac{1}{2}M_p^2\left(\left(2\dot{H} + 3H^2\right)\phi^2 + 4H\phi\dot{\phi} + 2\phi\ddot{\phi} + 2\dot{\phi}^2\right)\right] \tag{8}$$

$$-3M_p^2\left(2H^2 + \dot{H}\right)\phi + \frac{3H^2}{\widetilde{M}^2}\ddot{\phi} + 3H\left(\frac{3H^2}{\widetilde{M}^2} + \frac{2\dot{H}}{\widetilde{M}^2}\right)\dot{\phi} + V'(\phi) = 0. \tag{9}$$

The slow-roll parameters in this model are defined as

$$\epsilon \equiv -\frac{\dot{H}}{H^2}, \qquad \eta \equiv -\frac{1}{H}\frac{\ddot{H}}{\dot{H}}. \tag{10}$$

To have inflationary phase, ϵ and η should satisfy slow-roll conditions ($\epsilon \ll 1, \eta \ll 1$). In our setup, we find the following result:

$$\epsilon = \left[1 + \frac{\phi^2}{2} - \frac{\dot{\phi}^2}{2\widetilde{M}^2 M_p^2}\right]^{-1} \cdot \left[\frac{3\dot{\phi}^2}{2\widetilde{M}^2 M_p^2} + \frac{\phi\dot{\phi}}{2H} + \frac{\ddot{\phi}}{H\dot{\phi}}\left(\frac{\phi\dot{\phi}}{2H} - \frac{\dot{\phi}^2}{\widetilde{M}^2 M_p^2}\right)\right] \tag{11}$$

and

$$\eta = -2\epsilon - \frac{\dot{\epsilon}}{H\epsilon}. \tag{12}$$

Within the slow-roll approximation, (7), (8), and (9) can be written, respectively, as

$$H^2 \simeq \frac{1}{3M_p^2}\left[-\frac{3}{2}M_p^2 H^2\phi^2 + V(\phi)\right], \tag{13}$$

$$\dot{H} \simeq -\frac{1}{2M_p^2}\left[\frac{3H^2\dot{\phi}^2}{\widetilde{M}^2} - M_p^2 H\phi\dot{\phi} + M_p^2\dot{H}\phi^2\right], \tag{14}$$

and

$$-6M_p^2H^2\phi+\frac{9H^3\dot{\phi}}{\widetilde{M}^2}+V'(\phi)\simeq 0. \quad (15)$$

The number of e-folds during inflation is defined as

$$\mathcal{N}=\int_{t_{hc}}^{t_e}H\,dt, \quad (16)$$

where t_{hc} and t_e are time of horizon crossing and end of inflation, respectively. The number of e-folds in the slow-roll approximation in our setup can be expressed as follows:

$$\mathcal{N}\simeq\int_{\phi_{hc}}^{\phi_e}\frac{V(\phi)\,d\phi}{M_p^2\left(1+(1/2)\phi^2\right)\left[2M_p^2\widetilde{M}^2\phi-M_p^2\widetilde{M}^2\left(V'(\phi)/V(\phi)\right)\left(1+(1/2)\phi^2\right)\right]}. \quad (17)$$

After providing the basic setup of the model, for testing cosmological viability of this extended model, we treat the perturbations in comparison with observation.

3. Second-Order Action: Linear Perturbations

In this section, we study linear perturbations around the homogeneous background solution. To this end, the first step is expanding the action (1) up to the second order in small fluctuations. It is convenient to work in the ADM formalism given by [38]

$$ds^2=-N^2dt^2+h_{ij}\left(N^idt+dx^i\right)\left(N^jdt+dx^j\right), \quad (18)$$

where N^i is the shift vector and N is the lapse function. We expand the lapse function and shift vector to $N=1+2\Phi$ and $N^i=\delta^{ij}\partial_j\Upsilon$, respectively, where Φ and Υ are three-scalars. Also, $h_{ij}=a^2(t)[(1+2\Psi)\delta_{ij}+\gamma_{ij}]$, where Ψ is spatial curvature perturbation and γ_{ij} is shear three-tensor which is traceless and symmetric. In the rest of our study, we choose $\delta\Phi=0$ and $\gamma_{ij}=0$. By taking into account the scalar perturbations in linear-order, the metric (18) is written as (see, for instance, [39])

$$ds^2=-(1+2\Phi)\,dt^2+2\partial_i\Upsilon dt dx^i + a^2(t)(1+2\Psi)\,\delta_{ij}dx^idx^j. \quad (19)$$

Now by replacing metric (19) in action (1) and expanding the action up to the second order in perturbations, we find (see, for instance, [40, 41])

$$\begin{aligned}S^{(2)}=\int dt dx^3a^3\Bigg[&-\frac{3}{2}\left(M_p^2\phi^2-\frac{\dot{\phi}^2}{\widetilde{M}^2}\right)\dot{\Psi}^2 \\ &+\frac{1}{a^2}\left(\left(M_p^2\phi^2-\frac{\dot{\phi}^2}{\widetilde{M}^2}\right)\dot{\Psi}\right. \\ &\left.-\left(M_p^2H\phi^2+M_p^2\phi\dot{\phi}-\frac{3H\dot{\phi}^2}{\widetilde{M}^2}\right)\Phi\right)\partial^2\Upsilon \\ &-\frac{1}{a^2}\left(M_p^2\phi^2-\frac{\dot{\phi}^2}{\widetilde{M}^2}\right)\Phi\partial^2\Psi+3\left(M_p^2H\phi^2\right. \\ &\left.+M_p^2\phi\dot{\phi}-\frac{3H\dot{\phi}^2}{\widetilde{M}^2}\right)\Phi\dot{\Psi}+3H\left(-\frac{1}{2}M_p^2H\phi^2\right. \\ &\left.-M_p^2\phi\dot{\phi}+\frac{3H\dot{\phi}^2}{\widetilde{M}^2}\right)\Phi^2+\frac{1}{2a^2}\left(M_p^2\phi^2+\frac{\dot{\phi}^2}{\widetilde{M}^2}\right) \\ &\cdot(\partial\Psi)^2\Bigg].\end{aligned} \quad (20)$$

By variation of action (20) with respect to N and N^i we find

$$\Phi=\frac{M_p^2\phi^2-\dot{\phi}^2/\widetilde{M}^2}{M_p^2H\phi^2+M_p^2\phi\dot{\phi}-3H\dot{\phi}^2/\widetilde{M}^2}\dot{\Psi}, \quad (21)$$

$$\begin{aligned}\partial^2\Upsilon=&\frac{2a^2}{3}\cdot\frac{\left(-(9/2)M_p^2H^2\phi^2-9M_p^2H\phi\dot{\phi}+27H^2\dot{\phi}^2/\widetilde{M}^2\right)}{\left(M_p^2H\phi^2+M_p^2\phi\dot{\phi}-3H\dot{\phi}^2/\widetilde{M}^2\right)} \\ &+3\dot{\Psi}a^2-\frac{M_p^2\phi^2-\dot{\phi}^2/\widetilde{M}^2}{M_p^2H\phi^2+M_p^2\phi\dot{\phi}-3H\dot{\phi}^2/\widetilde{M}^2}\dot{\Psi}.\end{aligned} \quad (22)$$

Finally the second-order action can be rewritten as follows:

$$S^{(2)}=\int dt dx^3a^3\vartheta_s\left[\dot{\Psi}^2-\frac{c_s^2}{a^2}(\partial\Psi)^2\right] \quad (23)$$

where by definition

$$\vartheta_s\equiv 6\frac{\left(M_p^2\phi^2-\dot{\phi}^2/\widetilde{M}^2\right)^2\left(-(1/2)M_p^2H^2\phi^2-M_p^2H\phi\dot{\phi}+(3/\widetilde{M}^2)H^2\dot{\phi}^2\right)}{\left(M_p^2H\phi^2+M_p^2\phi\dot{\phi}-(3/\widetilde{M}^2)H\dot{\phi}^2\right)^2}+3\left(\frac{1}{2}M_p^2\phi^2-\frac{1}{2\widetilde{2}}\dot{\phi}^2\right) \quad (24)$$

and

$$\begin{aligned}
c_s^2 \equiv \frac{3}{2} \Bigg\{ \bigg(& M_p^2\phi^2 - \frac{\dot{\phi}^2}{\widetilde{M}^2} \bigg)^2 \bigg(M_p^2 H\phi^2 + M_p^2\phi\dot{\phi} \\
& - \frac{3H\dot{\phi}^2}{\widetilde{M}^2} \bigg) H - \bigg(M_p^2 H\phi^2 + M_p^2\phi\dot{\phi} - \frac{3H\dot{\phi}^2}{\widetilde{M}^2} \bigg)^2 \\
& \cdot \bigg(M_p^2\phi^2 - \frac{\dot{\phi}^2}{\widetilde{M}^2} \bigg) 4 \bigg(M_p^2\phi^2 - \frac{\dot{\phi}^2}{\widetilde{M}^2} \bigg) \bigg(M_p^2\phi\dot{\phi} \\
& - \frac{\dot{\phi}\ddot{\phi}}{\widetilde{M}^2} \bigg) \bigg(M_p^2 H\phi^2 + M_p^2\phi\dot{\phi} - \frac{3H\dot{\phi}^2}{\widetilde{M}^2} \bigg) - \bigg(M_p^2 \\
& - \frac{\dot{\phi}^2}{\widetilde{M}^2} \bigg)^2 \bigg(M_p^2\dot{H}\phi^2 + 2M_p^2 H\phi\dot{\phi} M_p^2\dot{\phi}^2 + M_p^2\phi\ddot{\phi} \\
& - \frac{3\dot{H}\dot{\phi}^2}{\widetilde{M}^2} - \frac{6}{\widetilde{M}^2} H\dot{\phi}\ddot{\phi} \bigg) \Bigg\} \Bigg\{ 9 \bigg[\frac{1}{2} M_p^2\phi^2 - \frac{\dot{\phi}^2}{2\widetilde{M}^2} \bigg] \\
& \cdot \bigg[4 \bigg(\frac{1}{2} M_p^2\phi^2 - \frac{\dot{\phi}^2}{2\widetilde{M}^2} \bigg) \\
& \cdot \bigg(-\frac{1}{2} M_p^2 H^2\phi^2 - M_p^2 H\phi\dot{\phi} + \frac{3}{M^{\tilde{2}} H^2\dot{\phi}^2} \bigg) \\
& + \bigg(M_p^2 H\phi^2 + M_p^2\phi\dot{\phi} - \frac{3H\dot{\phi}^2}{\widetilde{M}^2} \bigg)^2 \bigg] \Bigg\}^{-1}.
\end{aligned} \tag{25}$$

In order to obtain quantum perturbations Ψ, we can find equation of motion of the curvature perturbation by varying action (23) which follows

$$\ddot{\Psi} + \left(3H + \frac{\dot{\vartheta}_s}{\vartheta_s} \right) + \frac{c_s^2 k^2}{a^2} \Psi = 0. \tag{26}$$

By solving the above equation up to the lowest order in slow-roll approximation, we find

$$\Psi = \frac{iH \exp(-ic_s k\tau)}{2c_s^{3/2}\sqrt{k^3}\vartheta_s} (1 + ic_s k\tau). \tag{27}$$

By using the two-point correlation functions, we can study power spectrum of curvature perturbation in this setup. We find two-point correlation function by obtaining vacuum expectation value at the end of inflation. We define the power spectrum P_s, as

$$\begin{aligned}
& \langle 0 \mid \Psi(0, \mathbf{k}_1) \Psi(0, \mathbf{k}_2) \mid 0 \rangle \\
& \quad = \frac{2\pi^2}{k^3} P_s (2\pi)^3 \delta^3 (\mathbf{k}_1 + \mathbf{k}_2),
\end{aligned} \tag{28}$$

where

$$P_s = \frac{H^2}{8\pi^2 \vartheta_s c_s^3}. \tag{29}$$

The spectral index of scalar perturbations is given by (see [42–44] for more details on the cosmological perturbations in generalized gravity theories and also inflationary spectral index in these theories)

$$n_s - 1 = \left. \frac{d \ln P_s}{d \ln k} \right|_{c_s k = aH} = -2\epsilon - \delta_F - \eta_s - S \tag{30}$$

where by definition

$$\begin{aligned}
\delta_F &= \frac{\dot{f}}{H(1+f)}, \\
\eta_s &= \frac{\dot{\epsilon}_s}{H\epsilon_s}, \\
S &= \frac{\dot{c}_s}{Hc_s}
\end{aligned} \tag{31}$$

also

$$\epsilon_s = \frac{\vartheta_s c_s^2}{M_{pl}^2 (1+f)}. \tag{32}$$

We obtain finally

$$\begin{aligned}
n_s - 1 = & -2\epsilon - \frac{1}{H} \frac{d \ln c_s}{dt} \\
& - \frac{1}{H} \frac{d \ln \left[2H \left(1 + \phi^2/2 \right) \epsilon + \phi\dot{\phi} \right]}{dt},
\end{aligned} \tag{33}$$

which shows the scale-dependence of perturbations due to deviation of n_s from 1.

Now we study tensor perturbations in this setup. To this end, we write the metric as follows:

$$ds^2 = -dt^2 + a(t)^2 \left(\delta_{ij} + T_{ij} \right) dx^i dx^j, \tag{34}$$

where T_{ij} is a spatial shear 3-tensor which is transverse and traceless. It is convenient to write T_{ij} in terms of two polarization modes, as follows:

$$T_{ij} = T_+ e_{ij}^+ + T^\times e_{ij}^\times, \tag{35}$$

where e_{ij}^+ and $e_{ij}^\times$ are the polarization tensors. In this case, the second-order action for the tensor mode can be written as

$$S_T = \int dt dx^3 a^3 \vartheta_T \left[\dot{T}_{(+,\times)}^2 - \frac{c_T^2}{a^2} \left(\partial T_{(+,\times)} \right)^2 \right], \tag{36}$$

where by definition

$$\vartheta_T \equiv \frac{1}{8} \left(M_p^2 \phi^2 - \frac{\dot{\phi}^2}{\widetilde{M}^2} \right) \tag{37}$$

and

$$c_T^2 \equiv \frac{\widetilde{M}^2 M_p^2 \phi^2 + \dot{\phi}^2}{\widetilde{M}^2 M_p^2 \phi^2 - \dot{\phi}^2}. \tag{38}$$

Now, the amplitude of tensor perturbations is given by

$$P_T = \frac{H^2}{2\pi^2 \vartheta_T c_T^3}, \tag{39}$$

where we have defined the tensor spectral index as

$$n_T \equiv \left.\frac{d\ln P_T}{d\ln k}\right|_{c_T k = aH} = -2\epsilon - \delta_F. \tag{40}$$

By using above equations, we get finally

$$n_T = -2\epsilon - \frac{\phi\dot{\phi}}{H\left(1+\phi^2/2\right)}. \tag{41}$$

The tensor-to-scalar ratio as an important observational quantity in our setup is given by

$$\begin{aligned} r = \frac{P_T}{P_s} &= 16c_s\left(\epsilon + \frac{\phi\dot{\phi}}{2H\left(1+\phi^2/2\right)} + O\left(\epsilon^2\right)\right) \\ &\simeq -8c_s n_T \end{aligned} \tag{42}$$

which yields the standard consistency relation.

4. Third-Order Action: Non-Gaussianity

Since a two-point correlation function of the scalar perturbations gives no information about possible non-Gaussian feature of distribution, we study higher-order correlation functions. A three-point correlation function is capable of giving the required information. For this purpose, we should expand action (1) up to the third order in small fluctuations around the homogeneous background solutions. In this respect, we obtain

$$\begin{aligned} S^{(3)} = \int dt dx^3 a^3 \Bigg\{ & 3\Phi^3\left[M_p^2 H^2\left(1+\frac{\phi^2}{2}\right) + M_p^2 H\phi\dot{\phi} - \frac{5}{\widetilde{M}^2}H^2\dot{\phi}^2\right] \\ & + \Phi^2\left[9\Psi\left(-\frac{1}{2}M_p^2\phi^2 - M_p^2 H\phi\dot{\phi} + \frac{3}{\widetilde{M}}H^2\dot{\phi}^2\right)\right. \\ & + 6\dot{\Psi}\left(-M_p^2 H\left(1+\frac{\phi^2}{2}\right) - \frac{1}{2}M_p^2\phi\dot{\phi}\frac{3}{\widetilde{M}^2}H\dot{\phi}^2\right) \\ & - \frac{\dot{\phi}^2}{\widetilde{M}^2 a^2}\partial^2\Psi - \frac{2}{a^2} \\ & \left.\cdot\,\partial^2\Upsilon\left(-M_p^2 H\left(1+\frac{\phi^2}{2}\right) - \frac{1}{2}M_p^2\phi\dot{\phi}\frac{3}{\widetilde{M}^2}H\dot{\phi}^2\right)\right] \\ & + \Phi\left[\frac{1}{a^2}\left(-M_p^2 H\phi^2 - M_p^2\phi\dot{\phi} + \frac{3H\dot{\phi}^2}{\widetilde{M}^2}\right)\partial_i\Psi\partial_i\Upsilon\right. \\ & - 9\left(-M_p^2 H\phi^2 - M_p^2\phi\dot{\phi} + \frac{3H\dot{\phi}^2}{\widetilde{M}^2}\right)\dot{\Psi}\Psi \\ & + \frac{1}{2a^4}\left(M_p^2\left(1+\frac{\phi^2}{2}\right) + \frac{3}{2}\frac{\dot{\phi}^2}{\widetilde{M}^2}\right) \\ & \cdot\left(\partial_i\partial_j\Upsilon\partial_i\partial_j\Upsilon - \partial^2\Upsilon\partial^2\Upsilon\right) \\ & + \frac{1}{a^2}\left(-M_p^2 H\phi^2 - M_p^2\phi\dot{\phi} + \frac{3H\dot{\phi}^2}{\widetilde{M}^2}\right)\Psi\partial^2\Upsilon \\ & + \frac{4}{2a^2}\left(M_p^2\left(1+\frac{\phi^2}{2}\right) + \frac{3}{2}\frac{\dot{\phi}^2}{\widetilde{M}^2}\right)\dot{\Psi}\partial^2\Upsilon \\ & + \frac{1}{a^2}\left(-M_p^2\phi^2 + \frac{\dot{\phi}}{\widetilde{M}^2}\right)\Psi\partial^2\Psi \\ & + \frac{1}{2a^2}\left(-M_p^2\phi^2 + \frac{\dot{\phi}}{\widetilde{M}^2}\right)(\partial\Psi)^2 \\ & \left. - 6\left(M_p^2\left(1+\frac{\phi^2}{2}\right) + \frac{3}{2}\frac{\dot{\phi}^2}{\widetilde{M}^2}\right)\dot{\Psi}^2\right] + \frac{1}{2a^2}\left(M_p^2\phi^2\right. \\ & \left. + \frac{\dot{\phi}^2}{\widetilde{M}^2}\right)\Psi(\partial\Psi)^2 + \frac{9}{2}\left(-M_p^2\phi^2 + \frac{\dot{\phi}}{\widetilde{M}^2}\right)\dot{\Psi}^2\Psi \\ & - \frac{1}{a^2}\left(-M_p^2\phi^2 + \frac{\dot{\phi}}{\widetilde{M}^2}\right)\dot{\Psi}\partial_i\Psi\partial_i\Upsilon - \frac{1}{a^2}\left(-M_p^2\phi^2\right. \\ & \left. + \frac{\dot{\phi}}{\widetilde{M}^2}\right)\dot{\Psi}\Psi\partial^2\Upsilon - \frac{3}{4a^4}\Psi\left(-M_p^2\phi^2 + \frac{\dot{\phi}}{\widetilde{M}^2}\right) \\ & \cdot\left(\partial_i\partial_j\Upsilon\partial_i\partial_j\Upsilon - \partial^2\Upsilon\partial^2\Upsilon\right) + \frac{1}{a^4}\left(-M_p^2\phi^2 + \frac{\dot{\phi}}{\widetilde{M}^2}\right) \\ & \cdot\,\partial_i\Psi\partial_i\Upsilon\partial^2\Upsilon\Bigg\} \end{aligned} \tag{43}$$

We use (21) and (22) for eliminating Φ and Υ in this relation. For this end, we introduce the quantity χ as follows:

$$\begin{aligned} \Upsilon = {} & \frac{M_p^2\widetilde{M}^2\phi^2 - \dot{\phi}^2}{\widetilde{M}^2 M_p^2\left(H\phi^2 + \phi\dot{\phi}\right) - 3H\dot{\phi}^2}\Psi \\ & + \frac{2\widetilde{M}^2 a^2\chi}{M_p^2\widetilde{M}^2\phi^2 - \dot{\phi}^2}, \end{aligned} \tag{44}$$

where

$$\partial^2\chi = \vartheta_s\dot{\Psi}. \tag{45}$$

Now the third-order action (43) takes the following form:

$$\begin{aligned} S^{(3)} = \int dt\, dx^3 a^3 \Bigg\{ & \left[-3M_p^2 c_s^{-2}\Psi\dot{\Psi}^2 + M_p^2 a^{-2}\Psi(\partial\Psi)^2\right. \\ & \left. + M_p^2 c_s^{-2} H^{-1}\dot{\Psi}^3\right]\left[\left(1+\frac{1}{4}\phi^2\right)\epsilon + \frac{5}{8}\frac{\phi\dot{\phi}}{H}\right] - 2\left(1\right. \\ & \left. + \frac{1}{4}\phi^2\right)^{-1}\left(\frac{5}{8}\frac{\phi\dot{\phi}}{c_s^2 H}\right)\dot{\Psi}\partial_i\Psi\partial_i\chi\Bigg\}. \end{aligned} \tag{46}$$

By calculating the three-point correlation function, we can study non-Gaussianity feature of the primordial perturbations. For the present model, we use the interaction picture in which the interaction Hamiltonian, H_{int}, is equal to the Lagrangian third-order action. The vacuum expectation value of curvature perturbations at $\tau = \tau_f$ is

$$\begin{aligned}\langle \Psi(\mathbf{k}_1)\Psi(\mathbf{k}_2)\Psi(\mathbf{k}_3)\rangle &= -i\int_{\tau_i}^{\tau_f} d\tau \langle 0 | \\ &\left[\Psi(\tau_f,\mathbf{k}_1)\Psi(\tau_f,\mathbf{k}_2)\Psi(\tau_f,\mathbf{k}_3), H_{int}(\tau)\right] | 0\rangle.\end{aligned} \tag{47}$$

By solving the above integral in Fourier space, we find

$$\begin{aligned}&\langle \Psi(\mathbf{k}_1)\Psi(\mathbf{k}_2)\Psi(\mathbf{k}_3)\rangle \\ &= (2\pi)^3\delta^3(\mathbf{k}_1+\mathbf{k}_2+\mathbf{k}_3)P_s^2F_\Psi(\mathbf{k}_1,\mathbf{k}_2,\mathbf{k}_3),\end{aligned} \tag{48}$$

where

$$F_\Psi(\mathbf{k}_1,\mathbf{k}_2,\mathbf{k}_3) = \frac{(2\pi)^2}{\prod_{i=1}^3 k_i^3}G_\Psi, \tag{49}$$

$$\begin{aligned}G_\Psi = &\left[\frac{3}{4}\left(\frac{2}{K}\Sigma_{i>j}k_i^2k_j^2 - \frac{1}{K^2}\Sigma_{i\neq j}k_i^2k_j^3\right)\right. \\ &+ \frac{1}{4}\left(\frac{1}{2}\Sigma_i k_i^3 + \frac{2}{K}\Sigma_{i>j}k_i^2k_j^2 - \frac{1}{K^2}\Sigma_{i\neq j}k_i^2k_j^3\right) \\ &\left.-\frac{3}{2}\left(\frac{(k_1k_2k_3)^2}{K^3}\right)\right]\left(1-\frac{1}{c_s^2}\right),\end{aligned} \tag{50}$$

and $K = \sum_i k_i$. Finally the nonlinear parameter f_{NL} is defined as follows:

$$f_{NL} = \frac{10}{3}\frac{G_\Psi}{\sum_{i=1}^3 k_i}. \tag{51}$$

Here we study non-Gaussianity in the orthogonal and the equilateral configurations [45, 46]. Firstly we should account G_Ψ in these configurations. To this end, we follow [19, 47, 48] to introduce a shape ζ_*^{equi} as $\zeta_*^{equi} = -(12/13)(3\zeta_1 - \zeta_2)$. In this manner we define the following shape which is orthogonal to ζ_*^{equi}

$$\zeta_*^{ortho} = -\frac{12}{14-13\beta}\left[\beta(3\zeta_1-\zeta_2)+3\zeta_1-\zeta_2\right], \tag{52}$$

where $\beta \simeq 1.1967996$. Finally, bispectrum (48) can be written in terms of ζ_*^{equi} and ζ_*^{ortho} as follows:

$$G_\Psi = G_1\zeta_*^{equi} + G_2\zeta_*^{ortho}, \tag{53}$$

where

$$G_1 = \frac{13}{12}\left[\frac{1}{24}\left(1-\frac{1}{c_s^2}\right)\right](2+3\beta) \tag{54}$$

and

$$G_2 = \frac{14-13\beta}{12}\left[\frac{1}{8}\left(1-\frac{1}{c_s^2}\right)\right]. \tag{55}$$

Now, by using (50)-(55) we obtain the amplitude of non-Gaussianity in the orthogonal and equilateral configurations, respectively, as

$$f_{NL}^{equi} = \frac{130}{36\sum_{i=1}^3 k_i^3}\left[\frac{1}{24}\left(\frac{1}{1-c_s^2}\right)\right](2+3\beta)\zeta_*^{equi}, \tag{56}$$

and

$$f_{NL}^{ortho} = \frac{140-130\beta}{36\sum_{i=1}^3 k_i^3}\left[\frac{1}{8}\left(1-\frac{1}{c_s^2}\right)\right]\zeta_*^{ortho}. \tag{57}$$

The equilateral and the orthogonal shape have a negative and a positive peak in $k_1 = k_2 = k_3$ limit, respectively [49]. Thus, we can rewrite the above equations in this limit as

$$f_{NL}^{equi} = \frac{325}{18}\left[\frac{1}{24}\left(\frac{1}{c_s^2}-1\right)\right](2+3\beta), \tag{58}$$

and

$$f_{NL}^{ortho} = \frac{10}{9}\left[\frac{1}{8}\left(1-\frac{1}{c_s^2}\right)\right]\left(\frac{7}{6}+\frac{65}{4}\beta\right), \tag{59}$$

respectively.

5. Confronting with Observation

The previous sections were devoted to the theoretical framework of this extended model. In this section, we compare our model with observational data to find some observational constraints on the model parameter space. In this regard, we introduce a suitable candidate for potential term in the action. (Note that in general λ has dimension related to the Planck mass. This can be seen easily by considering the normalization of ϕ via $V(\phi) = (1/n)\lambda(\phi/\phi_0)^n$ which indicates that λ cannot be dimensionless in general. When we consider some numerical values for λ in our numerical analysis, these values are in *"appropriate units"*.) We adopt $V(\phi) = (1/n)\lambda\phi^n$ which contains some interesting inflation models such as chaotic inflation. To be more specified, we consider a quartic potential with $n = 4$. Firstly we substitute this potential into (11) and then by adopting $\epsilon = 1$ we find the inflaton field's value at the end of inflation. Then by solving the integral (17), we find the inflaton field's value at the horizon crossing in terms of number of e-folds, N. Then we substitute ϕ_{hc} into (33), (42), (58), and (59). The resulting relations are the basis of our numerical analysis on the parameter space of the model at hand. To proceed with numerical analysis, we study the behavior of the tensor-to-scalar ratio versus the scalar spectral index. In Figure 1, we have plotted the tensor-to-scalar ratio versus the scalar spectral index for $N = 60$ in the background of Planck2015 data. The trajectory of result in this extended nonminimal inflationary model lies well in the confidence levels of Planck2015 observational data for viable spectral index and r. The amplitude of orthogonal configuration of non-Gaussianity versus the amplitude of equilateral configuration is depicted in Figure 2 for $N = 60$. We see that this extended nonminimal model, in some

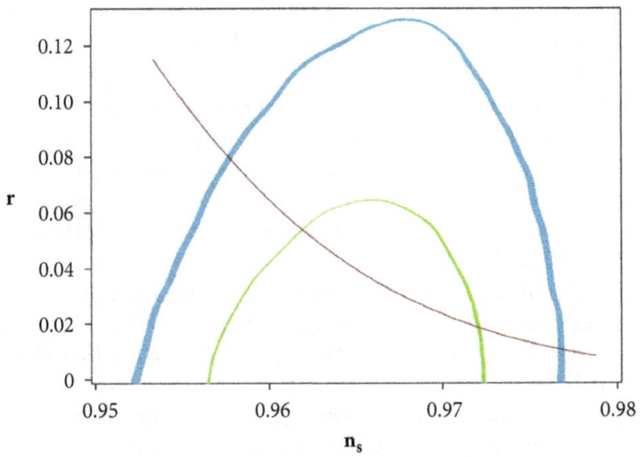

FIGURE 1: Tensor-to-scalar ratio versus the scalar spectral index in the background of Planck2015 TT,TE, and EE+lowP data.

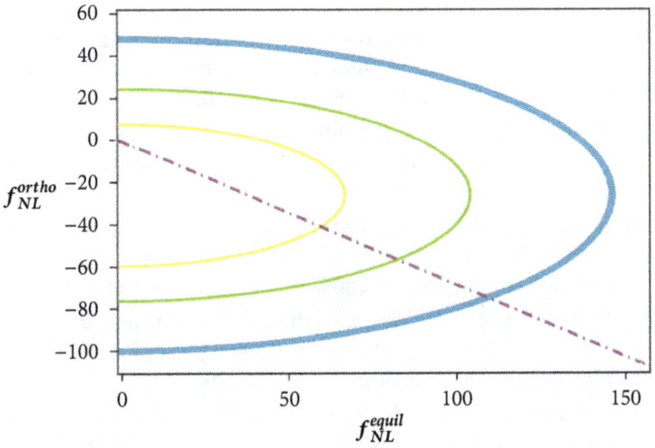

FIGURE 2: The amplitude of the orthogonal configuration versus the amplitude of the equilateral configuration of non-Gaussianity in the background of Planck2015 TTT, EEE, TTE, and EET data.

ranges of the parameter λ, is consistent with observation. If we restrict the spectral index to the observationally viable interval $0.95 < n_s < 0.97$, then λ is constrained to be in the interval $0.013 < \lambda < 0.095$ in *appropriate units*. If we restrict the equilateral configuration of non-Gaussianity to the observationally viable condition $-147 < f_{NL}^{equi} < 143$, then we find the constraint $\lambda < 0.1$ in our setup.

6. Summary and Conclusion

We studied an extended model of single field inflation where the inflaton and its derivatives are coupled to the background geometry. By focusing on the third-order action and nonlinear perturbations, we obtained observables of cosmological inflation, such as tensor-to-scalar ratio and the amplitudes of non-Gaussianities in this extended setup. By confronting the model's outcomes with observational data from Planck2015, we were able to constrain parameter space of the model. By adopting a quartic potential with $V(\phi) = (1/4)\lambda\phi^4$, restricting the model to realize observationally viable spectral index (or tensor-to-scalar ratio) imposes the constraint on coupling λ as $0.013 < \lambda < 0.095$. Also restricting the amplitude of equilateral amplitude of non-Gaussianity to the observationally supported value of $-147 < f_{NL}^{equi} < 143$ results in the constraint $\lambda < 0.1$ in appropriate units.

Conflicts of Interest

The authors declare that they have no conflicts of interest.

Acknowledgments

The work of K. Nozari has been supported financially by Research Institute for Astronomy and Astrophysics of Maragha (RIAAM) under research project no. 1/5750-1.

References

[1] V. F. Mukhanov and G. V. Chibisov, "Quantum fluctuations and a nonsingular universe," *JETP Letters*, vol. 33, pp. 549–553, 1981.

[2] A. H. Guth, "Inflationary universe: a possible solution to the horizon and flatness problems," *Physical Review D: Particles, Fields, Gravitation and Cosmology*, vol. 23, no. 2, pp. 347–356, 1981.

[3] A. Albrecht and P. J. Steinhardt, "Cosmology for grand unified theories with radiatively induced symmetry breaking," *Physical Review Letters*, vol. 48, no. 17, pp. 1220–1223, 1982.

[4] A. D. Linde, "A new inflationary universe scenario: A possible solution of the horizon, flatness, homogeneity, isotropy and primordial monopole problems," *Physics Letters B*, vol. 108, no. 6, pp. 389–393, 1982.

[5] A. A. Starobinsky, "Dynamics of phase transition in the new inflationary universe scenario and generation of perturbations," *Physics Letters B*, vol. 113, pp. 175–179, 1982.

[6] J. E. Lidsey, A. R. Liddle, E. W. Kolb, E. J. Copeland, T. Barreiro, and M. Abney, "Reconstructing the inflaton potential—an overview," *Reviews of Modern Physics*, vol. 69, no. 2, pp. 373–410, 1997.

[7] A. Liddle and D. Lyth, *Cosmological Inflation and Large-Scale Structure*, Cambridge University Press, 2000.

[8] E. Komatsu, K. M. Smith, and J. Dunkley, "Seven-year Wilkinson Microwave Anisotropy Probe (WMAP) observations: cosmological interpretation," *The Astrophysical Journal Supplement Series*, vol. 192, no. 2, p. 18, 2011.

[9] J. Martin, "The Observational Status of Cosmic Inflation after Planck," *The Cosmic Microwave Background*, pp. 41–134, 2016.

[10] J. Martin, "Inflation after Planck: and the winners are," *Cosmology and Nongalactic Astrophysics*, 2013.

[11] J. M. Maldacena, "Non-Gaussian features of primordial fluctuations in single field inflationary models," *Journal of High Energy Physics*, vol. 2003, no. 5, article 013, 2003.

[12] P. A. R. Ade, "Planck 2015 results. XX. Constraints on inflation," *Astrophysics: Cosmology and Nongalactic Astrophysics*, 2015.

[13] P. A. R. Ade, "Planck 2015 results. XIII. Cosmological parameters," *Astrophysics: Cosmology and Nongalactic Astrophysics*, 2015.

[14] P. A. R. Ade, "Planck 2015 results. XVII. Constraints on primordial non-Gaussianity," *Astrophysics: Cosmology and Nongalactic Astrophysics*, 2015.

[15] D. Baumann, "TASI lectures on inflation," *High Energy Physics - Theory*, 2009.

[16] Y. Wang, "Inflation, cosmic perturbations and non-gaussianities," *High Energy Physics - Theory*, 2013.

[17] A. De Felice and S. Tsujikawa, "Generalized Galileon cosmology," *Physical Review D: Particles, Fields, Gravitation and Cosmology*, vol. 84, no. 12, 2011.

[18] A. De Felice and S. Tsujikawa, "Primordial non-Gaussianities in general modified gravitational models of inflation," *Journal of Cosmology and Astroparticle Physics*, vol. 2011, no. 4, article 029, 2011.

[19] A. De Felice and S. Tsujikawa, "Shapes of primordial non-Gaussianities in the Horndeski's most general scalar-tensor theories," *Journal of Cosmology and Astroparticle Physics*, vol. 2013, no. 3, article no. 30, 2013.

[20] Q.-G. Huang and Y. Wang, "Large local non-Gaussianity from general ultra slow-roll inflation," *Journal of Cosmology and Astroparticle Physics*, vol. 2013, no. 6, article no. 35, 2013.

[21] V. Faraoni, "Inflation and quintessence with nonminimal coupling," *Physical Review D: Particles, Fields, Gravitation and Cosmology*, vol. 62, Article ID 023504, 2000.

[22] T. Futamase and K.-I. Maeda, "Chaotic inflationary scenario of the Universe with a nonminimally coupled inflaton field," *Physical Review D: Particles, Fields, Gravitation and Cosmology*, vol. 39, no. 2, pp. 399–404, 1989.

[23] D. S. Salopek, J. R. Bond, and J. M. Bardeen, "Designing density fluctuation spectra in inflation," *Physical Review D: Particles, Fields, Gravitation and Cosmology*, vol. 40, no. 6, pp. 1753–1788, 1989.

[24] R. Fakir and W. G. Unruh, "Improvement on cosmological chaotic inflation through nonminimal coupling," *Physical Review D: Particles, Fields, Gravitation and Cosmology*, vol. 41, no. 6, pp. 1783–1791, 1990.

[25] N. Makino and M. Sasaki, "The density perturbation in the chaotic inflation with non-minimal coupling," *Progress of Theoretical and Experimental Physics*, vol. 86, no. 1, pp. 103–118, 1991.

[26] J. Hwang and H. Noh, "COBE constraints on inflation models with a massive nonminimal scalar field," *Physical Review D: Particles, Fields, Gravitation and Cosmology*, vol. 60, no. 12, Article ID 063501, 1999.

[27] S. Tsujikawa and H. Yajima, "New constraints on multifield inflation with nonminimal coupling," *Physical Review D: Particles, Fields, Gravitation and Cosmology*, vol. 62, no. 12, 2000.

[28] C. Pallis and N. Toumbas, "Non-minimal sneutrino inflation, Peccei-Quinn phase transition and non-thermal leptogenesis," *Journal of Cosmology and Astroparticle Physics*, vol. 2011, no. 2, article 019, 2011.

[29] K. Nozari and S. Shafizadeh, "Non-minimal inflation revisited," *Physica Scripta*, vol. 2010, no. 8, Article ID 015901, 2010.

[30] C. P. Burgess, H. M. Lee, and M. Trott, "On Higgs inflation and naturalness," *Journal of High Energy Physics*, vol. 2010, no. 09 article 103, 2009.

[31] J. L. F. Barbón and J. R. Espinosa, "On the naturalness of Higgs inflation," *Physical Review D: Particles, Fields, Gravitation and Cosmology*, vol. 79, no. 8, Article ID 081302, 2009.

[32] C. P. Burgess, H. M. Lee, and M. Trott, "On Higgs inflation and naturalness," *Journal of High Energy Physics*, vol. 2010, no. 07, 2010.

[33] L. Amendola, "Cosmology with nonminimal derivative couplings," *Physics Letters B*, vol. 301, no. 2-3, pp. 175–182, 1993.

[34] C. Germani and A. Kehagias, "New model of inflation with nonminimal derivative coupling of standard model higgs boson to gravity," *Physical Review Letters*, vol. 105, no. 1, Article ID 011302, 2010.

[35] S. Tsujikawa, "Observational tests of inflation with a field derivative coupling to gravity," *Physical Review D: Particles, Fields, Gravitation and Cosmology*, vol. 85, Article ID 083518, 2012.

[36] H. M. Sadjadi and P. Goodarzi, "Reheating in non-minimal derivative coupling model," *Journal of Cosmology and Astroparticle Physics*, vol. 2013, no. 02, article 038, 2013.

[37] E. N. Saridakis and S. V. Sushkov, "Quintessence and phantom cosmology with nonminimal derivative coupling," *Physical Review D: Particles, Fields, Gravitation and Cosmology*, vol. 81, Article ID 083510, 2010.

[38] R. L. Arnowitt, S. Deser, and C. W. Misner, "Canonical variables for general relativity," *Physical Review A: Atomic, Molecular and Optical Physics*, vol. 117, no. 6, p. 1595, 1960.

[39] V. F. Mukhanov, H. A. Feldman, and R. H. Brandenberger, "Theory of cosmological perturbations," *Physics Reports*, vol. 215, no. 5-6, pp. 203–333, 1992.

[40] C. Cheung, A. L. Fitzpatrick, J. Kaplan, L. Senatore, and P. Creminelli, "The effective field theory of inflation," *Journal of High Energy Physics*, vol. 2008, no. 03, article no. 014, 2008.

[41] D. Seery and J. E. Lidsey, "Primordial non-Gaussianities in single-field inflation," *Journal of Cosmology and Astroparticle Physics*, vol. 2005, no. 6, article no. 003, 2005.

[42] J. Hwang and H. Noh, "Cosmological perturbations in generalized gravity theories," *Physical Review D: Particles, Fields, Gravitation and Cosmology*, vol. 54, no. 2, pp. 1460–1473, 1996.

[43] H. Noh and J. Hwang, "Inflationary spectra in generalized gravity: unified forms," *Physics Letters B*, vol. 515, no. 3-4, pp. 231–237, 2001.

[44] R. Myrzakulov, L. Sebastiani, and S. Vagnozzi, "Inflation in f(R, ϕ)-theories and mimetic gravity scenario," *The European Physical Journal C*, vol. 75, no. 9, 2015.

[45] D. Babich, P. Creminelli, and M. Zaldarriaga, "The shape of non-Gaussianities," *Journal of Cosmology and Astroparticle Physics*, no. 8, pp. 199–217, 2004.

[46] L. Senatore, M. K. Smith, and M. Zaldarriaga, "Non-Gaussianities in single field inflation and their optimal limits from the WMAP 5-year data," *Journal of Cosmology and Astroparticle Physics*, vol. 1, no. 28, 2010.

[47] J. R. Fergusson and E. P. S. Shellard, "Shape of primordial non-Gaussianity and the CMB bispectrum," *Physical Review D: Particles, Fields, Gravitation and Cosmology*, vol. 80, no. 4, Article ID 043510, 2009.

[48] C. T. Byrnes, "Lecture notes on non-Gaussianity," *Cosmology and Nongalactic Astrophysics*, 2014.

[49] X. Chen, M. X. Huang, S. Kachru, and G. Shiu, "Observational signatures and non-gaussianities of general single field inflation," *Journal of Cosmology and Astroparticle Physics*, vol. 2007, no. 1, article no. 02, 2007.

15

From Quantum Unstable Systems to the Decaying Dark Energy: Cosmological Implications

Aleksander Stachowski,[1] **Marek Szydłowski**,[1,2] **and Krzysztof Urbanowski**[3]

[1] *Astronomical Observatory, Jagiellonian University, Orla 171, 30-244 Kraków, Poland*
[2] *Mark Kac Complex Systems Research Centre, Jagiellonian University, Łojasiewicza 11, 30-348 Kraków, Poland*
[3] *Institute of Physics, University of Zielona Góra, Prof. Z. Szafrana 4a, 65-516 Zielona Góra, Poland*

Correspondence should be addressed to Marek Szydłowski; marek.szydlowski@uj.edu.pl

Academic Editor: Ricardo G. Felipe

We consider a cosmology with decaying metastable dark energy and assume that a decay process of this metastable dark energy is a quantum decay process. Such an assumption implies among others that the evolution of the Universe is irreversible and violates the time reversal symmetry. We show that if we replace the cosmological time t appearing in the equation describing the evolution of the Universe by the Hubble cosmological scale time, then we obtain time dependent $\Lambda(t)$ in the form of the series of even powers of the Hubble parameter H: $\Lambda(t) = \Lambda(H)$. Our special attention is focused on radioactive-like exponential form of the decay process of the dark energy and on the consequences of this type decay.

1. Introduction

In the explanation of the Universe, we encounter the old problem of the cosmological constant, which is related to understanding why the measured value of the vacuum energy is so small in comparison with the value calculated using quantum field theory methods [1]. Because of a cosmological origin of the cosmological constant one must also address another problem. Namely, it is connected with our understanding, with a question of not only why the vacuum energy is not only small, but also, as current Type Ia supernova observations to indicate, why the present mass density of the Universe has the same order of magnitude [2].

Both mentioned cosmological constant problems can be considered in the framework of the extension of the standard cosmological ΛCDM model in which the cosmological constant (naturally interpreted as related to the vacuum energy density) is running and its value is changing during the cosmic evolution.

Results of many recent observations lead to the conclusion that our Universe is in an accelerated expansion phase [3]. This acceleration can be explained as a result of a presence of dark energy. A detailed analysis of results of recent observations shows that there is a tension between local and primordial measurements of cosmological parameters [3]. It appears that this tension may be connected with dark energy evolving in time [4]. This paper is a contribution to the discussion of the nature of the dark energy. We consider the hypothesis that dark energy depends on time, $\rho_{\text{de}} = \rho_{\text{de}}(t)$, and it is metastable: We assume that it decays with the increasing time t to ρ_{bare}: $\rho_{\text{de}}(t) \longrightarrow \rho_{\text{bare}} \neq 0$ as $t \longrightarrow \infty$. The idea that vacuum energy decays was considered in many papers (see, e.g., [5, 6]). Shafieloo et al. [7] assumed that $\rho_{\text{de}}(t)$ decays according to the radioactive exponential decay law. Unfortunately, such an assumption is not able to reflect all the subtleties of evolution in the time of the dark energy and its decay process. It is because the creation of the Universe is a quantum process. Hence the metastable dark energy can be considered as the value of the scalar field at the false vacuum state and therefore the decay of the dark energy should be considered as a quantum decay process. The radioactive exponential decay law does not reflect correctly all phases of the quantum decay process. In general, analysing quantum decay processes one can distinguish the following phases [8, 9]: (i) the early time initial phase, (ii) the canonical or exponential phase (when the decay law has the exponential form), and (iii) the late time nonexponential phase. The first phase and the third one are missed when one considers the radioactive decay law only. Simply they are invisible to the radioactive exponential decay law. For example, the theoretical analysis of quantum decay processes shows that at late times the survival probability of the system considered in its initial state (i.e., the decay law) should tend to zero as $t \longrightarrow \infty$ much more slowly than any exponential function of time and that as a function of time it has the inverse power-like form at this regime of time [8, 10, 11]. So, all implications of the assumption that the decay process of the dark energy is a quantum decay process can be found only if we apply a quantum decay law to describe decaying metastable dark energy. This idea was used in [12], where the assumption made in [7] that $\rho_{\rm de}(t)$ decays according to the radioactive exponential decay law was improved by replacing that radioactive decay law by the survival probability $\mathcal{P}(t)$, that is, by the decay law derived assuming that the decay process is a quantum process.

This is the place where one has to emphasize that the use of the assumption that dark energy depends on time and is decaying during time evolution leads to the conclusion that such a process is irreversible and violates a time reversal symmetry. (Consequences of this effect will be analysed in next sections of this paper) Note that the picture of the evolving Universe, which results from the solutions of the Einstein equations completed with quantum corrections appearing as the effect of treating the false vacuum decay as a quantum decay process, is consistent with the observational data. The evolution starts from the early time epoch with the running $\Lambda(t)$ and then it goes to the final accelerating phase expansion of the Universe. In such a scenario the standard cosmological ΛCDM model emerges from the quantum false vacuum state of the Universe.

The paper is organised as follows: In Section 2 one finds a short introduction of formalism necessary for considering decaying dark energy as a quantum decay process. Cosmological implications of a decaying dark energy are considered in Section 3. Section 4 contains conclusions.

2. Decay of a Dark Energy as a Quantum Decay Process

In the quantum decay theory of unstable systems, properties of the survival amplitudes

$$\mathcal{A}(t) = \langle\phi|\phi(t)\rangle \tag{1}$$

are usually analysed. Here a vector $|\phi\rangle$ represents the unstable state of the system considered and $|\phi(t)\rangle$ is the solution of the Schrödinger equation

$$i\hbar\frac{\partial}{\partial t}|\phi(t)\rangle = \mathfrak{H}|\phi(t)\rangle. \tag{2}$$

The initial condition for (2) in the case considered is usually assumed to be

$$|\phi(t = t_0 \equiv 0)\rangle \overset{\text{def}}{=\!=} |\phi\rangle, \tag{3}$$

or equivalently

$$\mathcal{A}(0) = 1. \tag{4}$$

In (2) $\mathfrak{H}$ denotes the complete (full), self-adjoint Hamiltonian of the system. We have $|\phi(t)\rangle = \exp[-(i/\hbar)t\mathfrak{H}]|\phi\rangle$. It is not difficult to see that this property and hermiticity of H imply that

$$(\mathcal{A}(t))^* = \mathcal{A}(-t). \tag{5}$$

Therefore, the decay probability of an unstable state (usually called the decay law), i.e., the probability for a quantum system to remain at time t in its initial state $|\phi(0)\rangle \equiv |\phi\rangle$,

$$\mathcal{P}(t) \overset{\text{def}}{=\!=} |\mathcal{A}(t)|^2 \equiv \mathcal{A}(t)(\mathcal{A}(t))^*, \tag{6}$$

must be an even function of time [8]:

$$\mathcal{P}(t) = \mathcal{P}(-t). \tag{7}$$

This last property suggests that, in the case of the unstable states prepared at some instant t_0, say $t_0 = 0$, initial condition (3) for evolution equation (2) should be formulated more precisely. Namely, from (7), it follows that the probabilities of finding the system in the decaying state $|\phi\rangle$ at the instant, say $t = T \gg t_0 \equiv 0$, and at the instant $t = -T$ are the same. Of course, this can never occur. In almost all experiments in which the decay law of a given unstable subsystem system is investigated this particle is created at some instant of time, say t_0, and this instant of time is usually considered as the initial instant for the problem. From property (7) it follows that the instantaneous creation of the unstable subsystem system (e.g., a particle or an excited quantum level and so on) is practically impossible. For the observer, the creation of this object (i.e., the preparation of the state, $|\phi\rangle$, representing the decaying subsystem system) is practically instantaneous. What is more, using suitable detectors he/she is usually able to prove that it did not exist at times $t < t_0$. Therefore, if one looks for the solutions of Schrödinger equation (2) describing properties of the unstable states prepared at some initial instant t_0 in the system and if one requires these solutions to reflect situations described above, one should complete initial conditions (3), (4) for (2) by assuming additionally that

$$\begin{gathered}|\phi(t < t_0)\rangle = 0 \\ \text{or } \mathcal{A}(t)(t < t_0) = 0.\end{gathered} \tag{8}$$

Equivalently, within the problem considered, one can use initial conditions (3), (4) and assume that time t may vary from $t = t_0 > -\infty$ to $t = +\infty$ only, that is, that $t \in \mathbb{R}^+$.

Note that canonical (that is a classical radioactive) decay law $\mathcal{P}_c(t) = \exp[-t/\tau_0]$ (where τ_0 is a lifetime) does not satisfy property (7), which is valid only for the quantum decay law $\mathcal{P}(t)$. What is more, from (5) and (6) it follows that at very early times, i.e., at the Zeno times (see [8, 13]),

$$\left.\frac{\partial\mathcal{P}(t)}{\partial t}\right|_{t=0} = 0, \tag{9}$$

which implies that

$$\mathcal{P}(t) > e^{-t/\tau_0} \stackrel{\text{def}}{=\!=} \mathcal{P}_c(t) \quad \text{for } t \longrightarrow 0. \tag{10}$$

So at the Zeno time region the quantum decay process is much slower than any decay process described by the canonical (or classical) decay law $\mathcal{P}_c(t)$.

Now let us focus the attention on the survival amplitude $\mathcal{A}(t)$. An unstable state $|\phi\rangle$ can be modeled as wave packets using solutions of the following eigenvalue equation $\mathfrak{H}|E\rangle = E|E\rangle$, where $E \in \sigma_c(\mathfrak{H})$, and $\sigma_c(\mathfrak{H})$ denotes a continuum spectrum of $\mathfrak{H}$. Eigenvectors $|E\rangle$ are normalized as usual: $\langle E|E'\rangle = \delta(E - E')$. Using vectors $|E\rangle$ we can model an unstable state as the following wave-packet:

$$|\phi\rangle \equiv |\phi\rangle = \int_{E_{\min}}^{\infty} c(E)\,|E\rangle\, dE, \tag{11}$$

where expansion coefficients $c(E)$ are functions of the energy E and $E_{\min}$ is the lower bound of the spectrum $\sigma_c(\mathfrak{H})$ of $\mathfrak{H}$. The state $|\phi\rangle$ is normalized $\langle\phi|\phi\rangle = 1$, which means that it has to be $\int_{E_{\min}}^{\infty} |c(E)|^2 dE = 1$. Now using the definition of the survival amplitude $\mathcal{A}(t)$ and the expansion (11) we can find $\mathcal{A}(t)$, which takes the following form within the formalism considered:

$$\mathcal{A}(t) \equiv \mathcal{A}(t - t_0) = \int_{E_{\min}}^{\infty} \omega(E)\, e^{-iE(t-t_0)}\, dE, \tag{12}$$

where $\omega(E) \equiv |c(E)|^2 > 0$ and $\omega(E)dE$ is the probability to find the energy of the system in the state $|\phi\rangle$ between E and $E + dE$. The last relation (12) means that the survival amplitude $\mathcal{A}(t)$ is a Fourier transform of an absolute integrable function $\omega(E)$. If we apply the Riemann-Lebesgue Lemma to integral (12) then one concludes that there must be $\mathcal{A}(t) \longrightarrow 0$ as $t \longrightarrow \infty$. This property and relation (12) are an essence of the Fock–Krylov theory of unstable states [14, 15].

As it is seen from (12), the amplitude $\mathcal{A}(t)$ and thus the decay law $\mathcal{P}(t)$ of the unstable state $|\phi\rangle$ are completely determined by the density of the energy distribution $\omega(E)$ for the system in this state [14, 15] (see also [8, 10, 11, 16–21]).

In the general case the density $\omega(E)$ possesses properties analogous to the scattering amplitude; i.e., it can be decomposed into a threshold factor, a pole-function $P(E)$ with a simple pole, and a smooth form factor $F(E)$. There is $\omega(E) = \Theta(E - E_{\min})(E - E_{\min})^{\alpha_l} P(E)F(E)$, where α_l depends on the angular momentum l through $\alpha_l = \alpha + l$ [8] (see equation (6.1) in [8]), $0 \le \alpha < 1$) and $\Theta(E)$ is a step function: $\Theta(E) = 0$ for $E \le 0$ and $\Theta(E) = 1$ for $E > 0$. The simplest choice is to take $\alpha = 0$, $l = 0$, $F(E) = 1$ and to assume that $P(E)$ has a Breit–Wigner (BW) form of the energy distribution density. (The mentioned Breit–Wigner distribution was found when the cross section of slow neutrons was analysed [22]) It turns out that the decay curves obtained in this simplest case are very similar in form to the curves calculated for the above described more general $\omega(E)$ (see [16] and analysis in [8]). So to find the most typical properties of the decay process it is sufficient to make the relevant calculations for $\omega(E)$ modeled by the Breit–Wigner distribution of the energy density $\omega(E) \equiv \omega_{\text{BW}}(E) \stackrel{\text{def}}{=\!=} (N/2\pi)\Theta(E - E_{\min})(\Gamma_0/((E - E_0)^2 + (\Gamma_0/2)^2))$, where N is a normalization constant. The parameters E_0 and Γ_0 correspond to the energy of the system in the unstable state and its decay rate at the exponential (or canonical) regime of the decay process. $E_{\min}$ is the minimal (the lowest) energy of the system. Inserting $\omega_{\text{BW}}(E)$ into formula (12) for the amplitude $\mathcal{A}(t)$ and assuming for simplicity that $t_0 = 0$, after some algebra, one finds that

$$\mathcal{A}(t) = \frac{N}{2\pi} e^{-(i/\hbar)E_0 t} I_\beta\left(\frac{\Gamma_0 t}{\hbar}\right), \tag{13}$$

where

$$I_\beta(\tau) \stackrel{\text{def}}{=\!=} \int_{-\beta}^{\infty} \frac{1}{\eta^2 + 1/4} e^{-i\eta\tau} d\eta. \tag{14}$$

Here $\tau = \Gamma_0 t/\hbar \equiv t/\tau_0$, τ_0 is the lifetime, $\tau_0 = \hbar/\Gamma_0$, and $\beta = (E_0 - E_{min})/\Gamma_0 > 0$. The integral $I_\beta(\tau)$ has the following structure:

$$I_\beta(\tau) = I_\beta^{\text{pole}}(\tau) + I_\beta^{L}(\tau), \tag{15}$$

where

$$I_\beta^{\text{pole}}(\tau) = \int_{-\infty}^{\infty} \frac{1}{\eta^2 + 1/4} e^{-i\eta\tau} d\eta \equiv 2\pi e^{-\tau/2}, \tag{16}$$

and

$$I_\beta^{L}(\tau) = -\int_{+\beta}^{\infty} \frac{1}{\eta^2 + 1/4} e^{+i\eta\tau} d\eta. \tag{17}$$

(The integral $I_\beta^L(\tau)$ can be expressed in terms of the integral-exponential function [23–26] (for a definition, see [27, 28])) The result (15) means that there is a natural decomposition of the survival amplitude $\mathcal{A}(t)$ into two parts:

$$\mathcal{A}(t) = \mathcal{A}_c(t) + \mathcal{A}_L(t), \tag{18}$$

where

$$\mathcal{A}_c(t) = \frac{N}{2\pi} e^{-(i/\hbar)E_0 t} I_\beta^{\text{pole}}\left(\frac{\Gamma_0 t}{\hbar}\right) \equiv N e^{-(i/\hbar)E_0 t} e^{-\Gamma_0 t/2}, \tag{19}$$

and

$$\mathcal{A}_L(t) = \frac{N}{2\pi} e^{-(i/\hbar)E_0 t} I_\beta^{L}\left(\frac{\Gamma_0 t}{\hbar}\right), \tag{20}$$

and $\mathcal{A}_c(t)$ is the canonical part of the amplitude $\mathcal{A}(t)$ describing the pole contribution into $\mathcal{A}(t)$ and $\mathcal{A}_L(t)$ represents the remaining part of $\mathcal{A}(t)$.

From decomposition (18) it follows that in the general case within the model considered the survival probability (6) contains the following parts:

$$\begin{aligned}\mathcal{P}(t) &= |\mathcal{A}(t)|^2 \equiv |\mathcal{A}_c(t) + \mathcal{A}_L(t)|^2 \\ &= |\mathcal{A}_c(t)|^2 + 2\mathfrak{R}\left[\mathcal{A}_c(t)(\mathcal{A}_L(t))^*\right] + |\mathcal{A}_L(t)|^2.\end{aligned} \tag{21}$$

This last relation is especially useful when one looks for a contribution of late time properties of the quantum unstable system to the survival amplitude.

The late time form of the integral $I_\beta^L(\tau)$ and thus the late time form of the amplitude $\mathscr{A}_L(t)$ can be relatively easy to find using analytical expression for $\mathscr{A}_L(t)$ in terms of the integral-exponential functions or simply performing the integration by parts in (17). One finds for $t \longrightarrow \infty$ (or $\tau \longrightarrow \infty$) that the leading term of the late time asymptotic expansion of the integral $I_\beta^L(\tau)$ has the following form:

$$I_\beta^L(\tau) \simeq -\frac{i}{\tau} \frac{e^{i\beta\tau}}{\beta^2 + 1/4} + \dots, \quad (\tau \longrightarrow \infty). \tag{22}$$

Thus inserting (22) into (20) one can find late time form of $\mathscr{A}_L(t)$.

As was mentioned we consider the hypothesis that a dark energy depends on time, $\rho_{de} = \rho_{de}(t)$, and decays with the increasing time t to ρ_{bare}: $\rho_{de}(t) \longrightarrow \rho_{bare} \neq 0$ as $t \longrightarrow \infty$. We assume that it is a quantum decay process. The consequence of this assumption is that we should consider $\rho_{de}(t_0)$ (where t_0 is the initial instant) as the energy of an excited quantum level (e.g., corresponding to the false vacuum state) and the energy density ρ_{bare} as the energy corresponding to the true lowest energy state (the true vacuum) of the system considered. Our hypothesis means that $(\rho_{de}(t) - \rho_{bare}) \longrightarrow 0$ as $t \longrightarrow \infty$. As it was said we assumed that the decay process of the dark energy is a quantum decay process: From the point of view of the quantum theory of decay processes this means that $\lim_{t\to\infty}(\rho_{de}(t) - \rho_{bare}) = 0$ according to the quantum mechanical decay law. Therefore if we define

$$\widetilde{\rho}_{de}(t) \stackrel{\text{def}}{=\!=} \rho_{de}(t) - \rho_{bare}, \tag{23}$$

our assumption means that the decay law for $\widetilde{\rho}_{de}(t)$ has the following form (see [12]):

$$\begin{aligned}\widetilde{\rho}_{de}(t) &= \widetilde{\rho}_{de}(t_0)\,\mathscr{P}(t) \equiv \widetilde{\rho}_{de}(t_0) \\ &\cdot \left(|\mathscr{A}_c(t)|^2 + 2\Re\left[\mathscr{A}_c(t)(\mathscr{A}_L(t))^*\right] + |\mathscr{A}_L(t)|^2\right),\end{aligned} \tag{24}$$

where $\mathscr{P}(t)$ is given by relation (6), or, equivalently, our assumption means that the decay law for $\widetilde{\rho}_{de}(t)$ has the following form (compare [12]):

$$\begin{aligned}\rho_{de}(t) &\equiv \rho_{bare} + \widetilde{\rho}_{de}(t_0) \\ &\cdot \left(|\mathscr{A}_c(t)|^2 + 2\Re\left[\mathscr{A}_c(t)(\mathscr{A}_L(t))^*\right] + |\mathscr{A}_L(t)|^2\right),\end{aligned} \tag{25}$$

where $\widetilde{\rho}_{de}(t_0) = (\rho_{de}(t_0) - \rho_{bare})$ and $\mathscr{P}(t)$ is replaced by (21). Taking into account the standard relation between ρ_{de} and the cosmological constant Λ we can write

$$\begin{aligned}\Lambda_{eff}(t) &\equiv \Lambda_{bare} + \widetilde{\Lambda}(t_0) \\ &\cdot \left(|\mathscr{A}_c(t)|^2 + 2\Re\left[\mathscr{A}_c(t)(\mathscr{A}_L(t))^*\right] + |\mathscr{A}_L(t)|^2\right),\end{aligned} \tag{26}$$

where $\widetilde{\Lambda}(t_0) \equiv \widetilde{\Lambda}_0 = (\Lambda(t_0) - \Lambda_{bare})$. Thus within the considered case using definition (6) or relation (21) we can determine changes in time of the dark energy density $\rho_{de}(t)$ (or running $\Lambda(t)$) knowing the general properties of survival amplitude $\mathscr{A}(t)$.

The above described approach is self-consistent if we identify $\rho_{de}(t_0)$ with the energy E_0 of the unstable system divided by the volume V_0 (where V_0 is the volume of the system at $t = t_0$): $\rho_{de}(t_0) \equiv \rho_{de}^{qft} \stackrel{\text{def}}{=\!=} \rho_{de}^0 = E_0/V_0$ and $\rho_{bare} = E_{min}/V_0$. Here ρ_{de}^{qft} is the vacuum energy density calculated using quantum field theory methods. In such a case

$$\beta = \frac{E_0 - E_{min}}{\Gamma_0} \equiv \frac{\rho_{de}^0 - \rho_{bare}}{\gamma_0} > 0, \tag{27}$$

(where $\gamma_0 = \Gamma_0/V_0$), or equivalently $\Gamma_0/V_0 \equiv (\rho_{de}^0 - \rho_{bare})/\beta$.

3. Cosmological Implications of Decaying Vacuum

Let us consider cosmological implications of the parameter Λ with the time parameterized decaying part, derived in the previous section, in the form

$$\Lambda \equiv \Lambda_{eff}(t) = \Lambda_{bare} + \delta\Lambda(t), \tag{28}$$

where $\delta\Lambda(t)$ describes quantum corrections and it is given by a series with respect to $1/t$; i.e.,

$$\delta\Lambda(t) = \sum_{n=1}^{\infty} \alpha_{2n} \left(\frac{1}{t}\right)^{2n}, \tag{29}$$

where t is the cosmological scale time and the functions $\Lambda_{eff}(t)$ and $\delta\Lambda(t)$ have a reflection symmetry with respect to the cosmological time $\delta\Lambda(-t) = \delta\Lambda(t)$. The next step in deriving dynamical equations for the evolution of the Universe is to consider this parameter as a source of gravity which contributes to the effective energy density; i.e.,

$$3H(t)^2 = \rho_m(t) + \rho_{de}(t), \tag{30}$$

where $\rho_{de}(t)$ is identified as the energy density of the quantum decay process of vacuum

$$\rho_{de}(t) = \Lambda_{bare} + \delta\Lambda(t). \tag{31}$$

In this paper, we assume that $c = 8\pi G = 1$. The Einstein field equation for the FRW metric reduces to

$$\begin{aligned}\frac{dH(t)}{dt} &= -\frac{1}{2}\left(\rho_{eff}(t) + p_{eff}(t)\right) \\ &= -\frac{1}{2}\left(\rho_m(t) + 0 + p_{de}(t) - \rho_{de}(t)\right),\end{aligned} \tag{32}$$

where $\rho_{eff} = \rho_m + \rho_{de}$, $p_{eff} = 0 + p_{de}$, or

$$\frac{dH(t)}{dt} = -\frac{1}{2}\rho_m(t) = -\frac{1}{2}\left(3H(t)^2 - \Lambda_{bare} - \delta\Lambda(t)\right). \tag{33}$$

Szydlowski et al. [12] considered the radioactive-like decay of metastable dark energy. For the late time, this decay process has three consecutive phases: the phase of radioactive decay, the phase of damping oscillations, and finally the phase of power law decaying. When $\beta > 0$ for $t > (\hbar/\Gamma_0)(2\beta/(\beta^2 + 1/4))$, dark energy can be described in the following form (see (25) and [12]):

$$\rho_{de}(t) \approx \rho_{bare} + \epsilon\left(4\pi^2 e^{-(\Gamma_0/\hbar)t} + \frac{4\pi e^{-(\Gamma_0/2\hbar)t}\sin(\beta(\Gamma_0/\hbar)t)}{(1/4+\beta^2)(\Gamma_0/\hbar)t} + \frac{1}{((1/4+\beta^2)(\Gamma_0/\hbar)t)^2}\right), \quad (34)$$

where ϵ, Γ_0, and β are model parameters. Equation (34) results directly from (25): One only needs to insert (22) into formula for $\mathcal{A}_L(t)$ and result (19) instead of $\mathcal{A}_c(t)$ into (25). In this paper, we consider the first phase of decay process, in other words, the phase of radioactive (exponential) decay.

The model with the radioactive (exponential) decay of dark energy was investigated by Shafieloo et al. [7]. During the phase of the exponential decay of the vacuum

$$\frac{d\delta\Lambda(t)}{dt} = A\delta\Lambda(t), \quad (35)$$

where $A = \text{const} < 0$ ($\delta\Lambda(t)$ is decaying).

The set of equations (33) and (35) constitute a two-dimensional closed autonomous dynamical system in the form

$$\frac{dH(t)}{dt} = -\frac{1}{2}\left(3H(t)^2 - \Lambda_{bare} - \delta\Lambda(t)\right),$$
$$\frac{d\delta\Lambda(t)}{dt} = A\delta\Lambda(t). \quad (36)$$

System (36) has the time dependent first integral in the form

$$\rho_m(t) = 3H(t)^2 - \Lambda_{bare} - \delta\Lambda(t). \quad (37)$$

At the finite domain, system (36) possesses only one critical point representing the standard cosmological model (the running part of Λ vanishes, i.e., $\delta\Lambda(t) = 0$).

System (36) can be rewritten in variables

$$x = \frac{\delta\Lambda(t)}{3H_0^2},$$
$$y = \frac{H(t)}{H_0} \quad (38)$$

where H_0 is the present value of the Hubble function. Then

$$\frac{dx}{d\sigma} = \frac{A}{H_0}x$$
$$\frac{dy}{d\sigma} = -\frac{1}{2}\left(3y^2 - 3\Omega_{\Lambda_{bare}} - 3x\right), \quad (39)$$

where $\Omega_{\Lambda_{bare}} = \Lambda_{bare}/3H_0^2$ and $\sigma = H_0 t$ are a new reparametrized time. The phase portrait of system (39) is shown in Figure 1.

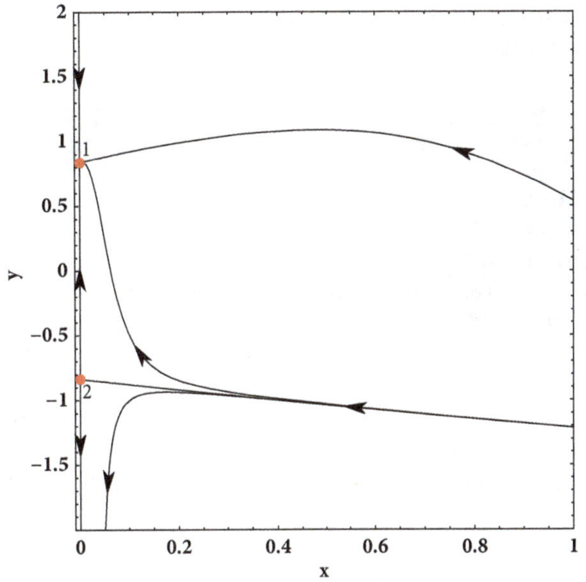

FIGURE 1: The phase portrait of system (39). Critical point 1 ($x = 0, y = \sqrt{\Lambda_{bare}}/\sqrt{3}H_0$) is the stable node and critical point 2 ($x = 0, y = -\sqrt{\Lambda_{bare}}/\sqrt{3}H_0$) is the saddle. These critical points represent the de Sitter universes. Here, H_0 is the present value of the Hubble function. The value of A is assumed as $-H_0$. Note that the phase portrait is not symmetric under reflection $H \longrightarrow -H$. While critical point 1 is a global attractor, only a unique separatrix reaches critical point 2.

Szydlowski et al. [12] demonstrated that the contribution of the energy density of the decaying quantum vacuum possesses three disjoint phases during the cosmic evolution. The phase of exponential decay like in the radioactive decay processes is long phase in the past and future evolution. Our estimation of model parameter shows that we are living in the Universe with the radioactive decay of the quantum vacuum.

It is interesting that, during this phase, the Universe violates the reflection symmetry of the time: $t \longrightarrow -t$. In cosmology and generally in physics there is a fundamental problem of the origin of irreversibility in the Universe [29]. Note that in our model irreversibility is a consequence of the radioactive decay of the quantum vacuum.

If we considered radioactivity (in which the time reversal symmetry is broken) then a direction of decaying vacuum is in accordance with the thermodynamical arrow of time. First, note that it is in some sense very natural that the dynamics of the Universe is in fact irreversible when the full quantum evolution is taken into account. Therefore, radioactive decay of vacuum irreversibility has a thermodynamic interpretation as far as the evolution of the Universe is concerned: in horizon thermodynamics the area of the cosmological horizon is interpreted as (beginning proportional to) the entropy, i.e., the Hawking entropy. In a system where Λ decays as a result of irreversible quantum processes we obtain the very natural conclusion that the entropy of the Universe grows, in many cases without an upper bound [30, 31].

In the general parameterization (29), of course, the symmetry of changing $t \longrightarrow -t$ is present and this symmetry is also in a one-dimensional nonautonomous dynamical system describing the evolution of the Universe:

$$\frac{dH(t)}{dt} = -\frac{1}{2}\left(3H(t)^2 - \Lambda_{\text{bare}} - \sum_{n=1}^{\infty} \alpha_{2n} t^{-2n}\right). \quad (40)$$

In cosmology, especially in quantum cosmology, the analysis of the concept of time seems to be the key for the construction of an adequate quantum gravity theory, which we would like to apply to the description of early Universe.

The good approximation of (40) is to replace in it the cosmological time by the Hubble cosmological scale time

$$t_{\text{H}} = \frac{1}{H}. \quad (41)$$

In consequence, parameterization (29) can be rewritten in the new form

$$\delta\Lambda(t) = \delta\Lambda(H(t)) = \sum_{n=1}^{\infty} \alpha_{2n} H(t)^{2n}. \quad (42)$$

After putting this form into (40), we obtain dynamical system in an autonomous form with the preserved symmetry of time $t \longrightarrow -t$, $H \longrightarrow -H$. In Figure 2 presents a diagram of the evolution of the Hubble function obtained from the following one-dimensional dynamical system:

$$\frac{dH(t)}{dt} = -\frac{1}{2}\left(3H(t)^2 - \Lambda_{\text{bare}} - \sum_{n=1}^{\infty} \alpha_{2n} H(t)^{2n}\right). \quad (43)$$

For comparison, the evolution of the Hubble functions derived in the ΛCDM model, model (40), and model (43) are presented in Figure 3. For the existence of the de Sitter global attractor as $t \longrightarrow \infty$ asymptotically a contribution coming from the decaying part of $\delta\Lambda(H(t)) = \sum_{n=1}^{\infty} \alpha_{2n} H(t)^{2n}$ should be vanishing.

This condition guarantees for us a consistency of our model with astronomical observations of the accelerating phase of the Universe [3].

If all parameters α_{2n} for $n > 1$ equal zero then the Hubble parameter is described by the following formula:

$$H(a) = \pm\sqrt{\frac{\rho_{\text{m},0} a^{\alpha_{21}-3} + \Lambda_{\text{bare}}}{3 - \alpha_{21}}}, \quad (44)$$

or

$$H(z) = \pm\sqrt{\frac{\rho_{\text{m},0} (1+z)^{3-\alpha_{21}} + \Lambda_{\text{bare}}}{3 - \alpha_{21}}}, \quad (45)$$

where $z = a^{-1} - 1$ is redshift. From (44) we can obtain the following formula for the expanding Universe:

$$a(t) = \left(\frac{\rho_{\text{m},0}}{\Lambda_{\text{bare}}} \sinh\left(\frac{\sqrt{(3-\alpha_{21})\Lambda_{\text{bare}}}}{2} t\right)\right)^{2/(3-\alpha_{21})}. \quad (46)$$

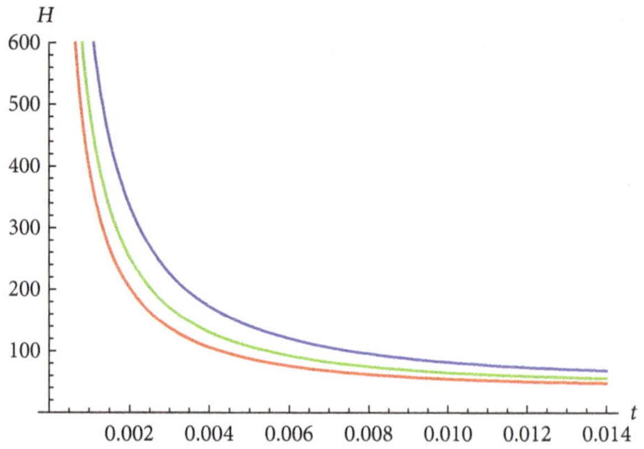

FIGURE 2: The diagram of the evolution of the Hubble function with respect to the cosmological time t, which is described by (43) with $\alpha_{21} \neq 0$ and $\alpha_{2n} = 0$ for every $n > 1$. For illustration, two example values of the parameter α_{21} = are chosen: −1 and −2. The top blue curve describes the evolution of the Hubble function in the ΛCDM model. The middle curve describes one for $\alpha_{21} = -1$ and the bottom red curve describes one for $\alpha_{21} = -2$. The Hubble function is expressed in km/s Mpc and the cosmological time t is expressed in s Mpc/km.

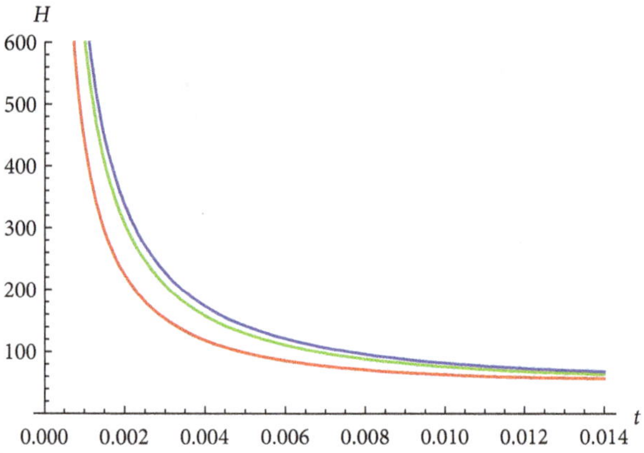

FIGURE 3: The diagram of the evolution of the Hubble function with respect to the cosmological time t, which is described by (40) and (43) with $\alpha_{21} \neq 0$ and $\alpha_{2n} = 0$ for every $n > 1$. For illustration, the value of the parameter α_{21} = is chosen as −0.3. The top blue curve describes the evolution of the Hubble function in the ΛCDM model. The middle curve describes one for (43) and the bottom red curve describes one for (40). The Hubble function is expressed in km/s Mpc and the cosmological time t is expressed in s Mpc/km. Note that these models are not qualitatively different.

Figure 4 presents the evolution of the scale factor, which is described by (46). Eq. (46) gives us the following formula:

$$H(t) = \sqrt{\frac{\Lambda_{\text{bare}}}{3 - \alpha_{21}}} \coth\left(\frac{1}{2}\sqrt{\Lambda_{\text{bare}}(3 - \alpha_{21})} t\right). \quad (47)$$

In the extension of Friedmann equation (37) matter is contributed as well as dark energy. The total energy-momentum tensor $T^{\mu\nu} = T^{\mu\nu}_{\text{m}} + T^{\mu\nu}_{\text{de}}$ is of course conserved. However, between the matter and dark energy sectors exist an interaction—the energy density is transferred between

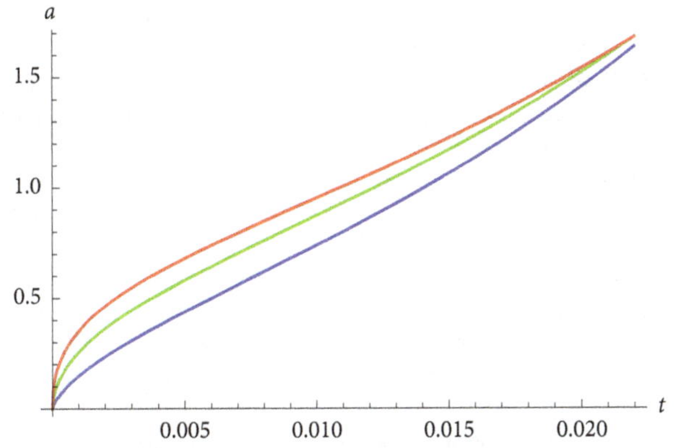

FIGURE 4: The diagram of the evolution of the scale factor with respect to the cosmological time t, which is described by (46). For illustration, two example values of the parameter α_{21} = are chosen: −1 and −2. The bottom blue curve describes the evolution of the scale factor in the ΛCDM model. The middle curve describes one for $\alpha_{21} = -1$ and the top red curve describes one for $\alpha_{21} = -2$. The cosmological time t is expressed in s Mpc/km.

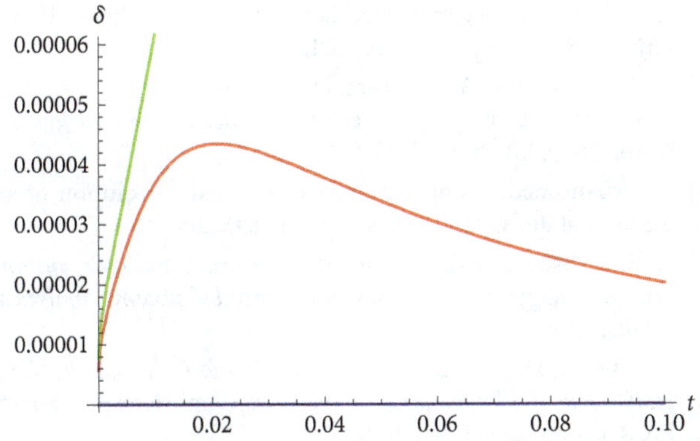

FIGURE 5: The diagram of the evolution of the parameter δ with respect to the cosmological time t. For illustration, two example values of the parameter A are chosen: $A = -100$km/s Mpc (the top green curve) and $A = -200$km/s Mpc (the middle red curve). For comparison the ΛCDM model with the parameter $\delta = 0$ is represented by the bottom blue curve. Here, the value of the parameter B is equal to 1. The cosmological time t is expressed in s Mpc/km.

these sectors. This process can be described by the system of equations

$$\frac{d\rho_{\mathrm{m}}(t)}{dt} + 3H(t)\rho_{\mathrm{m}}(t) = -\frac{d\rho_{\mathrm{de}}(t)}{dt} = -\frac{d\Lambda_{\mathrm{eff}}(t)}{dt},$$
$$\frac{d\rho_{\mathrm{de}}(t)}{dt} = \frac{d\Lambda_{\mathrm{eff}}(t)}{dt}, \tag{48}$$

where it is assumed that pressure of matter $p_{\mathrm{m}} = 0$ and $p_{\mathrm{de}} = -\rho_{\mathrm{de}}$. The time variability of the matter and energy density of decaying vacuum are demonstrated in Figure 5.

In the special case of radioactive decay of vacuum (48) reduces to

$$\frac{d\rho_{\mathrm{m}}(t)}{dt} + 3H(t)\rho_{\mathrm{m}}(t) = -ABe^{At} = -A\delta\Lambda(t),$$
$$\frac{d\rho_{\mathrm{de}}(t)}{dt} = A\delta\Lambda(t) \tag{49}$$

or

$$\frac{1}{a(t)^3}\frac{d}{dt}\left(a(t)^3\rho_{\mathrm{m}}(t)\right) = -ABe^{At} = -A\delta\Lambda(t) \Longrightarrow$$
$$\rho_{\mathrm{m}}(t)\,a(t)^3 = \rho_{\mathrm{m},0}a_0^3 - \int ABe^{At}a(t)^3\,dt, \tag{50}$$
$$\frac{d\rho_{\mathrm{de}}(t)}{dt} = A\delta\Lambda(t).$$

In the case of the interaction between matter and decaying dark energy, the natural consequence of conservation of the total energy-momentum tensor $T^{\mu\nu}$ is a modification of the standard formula for scaling matter.

Let $\rho_{\mathrm{m}}(t) = \rho_{m,0}a^{-3+\delta(t)}$, where $\delta(t)$ is a deviation from the canonical scaling of dust matter [32, 33]. Then we have

$$\delta(t) = \frac{\ln\left(\rho_{\mathrm{m}}(t)/\rho_{\mathrm{m},0}\right)}{\ln a(t)} + 3. \tag{51}$$

4. Conclusions

From our investigation of cosmological implications of effects of the quantum decay of metastable dark energy, one can derive following results:

(i) The cosmological models with the running cosmological parameter can be included in the framework of some extension of Friedmann equation. The new ingredient in the comparison with the standard cosmological model (ΛCDM model) is that the total energy-momentum tensor is conserved and the interaction takes place between the matter and dark energy sectors. In consequence the canonical scaling law $\rho_{\mathrm{m}} \propto a^{-3}$ is modified. Because $\Lambda(t)$ is decaying ($d\Lambda/dt < 0$) energy of matter in the comoving volume $\propto a^3$ is growing with time.

(ii) We have found that the appearance of the universal exponential contribution in energy density of the decaying vacuum can explain the irreversibility of the cosmic evolution. While the reversibility $t \longrightarrow -t$ is still present in the dynamical equation describing the evolutional scenario, in the first phase of radioactive decay, this symmetry is violated.

(iii) We have also compared the time evolution of the Hubble function in the model under consideration (where $\Lambda(t)$ is parameterized by the cosmological time) with Sola et al. [34] parameterization by the Hubble function. Note that both parameterizations coincide if time t is replaced by the Hubble scale time $t_H = 1/H$. If the evolution of the Universe is invariant in the scale, i.e., the scale factor a is changing in power law, then this correspondence is exact.

Conflicts of Interest

The authors declare that they have no conflicts of interest.

Acknowledgments

The authors are very grateful to Dr. Tommi Markkanen for suggestion of conclusion of their paper and indication of important references in the context of irreversibility of quantum decay process and thermodynamic arrow of time.

References

[1] S. Weinberg, *Reviews of Modern Physics*, vol. 61, pp. 1–23, 1989.

[2] S. Weinberg, "The cosmological constant problems," in *Sources and Detection of Dark Matter and Dark Energy in the Universe*, pp. 18–26, Springer-Verlag Berlin Heidelberg, New York, NY, USA, 2001.

[3] P. A. R. Ade et al., "Planck2015 results. XIV. Dark energy and modified gravity," *Astronomy & Astrophysics*, vol. 594, no. A14, p. 31, 2016.

[4] E. Di Valentino, E. V. Linder, and A. Melchiorri, "A Vacuum Phase Transition Solves H_0 Tension," *Physical Review D*, vol. 97, p. 043528, 2018.

[5] L. M. Krauss and J. Dent, "The Late Time Behavior of False Vacuum Decay: Possible Implications for Cosmology and Metastable Inflating States," *Physical Review Letters*, vol. 100, p. 171301, 2008.

[6] L. M. Krauss, J. Dent, and G. D. Starkman, "Late Time Decay of the False Vacuum, Measurement, and Quantum Cosmology," *International Journal of Modern Physics D*, vol. 17, p. 2501, 2008.

[7] A. Shafieloo, D. K. Hazra, V. Sahni, and A. A. Starobinsky, "Metastable Dark Energy with Radioactive-like Decay," *Monthly Notices of the Royal Astronomical Society*, vol. 473, p. 2760, 2018.

[8] L. Fonda, G. C. Ghirardi, and A. Rimini, "Decay theory of unstable quantum systems," *Reports on Progress in Physics*, vol. 41, p. 587, 1978.

[9] M. Peshkin, A. Volya, and V. Zelevinsky, "Non-exponential and oscillatory decays in quantum mechanics," *Europhysics Letters*, vol. 107, no. 4, p. 40001, 2014.

[10] L. A. Khalfin, "Contribution to the decay theory of a quasi-stationary state," *Journal of Experimental and Theoretical Physics*, vol. 33, p. 1371, 1958.

[11] L. A. Khalfin, "Contribution to the decay theory of a quasi-stationary state," *Soviet Physics—JETP*, vol. 6, p. 1053, 1958.

[12] M. Szydlowski, A. Stachowski, and K. Urbanowski, "Quantum mechanical look at the radioactive-like decay of metastable dark energy," *European Physical Journal C*, vol. 77, p. 902, 2017.

[13] K. Urbanowski, "Early-time properties of quantum evolution," *Physical Review A*, vol. 50, p. 2847, 1994.

[14] N. S. Krylov and V. A. Fock, "O dvuh osnovnyh tolkovanijah sootnosenija neopredelenosti dla energii i vremeni," *Journal of Experimental and Theoretical Physics*, vol. 17, p. 93, 1947 (Russian).

[15] V. A. Fock, *Fundamentals of quantum mechanics*, Mir Publishers, Moscow, 1978.

[16] N. G. Kelkar and M. Nowakowski, "No classical limit of quantum decay for broad states," *Journal Of Physics A*, vol. 43, p. 385308, 2010.

[17] J. Martorell, J. G. Muga, and D. W. L. Sprung, "Quantum post-exponential decay," in *Time in Quantum Mechanics - Vol. 2*, J. G. Muga, A. Ruschhaupt, and A. del Campo, Eds., vol. 789 of *Lecture Notes in Physics*, pp. 239–275, Springer-Verlag Berlin Heidelberg, Germany, 2009.

[18] E. Torrontegui, J. Muga, J. Martorell, and D. Sprung, "Quantum decay at long times," in *Unstable States in the Continuous Spectra, Part I: Analysis, Concepts, Methods, and Results*, C. A. Nicolaides and E. Brandas, Eds., vol. 60 of *Advances in Quantum Chemistry*, pp. 485–535, Elsevier, 2010.

[19] G. Garcia-Calderon, R. Romo, and J. Villavicencio, "Survival probability of multibarrier resonance systems: exact analytical approach," *Physical Review B*, vol. 76, p. 035340, 2007.

[20] F. Giraldi, "Logarithmic decays of unstable states," *European Physical Journal D*, vol. 69, p. 5, 2015.

[21] F. Giraldi, "Logarithmic decays of unstable states II," *European Physical Journal D*, vol. 70, p. 229, 2016.

[22] G. Breit and E. Wigner, "Capture of slow neutrons," *Physical Review*, vol. 49, p. 519, 1936.

[23] K. M. Sluis and E. A. Gislason, "Decay of a quantum-mechanical state described by a truncated Lorentzian energy distribution," *Physical Review A*, vol. 43, p. 4581, 1991.

[24] K. Urbanowski, "A quantum long time energy red shift: a contribution to varying α theories," *European Physical Journal C*, vol. 58, p. 151, 2008.

[25] K. Urbanowski, "Long time properties of the evolution of an unstable state," *Open Physics*, vol. 7, p. 696, 2009.

[26] K. Raczynska and K. Urbanowski, *Survival amplitude, instantaneous energy and decay rate of an unstable system: Analytical results*, 2018.

[27] W. J. Olver, D. W. Lozier, R. F. Boisvert, and C. W. Clark, *NIST Handbook of Mathematical Functions*, Cambridge University Press, Cambridge, USA, 2010.

[28] M. Abramowitz and I. A. Stegun, *Handbook of Mathematical Functions with Formulas, Graphs, and Mathematical Tables*, vol. 55 of *Applied Mathematics Series*, National Bureau of Standards, Washington, DC, USA, 1964.

[29] H. D. Zeh, *The Physical Basis of the Direction of Time*, Springer-Verlag Berlin Heidelberg, Germany, 5th edition, 2007.

[30] K. Freese, F. C. Adams, J. A. Frieman, and E. Mottola, "Cosmology with decaying vaccum energy," *Nuclear Physics B*, vol. 287, p. 797, 1987.

[31] T. Markkanen, "De Sitter Stability and Coarse Graining," *European Physical Journal C*, vol. 78, p. 97, 2018.

[32] I. L. Shapiro, J. Sola, C. Espana-Bonet, and P. Ruiz-Lapuente, "Variable cosmological constant as a Planck scale effect," *Physics Letters B*, vol. 574, p. 149, 2003.

[33] C. Espana-Bonet, P. Ruiz-Lapuente, I. L. Shapiro, and J. Sola, "Testing the running of the cosmological constant with Type Ia Supernovae at high z," *Journal of Cosmology and Astroparticle Physics*, vol. 402, p. 6, 2004.

[34] I. L. Shapiro and J. Sola, "On the possible running of the cosmological "constant"," *Physics Letters B*, vol. 682, p. 105, 2009.

16

Localization of Energy-Momentum for a Black Hole Spacetime Geometry with Constant Topological Euler Density

Irina Radinschi,[1] **Theophanes Grammenos**,[2] **Farook Rahaman**,[3] **Andromahi Spanou**,[4] **Marius Mihai Cazacu**,[1] **Surajit Chattopadhyay**,[5] **and Antonio Pasqua**[6]

[1] *Department of Physics "Gh. Asachi" Technical University, Iasi 700050, Romania*
[2] *Department of Civil Engineering, University of Thessaly, 383 34 Volos, Greece*
[3] *Department of Mathematics, Jadavpur University, Kolkata 700 032, West Bengal, India*
[4] *School of Applied Mathematics and Physical Sciences, National Technical University of Athens, 157 80 Athens, Greece*
[5] *Department of Mathematics, Amity University, Kolkata 700135, India*
[6] *Department of Physics, University of Trieste, 34127 Trieste, Italy*

Correspondence should be addressed to Irina Radinschi; radinschi@yahoo.com

Academic Editor: Elias C. Vagenas

The evaluation of the energy-momentum distribution for a new four-dimensional, spherically symmetric, static and charged black hole spacetime geometry with constant nonzero topological Euler density is performed by using the energy-momentum complexes of Einstein and Møller. This black hole solution was recently developed in the context of the coupled Einstein–nonlinear electrodynamics of the Born-Infeld type. The energy is found to depend on the mass M and the charge q of the black hole, the cosmological constant Λ, and the radial coordinate r, while in both prescriptions all the momenta vanish. Some limiting and particular cases are analyzed and discussed, illustrating the rather extraordinary character of the spacetime geometry considered.

1. Introduction

The issue of energy-momentum localization systematised researchers' work in a special way. Looking deeply into the problem, it was clear that the main difficulty consists in the lack of a proper definition for the energy density of gravitational backgrounds. In this light, much research work has been done over the last years concerning the best tools used for the energy-momentum localization. A brief survey points out the leading role played notably by super-energy tensors [1–4], quasi-local expressions [5–10], and the famous energy-momentum complexes of Einstein [11, 12], Landau-Lifshitz [13], Papapetrou [14], Bergmann-Thomson [15], Møller [16], and Weinberg [17]. Among the aforementioned computational tools, the energy-momentum complexes have been proven to be interesting and useful as well due to the diverse and numerous reasonable expressions that can be obtained by their application. Some observations are in order here. First, according to their underlying mathematical mechanism, their construction involves the use of two parts, one for the matter and the other for the gravitational field. Second, despite the fact that the energy-momentum complexes allow one to obtain many interesting and physically meaningful results for different space-time geometries, their construction is connected to an inherent central problem, namely, their coordinate dependence. As it is well-known from the relevant literature, this problem has found a solution in the case of the Møller energy-momentum complex. Indeed, the calculations for the energy-momentum of a given gravitational background in the Møller prescription enable the use of any coordinates such that the energy density component transforms as a four-vector density under purely spatial coordinate transformations for metrics with a line element of the form $ds^2 = g_{00}dt^2 - g_{ij}dx^i dx^j$. As for the other energy-momentum complexes, the Schwarzschild Cartesian coordinates and the Kerr-Schild

Cartesian coordinates have to be utilised for the calculations providing physically reasonable results for the cases of space-time geometries in $(3+1)$, $(2+1)$, and $(1+1)$ dimensions (see, e.g., [18–34] and references therein). At this point it should be noticed that, in order to avoid the coordinate dependence, an alternative method for the computation of the energy and momentum distributions is provided by the teleparallel equivalent of general relativity and certain modified versions of the teleparallel theory (see, e.g., [35–40] and references therein).

Regarding the Einstein, Landau-Lifshitz, Papapetrou, Bergmann-Thomson, Weinberg, and Møller energy-momentum complexes there is an agreement with the definition of the quasi-local mass introduced by Penrose [41] and developed by Tod [42] for some gravitational backgrounds. We point out that some rather recent works show that several energy-momentum complexes "provide the same results" for any metric of the Kerr-Schild class and indeed even for solutions that are more general than those of the Kerr-Schild class (see, e.g., [43, 44] and the interesting article [45] on the subject). Further, the entire historical development of the energy-momentum complexes that started with the formulation of their definitions also includes the attempts made for their rehabilitation [46–49]. In this sense, perhaps the most interesting issue was the fact that different energy-momentum complexes yield the same results for the energy-momentum distribution in the case of various gravitating systems.

The present paper has the following structure: in Section 2 we describe the new class of four-dimensional spherically symmetric, static, and charged black hole solutions with constant nonzero topological Euler density which we will consider. Section 3 focuses on the presentation of the Einstein and Møller prescriptions used for performing the calculations. In Section 4 we present the calculations and the results obtained for the energy and momentum distributions. Finally, in the discussion provided in Section 5, we give a brief description of the results obtained as well as some limiting and particular cases. Throughout we use geometrized units ($c = G = 1$) and the signature chosen is $(+,-,-,-)$. Further, the calculations are performed by using the Schwarzschild Cartesian coordinates $\{t, x, y, z\}$ for the Einstein energy-momentum complex and the Schwarzschild coordinates $\{t, r, \theta, \varphi\}$ for the Møller energy-momentum complex. Finally, Greek indices run from 0 to 3, while Latin indices range from 1 to 3.

2. Description of the New Black Hole Solution with Constant Topological Euler Density

This section deals with the presentation of the new four-dimensional spherically symmetric, static, and charged black hole solution with constant topological Euler density [50] examined in the present study. Connecting geometry with topology, from the generalised Gauss–Bonnet theorem (see, e.g., [51]) applied to four dimensions the Euler–Poincaré characteristic is obtained by the integral of the Euler density

$$\mathscr{G} = \frac{1}{32\pi^2}\left(R^{\kappa\lambda\mu\nu}R_{\kappa\lambda\mu\nu} - 4R^{\kappa\lambda}R_{\kappa\lambda} + R^2\right), \qquad (1)$$

where $R_{\kappa\lambda\mu\nu}$ is the Riemann curvature tensor, $R_{\kappa\lambda}$ is the Ricci tensor, and R is the Ricci scalar (often the topological Euler density is given without the factor $1/32\pi^2$ (see, e.g., [52]). In fact, the terms in the parenthesis constitute the so-called "quadratic Gauss-Bonnet term" in the Lovelock gravity Lagrangian). For a general, spherically symmetric and static geometry described by the line element

$$ds^2 = f(r)\,dt^2 - f(r)^{-1}\,dr^2 - r^2\left(d\theta^2 + \sin^2\theta d\varphi^2\right) \qquad (2)$$

(1) becomes

$$\mathscr{G} = \frac{4}{r^2}\left\{f'(r)^2 + \left[f(r) - 1\right]f''(r)\right\}, \qquad (3)$$

where the constant $32\pi^2$ has been absorbed in $\mathscr{G}$. For constant topological density $\mathscr{G} = \alpha \neq 0$, (3) gives for the metric function

$$f(r) = 1 \pm \left(1 - 2A + Br + \frac{\alpha r^4}{24}\right)^{1/2}, \qquad (4)$$

with A, B arbitrary constants. In what follows, we will keep only the negative sign of (4), as it is the one leading to black hole solutions.

The new solution derived in [50] is based on the coupling of gravity to nonlinear electromagnetic fields as described by the nonlinear generalisation of Maxwell's electrodynamics according to the Born–Infeld theory. Thus, in the chosen case of electrovacuum, the radial electric field,

$$E(r) = \frac{r^2}{4q}\left(4R^{\mu\nu}R_{\mu\nu} - R^2\right)^{1/2} \qquad (5)$$

where q is the electric charge, solves the Einstein–nonlinear electrodynamics coupled system with the Ricci tensor $R_{\mu\nu}$ and the Ricci scalar R calculated by using the line element (2) and the metric function (4). In fact, with the values $A = 1/2 + q^2\Lambda/3$, $B = 4M\Lambda/3$, and $\alpha = 8\Lambda^2/3$, with Λ being the cosmological constant and M being the mass of the black hole, the line element (2) with the metric function (4) becomes

$$\begin{aligned} ds^2 = {} & \left[1 - \sqrt{\frac{4M\Lambda r}{3} + \frac{\Lambda^2 r^4}{9} - \frac{2q^2\Lambda}{3}}\right]dt^2 \\ & - \left[1 - \sqrt{\frac{4M\Lambda r}{3} + \frac{\Lambda^2 r^4}{9} - \frac{2q^2\Lambda}{3}}\right]^{-1}dr^2 \\ & - r^2\left(d\theta^2 + \sin^2\theta d\varphi^2\right), \end{aligned} \qquad (6)$$

describing a Reissner–Nordström–de Sitter black hole space-time geometry.

Now one can distinguish between two different cases, namely, the massive case ($M \neq 0$) and the massless case ($M = 0$). In the first case, (6), when $q = 0$ and $\Lambda > 0$, shows that the geometry is regular everywhere except at the origin $r = 0$. However, this case has no particular interest from the electrodynamic viewpoint. Black hole solutions with zero mass were proposed as a conjecture by A. Strominger [53] in order to explain conifold singularities in the context of string theory. In particular, the ten-dimensional IIA (resp., IIB) string theory admits black D2- (resp., D3-) brane solutions with a mass proportional to their area. After applying a Calabi-Yau compactification these solutions may wrap around minimal 2-surfaces (resp., 3-surfaces) in the Calabi-Yau space and they appear as four-dimensional black holes. As the area of the surface around which they wrap is let to go to zero, the corresponding extremal black holes become massless, topologically stable, structures. In fact, the existence of stable black hole solutions with zero ADM mass was shown in [54] although their relation to the massless solutions suggested in [53] is still not clarified. Indeed, since then there has been an increasing interest in massless black hole solutions (see, e.g., [55, 56] and references therein). Recently, massless black hole solutions were obtained from the dyonic black hole solution of the Einstein–Maxwell–dilaton thery [57]. Although the black hole solution examined here does not originate from string theory we will proceed to the consideration of the massless ($M = 0$) case despite its physically dubious character.

Thus for $M = 0$, the line element (6) becomes

$$\begin{aligned} ds^2 = &\left[1-\sqrt{\frac{\Lambda^2 r^4}{9}-\frac{2q^2\Lambda}{3}}\right]dt^2 \\ &-\left[1-\sqrt{\frac{\Lambda^2 r^4}{9}-\frac{2q^2\Lambda}{3}}\right]^{-1}dr^2 \\ &-r^2\left(d\theta^2+\sin^2\theta d\varphi^2\right). \end{aligned} \tag{7}$$

Here, when $q = 0$ and $\Lambda \neq 0$ the de Sitter solution is obtained, while for $q \neq 0$ but $\Lambda = 0$ one gets the Minkowski solution. In the case $q \neq 0$ and $\Lambda < 0$, an event horizon exists and the spacetime is singular at the origin $r = 0$, while the electric field is everywhere regularisable. An overall detailed study of the black hole's behavior in the massive as well as in the massless case is presented in [50].

3. Einstein and Møller Energy-Momentum Complexes

In this section we outline the definitions of the Einstein and Møller energy-momentum complexes.

The expression for the Einstein energy-momentum complex [11] in the case of a (3+1)-dimensional gravitational background was later found to be given by

$$\theta_\nu^\mu = \frac{1}{16\pi}h_{\nu,\lambda}^{\mu\lambda}. \tag{8}$$

The Freud superpotentials $h_\nu^{\mu\lambda}$ in (8) are calculated by the compact formula (found by Landau and Lifshitz)

$$h_\nu^{\mu\lambda} = \frac{1}{\sqrt{-g}}g_{\nu\sigma}\left[-g\left(g^{\mu\sigma}g^{\lambda\kappa}-g^{\lambda\sigma}g^{\mu\kappa}\right)\right]_{,\kappa} \tag{9}$$

and satisfy the antisymmetric property

$$h_\nu^{\mu\lambda} = -h_\nu^{\lambda\mu}. \tag{10}$$

We notice that in the Einstein prescription the local conservation law is respected:

$$\theta_{\nu,\mu}^\mu = 0. \tag{11}$$

Consequently, the energy and momentum can be calculated in Einstein's prescription by

$$P_\nu = \iiint \theta_\nu^0 dx^1 dx^2 dx^3. \tag{12}$$

Here, θ_0^0 and θ_i^0 represent the energy and momentum density components, respectively.

Applying Gauss' theorem, the energy-momentum reads

$$P_\mu = \frac{1}{16\pi}\iint h_\mu^{0i} n_i dS, \tag{13}$$

with n_i being the outward unit normal vector over the surface dS. In (13) the component P_0 represents the energy.

According to [16], the Møller energy-momentum complex is

$$\mathcal{J}_\nu^\mu = \frac{1}{8\pi}M_{\nu,\lambda}^{\mu\lambda}, \tag{14}$$

with the Møller superpotentials $M_\nu^{\mu\lambda}$ given by

$$M_\nu^{\mu\lambda} = \sqrt{-g}\left(\frac{\partial g_{\nu\sigma}}{\partial x^\kappa}-\frac{\partial g_{\nu\kappa}}{\partial x^\sigma}\right)g^{\mu\kappa}g^{\lambda\sigma}. \tag{15}$$

The Møller superpotentials $M_\nu^{\mu\lambda}$ satisfy the antisymmetric property

$$M_\nu^{\mu\lambda} = -M_\nu^{\lambda\mu}. \tag{16}$$

As in the case of the Einstein prescription, in the Møller prescription the local conservation law is also satisfied:

$$\frac{\partial \mathcal{J}_\nu^\mu}{\partial x^\mu} = 0. \tag{17}$$

In (17) the component $\mathcal{J}_0^0$ represents the energy density and $\mathcal{J}_i^0$ gives the momentum density components.

For the Møller energy-momentum complex, the energy-momentum distributions are given by

$$P_\nu = \iiint \mathcal{J}_\nu^0 dx^1 dx^2 dx^3. \tag{18}$$

In particular, the energy distribution can be computed by

$$E = \iiint \mathcal{J}_0^0 dx^1 dx^2 dx^3. \tag{19}$$

Using Gauss' theorem one gets

$$P_\nu = \frac{1}{8\pi}\iint M_\nu^{0i} n_i dS, \tag{20}$$

where, again, n_i is the outward unit normal vector over the surface dS.

4. Energy and Momentum Distribution for the Black Hole Solution with Constant Topological Euler Density

To compute the energy and momenta with the Einstein energy-momentum complex, we have to transform the metric given by the line element (6) in Schwarzschild Cartesian coordinates by using the coordinate transformation $x = r\sin\theta\cos\varphi$, $y = r\sin\theta\sin\varphi$, $z = r\cos\theta$. Thus, we obtain a new form for the line element:

$$ds^2 = f(r)\,dt^2 - \left(dx^2 + dy^2 + dz^2\right) - \frac{f^{-1}(r) - 1}{r^2}(xdx + ydy + zdz)^2. \quad (21)$$

In Schwarzschild Cartesian coordinates for $\nu = 0, 1, 2, 3$ and $i = 1, 2, 3$ we find the following vanishing components of the superpotentials h_ν^{0i}:

$$h_1^{01} = h_1^{02} = h_1^{03} = 0,$$
$$h_2^{01} = h_2^{02} = h_2^{03} = 0, \quad (22)$$
$$h_3^{01} = h_3^{02} = h_3^{03} = 0.$$

In order to compute the nonvanishing components of the superpotentials in the Einstein prescription we use (9) and we obtain the following expressions:

$$h_0^{01} = \frac{2x}{r^2}\sqrt{\frac{4M\Lambda r}{3} + \frac{\Lambda^2 r^4}{9} - \frac{2q^2\Lambda}{3}}, \quad (23)$$

$$h_0^{02} = \frac{2y}{r^2}\sqrt{\frac{4M\Lambda r}{3} + \frac{\Lambda^2 r^4}{9} - \frac{2q^2\Lambda}{3}}, \quad (24)$$

$$h_0^{03} = \frac{2z}{r^2}\sqrt{\frac{4M\Lambda r}{3} + \frac{\Lambda^2 r^4}{9} - \frac{2q^2\Lambda}{3}}. \quad (25)$$

With the aid of the line element (21), the expression for the energy given by (13), and the expressions (23)-(25) for the superpotentials, we get the energy distribution for the examined black hole in the Einstein prescription:

$$E_E = \frac{r}{2}\sqrt{\frac{4M\Lambda r}{3} + \frac{\Lambda^2 r^4}{9} - \frac{2q^2\Lambda}{3}}. \quad (26)$$

In order to calculate the momentum components we employ (13) and (22) and performing the calculations we find that all the momenta vanish:

$$P_x = P_y = P_z = 0. \quad (27)$$

In the Møller prescription we perform the calculations in Schwarzschild coordinates $\{t, r, \theta, \varphi\}$ with the aid of the line element (6) and we find only one nonvanishing superpotential:

$$M_0^{01} = -\frac{1}{6}\frac{12M\Lambda + 4\Lambda^2 r^3}{\sqrt{12M\Lambda r + \Lambda^2 r^4 - 6q^2\Lambda}}r^2\sin\theta. \quad (28)$$

Table 1: Limiting behavior of the energy of the massive ($M \neq 0$) black hole solution in the Einstein and Møller prescriptions.

Energy	$r \longrightarrow \infty$	$q = 0$
E_E	∞	$\frac{r}{2}\sqrt{\frac{4M\Lambda r}{3} + \frac{\Lambda^2 r^4}{9}}$
E_M	$-\infty$	$-\frac{r^2}{12}\frac{12M\Lambda + 4\Lambda^2 r^3}{\sqrt{12M\Lambda r + \Lambda^2 r^4}}$

Using the above expression for the superpotential and the expression for the energy obtained from (20), we get the energy in the Møller prescription:

$$E_M = -\frac{r^2}{12}\frac{12M\Lambda + 4\Lambda^2 r^3}{\sqrt{12M\Lambda r + \Lambda^2 r^4 - 6q^2\Lambda}}. \quad (29)$$

Finally, all the momenta are found to be zero:

$$P_r = P_\theta = P_\varphi = 0. \quad (30)$$

5. Discussion

Our paper focuses on the analysis of the energy-momentum localization for a new four-dimensional, spherically symmetric, static, and charged black hole spacetime geometry with constant nonzero topological Euler density, given by the line element (6). The solution describes a Reissner–Nordström–de Sitter spacetime geometry as the result of the coupling of Einstein gravity with nonlinear electrodynamics of the Born–Infeld type. For $q = 0$ the solution has a near-de Sitter behavior, while for $\Lambda > 0$ the solution is regular everywhere except at the origin $r = 0$. To perform our study we use the Einstein and Møller energy-momentum complexes. The calculations provide the well-defined expressions (26) and (29) for the energy distribution in both prescriptions. These energy distributions depend on the mass M and the charge q of the black hole, the cosmological constant Λ, and the radial coordinate r, while in both energy-momentum complexes all the momenta vanish.

In order to study the limiting behavior of the energy distributions obtained by the Einstein and Møller prescriptions, we consider the energy for $r \longrightarrow \infty$ in the uncharged case $q = 0$ for the massive ($M \neq 0$) black hole and for $r \longrightarrow \infty$, $\Lambda = 0$, and $q = 0$ for the massless ($M = 0$) black hole.

Starting with the massive black hole ($M \neq 0$) the results for the limiting cases $r \longrightarrow \infty$ and $q = 0$ are presented in Table 1.

For the charged ($q \neq 0$) black hole without cosmological constant ($\Lambda = 0$), the spacetime geometry becomes the Minkowski geometry. If the black hole is uncharged ($q = 0$) and the cosmological constant is nonzero ($\Lambda > 0$), then the spacetime is regular everywhere except at the origin $r = 0$, as it is inferred from the calculation of the curvature invariants (Kretschmann and Ricci) in [50]. Indeed, the fact that, despite the regular behavior of the metric, the energy diverges at infinity in both prescriptions can possibly be attributed to the de Sitter like asymptotic behavior of the solution. The

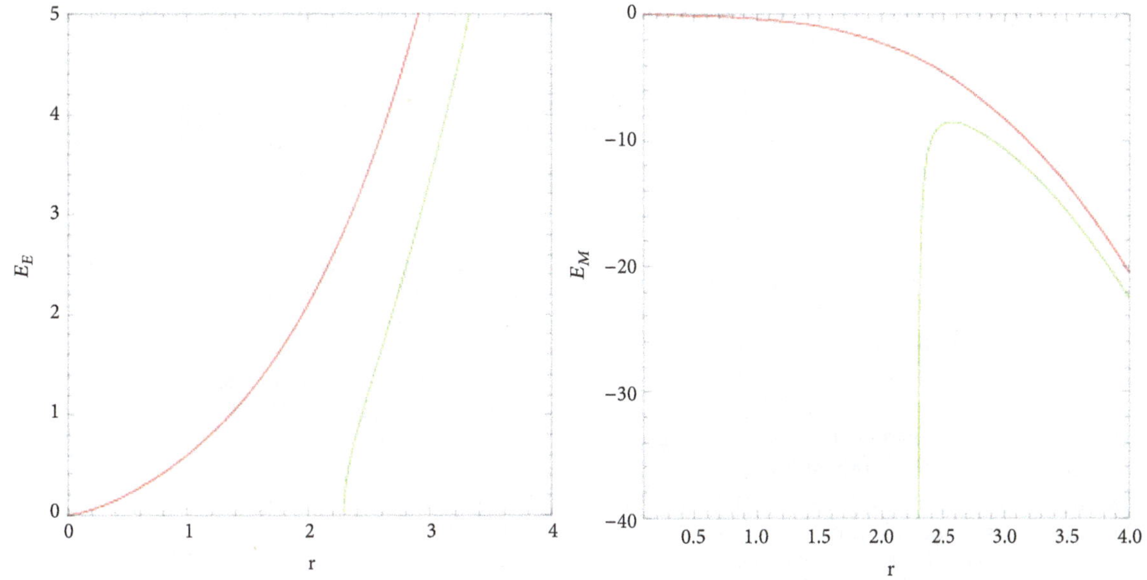

FIGURE 1: Evolution of the Einstein energy E_E (left) and of the Møller energy E_M (right) with respect to the r coordinate for the massive and charged black hole. In both cases we have chosen $M = 1$ and $q = 1$, while the red line and the green line correspond to $\Lambda = 1$ and $\Lambda = -1$, respectively.

evolution of the Einstein energy and of the Møller energy with respect to the r coordinate, as given by (26) and (29), is presented for $\Lambda > 0$ and $\Lambda < 0$ in Figure 1. In fact, the Møller energy given by (29) is negative for a large range of values for r and the parameters M, Λ, and q. The observed negativity of the energy might sound a note of warning regarding the physical interpretation of this result and, consequently, the merits of the Møller complex despite the physically acceptable results obtained by the latter in many other cases. Furthermore, if one considers the behavior of the energy regarding r, one concludes that the fourth-degree polynomial $12M\Lambda r + \Lambda^2 r^4 - 6q^2\Lambda$ in the denominator of (29) vanishes for several sets of values for the parameters M, Λ, and q with its roots consisting in two real and two complex conjugate solutions. As a result, at finite distances from the black hole the Einstein energy, given by (26), vanishes, while the Møller energy, given by (29), diverges. Equally interesting is the vanishing of the third-degree polynomial $12M\Lambda + 4\Lambda^2 r^3$ in the nominator of (29). In this case, one real and two complex conjugate solutions are obtained for every finite value of the mass parameter M and $\Lambda < 0$. As a result, the Møller energy vanishes at finite distances from the black hole. We have not found any plausible explanation for this rather pathological behavior of the energy in the two prescriptions, beside the fact that the black hole solution considered is quite peculiar. Some more light could be shed on this strange state of things through the comparison of the present results with the energy calculated, for example, by the Landau-Lifshitz and the Weinberg energy-momentum complexes for the metric considered, a task left for future work.

In the massless case ($M = 0$), when the black hole is charged ($q \neq 0$) and the cosmological constant is nonzero ($\Lambda \neq 0$), the spacetime geometry is described by the line element (7). In fact, an event horizon appears at $r_{\text{eh}} = [9/\Lambda^2 + 6q^2/\Lambda]^{1/4}$ for a negative cosmological constant $\Lambda < -3/(2q^2)$. Further, when $\Lambda > 0$, at the position $r_s = (6q^2/\Lambda)^{1/4}$ a singularity appears, while, when $\Lambda < 0$, this singularity can be avoided but there appears another singularity at $r = 0$ (see [50]). If this massless black hole has no charge ($q = 0$) but the cosmological constant in nonzero ($\Lambda \neq 0$), then a de Sitter spacetime geometry is obtained:

TABLE 2: Energy of the massless ($M = 0$) black hole solution in the Einstein and Møller prescriptions.

Energy	$q \neq 0, \Lambda \neq 0$	$q = 0, \Lambda \neq 0$
E_E	$\frac{r}{2}\sqrt{\frac{\Lambda^2 r^4}{9} - \frac{2q^2\Lambda}{3}}$	$\frac{r}{2}\sqrt{\frac{\Lambda^2 r^4}{9}}$
E_M	$-\frac{1}{3}\frac{\Lambda^2 r^5}{\sqrt{\Lambda^2 r^4 - 6q^2\Lambda}}$	$-\frac{1}{3}\frac{\Lambda^2 r^5}{\sqrt{\Lambda^2 r^4}}$

$$ds^2 = \left(1 - \sqrt{\frac{\Lambda^2 r^4}{9}}\right) dt^2 - \left(1 - \sqrt{\frac{\Lambda^2 r^4}{9}}\right)^{-1} dr^2 - r^2\left(d\theta^2 + \sin^2\theta d\varphi^2\right), \quad (31)$$

with a cosmological horizon appearing at $r_{\text{ch}} = \sqrt{3/\Lambda}$. The energy of the massless black hole for the charged and the uncharged case, computed in the Einstein and Møller prescriptions, is presented in Table 2.

In Table 3 we present the limiting behavior of the Einstein energy and the Møller energy for the charged and the uncharged massless ($M = 0$) black hole as $r \longrightarrow \infty$. In both cases the cosmological constant is nonzero.

Table 3: Limiting behavior of the energy of the massless black hole solution in the Einstein and Møller prescriptions.

Energy	$q \neq 0, \Lambda \neq 0, r \longrightarrow \infty$	$q = 0, \Lambda \neq 0, r \longrightarrow \infty$
E_E	∞	∞
E_M	$-\infty$	$-\infty$

According to the obtained results, we come to the conclusion that the Einstein and Møller energy-momentum complexes may provide an instructive tool for the study of the energy-momentum localization of gravitating systems, although in this work we cannot reach a definite answer to the problem of localization. That being said, the investigation of the problem of the energy-momentum localization in the context of the black hole solution considered here, by applying other energy-momentum complexes as well as the teleparallel equivalent of general relativity (TEGR), is planned as a future perspective.

Conflicts of Interest

The authors declare that they have no conflicts of interest.

Acknowledgments

Farook Rahaman and Surajit Chattopadhyay are grateful to the Inter-University Centre for Astronomy and Astrophysics (IUCAA), India, for providing the Associateship Programme. Farook Rahaman is thankful to DST, Govt. of India, for providing financial support under the SERB programme. Irina Radinschi is indebted to Prof. Rodica Tudorache and Prof. Dorel Fetcu from the Department of Mathematics and Informatics of the "Gheorghe Asachi" Technical University, Iasi, Romania, for their invaluable assistance.

References

[1] L. Bel, "Définition d'une densité d'énergie et d'un état de radiation totale généralisée," *Comptes Rendus de l'Académie des Sciences*, vol. 246, pp. 3015–3018, 1958.

[2] I. Robinson, "On the Bel—Robinson tensor," *Classical and Quantum Gravity*, vol. 14, no. 1A, pp. A331–A333, 1997.

[3] M. A. G. Bonilla and J. M. M. Senovilla, "Some properties of the BEL and BEL-Robinson tensors," *General Relativity and Gravitation*, vol. 29, no. 1, pp. 91–116, 1997.

[4] J. M. Senovilla, "Super-energy tensors," *Classical and Quantum Gravity*, vol. 17, no. 14, pp. 2799–2841, 2000.

[5] J. D. Brown and J. W. York Jr., "Quasilocal energy and conserved charges derived from the gravitational action," *Physical Review D: Particles, Fields, Gravitation and Cosmology*, vol. 47, no. 4, pp. 1407–1419, 1993.

[6] S. A. Hayward, "Quasilocal gravitational energy," *Physical Review D: Particles, Fields, Gravitation and Cosmology*, vol. 49, no. 2, pp. 831–839, 1994.

[7] C.-M. Chen and J. M. Nester, "Quasilocal quantities for general relativity and other gravity theories," *Classical and Quantum Gravity*, vol. 16, no. 4, pp. 1279–1304, 1999.

[8] C.-C. M. Liu and S.-T. Yau, "Positivity of quasilocal mass," *Physical Review Letters*, vol. 90, no. 23, Article ID 231102, 2003.

[9] L. Balart, "Quasilocal energy, Komar charge and horizon for regular black holes," *Physics Letters B*, vol. 687, no. 4-5, pp. 280–285, 2010.

[10] L. B. Szabados, "Quasi-local energy-momentum and angular momentum in general relativity," *Living Reviews in Relativity*, vol. 12, article 4, 2009.

[11] A. Einstein, "On the general theory of relativity," *Sitzungsberichte der Königlich Preussischen Akademie der Wissenschaften*, vol. 47, pp. 778–786, 1915, Addendum: Sitzungsberichte der Königlich Preussischen Akademie der Wissenschaften., vol. 47, p. 799, 1915.

[12] A. Trautman, "Conservation Laws in General Relativity," in *Gravitation: An Introduction to Current Research*, L. Witten, Ed., pp. 169–198, John Wiley & Sons, New York, NY, USA, 1962.

[13] L. D. Landau and E. M. Lifshitz, *The Classical Theory of Fields*, Addison-Wesley, Reading, Mass, USA, 1951.

[14] A. Papapetrou, "Einstein's theory of gravitation and flat space," *Proceedings of the Royal Irish Academy*, vol. 52, pp. 11–23, 1948.

[15] P. G. Bergmann and R. Thomson, "Spin and angular momentum in general relativity," *Physical Review A: Atomic, Molecular and Optical Physics*, vol. 89, no. 2, pp. 400–407, 1953.

[16] C. Møller, "On the localization of the energy of a physical system in the general theory of relativity," *Annals of Physics*, vol. 4, no. 4, pp. 347–371, 1958.

[17] S. Weinberg, *Gravitation and Cosmology: Principles and Applications of General Theory of Relativity*, John Wiley & Sons, New York, NY, USA, 1972.

[18] A. K. Sinha, G. K. Pandey, A. K. Bhaskar et al., "Effective gravitational mass of the Ayón-Beato and García metric," *Modern Physics Letters A*, vol. 30, no. 25, 1550120, 12 pages, 2015.

[19] S. K. Tripathy, B. Mishra, G. K. Pandey, A. K. Singh, T. Kumar, and S. S. Xulu, "Energy and momentum of Bianchi type VI h universes," *Advances in High Energy Physics*, vol. 2015, Article ID 705262, 8 pages, 2015.

[20] M. Saleh, B. B. Thomas, and T. C. Kofane, "Energy distribution and thermodynamics of the quantum-corrected Schwarzschild black hole," *Chinese Physics Letters*, vol. 34, no. 8, Article ID 080401, 2017.

[21] P. K. Sahoo, K. L. Mahanta, D. Goit et al., "Einstein energy-momentum complex for a phantom black hole metric," *Chinese Physics Letters*, vol. 32, no. 2, Article ID 020402, 2015.

[22] I.-C. Yang, "Some characters of the energy distribution for a charged wormhole," *Chinese Journal of Physics*, vol. 53, no. 6, Article ID 110108, pp. 1–4, 2015.

[23] I. Radinschi and T. Grammenos, "Møller's energy-momentum complex for a spacetime geometry on a noncommutative curved D3-brane," *International Journal of Theoretical Physics*, vol. 47, no. 5, pp. 1363–1372, 2008.

[24] I.-C. Yang, C.-L. Lin, and I. Radinschi, "The energy of regular black hole in general relativity coupled to nonlinear electrodynamics," *International Journal of Theoretical Physics*, vol. 48, no. 1, pp. 248–255, 2009.

[25] E. C. Vagenas, "Energy distribution in 2D stringy black hole backgrounds," *International Journal of Modern Physics A*, vol. 18, no. 31, pp. 5781–5794, 2003.

[26] I. Radinschi, F. Rahaman, and A. Banerjee, "On the energy of Hořava-Lifshitz black holes," *International Journal of Theoretical Physics*, vol. 50, no. 9, pp. 2906–2916, 2011.

[27] I. Radinschi, F. Rahaman, T. Grammenos, and S. Islam, "Einstein and Møller Energy-Momentum Complexes for a New Regular Black Hole Solution with a Nonlinear Electrodynamics Source," *Advances in High Energy Physics*, vol. 2016, Article ID 9049308, 2016.

[28] I. Radinschi, T. Grammenos, F. Rahaman et al., "Energy-Momentum for a Charged Nonsingular Black Hole Solution with a Nonlinear Mass Function," *Advances in High Energy Physics*, vol. 2017, Article ID 7656389, 10 pages, 2017.

[29] M. Abdel-Megied and R. M. Gad, "Møller's Energy in the Kantowski-Sachs Space-Time," *Advances in High Energy Physics*, vol. 2010, Article ID 379473, 6 pages, 2010.

[30] M. Sharif and M. Azam, "Energy-momentum distribution: some examples," *International Journal of Modern Physics A*, vol. 22, no. 10, pp. 1935–1951, 2007.

[31] T. Multamäki, A. Putaja, E. C. Vagenas, and I. Vilja, "Energy-momentum complexes in $f(R)$ theories of gravity," *Classical and Quantum Gravity*, vol. 25, no. 7, Article ID 075017, 2008.

[32] L. Balart, "Energy distribution of (2+1)-dimensional black holes with nonlinear electrodynamics," *Modern Physics Letters A*, vol. 24, no. 34, pp. 2777–2785, 2009.

[33] A. M. Abbassi, S. Mirshekari, and A. H. Abbassi, "Energy-momentum distribution in static and nonstatic cosmic string space-times," *Physical Review D: Particles, Fields, Gravitation and Cosmology*, vol. 78, no. 6, Article ID 064053, 9 pages, 2008.

[34] J. Matyjasek, "Some remarks on the Einstein and Møller pseudotensors for static and spherically-symmetric configurations," *Modern Physics Letters A*, vol. 23, no. 8, pp. 591–601, 2008.

[35] J. W. Maluf, "The teleparallel equivalent of general relativity," *Annalen der Physik*, vol. 525, no. 5, pp. 339–357, 2013.

[36] R. M. Gad, "On teleparallel version of stationary axisymmetric solutions and their energy contents," *Astrophysics and Space Science*, vol. 346, no. 2, pp. 553–557, 2013.

[37] G. G. Nashed, "Energy and momentum of a spherically symmetric dilaton frame as regularized by teleparallel gravity," *Annalen der Physik*, vol. 523, no. 6, pp. 450–458, 2011.

[38] G. G. L. Nashed, "Braneworld black holes in teleparallel theory equivalent to general relativity and their Killing vectors, energy, momentum and angular momentum," *Chinese Physics B*, vol. 19, no. 2, p. 020401, 2010.

[39] M. Sharif and A. Jawad, "Energy contents of some well-known solutions in teleparallel gravity," *Astrophysics and Space Science*, vol. 331, no. 1, pp. 257–263, 2010.

[40] J. G. da Silva and S. C. Ulhoa, "On gravitational energy in conformal teleparallel gravity," *Modern Physics Letters A*, vol. 32, no. 21, 1750113, 10 pages, 2017.

[41] R. Penrose, "Quasilocal mass and angular momentum in general relativity," *Proceedings of the Royal Society A Mathematical, Physical and Engineering Sciences*, vol. 381, no. 1780, pp. 53–63, 1982.

[42] K. P. Tod, "Some examples of Penrose's quasi-local mass construction," *Proceedings of the Royal Society A Mathematical, Physical and Engineering Sciences*, vol. 388, no. 1795, pp. 457–477, 1983.

[43] J. M. Aguirregabiria, A. Chamorro, and K. S. Virbhadra, "Energy and angular momentum of charged rotating black holes," *General Relativity and Gravitation*, vol. 28, no. 11, pp. 1393–1400, 1996.

[44] S. S. Xulu, "Bergmann-Thomson energy-momentum complex for solutions more general than the Kerr-Schild class," *International Journal of Theoretical Physics*, vol. 46, no. 11, pp. 2915–2922, 2007.

[45] K. S. Virbhadra, "Naked singularities and Seifert's conjecture," *Physical Review D: Particles, Fields, Gravitation and Cosmology*, vol. 60, no. 10, Article ID 104041, 6 pages, 1999.

[46] G. Sun, C.-M. Chen, J.-L. Liu, and J. M. Nester, "An optimal choice of reference for the quasi-local gravitational energy and angular momentum," *Chinese Journal of Physics*, vol. 52, no. 1, part 1, pp. 111–125, 2014.

[47] C.-M. Chen and J. M. Nester, "A symplectic Hamiltonian derivation of the quasilocal energy-momentum for general relativity," *Gravitation & Cosmology*, vol. 6, no. 4, pp. 257–270, 2000.

[48] C.-M. Chen, J. M. Nester, and R.-S. Tung, "Gravitational energy for GR and Poincaré gauge theories: a covariant Hamiltonian approach," *International Journal of Modern Physics D: Gravitation, Astrophysics, Cosmology*, vol. 24, no. 11, 1530026, 73 pages, 2015.

[49] J. M. Nester, C. M. Chen, J.-L. Liu, and G. Sun, "A reference for the covariant Hamiltonian boundary term," in *Relativity and Gravitation-100 years after Einstein in Prague*, J. Bicák and T. Ledvinka, Eds., pp. 177–184, Springer, 2014.

[50] P. Bargueño and E. C. Vagenas, "Black holes with constant topological Euler density," *European Physics Letters*, vol. 115, no. 6, p. 60002, 2016.

[51] P. Gilkey, J. H. Park, and R. Vázquez-Lorenzo, *Aspects of Differential Geometry I*, chapter 3, Morgan & Claypool Publishers, 2015.

[52] M. Ammon and J. Erdmenger, *Gauge/Gravity Duality – Foundations and Applications*, Cambridge University Press, 2015.

[53] A. Strominger, "Massless black holes and conifolds in string theory," *Nuclear Physics. B. Theoretical, Phenomenological, and Experimental High Energy Physics. Quantum Field Theory and Statistical Systems*, vol. 451, no. 1-2, pp. 96–108, 1995.

[54] K. Behrndt, "About a class of exact string backgrounds," *Nuclear Physics. B. Theoretical, Phenomenological, and Experimental High Energy Physics. Quantum Field Theory and Statistical Systems*, vol. 455, no. 1-2, pp. 188–210, 1995.

[55] R. Emparan, "Massless black hole pairs in string theory," *Physics Letters. B. Particle Physics, Nuclear Physics and Cosmology*, vol. 387, no. 4, pp. 721–726, 1996.

[56] U. Nucamendi and D. Sudarsky, "Black holes with zero mass," *Classical and Quantum Gravity*, vol. 17, no. 19, pp. 4051–4058, 2000.

[57] P. Goulart, "Massless black holes and charged wormholes in string theory," 2016, https://arxiv.org/abs/1611.03164.

17

Particle Motion around Charged Black Holes in Generalized Dilaton-Axion Gravity

Susmita Sarkar,[1] Farook Rahaman,[1] Irina Radinschi,[2] Theophanes Grammenos,[3] and Joydeep Chakraborty[4]

[1] *Department of Mathematics, Jadavpur University, Kolkata 700 032, West Bengal, India*
[2] *Department of Physics, Gheorghe Asachi Technical University, 700050 Iasi, Romania*
[3] *Department of Civil Engineering, University of Thessaly, 383 34 Volos, Greece*
[4] *Department of Mathematics, Nagar College, P.O. Nagar, Dist. Mursidabad, West Bengal, India*

Correspondence should be addressed to Farook Rahaman; rahaman@associates.iucaa.in

Academic Editor: Edward Sarkisyan-Grinbaum

The behaviour of massive and massless test particles around asymptotically flat and spherically symmetric, charged black holes in the context of generalized dilaton-axion gravity in four dimensions is studied. All the possible motions are investigated by calculating and plotting the corresponding effective potential for the massless and massive particles as well. Further, the motion of massive (charged or uncharged) test particles in the gravitational field of charged black holes in generalized dilaton-axion gravity for the cases of static and nonstatic equilibrium is investigated by applying the Hamilton-Jacobi approach.

1. Introduction

Recently, scientists have focused their attention on the black hole solutions in various alternative theories of gravity, particularly theories of gravitation with background scalar and pseudoscalar fields. In the low energy effective action, usually string theory based-models are comprised of two massless scalar fields, the dilaton, and the axion (see, e.g., [1]). Sur, Das, and SenGupta [2] employed the dilaton and axion fields coupled to the electromagnetic field in a more generalized coupling with Einstein and Maxwell theory in four dimensions in the low energy action. Exploiting this new idea, they have found asymptotically flat and nonflat dilaton-axion black hole solutions. The vacuum expectation values of the various moduli of compactification are responsible for these couplings. These black hole solutions have been studied extensively in the literature; e.g., their thermodynamics has been investigated [3], thin-shell wormholes have been constructed from charged black holes in generalized dilaton-axion gravity [4], the energy of charged black holes in generalized dilaton-axion gravity has been calculated [5], the statistical entropy of a charged dilaton-axion black hole has been examined [6], and the superradiant instability of a dilaton-axion black hole under scalar perturbation has been investigated [7]. Among various properties of such black hole solutions, a subject of great interest is the study of the behaviour of a test particle in the gravitational field of such black holes.

In this paper, we study the behaviour of the time-like and null geodesics in the gravitational field of a charged black hole in generalized dilaton-axion gravity. The solution under study describes an asymptotically flat black hole and the motions of both massless and massive particles are analyzed. The effective potentials are calculated and plotted for various parameters in the cases of circular and radial geodesics. The motion of a charged test particle in the gravitational field of a charged black hole in generalized dilaton-axion gravity is also investigated using the Hamilton-Jacobi approach.

The present paper has the following structure: in Section 2 the charged black hole metric in generalized dilaton-axion gravity is presented. Section 3 focuses on the geodesic equation in the cases of massless particle motion ($L = 0$) and massive particle motion ($L = -1$). In Section 4 the effective potential is studied in both cases of the massless and the massive particle. Section 5 is devoted to the study of the motion of a test particle in static equilibrium as well as in nonstatic equilibrium. For the latter case, a chargeless ($e = 0$) and a charged test particle are considered. Finally, in Section 6, the results obtained in this paper are discussed.

2. Charged Black Hole Metric in Generalized Dilation-Axion Gravity

Recently, Sur, Das, and SenGupta [2] have discovered a new black hole solution for the Einstein-Maxwell scalar field system inspired by low energy string theory. In fact, they have considered a generalized action in which two scalar fields are minimally coupled to the Einstein-Hilbert-Maxwell field in four dimensions (in the Einstein frame, see, e.g., [8, 9]) having the form

$$I = \frac{1}{2\kappa}\int d^4x\sqrt{-g}\left[R - \frac{1}{2}\partial_\mu\varphi\partial^\mu\varphi - \frac{\omega(\varphi)}{2}\partial_\mu\zeta\partial^\mu\zeta - \alpha(\varphi,\zeta)F_{\mu\nu}F^{\mu\nu} - \beta(\varphi,\zeta)F_{\mu\nu} * F^{\mu\nu}\right], \tag{1}$$

where $\kappa = 8\pi G$, R is the curvature scalar, $F_{\mu\nu}$ describes the Maxwell field strength and φ, ζ are two massless scalar/pseudoscalar fields depending only on the radial coordinate r which are coupled to the Maxwell field through the functions α and β. Here, ζ acquires a nonminimal kinetic term of the form $\omega(\varphi)$ due to its interaction with φ (φ, ζ can be identified with the scalar dilaton field and the pseudoscalar axion field, respectively), while $*F^{\mu\nu} = (1/2)\varepsilon^{\mu\nu\kappa\lambda}F_{\kappa\lambda}$ is the Hodge-dual Maxwell field strength.

Indeed, with the action described by (1), a much wider class of black hole solutions has been found, whereby two types of metrics, asymptotically flat and asymptotically non-flat, for the black hole solutions have been obtained.

For our study we use the asymptotically flat solution to analyze the behaviour of massive and massless test particles around a spherically symmetric, charged black hole in generalized dilaton-axion gravity. The asymptotically flat metric considered is given by

$$ds^2 = -f(r)\,dt^2 + \frac{dr^2}{f(r)} + h(r)\,d\Omega^2, \tag{2}$$

with

$$f(r) = \frac{(r - r_-)(r - r_+)}{(r - r_0)^{(2-2n)}(r + r_0)^{2n}} \tag{3}$$

and

$$h(r) = \frac{(r + r_0)^{2n}}{(r - r_0)^{(2n-2)}}. \tag{4}$$

In (3) and (4), according to [2], in order to have nontrivial φ and ζ fields, the exponent n is a dimensionless constant strictly greater than 0 and strictly less than 1. The other various parameters are given as follows:

$$r_\pm = m_0 \pm \sqrt{m_0^2 + r_0^2 - \frac{1}{8}\left(\frac{K_1}{n} + \frac{K_2}{1-n}\right)}, \tag{5}$$

$$r_0 = \frac{1}{16m_0}\left(\frac{K_1}{n} - \frac{K_2}{1-n}\right), \tag{6}$$

$$m_0 = m - (2n-1)\,r_0, \tag{7}$$

$$K_1 = 4n\left[4r_0^2 + 2r_0(r_+ + r_-) + r_+r_-\right], \tag{8}$$

$$K_2 = 4(1-n)\,r_+r_-, \tag{9}$$

and

$$m = \frac{1}{16r_0}\left(\frac{K_1}{n} - \frac{K_2}{1-n}\right) + (2n-1)\,r_0, \tag{10}$$

where m is the mass of the black hole and $0 < n < 1$. The parameters r_+ and r_- determine the inner and outer event horizons, respectively. Also, for $r = r_0$, there is a curvature singularity and the parameters obey the condition $r_0 < r_- < r_+$.

3. Geodesic Equation

The geodesic equation for the metric (2) describing the motion in the plane $\theta = \pi/2$ is as follows [10]:

$$\left(\frac{dr}{d\tau}\right)^2 = Lf(r) + E^2 - \frac{J^2 f(r)}{h(r)}, \tag{11}$$

$$\frac{d\phi}{d\tau} = \frac{J}{h(r)}, \tag{12}$$

$$\frac{dt}{d\tau} = \frac{E}{f(r)}, \tag{13}$$

where L is known as the Lagrangian having the values 0 for a massless particle and -1 for a massive particle and E, J are constants identified as the energy per unit mass and the angular momentum, respectively.

Now we proceed to discuss the motion of the massless and the massive particle for the radial geodesic.

The radial geodesic equation ($J = 0$) is

$$\left(\frac{dr}{d\tau}\right)^2 = E^2 + Lf(r). \tag{14}$$

Using (13), (14) becomes

$$\left(\frac{dr}{dt}\right)^2 = (f(r))^2\left(1 + f(r)\frac{L}{E^2}\right). \tag{15}$$

Then, by inserting $f(r)$ from (3), (15) reads

$$\left(\frac{dr}{dt}\right)^2 = \left(\frac{(r - r_-)(r - r_+)}{(r - r_0)^{(2-2n)}(r + r_0)^{2n}}\right)^2 \cdot \left(1 + \frac{(r - r_-)(r - r_+)}{(r - r_0)^{(2-2n)}(r + r_0)^{2n}}\frac{L}{E^2}\right). \tag{16}$$

3.1. Massless Particle Motion ($L = 0$). For the motion of a massless particle the Lagrangian L vanishes. In this case the equation for the radial geodesic (16) becomes

$$\left(\frac{dr}{dt}\right)^2 = \left(\frac{(r - r_-)(r - r_+)}{(r - r_0)^{(2-2n)}(r + r_0)^{2n}}\right)^2. \tag{17}$$

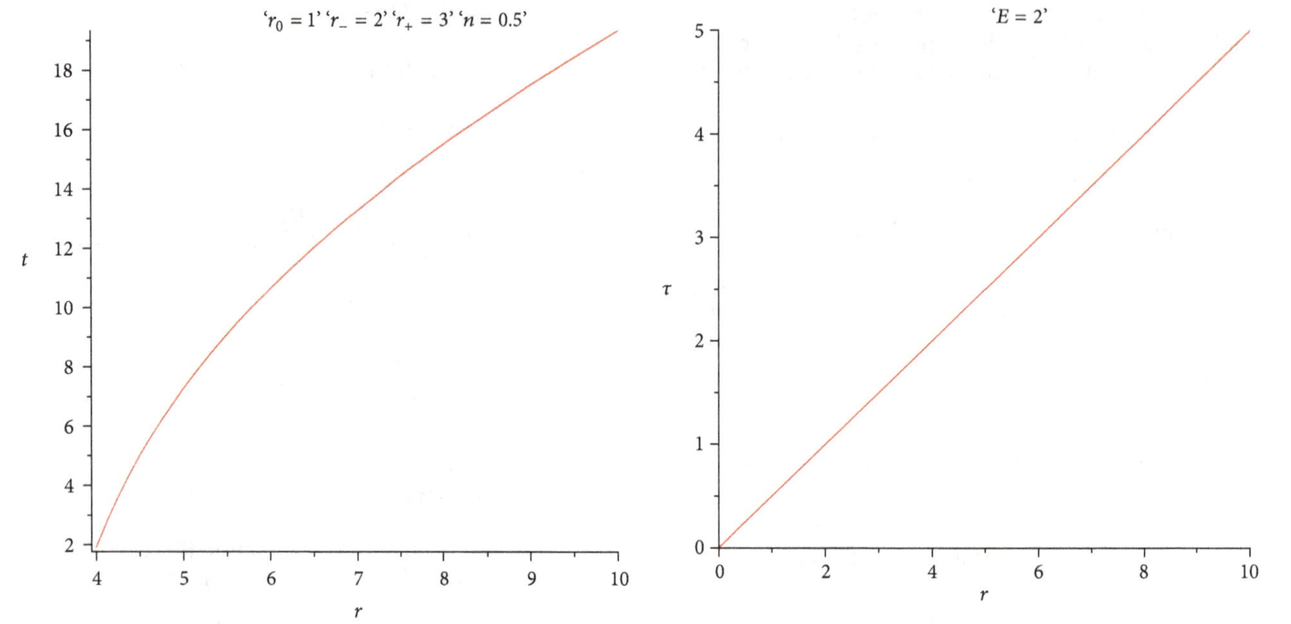

FIGURE 1: Graphs of $t - r$ (left) and $\tau - r$ (right) for a massless particle.

After integrating we get

$$\pm t = r + \frac{\ln(r - r_-) r^2}{r_- - r_+} - \frac{\ln(r - r_-) r_0^2}{r_- - r_+} - \frac{\ln(r - r_+) r_+^2}{r_- - r_+} + \frac{\ln(r - r_+) r_0^2}{r_- - r_+}. \tag{18}$$

Again from (14) we obtain for $L = 0$

$$\left(\frac{dr}{d\tau}\right)^2 = E^2, \tag{19}$$

from which we have a $\tau - r$ relationship

$$\pm\tau = \frac{r}{E}. \tag{20}$$

In Figure 1 (left) t is plotted with respect to the radial coordinate r and in Figure 1 (right) the proper time (τ) is plotted with respect to radial coordinate r for a massless particle.

3.2. Massive Particle Motion ($L = -1$). For a massive particle the Lagrangian L is -1 and from (14) and (15) we obtain for the motion of a massive particle the relationships between t and r and τ and r, respectively, as

$$\pm t = \int \frac{E dr}{\left((r - r_-)(r - r_+) / (r - r_0)^{(2-2n)} (r + r_0)^{2n}\right) \left(E^2 - (r - r_-)(r - r_+) / (r - r_0)^{(2-2n)} (r + r_0)^{2n}\right)^{1/2}} \tag{21}$$

$$\pm\tau = \int \frac{(r - r_0)^{(1-n)} (r + r_0)^n dr}{\left(E^2 (r - r_0)^{(2-2n)} (r + r_0)^{2n} - (r - r_-)(r - r_+)\right)^{1/2}}. \tag{22}$$

In Figure 2 the graphs of t with respect to the radial coordinate r (left) and of the proper time τ with respect to the radial coordinate r (right) for a massive particle are presented.

4. The Effective Potential

From the geodesic equation (11) we have

$$\frac{1}{2}\left(\frac{dr}{d\tau}\right)^2 = \frac{1}{2}\left[E^2 - f(r)\left(\frac{J^2}{h(r)} - L\right)\right]. \tag{23}$$

After comparing the above equation with the well known equation $(1/2)(dr/d\tau)^2 + V_{\text{eff}} = 0$, we obtain the following expression for the effective potential:

$$V_{\text{eff}} = -\frac{1}{2}\left[E^2 - f(r)\left(\frac{J^2}{h(r)} - L\right)\right]. \tag{24}$$

From (24) one can see that the effective potential depends on the energy per unit mass, E, and the angular momentum, J.

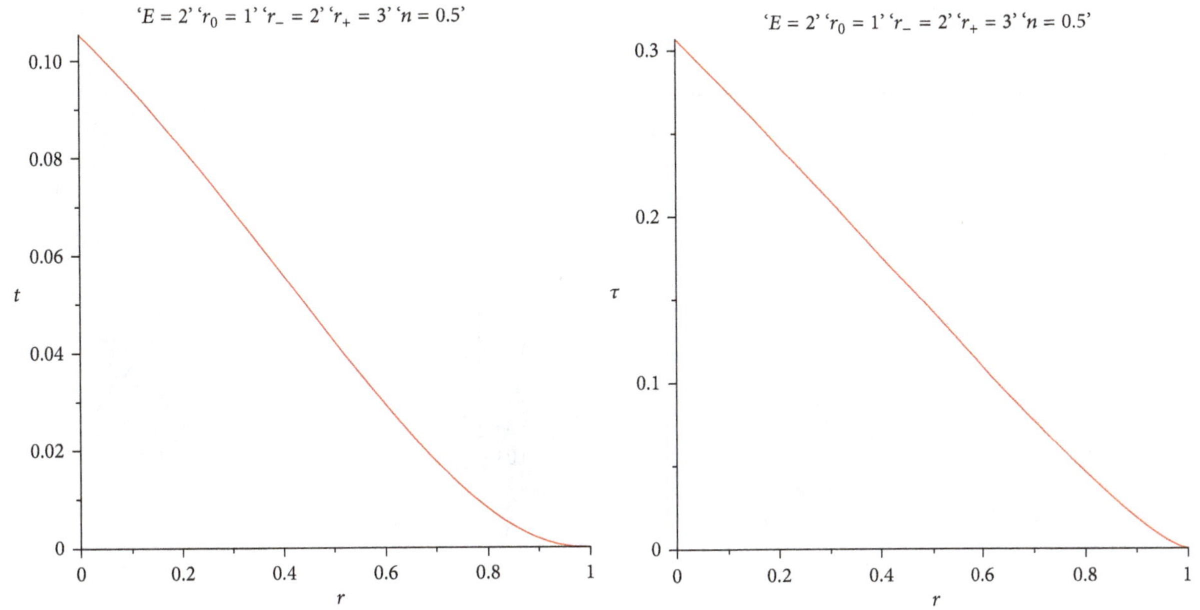

FIGURE 2: Graphs of $t-r$ (left) and $\tau - r$ (right) for a massive particle.

4.1. Massless Particle Case ($L = 0$). For the radial geodesics ($J = 0$), (24) yields

$$V_{\text{eff}} = -\frac{E^2}{2} \tag{25}$$

and the particle behaves like a free particle if $E = 0$.

Now we consider the circular geodesics ($J \neq 0$). The corresponding effective potential is given by

$$V_{\text{eff}} = -\frac{E^2}{2} + \frac{J^2}{2}\frac{(r-r_-)(r-r_+)}{(r+r_0)^{4n}}. \tag{26}$$

For $E \neq 0$ and $J = 0$ we infer from (26) that the effective potential does not depend on the charge and the mass of the black hole in generalized dilaton-axion gravity. The shape of the effective potential for $E \neq 0$ is shown in Figure 3 (left) for $J = 6$ (solid curve) and $J = 0$ (dotted line). We notice that for a nonzero value of J, the effective potential acquires a minimum, implying that stable circular orbits might exist. For $J = 0$, there are no stable circular orbits.

Now if we consider $E = 0$ and circular geodesics, i.e., $J \neq 0$, then, from (26), it is clear that the roots of the effective potential are the same as the horizon values. Further, the effective potential is negative between its two roots, i.e., between the horizons. Hence, since the effective potential has a minimum value stable circular orbits must exist, a conclusion that is confirmed in Figure 3 (right).

4.2. Massive Particle Case ($L = -1$). The effective potential for the massive particle is obtained from (24) as

$$V_{\text{eff}} = -\frac{1}{2}\left[E^2 - f(r)\left(\frac{J^2}{h(r)} + 1\right)\right]. \tag{27}$$

Now, for the radial geodesics with $J = 0$, $E = 0$, the above equation yields

$$V_{\text{eff}} = \frac{1}{2}\frac{(r-r_-)(r-r_+)}{(r-r_0)^{(2-2n)}(r+r_0)^{2n}}. \tag{28}$$

From (28) we notice that the solutions for the effective potential coincide with the horizon values for radial geodesics with $E = 0$, which is demonstrated graphically in Figure 4 (left). Further, from Figure 4 (left), one can see that the motion of the particle is bounded in the interior region of the black hole. The behaviour of the effective potential for $E \neq 0$ is depicted in the bottom part of Figure 4 (left). In this case, we also deduce that a bound orbit is possible for the massive particle.

Next we will consider the motion of a test particle with nonzero angular momentum. For $E = 0$, the roots of the effective potential coincide with the horizons (see Figure 4 (right)). Thus the particle is bounded in the interior region of the black hole.

5. Motion of a Test Particle

In this section we study the motion of a test particle of mass M and charge e in the gravitational field of a charged black hole in generalized dilaton-axion gravity. The Hamilton-Jacobi equation [11] is

$$g^{ik}\left(\frac{\partial S}{\partial x^i} + eA_i\right)\left(\frac{\partial S}{\partial x^k} + eA_k\right) + M^2 = 0, \tag{29}$$

where g_{ik}, A_i are the metric potential and the gauge potential, respectively, and S is Hamilton's standard characteristic

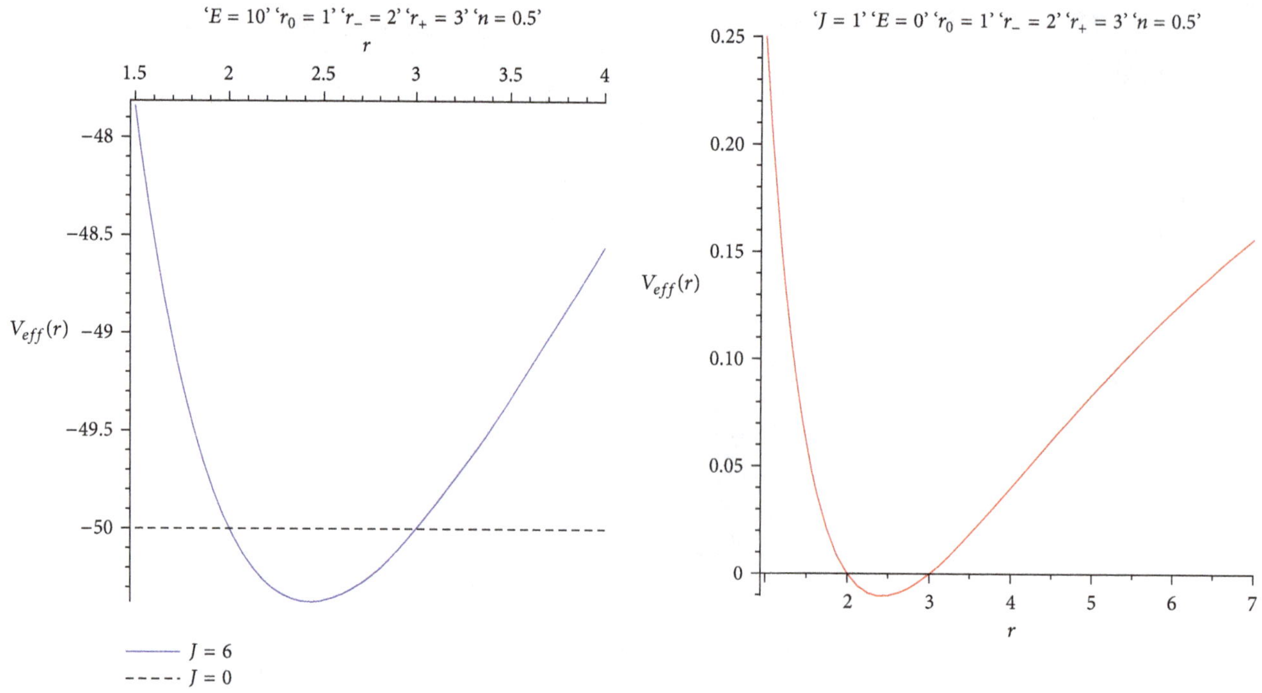

FIGURE 3: Graphs of $V_{\text{eff}} - r$ with $E = 10$, $J = 6$ and $J = 0$ (left) and $V_{\text{eff}} - r$ with $E = 0$ and $J = 1$ (right), for a massless particle.

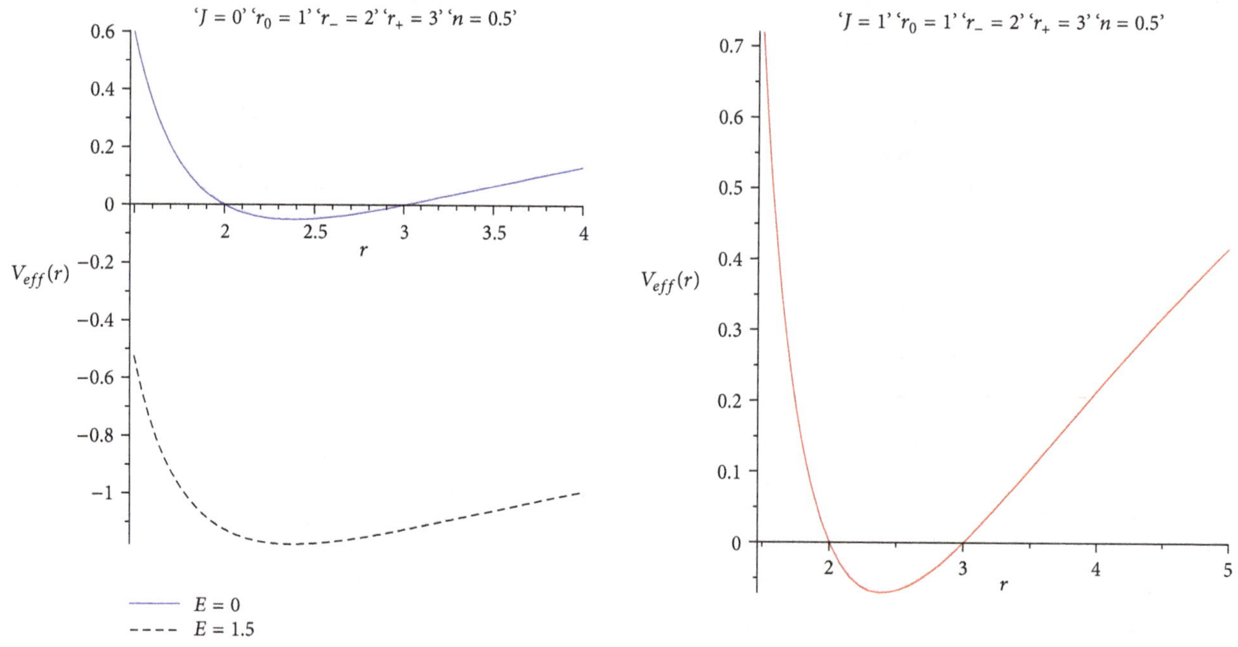

FIGURE 4: Graphs of $V_{\text{eff}} - r$ with $J = 0$, $E = 0$, and $E = 1.5$ (left) and $V_{\text{eff}} - r$ with $J = 1$ and $E = 0$ (right), for a massive particle.

function. The explicit form of the Hamilton-Jacobi equation for the line element (2) is

$$-\frac{1}{f(r)}\left(\frac{\partial s}{\partial t}+\frac{eQ}{r}\right)^2+f(r)\left(\frac{\partial S}{\partial r}\right)^2+\frac{1}{h(r)}\left(\frac{\partial S}{\partial \theta}\right)^2+\frac{1}{h(r)\sin^2\theta}\left(\frac{\partial S}{\partial \phi}\right)^2+M^2=0, \tag{30}$$

where Q is the charge of the black hole. To solve the above partial differential equation, let us assume a separable solution in the form

$$S(t,r,\theta,\phi)=-Et+S_1(r)+S_2(\theta)+J\phi, \tag{31}$$

where E and J are the energy and angular momentum of the particle, respectively. After some simplification we obtain

$$S_1(r) = \epsilon \int \left[\frac{(E - eQ/r)^2}{f^2} - \frac{p^2}{fh} - \frac{M^2}{f} \right]^{1/2}, \quad (32)$$

$$S_2(\theta) = \epsilon \int \left(p^2 - J^2 \operatorname{cosec}^2\theta\right)^{1/2}, \quad (33)$$

where $\epsilon = \pm 1$ and p is the separation constant also known as the momentum of the particle.

The radial velocity of the particle is given by

$$\frac{dr}{dt} = f^2 \left(E - \frac{eQ}{r}\right)^{-1} \cdot \left[\frac{1}{f^2}\left(E - \frac{eQ}{r}\right)^2 - \frac{p^2}{fh} - \frac{M^2}{f} \right]^{1/2}. \quad (34)$$

The turning points of the trajectory are obtained by the vanishing of the radial velocity, $dr/dt = 0$, which yields

$$\left(E - \frac{eQ}{r}\right)^2 - \frac{p^2 f}{h} - M^2 f = 0. \quad (35)$$

After solving this equation for E, we get

$$E = \frac{eQ}{r} + \sqrt{f}\left(\frac{p^2}{h} + M^2\right)^{1/2}. \quad (36)$$

The effective potential is obtained from the relation $V(r) = E/M$ as follows:

$$V = \frac{eQ}{Mr} + \sqrt{f}\left(\frac{p^2}{M^2 h} + 1\right)^{1/2}. \quad (37)$$

Using (3) and (4) the effective potential becomes

$$V(r) = \frac{eQ}{Mr} + \left(1 + \frac{p^2 (r - r_0)^{2n-2}}{M^2 (r + r_0)^{2n}}\right) \cdot \left(\frac{\sqrt{(r - r)(r - r_+)}}{(r - r_0)^{1-n}(r + r_0)^n}\right). \quad (38)$$

In the stationary system ($dV/dr = 0$) we obtain

$$\begin{aligned} &-\frac{eQ}{Mr^2} + \frac{1}{2}\frac{((r - r_-)(r - r_+))^{1/2}\left(p^2 (r - r_0)^{2n-2}(2n-2)/M^2 (r - r_0)(r + r_0)^{2n} - 2p^2 n (r - r_0)^{2n-2}/M^2 (r + r_0)^{2n+1}\right)}{\left(1 + p^2 (r - r_0)^{2n-2}/M^2 (r + r_0)^{2n}\right)^{1/2}(r - r_0)^{1-n}(r + r_0)^n} \\ &+ \frac{1}{2}\frac{(2r - r_+ - r_-)\left(1 + p^2 (r - r_0)^{2n-2}/M^2 (r + r_0)^{2n}\right)^{1/2}}{((r - r_-)(r - r_+))^{1/2}(r - r_0)^{1-n}(r + r_0)^n} \\ &- \frac{(1 - n)\left(1 + p^2 (r - r_0)^{2n-2}/M^2 (r + r_0)^{2n}\right)^{1/2}((r - r_-)(r - r_+))^{1/2}}{(r - r_0)^{2-n}(r + r_0)^n} \\ &- \frac{n\left(1 + p^2 (r - r_0)^{2n-2}/M^2 (r + r_0)^{2n}\right)^{1/2}((r - r_-)(r - r_+))^{1/2}}{(r - r_0)^{1-n}(r + r_0)^{n+1}} = 0. \end{aligned} \quad (39)$$

In order to use a more simplified equation and thus be able to visualize it by plotting its graph, one may select some specific value for n. Here we choose $n = 0.5$ (since $0 < n < 1$) and the simplified form of equation (39) is given by

$$\begin{aligned} \alpha(r) &:= \frac{2eQ(r^2 - r_0^2)^{1/2}}{Mr^2} \\ &+ \frac{2r((r - r_-)(r - r_+))^{1/2}}{(r^2 - r_0^2)}\left(\frac{p^2}{M^2 (r^2 - r_0^2)\beta} + \beta\right) \\ &- \frac{(2r - r_+ - r_-)\beta}{((r - r_-)(r - r_+))^{1/2}} = 0, \end{aligned} \quad (40)$$

where $\beta = (1 + p^2/M^2(r^2 - r_0^2))^{1/2}$.

5.1. Test Particle in Static Equilibrium. The momentum p must be zero in the static equilibrium system; thus from (40) we get

$$\begin{aligned} &\left(4e^2Q^2 - M^2\left(2r_- r_+ + r_+^2 - r_-^2\right)\right) r^8 \\ &+ \left(-4e^2Q^2 (r_- + r_+) \right. \\ &\left. + 4M^2\left(r_0^2 r_- + r_0^2 r_+ r_-^2 r_+ + r_- r_+^2\right)\right) r^7 \\ &+ \left(4e^2Q^2\left(r_- r_+ - 3r_0^2\right) \right. \\ &\left. - 2M^2\left(6r_0^2 r_- r_+ + r_0^2 r_-^2 + r_-^2 r_+^2 + r_0^2 r_+^2 - r_0^4\right)\right) r^6 \\ &+ \left(12e^2Q^2 r_0^2 (r_- + r_+) \right. \\ &\left. + 4M^2 r_0^2\left(r_+ r_-^2 + r_+^2 r_- + r_0^2 (r_+ + r_-)\right)\right) r^5 \end{aligned}$$

$$+ \left(12e^2Q^2r_0^2\left(r_0^2 - r_+r_-\right) - M^2r_0^4\left(2r_+r_- + r_+^2 + r_-^2\right)\right)r^4 - 12e^2Q^2r_0^4\left(r_+ + r_-\right) \cdot r^3 + 4e^2Q^2r_0^4\left(3r_+r_- - r_0^2\right)r^2 + 4e^2Q^2r_0^6\left(r_+ + r_-\right) \cdot r - 4e^2Q^2r_0^6r_+r_- = 0. \quad (41)$$

We notice that the last term of the above equation is negative. So, this equation has at least one positive real root. Consequently, a bound orbit is possible for the test particle, i.e., the test particle can be trapped by a charged black hole in generalized dilaton-axion gravity. In other words, a charged black hole in generalized dilaton-axion gravity exerts an attractive gravitational force on matter.

5.2. Test Particle in Nonstatic Equilibrium

5.2.1. Test Particle without Charge ($e = 0$). In this case (40) becomes

$$M^2\left(r_- + r_+\right)r^4 - 2\left(r_0^2M^2 + r_-r_+M^2 + p^2\right)r^3 + 3p^2\left(r_+ - r_-\right)r^2 + 2\left(r^4M^2 - r_0^2p^2 + r_0^2M^2r_+r_- - 2p^2r_+r_-\right)r + r_0^2\left(r_+p^2 + r_-p^2 - M^2r_0^2r_+ - M^2r_0^2r_-\right) = 0. \quad (42)$$

If $M^2r_0^2r_+ + M^2r_0^2r_- > r_+p^2 + r_-p^2$, one can see that the last term of the above equation is negative. Therefore, this equation must have at least one positive real root. Consequently, a bound orbit for the uncharged test particle is possible. If $M^2r_0^2r_+ + M^2r_0^2r_- = r_+p^2 + r_-p^2$, then (42) changes to a third-degree equation with two changes of sign. By Descarte's rule of sign, this equation must have either two positive roots or no positive roots at all. Thus, a bound orbit for the uncharged test particle may or may not be possible. For $M^2r_0^2r_+ + M^2r_0^2r_- < r_+p^2 + r_-p^2$, (42) has two changes of sign. Here, again a bound orbit for the uncharged test particle may or may not be possible.

5.2.2. Test Particle with Charge ($e \neq 0$). For a charged test particle with $n = 0.5$, the stationary system ($dV/dr = 0$) yields the form given in (40). As this equation is algebraically very complicated, we use the graph of $\alpha(r)$ in order to find out whether there exist any real positive roots. From Figure 5 one can see that, for different values of the test particle's charge, $\alpha(r)$ given by (40) does not intersect the r-axis. Hence, no real positive roots are possible. As a result, no bound orbit for the charged test particle is possible.

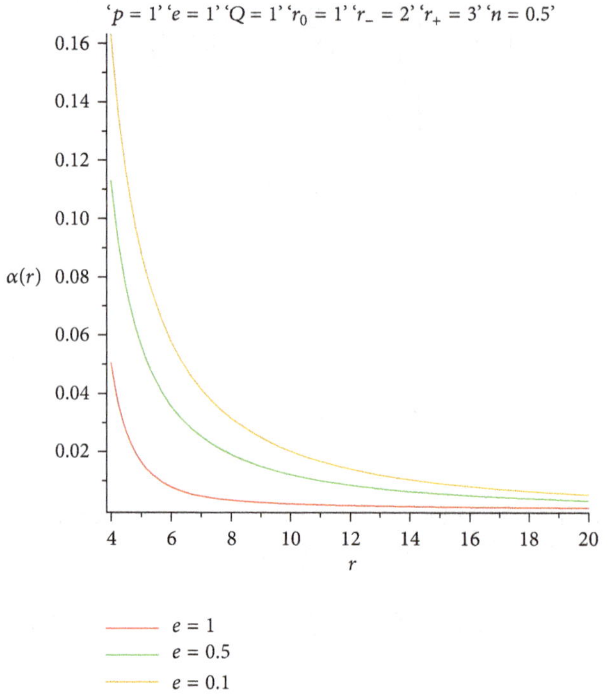

FIGURE 5: Graph of $\alpha(r)$ given in (40) for different values of the test particle's charge.

6. Discussion

In the present investigation, we have analyzed the behaviour of massless and massive particles in the gravitational field of a charged black hole in generalized dilaton-axion gravity in four dimensions. To this purpose, we have studied the motion of a massless particle ($L = 0$) and a massive particle ($L = -1$). We have plotted the graphs of t and the proper time τ with respect to the radial coordinate r. For the massless particle t increases nonlinearly with r (Figure 1 (left)), while the proper time τ increases linearly with r (Figure 1 (right)). In the case of the massive particle motion both t and τ decrease nonlinearly with r (Figure 2).

Further, studying the effective potential we ended up with (24) from which we conclude that V_{eff} depends on the energy per unit mass E and the angular momentum J. For a massless test particle we found that $V_{\text{eff}} = -E^2/2$ in the case of radial geodesics ($J = 0$), while if $E = 0$ the particle behaves like a free particle. In the case of circular geodesics ($J \neq 0$), V_{eff} is given by (26). In Figure 3 the behaviour of V_{eff} is presented for zero and nonzero E and various values of J. From these graphs it can be inferred that for $E \neq 0$ and $J \neq 0$ the effective potential V_{eff} has a minimum and circular orbits are possible, while for $E \neq 0$ and $J = 0$ no stable circular orbits can exist. In the cases $E = 0$ and $J \neq 0$ the effective potential V_{eff} changes sign two times between the horizons and stable circular orbits must exist.

From the examination of the calculated effective potential, (27) for a massive particle, we have considered the radial geodesics for the cases $J = 0$ and $E = 0$, $J = 0$ and $E \neq 0$, and $J \neq 0$ and $E = 0$. It is seen (Figure 4) that in the first case the roots of the effective potential coincide with the horizons' positions and the particle's orbit is bound in the black hole's interior. In the second case, a bound orbit is also possible. Finally, in the third case, i.e., when the particle's angular momentum does not vanish, the roots of the effective potential coincide again with the horizons' positions and the particle's orbit is bound again in the black hole's interior.

As a last step we have examined the motion of a massive and charged test particle in the gravitational field of a charged black hole in generalized dilaton-axion gravity by exploiting the Hamilton-Jacobi equation. The latter is set up for the space-time geometry considered and is analytically solved by applying additive separation of variables. As a result, the particle's radial velocity and the effective potential are determined in closed form. Then the case of static equilibrium is examined and it is found that the charged test particle may have a bound orbit; i.e., it can be trapped by a charged black hole in this context or, stated differently, the charged black hole exerts an attractive gravitational force upon the charged particle in generalized dilaton-axion gravity. In the case of nonstatic equilibrium, we have distinguished between an uncharged and a charged test particle. In the former case, conditions have been found for the possibility of existence of the particle's bound orbit. Finally, when the test particle carries a charge, it is seen graphically (Figure 5) that no bound orbit is possible.

An interesting perspective for future work would be the study of the motion of charged or uncharged test particles and the behaviour of geodesics for rotating black hole solutions or for black hole solutions in more than four space-time dimensions in the context of generalized dilaton-axion gravity.

Conflicts of Interest

The authors declare that they have no conflicts of interest.

Acknowledgments

Farook Rahaman would like to thank the authorities of the Inter-University Centre for Astronomy and Astrophysics, Pune, India, for providing research facilities. Farook Rahaman and Susmita Sarkar are also grateful to DST-SERB (Grant No.: EMR/2016/000193) and UGC (Grant No.: 1162/(sc)(CSIR-UGC NET, DEC 2016)), Govt. of India, for financial support, respectively.

References

[1] T. Ortín, *Gravity and Strings*, Chapter 16, Cambridge University Press, 2004.

[2] S. Sur, S. Das, and S. SenGupta, "Charged black holes in generalized dilaton-axion gravity," *Journal of High Energy Physics*, vol. 10, article no. 064, 2005.

[3] T. Ghosh and S. SenGupta, "Thermodynamics of dilation-axion black holes," *Physical Review D*, vol. 78, no. 12, Article ID 124005, 2008.

[4] A. A. Usmani, Z. Hasan, F. Rahaman, S. A. Rakib, S. Ray, and P. K. Kuhfittig, "Thin-shell wormholes from charged black holes in generalized dilaton-axion gravity," *General Relativity and Gravitation*, vol. 42, no. 12, pp. 2901–2912, 2010.

[5] I. Radinschi, F. Rahaman, and A. Ghosh, "On the energy of charged black holes in generalized dilaton-axion gravity," *International Journal of Theoretical Physics*, vol. 49, no. 5, pp. 943–956, 2010.

[6] Z. M. Yang, X.-L. Li, and Y. Gao, "Entanglement entropy of charged dilaton-axion black hole and quantum isolated horizon," *European Physical Journal Plus*, vol. 131, no. 9, p. 304, 2016.

[7] T. Ghosh and S. SenGupta, "Tunneling across dilaton-axion black holes," *European Physics Letters*, vol. 120, no. 5, Article ID 50003, 2017.

[8] A. S. Bhatia and S. Sur, "Dynamical system analysis of dark energy models in scalar coupled metric-torsion theories," *International Journal of Modern Physics D*, vol. 26, no. 13, Article ID 1750149, 2017.

[9] S. Sur and A. S. Bhatia, "Weakly dynamic dark energy via metric-scalar couplings with torsion," *Journal of Cosmology and Astroparticle Physics*, vol. 2017, no. 7, p. 39, 2017.

[10] M. Kalam, F. Rahaman, and S. Mondal, "Particle motion around tachyon monopole," *General Relativity and Gravitation*, vol. 40, no. 9, pp. 1849–1861, 2008.

[11] S. Chakraborty and M. F. Rahaman, "Motion of test particles around gauge monopoles or near cosmic strings considering semiclassical gravitational effects," *International Journal of Modern Physics D*, vol. 9, no. 2, pp. 155–159, 2000.

18

From de Sitter to de Sitter: Decaying Vacuum Models as a Possible Solution to the Main Cosmological Problems

G. J. M. Zilioti,[1] R. C. Santos,[2] and J. A. S. Lima[3]

[1]*Universidade Federal do ABC (UFABC), Santo André, 09210-580 São Paulo, Brazil*
[2]*Departamento de Física, Universidade Federal de São Paulo (UNIFESP), 09972-270 Diadema, SP, Brazil*
[3]*Departamento de Astronomia, Universidade de São Paulo (IAGUSP), Rua do Matão 1226, 05508-900 São Paulo, Brazil*

Correspondence should be addressed to J. A. S. Lima; jas.lima@iag.usp.br

Academic Editor: Marek Szydlowski

Decaying vacuum cosmological models evolving smoothly between two extreme (very early and late time) de Sitter phases are able to solve or at least to alleviate some cosmological puzzles; among them we have (i) the singularity, (ii) horizon, (iii) graceful-exit from inflation, and (iv) the baryogenesis problem. Our basic aim here is to discuss how the coincidence problem based on a large class of running vacuum cosmologies evolving from de Sitter to de Sitter can also be mollified. It is also argued that even the cosmological constant problem becomes less severe provided that the characteristic scales of the two limiting de Sitter manifolds are predicted from first principles.

1. Introduction

The present astronomical observations are being successfully explained by the so-called cosmic concordance model or Λ_0CDM cosmology [1]. However, such a scenario can hardly provide by itself a definite explanation for the complete cosmic evolution involving two unconnected accelerating inflationary regimes separated by many aeons. Unsolved mysteries include the predicted existence of a space-time singularity in the very beginning of the Universe, the "graceful-exit" from primordial inflation, the baryogenesis problem, that is, the matter-antimatter asymmetry, and the cosmic coincidence problem. Last but not least, the scenario is also plagued with the so-called cosmological constant problem [2].

One possibility for solving such evolutionary puzzles is to incorporate energy transfer among the cosmic components, as what happens in decaying or running vacuum models or, more generally, in the interacting dark energy cosmologies. Here we are interested in the first class of models because the idea of a time-varying vacuum energy density or $\Lambda(t)$-models ($\rho_\Lambda \equiv \Lambda(t)/8\pi G$) in the expanding Universe is physically more plausible than the current view of a strict constant Λ [3–13].

The cosmic concordance model suggests strongly that we live in a flat, accelerating Universe composed of $\sim$ 1/3 of matter (baryons + dark matter) and $\sim$ 2/3 of a constant vacuum energy density. The current accelerating period ($\ddot{a} > 0$) started at a redshift $z_a \sim 0.69$ or equivalently when $2\rho_\Lambda = \rho_m$. Thus, it is remarkable that the constant vacuum and the time-varying matter-energy density are of the same order of magnitude just by now thereby suggesting that we live in a very special moment of the cosmic history. This puzzle ("why now"?) has been dubbed by the cosmic coincidence problem (CCP) because of the present ratio $\Omega_m/\Omega_\Lambda \sim \mathcal{O}(1)$, but it was almost infinite at early times [14, 15]. There are many attempts in the literature to solve such a mystery, some of them closely related to interacting dark energy models [16–18].

Recently, a large class of flat nonsingular FRW type cosmologies, where the vacuum energy density evolves like a truncated power-series in the Hubble parameter H, has been discussed in the literature [19–22] (its dominant term behaves like $\rho_\Lambda(H) \propto H^{n+2}, n > 0$). Such models has some interesting features; among them, there are (i) a new mechanism for inflation with no "graceful-exit" problem, (ii) the late time expansion history which is very close to the cosmic concordance model, and (iii) a smooth link between the initial and final de Sitter stages through the radiation and matter dominated phases.

In this article we will show in detail how the coincidence problem is also alleviated in the context of this class of decaying vacuum models. In addition, partially based on previous works, we also advocate here that a generic running vacuum cosmology providing a complete cosmic history evolving between two extreme de Sitter phases is potentially able to mitigate several cosmological problems.

2. The Model: Basic Equations

The Einstein equations, $G^{\mu\nu} = 8\pi G\ [T^{\mu\nu}_{(\Lambda)} + T^{\mu\nu}_{(T)}]$, for an interacting vacuum-matter mixture in the FRW geometry read [19, 20]

$$8\pi G\ \rho_T + \Lambda(H) = 3H^2, \tag{1}$$

$$8\pi G\ p_T - \Lambda(H) = -2\dot{H} - 3H^2, \tag{2}$$

where $\rho_T = \rho_M + \rho_R$ and $p = p_M + p_R$ are the total energy density and pressure of the material medium formed by nonrelativistic matter and radiation. Note that the bare Λ appearing in the geometric side was absorbed on the matter-energy side in order to describe the effective vacuum with energy density $\rho_\Lambda = -p_\Lambda \equiv \Lambda(H)/8\pi G$. Naturally, the time dependence of Λ is provoked by the vacuum energy transfer to the fluid component. In this context, the total energy conservation law, $u_\mu[T^{\mu\nu}_{(\Lambda)} + T^{\mu\nu}_{(T)}]_{;\nu} = 0$, assumes the following form:

$$\dot{\rho}_T + 3H(\rho_T + p_T) = -\dot{\rho}_\Lambda \equiv -\frac{\dot{\Lambda}}{8\pi G}. \tag{3}$$

What about the behavior of $\dot{\Lambda}$*?* Assuming that the created particles have zero chemical potential and that the vacuum fluid behaves like a condensate carrying no entropy, as what happens in the Landau-Tisza two-fluid description employed in helium superfluid dynamics[23], it has been shown that $\dot{\Lambda} < 0$ as a consequence of the second law of thermodynamics [10], that is, the vacuum energy density diminishes in the course of the evolution. Therefore, in what follows we consider that the coupled vacuum is continuously transferring energy to the dominant component (radiation or nonrelativistic matter components). Such a property defines precisely the physical meaning of decaying or running vacuum cosmologies in this work.

Now, by combining the above field equation it is readily checked that

$$\dot{H} + \frac{3(1+\omega)}{2}H^2 - \frac{1+\omega}{2}\Lambda(H) = 0, \tag{4}$$

where the equation of state $p_T = \omega\rho_T$ ($\omega \geq 0$) was used. The above equations are solvable only if we know the functional form of $\Lambda(H)$.

The decaying vacuum law adopted here was first proposed based on phenomenological grounds [7–9, 11] and later on suggested by the renormalization group approach techniques applied to quantum field theories in curved space-time [24]. It is given by

$$\Lambda(H) \equiv 8\pi G\rho_\Lambda = c_0 + 3\nu H^2 + \alpha\frac{H^{n+2}}{H_I{}^n}, \tag{5}$$

where H_I is an arbitrary time scale describing the primordial de Sitter era (the upper limit of the Hubble parameter), ν and α are dimensionless constants, and c_0 is a constant with dimension of $[H]^2$.

In a point of fact, the constant α above does not represent a new degree of freedom. It can be determined with the proviso that, for large values of H, the model starts from a de Sitter phase with $\rho = 0$ and $\Lambda_I = 3H_I^2$. In this case, from (5) one finds $\alpha = 3(1-\nu)$ because the first two terms there are negligible in this limit [see Eq. (1) in [9] for the case $n = 1$ and [11] for a general n]. The constant c_0 can be fixed by the time scale of the final de Sitter phase. For $H << H_I$ we also see from (4) that $c_0 = 3(1-\nu)H_F^2$, where H_F characterizes the final de Sitter stage (see (6) and (8)). Hence, the phenomenological law (5) assumes the final form:

$$\Lambda(H) = 3(1-\nu)H_F^2 + 3\nu H^2 + 3(1-\nu)\frac{H^{n+2}}{H_I{}^n}. \tag{6}$$

This is an interesting 3-parameter phenomenological expression. It depends on the arbitrary dimensionless constant ν and also the two extreme Hubble parameters (H_I, H_F) describing the primordial and late time inflationary phases, respectively. Current observations imply that the value of ν is very small, $|\nu| \sim 10^{-6} - 10^{-3}$ [25, 26]. More interestingly, the analytical results discussed below remain valid even for $\nu = 0$. In this case, we obtain a sort of minimal model defined only by a pair of physical time scales, H_I and H_F, determining the entire evolution of the Universe. As we shall see, the possible existence of these two extreme de Sitter regimes suggests a different perspective to the cosmological constant problem.

By inserting the above expression into (3) we obtain the equation of motion:

$$\dot{H} + \frac{3(1+\omega)(1-\nu)}{2}H^2\left[1 - \frac{H_F^2}{H^2} - \frac{H^n}{H_I^n}\right] = 0. \tag{7}$$

In principle, all possible de Sitter phases here are simply characterized by a constant Hubble parameter (H_C) satisfying the conditions $\dot{H} = \rho = p = 0$ and $\Lambda = 3H_C^2$. For all physically relevant values of ν and ω in the present context, we see that the condition $\dot{H} = 0$ is satisfied whether the possible values of H_C are constrained by the algebraic equation involving the arbitrary (initial and final) de Sitter vacuum scales H_I and H_F:

$$H_C^{n+2} - H_I^n H_C^2 + H_F^2 H_I^n = 0. \tag{8}$$

In particular, for $n = 2$, the value preferred from the covariance of the action, the exact solution is given by

$$H_C^2 = \frac{H_I^2}{2} \pm \frac{H_I^2}{2}\sqrt{1 - \frac{4H_F^2}{H_I^2}}, \tag{9}$$

and since $H_F << H_I$ we see that the two extreme scaling solutions for $n = 2$ are $H_{1C} = H_I$ and $H_{2C} = H_F$. However, we also see directly from (8) that the condition $H_F << H_I$, also guarantees that such solutions are valid regardless of the values of n. In certain sense, since H_0 is only the present day expansion rate, characterizing a quite casual stage of the recent evolving Universe, probably, it is not the interesting scale to be a priori predicted. In what follows we consider that the pair of extreme de Sitter scales (H_I, H_F) are the physically relevant quantities. This occurs because different from H_0, the expanding de Sitter rates are associated with very specific limiting manifolds. For instance, it is widely known that de Sitter spaces are static when written in a suitable coordinate system. Besides the discussion on the coincidence problem (see next section), a new idea to be advocated here is that the prediction of such scales, at least in principle, should be an interesting theoretical target. Their first principles prediction would open a new and interesting route to investigate the cosmological constant problem.

The solutions for the Hubble parameter describing analytically the transitions vacuum-radiation ($\omega = 1/3$) and matter-vacuum ($\omega = 0$) can be expressed in terms of the scale factor, the couple of scales (H_I, H_F), and free parameters (ν, n):

$$H = \frac{H_I}{\left[1 + Ca^{2n(1-\nu)}\right]^{1/n}}, \quad (10)$$

$$H = H_F\left[Da^{-3(1-\nu)} + 1\right]^{1/2}. \quad (11)$$

We remark that the transition radiation-matter is like that in the standard cosmic concordance model. The only difference is due to the small ν parameter that can be fixed to be zero (minimal model). Indeed, if one fixes $\nu = 0$, the matter-vacuum transition is exactly the same one appearing in the flat ΛCDM model. As we shall see below, the final scale H_F can be expressed as a simple function of H_0, ν, and Ω_Λ. Naturally, the existence of such an expression is needed in order to compare with the present observations. However, it cannot be used to hide the special meaning played by H_F in a possible solution of the cosmological constant problem.

3. Alleviating the Coincidence Problem

The so-called coincidence problem is very well known. It comes from the fact that the matter-energy density of the nonrelativistic components (baryons + dark matter) decreases as the Universe expands while the vacuum energy density (ρ_{Λ_0}) is always constant in the cosmic concordance model (Λ_0CDM). This happens also because the energy densities of the radiation ρ_γ (CMB photons) and neutrinos (ρ_ν) are negligible today. Thus, in a broader perspective, one may also say that the ratio $(\rho_M + \rho_R)/\rho_{\Lambda_0}$, where $\rho_R = \rho_\gamma + \rho_\nu$, was almost infinite at early times, but it is nearly of the order unity today.

The current fine-tuning behind the coincidence problem can also be readily defined in terms of the corresponding density parameters, since ($\Omega_{\Lambda_0} \sim 0.7$ and $\Omega_M + \Omega_R \sim 0.3$), so that the ratio is of the order unity some 14 billion years later.

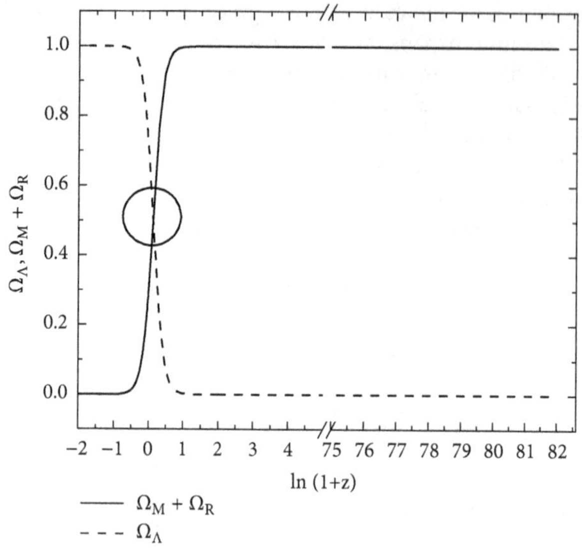

FIGURE 1: Standard coincidence problem in the cosmic concordance model (Λ_0CDM). Solid and dashed lines represent the evolution of the vacuum (Ω_{Λ_0}) and total matter-radiation ($\Omega_M + \Omega_R$) density parameters. The circle marks the low (and unique!) redshift presenting the extreme coincidence between the density parameter of the vacuum and material medium. Note that the model discussed here is fully equivalent to Λ_0CDM when the time dependent corrections in the decay $\Lambda(t)$ expression are neglected [see (5)].

In Figure 1 we display the standard view of the coincidence problem in terms of the corresponding density parameters: $\Omega_M = \Omega_b + \Omega_{cdm}$ (baryons + cold dark matter) and $\Omega_R = \Omega_\gamma + \Omega_\nu$ (CMB photons + relic neutrinos). As one may conclude from the figure, the ratio was practically infinite at very high redshifts, that is, at the early Universe (say, roughly at the Planck time). However, both densities are nearly coincident at present. The ratio $(\Omega_R + \Omega_M)/\Omega_{\Lambda_0} \sim 1$) is at low redshifts. Note also that, in the far future, that is, very deep in the de Sitter stage, the ratio approaches zero or equivalently the inverse ratio is almost infinite because the vacuum component becomes fully dominant.

A natural way to solve this puzzle is to assume that the vacuum energy density must vary in the course of the expansion. As shown in the previous section, the characteristic scales of the $\Lambda(t)$ model specify the evolution during the extreme de Sitter phases: the primordial vacuum solution with $Ca^{2n(1-\nu)} << 1$ and $H = H_I$ behaves like a "repeller" in the distant past, while the final vacuum solution for $a >> 1$, that is, $Da^{3(1-\nu)} \longmapsto 0$ and $H = H_F$, is an attractor in the distant future.

The arbitrary integration constants C and D are also easily determined. The constant C can be fixed by the end of the primordial inflation ($\ddot{a} = 0$) or equivalently $\rho_\Lambda = \rho_R$. This means that $C = a_{(eq)}^{-2n(1-\nu)}/(1-2\nu)$ [$a_{(eq)}$ corresponds to the value of the scale factor at vacuum-radiation equality]. In terms of the present day observable quantities we also find $D = \Omega_{M0}/(\Omega_{\Lambda 0} - \nu)$ and $H_F = H_0\sqrt{\Omega_{\Lambda 0} - \nu}/\sqrt{1-\nu}$. For $\nu = 0$ and $\Omega_{\Lambda 0} \sim 0.7$ one finds $H_F \sim 0.83H_0$, as expected a little smaller than H_0. The small observable parameter

$\nu < 10^{-3}$ quantifies the difference between the late time decaying vacuum model and the cosmic concordance cosmology; namely,

$$H = \frac{H_0}{\sqrt{1-\nu}} \left[\Omega_{M0} a^{-3(1-\nu)} + 1 - \Omega_{M0} - \nu\right]^{1/2}. \tag{12}$$

As remarked above, the $H(a)$ expression of the standard ΛCDM model is fully recovered for $\nu = 0$.

The solution of the coincidence problem in the present framework can be demonstrated as follows. The density parameters of the vacuum and material medium are given by

$$\Omega_\Lambda \equiv \frac{\Lambda(H)}{3H^2} = \nu + (1-\nu)\frac{H_F^2}{H^2} + (1-\nu)\frac{H^n}{H_I^n}, \tag{13}$$

$$\Omega_T \equiv 1 - \Omega_\Lambda = 1 - \nu - (1-\nu)\frac{H_F^2}{H^2} - (1-\nu)\frac{H^n}{H_I^n}. \tag{14}$$

Such results are a simple consequence of expression (6) for $\Lambda(H)$ and constraint Friedman equation (1). Note that $\Omega_T \equiv \Omega_M + \Omega_R$ is always describing the dominant component, either the nonrelativistic matter ($\omega = 0$) or radiation ($\omega = 1/3$).

The density parameters of the vacuum and material medium are equal in two different epochs specifying the dynamic transition between the distinct dominant components. These specific moments of time will be characterized here by Hubble parameters H_1^{eq} and H_2^{eq}. The first equality (vacuum-radiation, $\rho_\Lambda = \rho_R$) occurs just at the end of the first accelerating stage ($\ddot{a} = 0$), that is, when $H_1^{eq} = [(1-2\nu)/2(1-\nu)]^{1/n} H_I$, while the second one is at low redshifts when $H_2^{eq} = [2(1-\nu)/(1-2\nu)]^{1/2} H_F$. Note that such results are also valid for the minimal model by taking $\nu = 0$. In particular, inserting $\nu = 0$ in the first expression above we find $H_1^{eq} = H_I/2^{1/n}$. The scale H_2^{eq} can also be determined in terms of H_0. By adding the result $H_F \sim 0.83 H_0$ we find for $\nu = 0$ that $H_2^{eq} \sim 1.18 H_0$, which is higher than H_0, as should be expected for the matter-vacuum transition.

Naturally, the existence of two subsequent equalities on the density parameter suggests a solution to the coincidence problem. Neglecting terms of the order of 10^{-120} and 10^{-60n} in above expressions, it is easy to demonstrate the following results:

(1) $\lim_{H \to H_I} \Omega_\Lambda = 1$ and $\lim_{H \to H_I} \Omega_T = 0$,

(2) $\lim_{H \to H_F} \Omega_\Lambda = 1$ and $\lim_{H \to H_F} \Omega_T = 0$.

The meaning of the above results is quite clear. The density parameters of the vacuum and material components (radiation + matter) perform a cycle, that is, Ω_Λ, and $\Omega_M + \Omega_R$ are periodic in the long run.

In Figure 2, we show the complete evolution of the vacuum and matter-energy density parameters for this class of decaying vacuum model. Different from Figure 1 we observe that the values of Ω_Λ and $\Omega_M + \Omega_R$ are cyclic in the long run.

These parameters start and finish the evolution satisfying the above limits. The physical meaning of such evolution is also remarkable. For any value of $n > 0$, the model starts as a pure unstable vacuum de Sitter phase with $H = H_I$ (in the beginning there is no matter or radiation, $\Omega_\Lambda = 1, \Omega_M + \Omega_R = 0$). The vacuum decays and the model evolves smoothly to a quasi-radiation phase parametrized by the small ν-parameter.

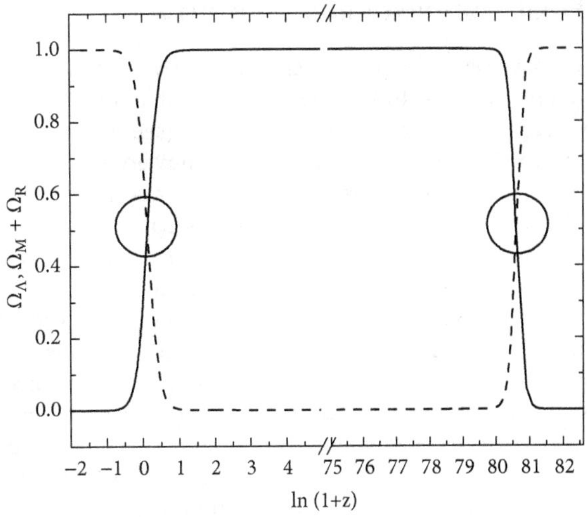

FIGURE 2: Solution of the coincidence problem in running vacuum cosmologies. The right graphic is our model; the left is ΛCDM. Solid and dashed lines represent the evolution of the vacuum (Ω_Λ) and total matter-radiation ($\Omega_M + \Omega_R$) density parameters for n=2, $\nu = 10^{-3}$, and $H_I/H_0 = 10^{60}$. The late time coincidence between the density parameter of the vacuum and material medium (left circle) has already occurred at very early times (right circle). Note also that the values 5 and 75 in the horizontal axis were glued in order to show the complete evolution (the suppressed part presents exactly the same behavior). Different values of n change slightly the value of the redshift for which $\Omega_\Lambda = \Omega_M + \Omega_R$ at the very early Universe (see also discussion in the text).

The circles show the redshifts for which $\Omega_\Lambda = \Omega_M + \Omega_R$. Of course, the existence of two equality solutions alleviates the cosmic coincidence problem.

The robustness of the solution must also be commented on. It holds not only for any value of $n > 0$ but also for $\nu = 0$. In the latter case, the primordial nonsingular vacuum state deflates directly to the standard FRW radiation phase. Later on, the transition from radiation to matter-vacuum dominated phase also occurs, thereby reproducing exactly the matter-vacuum transition of the standard Λ_0CDM model.

The "irreversible entropic cycle" from initial Sitter (H_I) to the late time de Sitter stage is completed when the Hubble parameter approaches its small final value ($H \longmapsto H_F$). The de Sitter space-time that was a "repeller" (unstable solution) at very early times ($z \longrightarrow \infty$) becomes an attractor in the distant future ($z \longrightarrow -1$) driven by the incredibly low energy scale H_F which is associated with the late time vacuum energy density, $\rho_M \longrightarrow 0, \rho_{\Lambda F} \propto H_F^2$.

Like the above solution to the coincidence problem, some cosmological puzzles can also be resolved along the same lines because the time behavior of the present scenario even fixing $\alpha = 1 - \nu$ has been proven here to be exactly the one discussed in [20] (see also [9] for the case $n = 1$).

4. Final Comments and Conclusion

As we have seen, the phenomenological $\Lambda(t)$-term provided a possible solution to the coincidence problem because the ratio Ω_M/Ω_Λ is periodic in long run (see Figure 2). In other words, the coincidence is not a novelty exclusive of the current epoch (low redshifts) since it also happened in the very early Universe at extremely high redshifts. In this framework, such a result seems to be robust because it is not altered even to the minimal model, that is, for $\nu = 0$.

It should also be stressed that the alternative complete cosmological scenario (from de Sitter to de Sitter) is not a singular attribute of decaying vacuum models. For instance, it was recently proved that at the background level such models are equivalent to gravitationally induced particle production cosmologies [27, 28] by identifying $\Lambda(t) \equiv \rho\Gamma/3H$, where Γ is the gravitational particle production rate. In a series of papers [29, 30], the dynamical equivalence of such scenario at late times with the cosmic concordance model was also discussed. It is also interesting that such a reduction of the dark sector can mimic the cosmic concordance model (Λ_0CDM) at both the background and perturbative levels [31, 32]. In principle, this means that alternative scenarios evolving smoothly between two extreme de Sitter phases are also potentially able to provide viable solutions of the main cosmological puzzles. However, different from $\Lambda(t)$-cosmologies, such alternatives are unable to explain the cosmological constant problem with this extreme puzzle becoming restricted to the realm of quantum field theory.

At this point, in order to compare our results with alternative models also evolving between two extreme de Sitter stages, it is interesting to review briefly how the main cosmological problems are solved (or alleviated) within this class of models driven by a pure decaying vacuum initial state:

(i) *Singularity:* the space-time in the distant past is a nonsingular de Sitter geometry with an arbitrary energy scale H_I. In order to agree with the semiclassical description of gravity, the arbitrary scale H_I must be constrained by the upper limit $H_I \leq 10^{19}$ GeV (Planck energy) in natural units or equivalently based on general relativity is valid only for times greater than the Plank time, $H_I^{-1} \geq 10^{-43}$ sec.

(ii) *Horizon problem:* the ansatz (6) can mathematically be considered as the simplest decaying vacuum law which destabilizes the initial de Sitter configuration. Actually, in such a model the Universe begins as a steady-state cosmology, $R \sim e^{H_I t}$. Since the model is nonsingular, it is easy to show that the horizon problem is naturally solved in this context (see, for instance, [22]).

(iii) *"Graceful-Exit" from inflation:* the transition from the early de Sitter to the radiation phase is smooth and driven by (10). The first coincidence of density parameters happens for $H = H_1^{eq}, \rho_\Lambda = \rho_R$, and $\ddot{a} = 0$, that is, when the first inflationary period ends (see Figure 2). All the radiation entropy ($S_0 \sim 10^{88}$, in dimensionless units) and matter-radiation content now observed were generated during the early decaying vacuum process (see [21] for the entropy produced in the case $n = 2$). For an arbitrary $n > 0$, the exit of inflation and the entropy production had also already been discussed [22]. Some possible curvature effects were also analyzed [33].

(iv) *Baryogenesis problem:* recently, it was shown that the matter-antimatter asymmetry can also be induced by a derivative coupling between the running vacuum and a nonconserving baryon current [34, 35]. Such an ingredient breaks dynamically CPT thereby triggering baryogenesis through an effective chemical potential (for a different but related approach see [36]). Naturally, baryogenesis induced by a running vacuum process has at least two interesting features: (i) the variable vacuum energy density is the same ingredient driving the early accelerating phase of the Universe and it also controls the baryogenesis process; (ii) the running vacuum is always accompanied by particle production and entropy generation [8, 10, 22]. This nonisentropic process is an extra source of T-violation (beyond the freeze-out of the B-operator) which as first emphasized by Sakharov [37] is a basic ingredient for successful baryogenesis. In particular, for $\nu = 0$ it was found that the observed B-asymmetry ordinarily quantified by the η parameter

$$5.7 \times 10^{-10} < \eta < 6.7 \times 10^{-10} \quad (15)$$

can be obtained for a large range of the relevant parameters (H_I, n) of the present model [34, 35]. Thus, as remarked before, the proposed running vacuum cosmology may also provide a successful baryogenesis mechanism.

(v) *de Sitter Instability and the future of the Universe:* another interesting aspect associated with the presence of two extreme Sitter phases as discussed here is the intrinsic instability of such space-time. Long time ago, Hawking showed that the space-time of a static black hole is thermodynamically unstable to macroscopic fluctuation in the temperature of the horizon [38]. Later on, it was also demonstrated by Mottola [39] based on the validity of the generalized second law of thermodynamics that the same arguments used by Hawking in the case of black holes remain valid for the de Sitter space-time. In the case of the primordial de Sitter phase, described here by the characteristic scale H_I, such an instability is dynamically described by solution (10) for $H(a)$. As we know, it behaves like a "repeller" driving the model to the radiation phase. However, the instability result in principle must also be valid to the final de Sitter stage which behaves like an attractor. In this way, once the final de Sitter phase is reached, the space-time would evolve to an energy scale smaller than H_F thereby starting a new evolutionary "cycle" in the long run.

(vi) *Cosmological constant problem:* it is known that phenomenological decaying vacuum models are unable to solve this conundrum [22, 34]. The basic reason seems to be related to the clear impossibility to predict the present day value of the vacuum energy density (or equivalently the value of H_0) from first principles. However, the present phenomenological approach can provide a new line of inquiry in the search for alternative (first principle) solutions for this remarkable puzzle. In this concern, we notice that the minimal model discussed here depends only on two relevant physical scales (H_F, H_I) which are associated with the extreme de Sitter phases. The existence of such scales implies that the ratio between the late and very early vacuum energy densities $\rho_{\Lambda F}/\rho_{\Lambda I} = (H_F/H_I)^2$ does not depend explicitly on the Planck mass. Indeed, the gravitational constant (in natural units, $G = M_{Planck}^{-2}$) arising in the expressions of the early and late time vacuum energy densities cancels out in the above ratio. Since $H_F \sim 10^{-42}$GeV, by assuming that $H_I \sim 10^{19}$GeV (the cutoff of classical theory of gravity), one finds that the ratio $\rho_{\Lambda F}/\rho_{\Lambda I} \sim 10^{-122}$, as suggested by some estimates based on quantum field theory, a result already obtained in some nonsingular decaying vacuum models [19]. In this context, the open new perspective is related to the search for a covariant action principle where both scales arise naturally. One possibility is related to models whose theoretical foundations are based on modified gravity theories like $F(R), F(R,T), etc$ [see, for instance, [40, 41]].

The results outlined above suggest that decaying vacuum models phenomenologically described by $\Lambda(t)$-cosmologies may be considered an interesting alternative to the mixing scenario formed by the standard ΛCDM plus inflation. However, although justified from different viewpoints, the main difficulty of such models seems to be a clear-cut covariant Lagrangian description.

Conflicts of Interest

The authors declare that there are no conflicts of interest regarding the publication of this paper.

Acknowledgments

G. J. M. Zilioti is grateful for a fellowship from CAPES (Brazilian Research Agency), R. C. Santos acknowledges the INCT-A project, and J. A. S. Lima is partially supported by CNPq, FAPESP (LLAMA project), and CAPES (PROCAD 2013). The authors are grateful to Spyros Basilakos, Joan Solà, and Douglas Singleton for helpful discussions.

References

[1] P. A. R. Ade et al., *Planck Collaboration A&A*, vol. 594, pp. 1502–1589, 2016.

[2] S. Weinberg, "The cosmological constant problem," *Reviews of Modern Physics*, vol. 61, no. 1, pp. 1-23, 1989.

[3] M. Bronstein, "On the expanding universe," *Physikalische Zeitschrift der Sowjetunion*, vol. 3, pp. 73–82, 1933.

[4] M. Özer and M. Taha, "A possible solution to the main cosmological problems," *Physics Letters B*, vol. 171, no. 4, pp. 363–365, 1986.

[5] M. Ozer and M. O. Taha, "A model of the universe free of cosmological problems," *Nuclear Physics*, vol. 287, p. 776, 1987.

[6] K. Freese, F. C. Adams, J. A. Frieman, and E. Mottola, "Cosmology with decaying vacuum energy," *Nuclear Physics A*, vol. 287, p. 797, 1987.

[7] J. C. Carvalho, J. A. Lima, and I. Waga, " Cosmological consequences of a time-dependent ," *Physical Review D: Particles, Fields, Gravitation and Cosmology*, vol. 46, no. 6, pp. 2404–2407, 1992.

[8] I. Waga, "Decaying vacuum flat cosmological models - Expressions for some observable quantities and their properties," *The Astrophysical Journal*, vol. 414, p. 436, 1993.

[9] J. A. Lima and J. M. Maia, "Deflationary cosmology with decaying vacuum energy density," *Physical Review D: Particles, Fields, Gravitation and Cosmology*, vol. 49, no. 10, pp. 5597–5600, 1994.

[10] J. A. Lima, "Thermodynamics of decaying vacuum cosmologies," *Physical Review D: Particles, Fields, Gravitation and Cosmology*, vol. 54, no. 4, pp. 2571–2577, 1996.

[11] J. M. F. Maia, *Some Applications of Scalar Fields in Cosmology [Ph.D. thesis]*, University of São Paulo, Brazil, 2000.

[12] M. Szydlowski, A. Stachowski, and K. Urbanowski, "Cosmology with a Decaying Vacuum Energy Parametrization Derived from Quantum Mechanics," *Journal of Physics: Conference Series*, vol. 626, no. 1, 2015.

[13] M. Szydłowski, A. Stachowski, and K. Urbanowski, "Quantum mechanical look at the radioactive-like decay of metastable dark energy," *The European Physical Journal C*, vol. 77, no. 12, 2017.

[14] P. J. Steinhardt, L. Wang, and I. Zlatev, "Cosmological tracking solutions," *Physical Review D: Particles, Fields, Gravitation and Cosmology*, vol. 59, no. 12, Article ID 123504, 1999.

[15] P. J. Steinhardt, *Critical Problems in Physics*, V. L. Fitch and D. R. Marlow, Eds., Princeton University Press, Princeton, N. J, 1999.

[16] S. Dodelson, M. Kaplinghat, and E. Stewart, "Solving the Coincidence Problem: Tracking Oscillating Energy," *Physical Review Letters*, vol. 85, no. 25, pp. 5276–5279, 2000.

[17] W. Zimdahl, D. Pavón, and L. P. Chimento, "Interacting quintessence," *Physics Letters B*, vol. 521, no. 3-4, pp. 133–138, 2001.

[18] A. Barreira and P. P. Avelino, *Physical Review D: Particles, Fields, Gravitation and Cosmology*, vol. 84, no. 8, 2011.

[19] J. A. S. Lima, S. Basilakos, and J. Solá, "Expansion History with Decaying Vacuum: A Complete Cosmological Scenario," *Monthly Notices of the Royal Astronomical Society*, vol. 431, p. 923, 2013.

[20] E. L. D. Perico, J. A. S. Lima, S. Basilakos, and J. Solà, "Complete cosmic history with a dynamical Λ=Λ(H) term," *Physical Review D: Particles, Fields, Gravitation and Cosmology*, vol. 88, Article ID 063531, 2013.

[21] J. A. Lima, S. Basilakos, and J. Solà, "Nonsingular decaying vacuum cosmology and entropy production," *General Relativity and Gravitation*, vol. 47, no. 4, article 40, 2015.

[22] J. A. S. Lima, S. Basilakos, and J. Solà, "Thermodynamical aspects of running vacuum models," *The European Physical Journal C*, vol. 76, no. 4, article 228, 2016.

[23] L. Landau and E. Lifshitz, *Journal of Fluid Mechanics*, Pergamon Press, 1959.

[24] I. L. Shapiro and J. Solà, "Scaling behavior of the cosmological constant and the possible existence of new forces and new light degrees of freedom," *Physics Letters B*, vol. 475, no. 3-4, pp. 236–246, 2000.

[25] S. Basilakos, D. Polarski, and J. Solà, "Generalizing the running vacuum energy model and comparing with the entropic-force models," *Physical Review D: Particles, Fields, Gravitation and Cosmology*, vol. 86, Article ID 043010, 2012.

[26] A. Gomez-Valent and J. Solà, "Vacuum models with a linear and a quadratic term in H: structure formation and number counts analysis," *Monthly Notices of the Royal Astronomical Society*, vol. 448, p. 2810, 2015.

[27] I. Prigogine, J. Geheniau, E. Gunzig, and P. Nardone, "Thermodynamics and cosmology," *General Relativity and Gravitation*, vol. 21, p. 767, 1989.

[28] M. O. Calvão, J. A. S. Lima, and I. Waga, "On the thermodynamics of matter creation in cosmology," *Physics Letters A*, vol. 162, no. 3, pp. 223–226, 1992.

[29] J. A. S. Lima, J. F. Jesus, and F. A. Oliveira, "CDM accelerating cosmology as an alternative to ΛCDM model," *Journal of Cosmology and Astroparticle Physics*, vol. 11, article 027, 2010.

[30] J. A. S. Lima, S. Basilakos, and F. E. M. Costa, "New cosmic accelerating scenario without dark energy," *Physical Review D: Particles, Fields, Gravitation and Cosmology*, vol. 86, Article ID 103534, 2012.

[31] R. O. Ramos, M. Vargas dos Santos, and I. Waga, "Matter creation and cosmic acceleration," *Physical Review D: Particles, Fields, Gravitation and Cosmology*, vol. 89, no. 8, Article ID 083524, 2014.

[32] R. O. Ramos, M. Vargas dos Santos, and I. Waga, "Degeneracy between CCDM and ΛCDM cosmologies," *Physical Review D*, vol. 90, Article ID 127301, 2014.

[33] P. Pedram, M. Amirfakhrian, and H. Shababi, "On the (2+1)-dimensional Dirac equation in a constant magnetic field with a minimal length uncertainty," *International Journal of Modern Physics D*, vol. 24, no. 2, Article ID 1550016, 8 pages, 2015.

[34] J. A. S. Lima and D. Singleton, "Matter-Antimatter Asymmetry Induced by a Running Vacuum Coupling," *The European Physical Journal C*, vol. 77, p. 855, 2017.

[35] J. A. S. Lima and D. Singleton, "Matter-antimatter asymmetry and other cosmological puzzles via running vacuum cosmologies," *International Journal of Modern Physics D*, 2018.

[36] V. K. Oikonomou, S. Pan, and R. C. Nunes, "Gravitational Baryogenesis in Running Vacuum models," *International Journal of Modern Physics A*, vol. 32, no. 22, Article ID 1750129, 2017.

[37] A. Sakharov, "Violation of CP invariance, C asymmetry, and baryon asymmetry of the universe," *JETP Letters*, vol. 5, p. 24, 1967.

[38] S. W. Hawking, "Black holes and thermodynamics," *Physical Review D: Particles, Fields, Gravitation and Cosmology*, vol. 13, no. 2, pp. 191–197, 1976.

[39] E. Mottola, "Thermodynamic instability of de Sitter space," *Physical Review D: Particles, Fields, Gravitation and Cosmology*, vol. 33, no. 6, pp. 1616–1621, 1986.

[40] T. P. Sotiriou and V. Faraoni, "*f(R)* theories of gravity," *Reviews of Modern Physics*, vol. 82, no. 1, article 451, 2010.

[41] T. Harko, F. S. N. Lobo, S. Nojiri, and S. D. Odintsov, "*F(R, T)* gravity," *Physical Review D: Particles, Fields, Gravitation and Cosmology*, vol. 84, no. 2, Article ID 024020, 2011.

19

Simplified Dark Matter Models

Enrico Morgante

Deutsches Elektronen-Synchrotron DESY, Notkestraße 85, 22607 Hamburg, Germany

Correspondence should be addressed to Enrico Morgante; enrico.morgante@desy.de

Academic Editor: Farinaldo Queiroz

I review the construction of simplified models for dark matter searches. After discussing the philosophy and some simple examples, I turn the attention to the aspect of the theoretical consistency and to the implications of the necessary extensions of these models.

1. Introduction

Producing and studying the properties of the dark matter (DM) particles at the LHC are an extremely exciting possibility that would open the door to a new understanding of the interplay between astrophysics, cosmology, and particle physics. Essentially all the naturalness-inspired scenarios can accommodate the presence of a good dark matter candidate: a neutral and very long-lived particle that was copiously produced in the early universe and then lost thermal contact with the SM (if it ever occurred) leaving a relic density $\Omega_{\mathrm{DM}} \sim 0.26$ of cold particles. The fact that such a stable weakly interacting massive particle with a mass around the weak scale has automatically a relic abundance close to the measured one is a remarkable property which is often dubbed *WIMP miracle*. The LHC is a perfectly suited machine to look for this kind of particles, and current bounds from ATLAS and CMS complement those from direct and indirect searches.

A key task in these studies is that of choosing a theoretical framework to compare with data and compare the results of different experiments. Given the plethora of particle physics models beyond the SM providing a WIMP candidate, it is highly desirable to study the signatures of this DM candidate in a model-independent way. In the early stages of the LHC, this was achieved by means of the effective field theory approach (EFT). In this framework, the Standard Model (SM) is complemented by a set of non-renormalizable operators that parametrize the interaction of the DM particle with SM fields in terms of one effective scale Λ and of the DM mass m_χ [1]. The EFT approach has proven to be very useful in the analysis of LHC Run I data [1–15], because of the great advantage of giving bounds that are as model-independent as possible: for a given choice of the spin of the DM particle, the number of operators that can couple it to the SM and may give interesting signatures at the LHC is limited, for a fixed mass dimension. Since direct and indirect detection of WIMPs, as well as WIMP production at the LHC, all require an interaction of the WIMPs with the SM particles, and such an interaction may be generated by the same operator, the EFT approach has the additional advantage of facilitating the analysis of the correlations between the various kinds of experiments.

The important drawback of the EFT description is its intrinsic energy limitation. At energies larger than some cutoff Λ, the contribution of higher dimension operators to the computation of scattering amplitudes becomes comparable to the lower ones, signalling the breakdown of perturbativity. More in particular, if the EFT is seen as the low energy limit of a theory with a mediator of mass M which is above the energy scale probed by the experiment, the cutoff is obtained as $\Lambda^2 \sim M^2/g^2$, where g is some combination of the coupling constants, and the theory is valid up to a momentum exchange $p^2 \lesssim M^2 \sim \Lambda^2$, where we have assumed $g \sim 1$. In direct and indirect detection this constraint is typically satisfied thanks to the low velocity of the incoming particle. On the other hand, the momentum exchanged in the partonic interactions at the LHC is of order few TeV, larger than the values of Λ that can be excluded within the EFT framework, making the naïve EFT bounds unreliable except for values of the couplings close to the perturbative bound $g \lesssim 4\pi$ [16–18]. In principle this does not mean that the EFT approach is not useful. Recasting procedures can be adopted to rederive bounds considering only a fraction of the events in the simulation that correspond to those which fulfil the requirement on the momentum [17, 19]. Clearly, the new bounds would be much weaker, but their simplicity still suggests that they should not be disregarded.

Partly in response to the problems of EFTs, and partly inspired by their rich phenomenological implications, in more recent years the LHC community has turned its attention to the tool-kit of simplified models. Such models are characterized by the most important state mediating the interaction of the DM particle with the SM, as well as the DM particle itself (see, for example, [14, 20–24] for early proposals). Including the effect of the mediator's propagator allows avoiding the energy limitation of the EFT, and simplified models are able to describe correctly the full kinematics of DM production at the LHC, at the price of a moderately increased number of parameters. As we are going to discuss below, the introduction of simplified models opens a new set of possibilities compared to the simpler EFT approach, while opening at the same time a set of new questions.

This paper is structured as follows. In Section 2, we are first going to describe the construction of DM simplified models from a bottom-up approach, providing some examples in Section 3. Then, in Section 4, we will point out the theoretical issues of such a construction, introducing a second generation of simplified models that have gained a lot of attention in recent times. Section 5 will contain our conclusions.

Thorough discussions about simplified DM models may be found in [25–30]. The second-generation models of Section 4 are discussed in [31, 32]. The discussion in this paper will be partly based on [33].

2. Philosophy of Simplified DM Models

As in the case of the EFT, the idea behind simplified models is to provide a good representation of possibly all realistic WIMP scenarios within the energy reach of the LHC, restricting to the smallest possible set of benchmark models, each with the minimal number of free parameters. Simplified models should be complete enough to give an accurate description of the physics at the scale probed by colliders, but at the same time they must have a limited number of new states and parameters.

The starting point is always the SM Lagrangian, complemented with a DM particle and a mediator that couples to it, through renormalizable operators, to quarks and gluons, which is necessary for the production of these states at a hadron collider. A coupling to other SM particles can be included as well and will add interesting experimental signatures to the model. In general, some simplifying assumptions can be made: for example, one can take all couplings to be equal, or the couplings to third generation's quarks to be dominant. Interactions that violate the accidental global symmetries of the SM must be handled with great care. Indeed, constraints on processes that violate these symmetries are typically very strong and may overcome those coming from DM searches or even rule out all of the interesting parameter space of the simplified model. For this reason, CP, lepton number, and baryon number conservation is typically assumed, together with minimal flavour violation (MFV). (Constraints on BSM models from CP and flavour violating observables are very strong, and the energy scale at which new physics may show up must be larger than tens of TeV in the best case, if the flavour structure of the model is generic. Minimal Flavour Violation is a way to reconcile these constraints with possible new physics at the TeV scale [34]. The basic idea is that the structure of flavour changing interactions must reproduce that of the SM. The SM is invariant under the flavour group $\mathcal{G}_F = SU(3)_q \times SU(3)_u \times SU(3)_d$, except a small breaking associated with the Yukawa matrices Y_u and Y_d. The invariance is restored if these matrices are regarded as "spurions" with transformation law $Y_u \sim (3, \overline{3}, 1)$ and $Y_d \sim (3, 1, \overline{3})$. Imposing MFV amounts to requiring that new physics is invariant under $\mathcal{G}_F$.) Even with this assumption, there are cases in which constraints from flavour physics may be stronger than those coming from mono-X searches [35] (see also [36] for a discussion of a non-minimally flavour violating dark sector).

Most simplified models of interest may be understood as the limit of a more general new physics scenario, where all new states but a few are integrated out because they have a mass larger than the energy scale reachable at the LHC or because they have no role in DM interactions with the SM. Similarly, in the limit where the mass of the mediator is very large, the EFT framework may be recovered by integrating out the mediator. On the contrary, there are new physics models which cannot be recast in terms of vanilla simplified models, typically because more than just one operator is active at the same time, possibly interfering with each other. The situation is summarized in Figure 1.

Even if this may sound obvious, we should stress that the correspondence between simplified models and EFT is not one to one. Simplified models that involve mediators of different spin nature may give rise to the same effective operator after a Fierz rotation, as pointed out in [19] with the example of a Majorana DM particle embedded in a Z' model or a SUSY-inspired model with coloured scalar mediators in the t-channel.

Even when a simple correspondence between the EFT and the simplified model is assumed, limits on the EFT cannot be readily translated onto the simplified model because of the possible resonant enhancement (that would make the limit stronger) or the typically softer missing energy spectrum (that would weaken the limit) [37]. Moreover, the different missing energy spectrum may require *ad hoc* optimization strategy by the experimental searches, and considering mediators of different mass requires different optimization for each case. Finally, models with a heavy mediator that would correspond to the EFT limit tend to predict a too large relic DM density (assuming no deviation from the standard cosmological history and no states other than the SM ones to annihilate into) and are therefore less appealing as a model of DM [38].

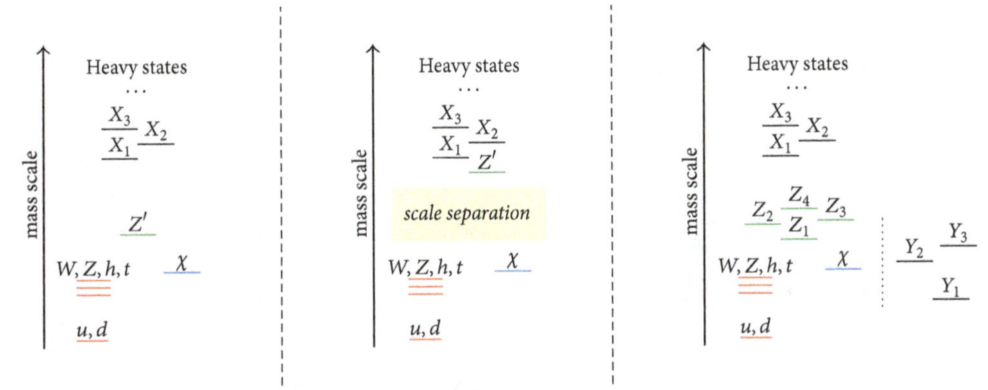

FIGURE 1: *Left:* a simplified model viewed as a sector of a more general new physics scenario. The SM is complemented by the DM particle χ and a mediator Z'. Other heavy states $X_1, X_2, X_3, \ldots$ may be integrated out because they are very heavy. *Centre:* in the case in which the mediator Z' itself has a very large mass, it may be integrated out as well and the interaction is mediated by effective operators. *Right:* in a generic setup the DM-SM interaction is mediated by a number of operators, possibly interfering with each other. Moreover, additional states $Y_1, Y_2, Y_3, \ldots$ that do not couple directly to DM may be present. Those states can constitute a new handle on the dark sector, other than DM. The 'less simplified' models described in Section 4 fall in this category.

From the point of view of LHC searches, the enlarged physical spectrum and parameter space of simplified models with respect to the EFT represents a challenge, and a greater variety of search channels is involved. While within the EFT approach the mono-X searches hold the stage, simplified models of DM can be constrained also with multi-jet + MET searches, with di-jet and di-leptons resonance searches and many others, depending on the degree of sophistication and on the ingredients of the model. Interestingly, many of these searches do not involve the DM particle, but only the mediator, and constraints are often stronger than the mono-X ones. On the other hand, the EFT is still a useful tool when dealing with strongly coupled theories, where a description in terms of a perturbative simplified model is not viable [39, 40].

3. First Generation of Simplified Models

3.1. s-Channel Mediators. We are now going to list a few examples of the simplified models of relevance for LHC searches, starting with those that include a fermionic DM χ (which for now we assume to be a Dirac spinor, but this is not necessary) and a mediator exchanged in the s-channel. The models under consideration are the following:

$$\mathscr{L}_V \supset \frac{1}{2} m_V^2 V_\mu V^\mu - m_\chi \bar{\chi}\chi - g_\chi V_\mu \bar{\chi}\gamma^\mu \chi - g_q^{ij} V_\mu \bar{q}_i \gamma^\mu q_j, \quad (1)$$

$$\mathscr{L}_A \supset \frac{1}{2} m_A^2 A_\mu A^\mu - m_\chi \bar{\chi}\chi - g_\chi A_\mu \bar{\chi}\gamma^\mu \gamma_5 \chi - g_q^{ij} A_\mu \bar{q}_i \gamma^\mu \gamma_5 q_j, \quad (2)$$

for a spin-1 mediator and

$$\mathscr{L}_S \supset -\frac{1}{2} m_S^2 S^2 - m_\chi \bar{\chi}\chi - y_\chi S \bar{\chi}\chi - y_q^{ij} S \bar{q}_i q_j + \text{h.c.}, \quad (3)$$

$$\mathscr{L}_P \supset -\frac{1}{2} m_P^2 P^2 - m_\chi \bar{\chi}\chi - i y_\chi P \bar{\chi}\gamma_5 \chi - i y_q^{ij} P \bar{q}_i \gamma_5 q_j + \text{h.c.}. \quad (4)$$

for a spin-0 mediator, where V, A, S, P stand for a vector, axial-vector, scalar, or a pseudoscalar mediator, respectively, $q = u, d$ and $i, j = 1, 2, 3$ are flavour indices. In the heavy m_{med} limit, the mediators can be integrated out, recovering the effective operators:

$$\begin{aligned} &\text{D1: } \frac{m_q}{\Lambda^3} \bar{\chi}\chi\ \bar{q}q \\ &\text{D4: } \frac{m_q}{\Lambda^3} \bar{\chi}\gamma^5 \chi\ \bar{q}\gamma^5 q \\ &\text{D5: } \frac{1}{\Lambda^2} \bar{\chi}\gamma_\mu \chi\ \bar{q}\gamma_\mu q \\ &\text{D8: } \frac{1}{\Lambda^2} \bar{\chi}\gamma_\mu \gamma^5 \chi\ \bar{q}\gamma_\mu \gamma^5 q \end{aligned} \quad (5)$$

where the nomenclature was first adopted in [1].

Let us briefly point out, and we will come back to this in Section 4, that the scalar and pseudo-scalar models of (3), (4) are not gauge invariant. This may lead to spurious results in processes where a W/Z boson is emitted. Moreover, in the axial-vector model perturbative unitarity is violated in a large portion of parameter space. We will return to these issues in Section 4.

Consistently with the MFV hypothesis, we force the couplings to be diagonal: $g_q^{ij} = g_q^i \delta^{ij}$. Moreover, we assume them to be flavour-blind in the (axial-)vector case and proportional to the SM Yukawa in the (pseudo)scalar ones:

$$\begin{aligned} &g_d^i = g_u^i \equiv g_q, \\ &y_q^{ij} \equiv y \frac{m_i}{v} \delta^{ij} \\ &\text{for } i = 1, 2, 3. \end{aligned} \quad (6)$$

In this way the spin-0 models have an enhanced coupling to the third generation's quarks, which makes the phenomenology quite different from the spin-1 models, both because of

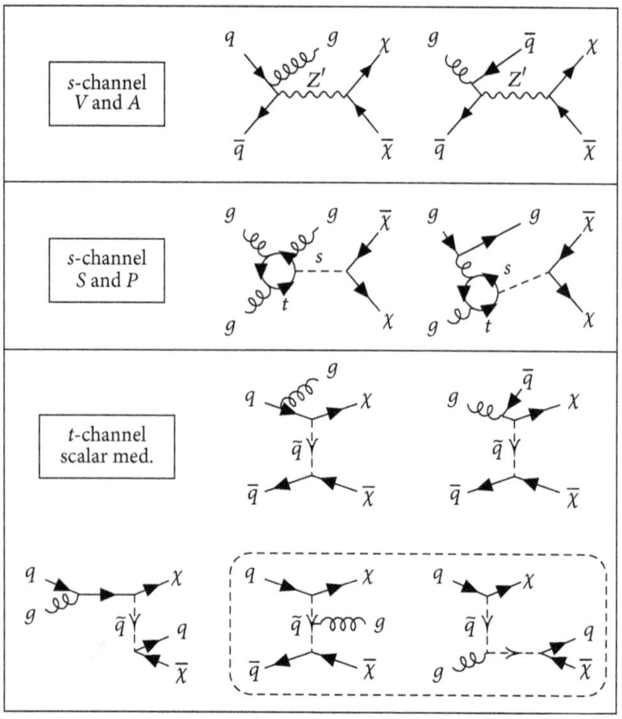

FIGURE 2: Feynman diagrams for the production of a DM pair in association with a quark or a gluon, leading to a mono-jet signature. Additional diagrams obtained by exchanging a fermion and a anti-fermion, as well as the ones obtained by permuting the gluon vertices in the loop in the (pseudo)scalar case, are neglected. The diagrams enclosed in the dashed box in the t-channel model are suppressed in the EFT limit.

the different production mechanism (gluon fusion with a top loop instead of $q\bar{q}$ annihilation) and because of the possibility of constraining such models with searches in b/t channels. As an example, Figure 2 shows the Feynman diagrams involved in the calculation of the mono-jet cross section.

In addition to the parameters of the Lagrangian, in the calculation of scattering amplitudes one must also include the decay width Γ of the mediator, which can be thought as a free parameter that encodes the unknown decay probability to other particles belonging to the dark sector. In this case, in the computation of the cross sections the couplings constants factor out and they affect only the normalization of the cross section through their product $g_q g_\chi$, while the spectra depend only on the masses m_{med}, m_χ. The problem with this approach is that, typically for large values of m_{med}, the benchmark value of Γ becomes smaller than the sum of the partial widths for decays into DM and quarks, which makes the choice unphysical [38]. To avoid the problem it is usually assumed that the mediator can not decay into particles other than the SM ones and, depending on its mass, the DM, and the width is computed accordingly as

$$\Gamma = \Gamma_\chi + \sum_f \Gamma_f + \Gamma_{gg} \tag{7}$$

This is usually referred to as the "minimal width assumption". Even if this choice eliminates one parameter, its drawback is that now the cross sections depend nontrivially on the couplings:

$$\sigma \propto \frac{g_\chi^2 g_q^2}{\left(s - m_{\text{med}}^2\right)^2 + m_{\text{med}}^2 \Gamma^2} \tag{8}$$

where Γ at the denominator depends on g_χ, g_q. Nevertheless, in most cases the dependence on the couplings is less important than the one on the masses, and so it is a good choice to fix the couplings and let the mass vary, thus presenting results as exclusion plots in the plane m_χ versus m_{med}. Following the recommendations of the LHC Dark Matter Working Group, useful benchmarks for the vector (V) and axial-vector (A) models are as follows [41, 42]:

$$\begin{aligned} V&: \begin{cases} g_\chi = 1, & g_q = 0.25, \quad g_\ell = 0 \\ g_\chi = 1, & g_q = 0.1, \quad g_\ell = 0.01 \end{cases} \\ A&: \begin{cases} g_\chi = 1, & g_q = 0.25, \quad g_\ell = 0 \\ g_\chi = 1, & g_q = 0.1, \quad g_\ell = 0.1 \end{cases} \end{aligned} \tag{9}$$

The exclusion lines that LHC draws have typically a simple structure. In MET+X searches in which DM is pair produced from the mediator and recoils against a SM particle (a photon, a hadronic jet or other) that is necessary to tag the event, the best sensitivity is obtained for $m_{\text{med}} > 2m_\chi$, where DM can be produced on resonance and the cross section is consequently enhanced. On the other hand, for $m_{\text{med}} < 2m_\chi$, the cross section is suppressed. (In the mono-jet channel, for $m_{\text{med}} \lesssim 2m_\chi$, the LHC at 14 TeV with 300 fb^{-1} is sensitive to $\mathcal{O}(1)$ couplings only for $m_\chi \lesssim \mathcal{O}(100\text{ GeV})$, while for $m_\chi \sim 1$ TeV it is sensitive only to couplings of order $g_\chi \cdot g_q \gtrsim 10$ [25]) Finally, for large m_{med} the EFT limit is recovered, but then again the constraining power is suppressed by the large m_{med}^4. Mono-jet limits in this region extend up to around 1.5 – 2 TeV, depending on the search, on the choice of vector or axial-vector mediator and to the values of the couplings (see, e.g., Figures 3 and 4 of [43] for the dependence of the m_{med} limit on the values g_q, g_χ). This gives rise to a typical triangular shape in the exclusion plots sketched in Figure 3 (see, e.g., [44] for a very recent example of such an exclusion plot in the mono-Z/W channel).

Sometimes, it proves very useful to show the constraints on s-channel simplified models in the plane $m_{\text{med}} - \Lambda$ for fixed m_χ, where $\Lambda = m_{\text{med}}/\sqrt{g_q g_\chi}$ is defined as the contact interaction scale. In this plane, the bounds have the typical shape shown in the right panel of Figure 3. For light mediator (Region I), DM production proceeds off-shell, and the cross section is suppressed (compared to the corresponding EFT result) by (m_{med}^4/s^2), where $\sqrt{s}$ can be estimated as $\min(\text{MET}^2, m_\chi^2)$. In Region II, the mediator is produced on-shell, and the cross section is enhanced. In this region the limit depends on the choice of the width Λ, as the cross section scales as $g_\chi^2 g_q^2/(m_{\text{med}}^2\Gamma^2)$. Finally, in Region III, the EFT limit is recovered.

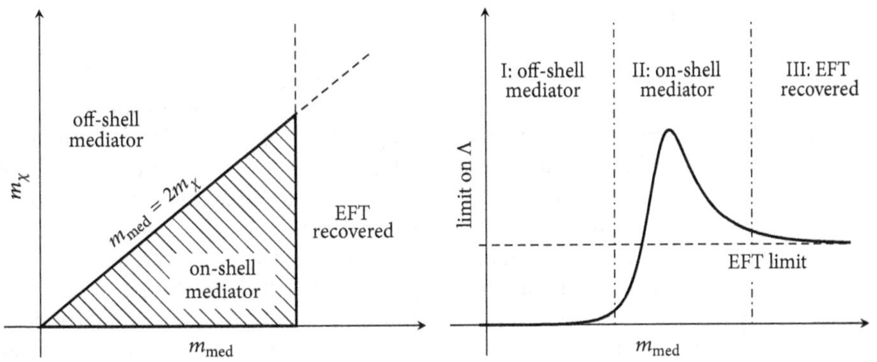

FIGURE 3: *Left*: sketch of the limits obtained from mono-X analysis in the $m_{\text{med}} - m_\chi$ plane (*left*, adapted from [32]) and in the $m_{\text{med}} - \Lambda$ plane (*right*, adapted from [37]).

Missing transverse energy searches are not the only handle that we have on simplified models. Searches for the mediator, for example, in the resonant di-jet channel, can lead to more stringent bounds in m_{med} for the same value of the coupling to quarks g_q.

Constraints on s-channel simplified models have been obtained by numerous groups, with a particular attention to the case of a (axial-)vector mediator, due to the problematic nature of the (pseudo)scalar models of (3), (4) (see the discussion in Section 4). Mono-jet constraints were discussed in [45–50]. A thorough comparison of mono-jet searches to di-jet searches, direct detection limits, dark matter overproduction in the early universe, and constraints from perturbative unitarity is performed in [43, 51–54]. In [55, 56] the problem of deriving limits for arbitrary values of the coupling constants starting from the benchmark ones is addressed. The case for a light DM (thus evading constraints from direct searches) is analysed in [22, 57, 58].

A very interesting phenomenology arises in the case where the couplings to third-generation quarks is larger than the couplings to the first two, as it is the case in the spin-0 models with MFV introduced above. Strong constraints on these models come from searches for one or two b-tagged jet + MET and $t\bar{t}$ + MET (see [59] for an early proposal within the EFT framework and [28, 48, 60–64] for a discussion in terms of simplified models). Summarizing, for light DM (m_χ = 1 GeV) current bounds obtained in the $t\bar{t}$ + MET channel can exclude couplings $g \gtrsim 1$ up to $m_{\text{med}} \lesssim 100$ GeV [63, 64], while this value can decrease to ≈ 0.5 at the end of the planned LHC runs [62]. These values are similar in magnitude to the ones obtained by a mono-jet analysis (see, e.g., [61]), which are much weaker than in the (axial-)vector case due to the assumed SM-Yukawa-like structure, which suppresses the coupling to light quarks. Still, LHC searches can provide the most stringent limit in some region of parameter space, complementing those coming from direct detection and from the relic abundance constraint and proving once more the importance of the complementarity of different probes [61].

3.2. t-Channel Mediators. Another interesting possibility is that of a coloured fermionic mediator with an interaction vertex between quarks and the WIMP resulting in a t-channel exchange, as with squark in supersymmetric models:

$$\mathcal{L} = \mathcal{L}_{\text{SM}} + \sum_i \left(g_L^i \overline{Q}_L^i \widetilde{Q}_L^i + g_u^i \overline{u}_R^i \widetilde{u}_R^i + g_d^i \overline{d}_R^i \widetilde{d}_R^i \right) \chi + \text{mass terms} + c.c., \quad (10)$$

where Q_L^i, u_R^i, d_R^i are the usual SM quarks, $\widetilde{Q}_L^i, \widetilde{u}_R^i, \widetilde{d}_R^i$ correspond to the respective scalar mediator (the squarks), and i represents a flavour index. Unlike the usual case in supersymmetry, here the WIMP χ can be taken to be either Dirac or Majorana fermion. This model is extensively analysed in [65–72]. While in the model above the flavour index is carried by the mediators, it could be the case that this index is assigned to the DM itself. This is the so-called "flavoured DM" scenario [73, 74]

As it can be easily understood, also in the case of t-channel mediators the phenomenology depends on the relative values of the couplings and of the masses involved. The minimal flavour violation hypothesis forces the couplings g^i and the masses of the scalars m_i to be equal. This assumption can be relaxed for the third generation. Particularly interesting is the case in which the coupling to the third generation is enhanced, and strong constraints come from searches for b, t quarks in the final state [25, 74–81]. A distinct phenomenology arises in the case of small couplings, in which the relic density is obtained by a out-of-equilibrium freeze-out mechanism and exotic collider signatures such as disappearing tracks and displaced vertices [80, 81].

Two interesting features of this model are worth listing that makes it qualitatively different from its low energy EFT limit. Firstly, being the squarks coloured, gluons may be emitted not only as initial state radiation but also from the mediator itself. This process is suppressed in the EFT limit by two powers of M_{med}, and this makes a large qualitative difference in the kinematic distribution within the simplified model and the corresponding operator. Secondly, when the mediator is light enough, its pair production becomes kinematically accessible, and an event like $p\,p \longrightarrow \widetilde{q}\,\widetilde{q} \longrightarrow q\chi\; q\chi$ leads to a di-jet + MET signature (or in general jets + MET, when additional jet radiation is taken into account).

Interestingly, this signature with two high-p_T jets leads to constraints stronger then the mono-jet one on a large portion of parameter space, except for the compressed region $|m_{\text{med}} - m_\chi| \ll m_{\text{med}}$ [70].

An interesting phenomenology arises in the case in which the t-channel mediators couple DM to both quarks and leptons, as in [82, 83]. Radiative corrections in this model strongly alter the spectrum of the Drell-Yan process $q\bar{q} \longrightarrow \ell^+\ell^-$. With this signal, one can probe "compressed regions" that are difficult to probe in direct searches for mediators like jets + MET searches. Moreover, these features can be used to make qualitative statements about dark matter's self-conjugation, mass, spin, and chirality of interactions.

3.3. Other Models. In addition to the models listed above, many interesting ones may be constructed that cannot be addressed here. Those include spin-2 mediators [84], t-channel fermionic mediators, fermiophobic scalar mediators [85], gluphylic mediator models [86–88], models with SM portals (Higgs or Z) [89], models of scalar DM [90], vector DM [91, 92], Higgs portal models with DM of diverse spin number [93–95], and others. A recent and comprehensive review is given in [30]. Interesting cosmological features of a model in which DM couples predominantly to the top quark are explored in [96]. In this case, it is possible to obtain the correct abundance with annihilations of DM particles to heavier states, at the tail of the velocity distribution.

4. Less Simplified Models

4.1. A Critical Look. The simplified models that we discussed so far can be viewed as an improvement of effective operators, where the effective scale Λ^4 is replaced with a propagator's denominator $(p^2 - M^2)^2 + \Gamma^2 M^2$ in order to avoid energy limitations and exploit resonant enhancement in the production cross section. This is one step above in a bottom-up approach. Nevertheless, the models described above suffer from other limitations and are not fully self consistent, as we are going to illustrate in a moment.

The reason why the theoretical consistency of the simplified models is important is twofold. On the one hand, violation of perturbative unitarity can lead to spuriously large predictions (e.g., in the process of W emission), leading to artificially strong bounds on the parameter space of the model. On the other hand, thinking of a full UV completion from which simplified models may descend, theoretical consistency is a necessary requirement at the level of the full theory. One could argue that this may not be the case for the simplified model, since other fields belonging to the full theory may restore the desired consistency. While this is generically true, these additional fields may add interesting ingredients to the phenomenology of the model, as they may produce new final states at the LHC and modify the annihilation cross section that enters the relic abundance calculation and the indirect detection fluxes, and their inclusion is therefore mandatory.

4.1.1. Gauge Non-Invariance of the (Pseudo-)Scalar Model. The scalar and pseudo-scalar models of (3), (4) are manifestly not gauge invariant. The problem comes with the Yukawas of the scalar mediator: assuming the DM is a singlet under the SM group, the invariance of the term $S\bar{\chi}\chi$ forces S to be a singlet as well, while for the Yukawas with the SM quarks $S\bar{q}q$ and $P\bar{q}\gamma_5 q$ the mediator should transform as a doublet. Clearly such a model can exist only as a consequence of EW symmetry breaking, and the DM-SM interaction has to be suppressed at low energy by some power of $v_{\text{EW}}/m_{\text{med}}$ [97, 98]. Clearly, a UV completion of the model is necessary.

As we will detail below, this issue can be fixed if the mediator is assumed to couple at tree level *only* to the DM and to the Higgs through the gauge invariant portal terms $S|H|^2$, $S^2|H|^2$, etc. that gives rise to a non-zero mixing angle after EW symmetry breaking [98–101]. This construction can be replicated for a pseudo-scalar, with the advantage of avoiding strong direct detection bounds, but at the price of introducing a new source of CP violation [102–104].

In [98], the model of (3) in compared to three possible consistent variations of it: first, the one in which the coupling of the scalar mediator to SM quarks is suppressed by $v_{\text{EW}}/m_{\text{med}}$; second, the model in which the mediator is replaced with the Higgs itself and the coupling to the DM particle is suppressed by the same factor; finally, the Higgs portal model including the $S-h$ mixing mentioned above. Naively, one could think that the S-mediator and the h mediator models can provide good approximations of the Higgs portal model. This is true only for large S mass ($m_S \gtrsim 1$ TeV for $m_\chi = 50$ GeV, or $m_S \gtrsim 5$ TeV for $m_\chi = 400$ GeV), in which case the Higgs portal resembles the model with the Higgs as a mediator, because the heavy scalar S can be integrated out. On the other hand, for lower masses the behaviour of the Higgs portal model descends from the interplay of the two mediating particles h and S, and the limits can be both stronger or weaker than the ones obtained in the two single-mediator models. In the case $m_\chi > m_h/2$, where the mixing angle is not constrained by the Higgs-to-invisibles branching ratio, the DM production cross section is dominated by S exchange, and in the heavy S limit a bound $M_\star \gtrsim 20$ GeV can be imposed, where

$$\frac{1}{M_\star^3} \approx \frac{\lambda \sin\alpha \cos\alpha}{v_H m_h^2}, \tag{11}$$

α being the mixing angle and λ the coupling of the scalar S to the DM particle [98].

A model of this kind naturally replicates the SM Yukawa-like structure of (3), (4) and displays promising experimental signatures (mono-jet, heavy quarks, mono-V, see [32] and references therein). Nevertheless, the mixing angle ϵ is strongly constrained by Higgs physics measurements (in particular the decay rate into invisible particles and the Higgs signal strength that are respectively enhanced and reduced by a non-zero mixing angle), and it does not add anything new to the LHC models' toolbox. An interesting option is to extend it to a two Higgs doublets model (2HDM) with the addition of a singlet scalar, evading all such constraints [101, 105].

4.1.2. Gauge Non-Invariance and Violation of Perturbative Unitarity. The vector and axial-vector simplified models of

(1), (2) are not in general invariant under the full SM gauge group $SU(3)_c \times SU(2)_L \times U(1)_Y$ but only under the unbroken subgroup $SU(3)_c \times U(1)_{e.m.}$. In particular, if the couplings to up and down quarks are different the mediator does not couple to the left handed quark doublet but to its two components separately, thus breaking gauge invariance. Similarly, the t-channel model of (10) is not gauge invariant unless the scalar mediator $\widetilde{Q}_L^i$ is charged and transforms as $(\mathbf{2}, -1/2)$ under $SU(2)_L \times U(1)_Y$. Violation of the electroweak gauge symmetry can lead to spuriously enhanced cross section for DM production with the initial state radiation of a W boson [97, 106]. This problem does not only affect the mono-W searches: W can indeed decay hadronically, enhancing the signal in the mono-jet search. For example, in the case of a vector mediator with opposite sign couplings to up and down quarks, this process dominates the mono-jet cross section for $\not{E}_T > 400$ GeV [107]. Similar issues should be present when considering Z or γ emission. In passing, this example shows that constraints descending from the internal consistency of the model cannot be neglected even when restricting to a particular MET search such as the mono-jet one that at first sight looks safe. Referring to the case of a vector mediator, different couplings of the up and down quarks can be made compatible with perturbative unitarity if an appropriate vertex WWZ' is added (where Z' is the new vector mediator), in similarity to what happens for the Z boson in the SM. In the t-channel model, instead, perturbative unitarity is restored if W emission from the charged mediator's line is included in the calculation.

4.1.3. Violation of Perturbative Unitarity with a s-Channel Axial-Vector Mediator. In the axial-vector model (2) the coupling to the longitudinal mode of the mediator is enhanced for heavy fermions by the ratio m_f/m_A. In particular, considering the elastic scattering of fermions (both SM fermions or DM) the perturbative unitarity bound on this model reads $m_f \lesssim m_{Z'}/(\sqrt{2}g_f^A)$ [108], where f may stand for both a SM fermion or the DM particle. Even if such a bound is satisfied, perturbative unitarity is still violated in the process of 2 fermions annihilation into $Z'Z'$, which is important for the calculation of the relic density and for indirect detection. In order to restore unitarity some additional ingredient has to be invoked. In particular, what violates unitarity is the longitudinal mode of the Z' boson; therefore the addition to the model of a scalar particle, invariant under the SM gauge group, that give rise to its mass via Higgs mechanism serves the purpose. In this case, the condition on the mass of the Z' would be the following [108]:

$$\sqrt{\pi}\frac{m_{Z'}}{g_{\mathrm{DM}}^A} \geq \max\left[m_s, \sqrt{2}m_{\mathrm{DM}}\right], \quad (12)$$

where m_s is the mass of the new scalar. At this point, it is clear that such an issue is not present in the vector model, because the mass of the mediator in that case can be obtained via a Stueckelberg mechanism without the need of additional particles (see [109] for further discussion).

4.1.4. Invariance of the SM Yukawas. Again referring to the axial-vector model of (2), if this is thought as a gauge extension of the SM, then the SM fermions must be charged under the new gauge symmetry (we restrict for simplicity to a $U(1)'$ theory, often referred to as a "Z' model"). Therefore, the Yukawa terms $H\overline{Q}_L d_R$ are gauge invariant only if the Higgs is charged as well, with $q_H = q_{qL} - q_{uR} = q_{dR} - q_{qL} = q_{eR} - q_{\ell L}$ (which is zero only if the SM fermions are vector-like under the new symmetry, as in the $B - L$ case) [108]. From this relation one sees that leptons have to be charged in this model, thus resulting in strong constraints from di-lepton searches. Another important consequence is that, after electroweak-symmetry breaking, the SM Z and the Z' have a non-zero mixing angle, and a tree level hZZ' vertex appears, with important phenomenological consequences and a complicate interplay between the two effects [110, 111].

4.1.5. Cancellation of Gauge Anomalies. If the interaction of DM with SM fermions is due to an extended gauge symmetry, in order for the theory to be consistent at the quantum level the charge assignment under the new gauge group cannot be generic. If all the fermions of the dark sector are uncharged under $SU(3)_c \times SU(2)_L \times U(1)_Y$, then the SM ones must have charges chosen in such a way to cancel the mixed anomalies of the dark gauge group with the SM. It can be shown that for a $U(1)'$ theory this forces the charges to be a linear combination of the SM hypercharge Y and of $B - L$ [112]. This implies that the mediator must couple to leptons, leading to tight constraints from resonance searches in the di-lepton channel (see, e.g., [110]).

Alternatively, additional heavy fermions, charged under the SM, may be added to the model. The mass of these fermions cannot be arbitrarily large: in order to cancel the anomalies, they must be chiral at least under the dark gauge group, and their mass is given by the vev of the dark Higgs. Fortunately, imposing the invariance of the SM Yukawa terms as discussed above in Section 4.1.4, it turns out that the gluon-gluon-Z' anomaly automatically cancels, and the new states need not to be coloured, reducing their impact on LHC searches [108]. On the other hand, they will enter the calculation of loop induced processes such as two photons decays that are not calculable in an anomalous theory and are relevant for indirect searches [110, 113]. (Anomalous models can be studied within a specific EFT framework, in which the required additional heavy fermions are integrated out, resulting in a set of effective Chern-Simons terms that must be added to the theory [114])

4.2. The $U(1)'$ *Model.* This model is one of the simplest possible extensions of the SM, in which the gauge group is enlarged by an additional $U(1)'$, spontaneously broken by the vev of a scalar field s, singlet under the SM, that gives mass to the dark gauge boson Z'. As mentioned above, interesting features come from this construction. First, the invariance of the SM Yukawas force the Higgs to be charged under $U(1)'$ whenever the charges of the left- and right-handed SM fermions differ from one another. Second, and consequently, the Higgs kinetic term includes interactions with the Z' boson, that induce after EW symmetry breaking a $ZZ'h$ that may have an impact on indirect detection [110]. Moreover, the dark Higgs s enriches the phenomenology, both at the LHC and in the calculation of annihilation rates. Finally, in consistent models that implement gauge anomaly cancellation the SM leptons must typically be charged and couple to the Z', which is therefore constrained by resonant di-lepton searches. (The coupling to leptons can be avoided by adding new fermions with nontrivial transformations under the SM gauge group such that all anomalies cancel [115], or in models where the DM carries a non-zero baryon number [116–119].) This constraint may be evaded if the Z' is lighter than the SM Z, thus escaping resonant searches [120].

A number of studies have addressed DM Z' models with respect to LHC, direct and indirect searches, as well as its cosmological implications [53, 121–126]. If DM is a Dirac fermion, its nonrelativistic scattering off nuclei is spin-independent, and direct detection constrain the DM mass to be larger than $\sim$ 1 TeV [125]. Such a strong constraint is lifted if the DM is a Majorana fermion, since the vector bilinear $\overline{\chi}\gamma_\mu\chi$ vanishes exactly and only the axial-vector one is left. In models in which the mediator couples to leptons with a coupling similar in magnitude to the one to quarks (as required by anomaly cancellation) resonant di-lepton searches forces the Z' to be heavier than $\sim$ 3 – 4 TeV, depending on the couplings and on the structure of fermionic charges [127]. Di-jet searches can exclude $m_{Z'} \lesssim 2.5$ TeV for a coupling $g_q \approx 0.25$, see [128]. Mono-jet searches are typically weaker than the di-jet ones, reaching $m_{Z'} \lesssim 1.5$ TeV for $g_q = 0.25$, $g_\chi = 1$ and for $m_\chi < m_{Z'}/2$ [129].

The dark Higgs becomes relevant when the annihilation rate is concerned. Indeed, the sZ' annihilation channel proceeds in s-wave and, therefore, it dominates the cross section for indirect detection and for the relic abundance calculation [130], together with the ss, $Z'Z'$, and Zh channels [130, 131]. In particular, the additional channels and operators increase the DM annihilation rate, thus reducing the relic abundance and alleviating constraints from DM overproduction in the early universe. Moreover, mono-dark-Higgs signals can be looked for at the LHC, with the typical signature being DM produced in association with a scalar resonance s decaying to a highly boosted $b\overline{b}$ pair [131]. In this case, the expected LHC sensitivity extends up to $m_{Z'} \lesssim 3.5$ TeV, $m_\chi \lesssim 600$ GeV for $m_s = 50$ GeV, or higher for heavier s. In general, such a model can be viewed as a model with two mediators that in some limit may reduce to a spin-1 or to a spin-0 mediator [53].

4.3. Two Higgs Doublets (plus One Singlet) Models and DM. Models in which the (pseudo-)scalar mediator that couples to DM obtains its coupling to SM quarks from mixing with a second Higgs doublet have received a significant attention recently. With respect to the scenario in which the singlet mixes directly with the SM Higgs discussed in Section 4.1.1, in this model the Higgs branching ratios and signal strength are not modified. Moreover, both a scalar and a pseudo-scalar mediator can be accommodated without adding new sources of CP violation. Indeed, in the presence of a second doublet, there are 8 spin-0 fields, three of which get 'eaten' by the $Z, W^\pm$ after symmetry breaking, thus leaving one charged scalar field, two neutral ones and one neutral pseudo-scalar, which the dark mediator can mix with. The pseudo-scalar case is particularly interesting in view of the fact that its low energy effective vertex $\overline{q}i\gamma_5 q\overline{\chi}i\gamma_5\chi$ leads to the both spin- and momentum-suppressed non relativistic interaction $(\vec{s}_\chi \cdot \vec{q})(\vec{s}_N \cdot \vec{q})$ [132, 133], on which constrains from direct detection are poor.

A thorough discussion of 2HDM (not related to DM) is given in [134]. The model of interest here, where the 2HDM is complemented with an additional pseudo-scalar mediator and with a DM fermion, is described in detail in [135]. In general, the model counts many free parameters, many of which can be fixed by requiring that Higgs and precision EW tests are not spoiled, and by the requirement of the stability of the scalar potential. In particular, one of the two Higgs states is assumed to have SM-like couplings, while the second doublet couples to SM vectors only at loop level: this is the so-called *alignment/decoupling limit*. Moreover, the neutral scalar, pseudo-scalar, and the charged components of the second Higgs doublet are assumed to have the same mass. A typical choice is then to fix the DM Yukawa coupling to 1, and the DM mass to a benchmark value of 10 GeV. The model then consists of 4 free parameters, which are typically chosen to be the mass of the pseudo-scalars m_a, m_A, the ratio of the VEVs $\tan\beta = v_1/v_2$ and the $a - A$ mixing angle θ.

In the context of DM, many search channels have been used to constrain this class of models. In particular, searches for mono-jet/$\gamma/Z/W/H$, non-resonant dijets, single top as well as $t\overline{t}$, $b\overline{b}$, $b\overline{b}Z$, all in association with missing transverse energy, have been explored [62, 101, 104, 105, 135–142]. As usual, constraints come also from visible channels (i.e., where no invisible DM particle is produced), as well as indirect detection and cosmology. Finally, all the usual concerns about 2HDM coming from flavour physics, EW precision tests, invisible Higgs decays, vacuum stability and perturbative unitarity apply too.

The Higgs' width to invisibles constrains is $m_a \gtrsim$ 100 GeV, while flavour constraints forces are $\tan\beta \gtrsim 1$. The most powerful searches are then the mono-Z and the mono-H ones that can exclude $\tan\beta$ up to 2 for values of m_a up to $200 - 350$ GeV, depending on the values of m_A and $\sin\theta$. For further details we refer the reader to [135].

Interestingly, since in such a model the coupling of DM to the heavy quarks is naturally enhanced, the Galactic Centre excess could be explained by a model which is testable at the LHC [139, 143].

Let us just mention another related possibility, which is the one of a second inert Higgs doublet, which does not couple to SM fermions except for its mixing with the Higgs. The lightest component of this doublet is a perfect candidate for scalar DM [144].

5. Conclusions

The question of which DM models should be adopted in defining new search strategies and in presenting experimental results is a pressing one, primarily for LHC searches. Simplified DM models are a possible answer to this question, living in between the effective operators approach (with a limited applicability at the LHC) and the realm of well-motivated BSM theories.

From a bottom-up viewpoint, the idea of simplified models is to expand the effective operators including mediator particles in the description, thus avoiding the energy limitations of the EFT approach and adding a richer phenomenology, new search channels, etc. In a top-down framework, instead, simplified models can be seen as a way to simplify the phenomenology of complex new physics models in such a way to restrict to the phenomena related to DM.

In order not to deal with unphysical results, the vanilla simplified models have to be supplied with additional constraints, couplings, and states, in a kind of second-order improvement. The typical consequence is that the strongest LHC constraints on the dark sector come from many possible observables other than DM production processes (as the mono-X searches) and di-jet searches (e.g., di-lepton resonances, mixing with Z boson and electroweak precision tests, Higgs width to invisibles, perturbative unitarity, and so on). This comes with no surprise, since the high energy reach of the LHC consent to explore a large variety of phenomena above the weak scale, without restricting to the lightest stable state of this new physics sector. This is quite the opposite with respect to what happens with direct and indirect searches, which are intrinsically limited to constrain the properties of the DM particle.

Theoretically consistent simplified models tend to loose part of their generality and to mimic richer BSM theories. For example, models containing a vector mediator and a dark Higgs may descend from gauged $\mathrm{U}(1)'$ constructions, while models featuring two Higgs doublets and a (pseudo-)scalar singlet resemble the Higgs sector of the NMSSM.

Simplified models cannot (or only partially) be viewed as an exhaustive toolbox to constrain all possible WIMP scenarios at once. For this reason, it is of extreme importance that the LHC collaborations publish their results on simple, search-specific, models in such a way that they are recastable for any other model (as it is for cut-and-count analyses). In turn, theoreticians should keep working in close contact with experimentalists in order to maximise the utility of the simplified models tool-kit. Finally, the use of (truncated) EFT should not be disregarded, since this is the most model-independent approach and it is economical from the point of view of the reduced dimensionality of its parameter space.

Conflicts of Interest

The author declares that there are no conflicts of interest regarding the publication of this paper.

Acknowledgments

The author is grateful to Michael Duerr and Davide Racco for the many useful comments on this manuscript.

References

[1] J. Goodman, M. Ibe, A. Rajaraman, W. Shepherd, T. M. Tait, and H. Yu, "Constraints on dark matter from colliders," *Physical Review D: Covering Particles, Fields, Gravitation, and Cosmology*, vol. 82, no. 11, 2010.

[2] M. Beltrán, D. Hooper, E. W. Kolb, Z. A. Krusberg, and T. M. Tait, "Maverick dark matter at colliders," *Physical Review D: Covering Particles, Fields, Gravitation, and Cosmology*, vol. 2010, no. 9, 2010.

[3] J. Goodman, M. Ibe, A. Rajaraman, W. Shepherd, T. M. Tait, and H. Yu, "Constraints on light Majorana dark matter from colliders," *High Energy Physics - Phenomenology*, vol. 695, no. 1-4, pp. 185–188, 2011.

[4] Y. Bai, P. J. Fox, and R. Harnik, "The Tevatron at the frontier of dark matter direct detection," *Physical Review D: Covering Particles, Fields, Gravitation, and Cosmology*, vol. 2010, no. 12, 2010.

[5] P. J. Fox, R. Harnik, J. Kopp, and Y. Tsai, "LEP shines light on dark matter," *Physical Review D: Covering Particles, Fields, Gravitation, and Cosmology*, vol. 84, 2011.

[6] A. Rajaraman, W. Shepherd, T. M. Tait, and A. M. Wijangco, "LHC bounds on interactions of dark matter," *Physical Review D: Covering Particles, Fields, Gravitation, and Cosmology*, vol. 84, no. 9, 2011.

[7] P. J. Fox, R. Harnik, J. Kopp, and Y. Tsai, "Missing energy signatures of dark matter at the LHC," *Physical Review D: Covering Particles, Fields, Gravitation, and Cosmology*, vol. 85, no. 5, 2012.

[8] I. M. Shoemaker and L. Vecchi, "Unitarity and monojet bounds on models for DAMA, CoGeNT, and CRESST-II," *Physical Review D: Covering Particles, Fields, Gravitation, and Cosmology*, vol. 86, no. 1, 2012.

[9] R. C. Cotta, J. L. Hewett, M. Le, and T. G. Rizzo, "Bounds on dark matter interactions with electroweak gauge bosons," *Physical Review D: Covering Particles, Fields, Gravitation, and Cosmology*, vol. 88, no. 11, 2013.

[10] H. K. Dreiner, M. Huck, M. Krämer, D. Schmeier, and J. Tattersall, "Illuminating dark matter at the ILC," *Physical Review D: Covering Particles, Fields, Gravitation, and Cosmology*, vol. 87, no. 7, 2013.

[11] Y. J. Chae and M. Perelstein, "Dark matter search at a linear collider: effective operator approach," *Journal of High Energy Physics*, vol. 5, no. 138, 2013.

[12] P. J. Fox and C. Williams, "Next-to-leading order predictions for dark matter production at hadron colliders," *Physical Review D: Particles, Fields, Gravitation and Cosmology*, vol. 87, no. 5, 2013.

[13] A. De Simone, A. Monin, A. Thamm, and A. Urbano, "On the effective operators for dark matter annihilations," *Journal of Cosmology and Astroparticle Physics*, no. 2, 2013.

[14] H. Dreiner, D. Schmeier, and J. Tattersall, "Contact interactions probe effective dark-matter models at the LHC," *EPL (Europhysics Letters)*, vol. 102, no. 5, 2013.

[15] J.-Y. Chen, E. W. Kolb, and L.-T. Wang, "Dark matter coupling to electroweak gauge and Higgs bosons: An effective field theory approach," *Physics of the Dark Universe*, vol. 2, no. 4, pp. 200–218, 2013.

[16] G. Busoni, A. De Simone, E. Morgante, and A. Riotto, "On the validity of the effective field theory for dark matter searches at the LHC," *Physics Letters B*, vol. 728, pp. 412–421, 2014.

[17] G. Busoni, A. De Simone, J. Gramling, E. Morgante, and A. Riotto, "On the validity of the effective field theory for dark matter searches at the LHC, part II: complete analysis for the s-channel," *Journal of Cosmology and Astroparticle Physics*, no. 6, 2014.

[18] G. Busoni, A. D. Simone, T. Jacques, E. Morgante, and A. Riotto, "On the Validity of the Effective Field Theory for Dark Matter Searches at the LHC, Part III: Analysis for the t-channel," *Journal of Cosmology and Astroparticle Physics*, vol. 2014, no. 1409, 2014.

[19] D. Racco, A. Wulzer, and F. Zwirner, "Robust collider limits on heavy-mediator Dark Matter," *Journal of High Energy Physics*, vol. 2015, no. 5, 2015.

[20] E. Dudas, Y. Mambrini, S. Pokorski, and A. Romagnoni, "(In)visible Z-prime and dark matter," *Journal of High Energy Physics*, vol. 2009, 2009.

[21] J. Goodman and W. Shepherd, "LHC Bounds on UV-Complete Models of Dark Matter," *High Energy Physics - Phenomenology*, 2011.

[22] H. An, X. Ji, and L. Wang, "Light dark matter and Z*ι* dark force at colliders," *Journal of High Energy Physics*, vol. 2012, no. 7, 2012.

[23] M. T. Frandsen, F. Kahlhoefer, A. Preston, S. Sarkar, and K. Schmidt-Hoberg, "LHC and Tevatron bounds on the dark matter direct detection cross-section for vector mediators," *Journal of High Energy Physics*, vol. 2012, no. 7, 2012.

[24] R. C. Cotta, A. Rajaraman, T. M. Tait, and A. M. Wijangco, "Particle physics implications and constraints on dark matter interpretations of the CDMS signal," *Physical Review D: Particles, Fields, Gravitation and Cosmology*, vol. 90, no. 1, 2014.

[25] J. Abdallah et al., "Simplified Models for Dark Matter and Missing Energy Searches at the LHC," *High Energy Physics - Phenomenology*, 2014.

[26] S. Malik, C. McCabe, H. Araujo, A. Belyaev, C. Boehm et al., "Interplay and Characterization of Dark Matter Searches at Colliders and in Direct Detection Experiments," *Physics of the Dark Universe*, vol. 9-10, pp. 51–58, 2015.

[27] J. Abdallah et al., "Simplified Models for Dark Matter Searches at the LHC," *Physics of the Dark Universe*, vol. 9-10, pp. 8–23, 2015.

[28] D. Abercrombie et al., "Dark Matter Benchmark Models for Early LHC Run-2 Searches: Report of the ATLAS/CMS Dark Matter Forum," 2015, https://arxiv.org/abs/1507.00966.

[29] A. De Simone and T. Jacques, "Simplified models vs. effective field theory approaches in dark matter searches," *The European Physical Journal C*, vol. 76, no. 7, 2016.

[30] G. Arcadi, M. Dutra, P. Ghosh et al., "The waning of the WIMP? A review of models, searches, and constraints," *The European Physical Journal C*, vol. 78, no. 3, 2018.

[31] A. Albert et al., "Towards the next generation of simplified Dark Matter models," *Physics of the Dark Universe*, vol. 16, pp. 49–70, 2017.

[32] F. Kahlhoefer, "Review of LHC dark matter searches," *International Journal of Modern Physics A*, vol. 32, 2017.

[33] E. Morgante, *Aspects of WIMP Dark Matter searches at colliders and other probes [Ph.D. thesis]*, Springer Theses, Geneva, Switzerland, 2017.

[34] G. D'Ambrosio, G. F. Giudice, G. Isidori, and A. Strumia, "Minimal flavor violation: An Effective field theory approach," *Nuclear Physics B*, vol. 645, pp. 155–187, 2002.

[35] M. J. Dolan, F. Kahlhoefer, C. McCabe, and K. Schmidt-Hoberg, "A taste of dark matter: flavour constraints on pseudoscalar mediators," *Journal of High Energy Physics*, vol. 2015, no. 3, 2015.

[36] P. Agrawal, M. Blanke, and K. Gemmler, "Flavored dark matter beyond Minimal Flavor Violation," *Journal of High Energy Physics*, vol. 2014, no. 10, 2014.

[37] O. Buchmueller, M. J. Dolan, and C. McCabe, "Beyond effective field theory for dark matter searches at the LHC," *Journal of High Energy Physics*, vol. 2014, no. 1, 2014.

[38] G. Busoni, A. D. Simone, T. Jacques, E. Morgante, and A. Riotto, "Making the most of the relic density for dark matter searches at the LHC 14 TeV Run," *Journal of Cosmology and Astroparticle Physics*, vol. 2015, 2015.

[39] S. Bruggisser, F. Riva, and A. Urbano, "Strongly Interacting Light Dark Matter," *SciPost Physics*, vol. 3, no. 3, 2017.

[40] S. Bruggisser, F. Riva, and A. Urbano, "The last gasp of dark matter effective theory," *Journal of High Energy Physics*, vol. 2016, no. 11, 2016.

[41] G. Busoni et al., "Recommendations on presenting LHC searches for missing transverse energy signals using simplified s-channel models of dark matter," 2016, https://arxiv.org/abs/1603.04156.

[42] A. Albert et al., "Recommendations of the LHC Dark Matter Working Group: Comparing LHC searches for heavy mediators of dark matter production in visible and invisible decay channels," 2017, https://arxiv.org/abs/1703.05703.

[43] A. Choudhury, K. Kowalska, L. Roszkowski, E. M. Sessolo, and A. J. Williams, "Less-simplified models of dark matter for direct detection and the LHC," *Journal of High Energy Physics*, vol. 2016, 2016.

[44] CMS collaboration, A. M. Sirunyan et al., "Search for new physics in final states with an energetic jet or a hadronically decaying W or Z boson and transverse momentum imbalance at $\sqrt{s} = 13$ TeV," *Physical Review D97*, vol. 092005, no. 9, 2018.

[45] O. Lebedev and Y. Mambrini, "Axial dark matter: The case for an invisible Z′," *Physics Letters B*, vol. 734, pp. 350–353, 2014.

[46] O. Buchmueller, M. J. Dolan, S. A. Malik, and C. McCabe, "Characterising dark matter searches at colliders and direct detection experiments: vector mediators," *Journal of High Energy Physics*, vol. 2015, no. 1, 2015.

[47] P. Harris, V. V. Khoze, M. Spannowsky, and C. Williams, "Constraining dark sectors at colliders: Beyond the effective theory approach," *Physical Review D: Particles, Fields, Gravitation and Cosmology*, vol. 91, no. 5, 2015.

[48] M. R. Buckley, D. Feld, and D. Gonçalves, "Scalar simplified models for dark matter," *Physical Review D: Particles, Fields, Gravitation and Cosmology*, vol. 91, no. 1, 2015.

[49] Q. Xiang, X. Bi, P. Yin, and Z. Yu, "Searches for dark matter signals in simplified models at future hadron colliders," *Physical Review D: Particles, Fields, Gravitation and Cosmology*, vol. 91, no. 9, 2015.

[50] P. Harris, V. V. Khoze, M. Spannowsky, and C. Williams, "Closing up on dark sectors at colliders: From 14 to 100 TeV," *Physical Review D: Particles, Fields, Gravitation and Cosmology*, vol. 93, no. 5, 2016.

[51] M. Fairbairn and J. Heal, "Complementarity of dark matter searches at resonance," *Physical Review D: Particles, Fields, Gravitation and Cosmology*, vol. 90, no. 11, 2014.

[52] M. Chala, F. Kahlhoefer, M. McCullough, G. Nardini, and K. Schmidt-Hoberg, "Constraining dark sectors with monojets and dijets," *Journal of High Energy Physics*, vol. 2015, no. 7, 2015.

[53] M. Duerr, F. Kahlhoefer, K. Schmidt-Hoberg, T. Schwetz, and S. Vogl, "How to save the WIMP: global analysis of a dark matter model with two s-channel mediators," *Journal of High Energy Physics*, vol. 2016, no. 9, 2016.

[54] M. Fairbairn, J. Heal, F. Kahlhoefer, and P. Tunney, "Constraints on Z*ι* models from LHC dijet searches and implications for dark matter," *Journal of High Energy Physics*, vol. 2016, no. 9, 2016.

[55] T. Jacques and K. Nordström, "Mapping monojet constraints onto simplified dark matter models," *Journal of High Energy Physics*, vol. 2015, no. 6, 2015.

[56] A. J. Brennan, M. F. McDonald, J. Gramling, and T. D. Jacques, "Collide and conquer: constraints on simplified dark matter models using mono-X collider searches," *Journal of High Energy Physics*, vol. 2016, no. 5, 2016.

[57] P. Gondolo, P. Ko, and Y. Omura, " Light dark matter in leptophobic ," *Physical Review D: Particles, Fields, Gravitation and Cosmology*, vol. 85, no. 3, 2012.

[58] H. An, R. Huo, and L.-T. Wang, "Searching for low mass dark portal at the LHC," *Physics of the Dark Universe*, vol. 2, no. 1, pp. 50–57, 2013.

[59] T. Lin, E. W. Kolb, and L. Wang, "Probing dark matter couplings to top and bottom quarks at the LHC," *Physical Review D: Particles, Fields, Gravitation and Cosmology*, vol. 88, no. 6, 2013.

[60] U. Haisch and E. Re, "Simplified dark matter top-quark interactions at the LHC," *Journal of High Energy Physics*, vol. 2015, no. 6, 2015.

[61] C. Arina, M. Backović, E. Conte et al., "A comprehensive approach to dark matter studies: exploration of simplified top-philic models," *Journal of High Energy Physics*, vol. 2016, no. 11, 2016.

[62] U. Haisch, P. Pani, and G. Polesello, "Determining the CP nature of spin-0 mediators in associated production of dark matter and $t\bar{t}$ pairs," *Journal of High Energy Physics*, vol. 2017, no. 2, 2017.

[63] CMS. collaboration, "Search for dark matter in association with a top quark pair at sqrt(s)=13 TeV," CMS-PAS-EXO-16-005.

[64] CMS collaboration, "Search for dark matter in association with a top quark pair at $\sqrt{s} = 13TeV$ in the dilepton channel," CMS-PAS-EXO-16-028.

[65] N. F. Bell, A. J. Galea, J. B. Dent, T. D. Jacques, L. M. Krauss, and T. J. Weiler, "Searching for dark matter at the LHC with a mono-Z," *Physical Review D: Particles, Fields, Gravitation and Cosmology*, vol. 86, no. 9, 2012.

[66] S. Chang, R. Edezhath, J. Hutchinson, and M. Luty, "Effective WIMPs," *Physical Review D: Particles, Fields, Gravitation and Cosmology*, vol. 89, no. 1, 2014.

[67] H. An, L. Wang, and H. Zhang, "Dark matter with t-channel mediator: a simple step beyond contact interaction," *Physical Review D: Particles, Fields, Gravitation and Cosmology*, vol. 2014, 2014.

[68] Y. Bai and J. Berger, "Fermion portal dark matter," *Journal of High Energy Physics*, vol. 2013, no. 11, 2013.

[69] A. DiFranzo, K. I. Nagao, A. Rajaraman, and T. M. Tait, "Simplified models for dark matter interacting with quarks," *Journal of High Energy Physics*, vol. 2013, no. 11, 2013.

[70] M. Papucci, A. Vichi, and K. M. Zurek, "Monojet versus the rest of the world I: t-channel models," *Journal of High Energy Physics*, vol. 2014, no. 11, 2014.

[71] M. Garny, A. Ibarra, S. Rydbeck, and S. Vogl, "Majorana dark matter with a coloured mediator: collider vs direct and indirect searches," *Journal of High Energy Physics*, vol. 2014, no. 6, 2014.

[72] M. Garny, A. Ibarra, and S. Vogl, "Signatures of Majorana dark matter with t-channel mediators," *International Journal of Modern Physics D*, vol. 24, 2015.

[73] P. Agrawal, B. Batell, D. Hooper, and T. Lin, "Flavored dark matter and the Galactic Center gamma-ray excess," *Physical Review D: Particles, Fields, Gravitation and Cosmology*, vol. 90, no. 6, 2014.

[74] B. Batell, T. Lin, and L. Wang, "Flavored dark matter and R-parity violation," *Journal of High Energy Physics*, vol. 2014, no. 1, 2014.

[75] J. Andrea, B. Fuks, and F. Maltoni, "Monotops at the LHC," *Physical Review D: Particles, Fields, Gravitation and Cosmology*, vol. 84, no. 7, 2011.

[76] J. F. Kamenik and J. Zupan, "Discovering dark matter through flavor violation at the LHC," *Physical Review D: Particles, Fields, Gravitation and Cosmology*, vol. 84, no. 11, 2011.

[77] A. Kumar and S. Tulin, "Top-flavored dark matter and the forward-backward asymmetry," *Physical Review D: Particles, Fields, Gravitation and Cosmology*, vol. 87, no. 9, 2013.

[78] J. Agram, J. Andrea, M. Buttignol, E. Conte, and B. Fuks, "Monotop phenomenology at the Large Hadron Collider," *Physical Review D: Particles, Fields, Gravitation and Cosmology*, vol. 89, no. 1, 2014.

[79] C. Kilic, M. D. Klimek, and J. Yu, "Signatures of top flavored dark matter," *Physical Review D: Particles, Fields, Gravitation and Cosmology*, vol. 91, no. 5, 2015.

[80] M. Garny, J. Heisig, B. Lülf, and S. Vogl, "Coannihilation without chemical equilibrium," *Physical Review D: Particles, Fields, Gravitation and Cosmology*, vol. 96, no. 10, 2017.

[81] M. Garny, J. Heisig, M. Hufnagel, and B. Lülf, "Top-philic dark matter within and beyond the WIMP paradigm," *Physical Review D: Particles, Fields, Gravitation and Cosmology*, vol. 97, no. 7, 2018.

[82] W. Altmannshofer, P. J. Fox, R. Harnik, G. D. Kribs, and N. Raj, "Dark matter signals in dilepton production at hadron colliders," *Physical Review D: Particles, Fields, Gravitation and Cosmology*, vol. 91, no. 11, 2015.

[83] R. M. Capdevilla, A. Delgado, A. Martin, and N. Raj, "Characterizing dark matter at the LHC in Drell-Yan events," *Physical Review D: Particles, Fields, Gravitation and Cosmology*, vol. 97, no. 3, 2018.

[84] S. Kraml, U. Laa, K. Mawatari, and K. Yamashita, "Simplified dark matter models with a spin-2 mediator at the LHC," *The European Physical Journal C*, vol. 77, no. 5, 2017.

[85] C. Englert, M. McCullough, and M. Spannowsky, "S-channel dark matter simplified models and unitarity," *Physics of the Dark Universe*, vol. 14, pp. 48–56, 2016.

[86] R. M. Godbole, G. Mendiratta, and T. M. Tait, "A simplified model for dark matter interacting primarily with gluons," *Journal of High Energy Physics*, vol. 2015, no. 8, 2015.

[87] R. M. Godbole, G. Mendiratta, A. Shivaji, and T. M. Tait, "Mono-jet signatures of gluphilic scalar dark matter," *Physics Letters B*, vol. 772, pp. 93–99, 2017.

[88] O. Ducu, L. Heurtier, and J. Maurer, "LHC signatures of a Z ′ mediator between dark matter and the SU(3) sector," *Journal of High Energy Physics*, vol. 2016, no. 3, 2016.

[89] G. Arcadi, Y. Mambrini, and F. Richard, "Z-portal dark matter," *Journal of Cosmology and Astroparticle Physics*, vol. 2015, 2015.

[90] GAMBIT collaboration, P. Athron et al., "Status of the scalar singlet dark matter model," *The European Physical Journal C*, vol. 77, pp. 568–1705, 2017.

[91] G. Arcadi, C. Gross, O. Lebedev, S. Pokorski, and T. Toma, "Evading direct dark matter detection in Higgs portal models," *Physics Letters B*, vol. 769, pp. 129–133, 2017.

[92] G. Arcadi, P. Ghosh, Y. Mambrini, M. Pierre, and F. S. Queiroz, "$Z^{'}$ portal to Chern-Simons Dark Matter," *Journal of Cosmology and Astroparticle Physics*, vol. 2017, 2017.

[93] X. He and J. Tandean, "New LUX and PandaX-II results illuminating the simplest Higgs-portal dark matter models," *Journal of High Energy Physics*, vol. 2016, no. 12, 2016.

[94] C. Chang, X. He, and J. Tandean, "Two-Higgs-doublet-portal dark-matter models in light of direct search and LHC data," *Journal of High Energy Physics*, vol. 2017, no. 4, 2017.

[95] C. Chang, X. He, and J. Tandean, "Exploring Spin-3/2 Dark Matter with Effective Higgs Couplings," *Physical Review D: Particles, Fields, Gravitation and Cosmology*, vol. 96, no. 7, 2017.

[96] A. Delgado, A. Martin, and N. Raj, "Forbidden dark matter at the weak scale via the top portal," *Physical Review D: Particles, Fields, Gravitation and Cosmology*, vol. 95, no. 3, 2017.

[97] N. F. Bell, Y. Cai, J. B. Dent, R. K. Leane, and T. J. Weiler, "Dark matter at the LHC: Effective field theories and gauge invariance," *Physical Review D: Particles, Fields, Gravitation and Cosmology*, vol. 92, no. 5, 2015.

[98] S. Baek, P. Ko, M. Park, W. Park, and C. Yu, "Beyond the dark matter effective field theory and a simplified model approach at colliders," *Physics Letters B*, vol. 756, pp. 289–294, 2016.

[99] S. Baek, P. Ko, and W. Park, "Search for the Higgs portal to a singlet fermionic dark matter at the LHC," *Journal of High Energy Physics*, vol. 2012, no. 2, 2012.

[100] S. Baek, P. Ko, W.-I. Park, and E. Senaha, "Vacuum structure and stability of a singlet fermion dark matter model with a singlet scalar messenger," *Journal of High Energy Physics*, vol. 2012, no. 11, 2012.

[101] N. F. Bell, G. Busoni, and I. W. Sanderson, "Self-consistent Dark Matter simplified models with an s-channel scalar mediator," *Journal of Cosmology and Astroparticle Physics*, vol. 2017, 2017.

[102] K. Ghorbani, "Fermionic dark matter with pseudo-scalar Yukawa interaction," *Journal of Cosmology and Astroparticle Physics*, vol. 1501, 2015.

[103] K. Ghorbani and L. Khalkhali, "Mono-Higgs signature in a fermionic dark matter model," *Journal of Physics G: Nuclear and Particle Physics*, vol. 44, no. 10, 2017.

[104] S. Baek, P. Ko, and J. Li, "Minimal renormalizable simplified dark matter model with a pseudoscalar mediator," *Physical Review D: Particles, Fields, Gravitation and Cosmology*, vol. 95, no. 7, 2017.

[105] D. Goncalves, P. A. N. Machado, and J. M. No, "Simplified Models for Dark Matter Face their Consistent Completions," *Physical Review D: Covering Particles, Fields, Gravitation, and Cosmology*, vol. 95, 2017.

[106] N. F. Bell, Y. Cai, and R. K. Leane, "Mono-W Dark Matter Signals at the LHC: Simplified Model Analysis," *Journal of Cosmology and Astroparticle Physics*, vol. 1601, 2016.

[107] U. Haisch, F. Kahlhoefer, and T. M. Tait, "On mono- W signatures in spin-1 simplified models," *Physics Letters B*, vol. 760, pp. 207–213, 2016.

[108] F. Kahlhoefer, K. Schmidt-Hoberg, T. Schwetz, and S. Vogl, "Implications of unitarity and gauge invariance for simplified dark matter models," *Journal of High Energy Physics*, vol. 2016, no. 2, 2016.

[109] N. F. Bell, Y. Cai, and R. K. Leane, "Impact of mass generation for spin-1 mediator simplified models," *Journal of Cosmology and Astroparticle Physics*, vol. 2017, 2017.

[110] T. Jacques, A. Katz, E. Morgante, D. Racco, M. Rameez, and A. Riotto, "Complementarity of DM searches in a consistent simplified model: the case of $Z^{'}$," *Journal of High Energy Physics*, vol. 2016, no. 10, 2016.

[111] Y. Cui and F. D'Eramo, "Surprises from complete vector portal theories: New insights into the dark sector and its interplay with Higgs physics," *Physical Review D: Particles, Fields, Gravitation and Cosmology*, vol. 96, no. 9, 2017.

[112] S. Weinberg, *The quantum theory of fields*, vol. 2 of *Modern applications*, Cambridge University Press, 2013.

[113] M. Duerr, P. Fileviez Pérez, and J. Smirnov, "Gamma lines from Majorana dark matter," *Physical Review D: Particles, Fields, Gravitation and Cosmology*, vol. 93, no. 2, 2016.

[114] A. Ismail, A. Katz, and D. Racco, "On dark matter interactions with the Standard Model through an anomalous Z*ı*," *Journal of High Energy Physics*, vol. 2017, no. 165, 2017.

[115] J. Ellis, M. Fairbairn, and P. Tunney, "Anomaly-free dark matter models are not so simple," *Journal of High Energy Physics*, no. 8, 2017.

[116] M. Duerr, P. Fileviez Perez, and M. B. Wise, "Gauge Theory for Baryon and Lepton Numbers with Leptoquarks," *Physical Review Letters*, vol. 11, 2013.

[117] M. Duerr and P. Fileviez Pérez, "Baryonic dark matter," *Physics Letters B*, vol. 732, pp. 101–104, 2014.

[118] P. Fileviez Pérez, S. Ohmer, and H. H. Patel, "Minimal theory for lepto-baryons," *Physics Letters B*, vol. 735, pp. 283–287, 2014.

[119] S. Ohmer and H. H. Patel, "Leptobaryons as Majorana dark matter," *Physical Review D: Particles, Fields, Gravitation and Cosmology*, vol. 92, no. 5, 2015.

[120] A. Alves, G. Arcadi, Y. Mambrini, S. Profumo, and F. S. Queiroz, "Augury of darkness: the low-mass dark $Z^{'}$ portal," *Journal of High Energy Physics*, vol. 2017, no. 4, 2017.

[121] A. Alves, S. Profumo, and F. S. Queiroz, "The dark Z' portal: direct, indirect and collider searches," *Journal of High Energy Physics*, vol. 2014, 2014.

[122] G. Arcadi, Y. Mambrini, M. H. Tytgat, and B. Zaldıvar, "Invisible $Z^{'}$ and dark matter: LHC vs LUX constraints," *Journal of High Energy Physics*, vol. 2014, no. 3, 2014.

[123] A. Alves, A. Berlin, S. Profumo, and F. S. Queiroz, "Dark Matter Complementarity and the $Z^{'}$ Portal," *Physical Review D*, vol. 92, 2015.

[124] K. Ghorbani and H. Ghorbani, "Two-portal dark matter," *Physical Review D: Particles, Fields, Gravitation and Cosmology*, vol. 91, no. 12, 2015.

[125] A. Alves, A. Berlin, S. Profumo, and F. S. Queiroz, "Dirac-fermionic dark matter in $U(1)_X$ models," *Journal of High Energy Physics*, vol. 76, pp. 1–33, 2015.

[126] G. Arcadi, M. D. Campos, M. Lindner, A. Masiero, and F. S. Queiroz, "Dark sequential Z' portal: Collider and direct detection experiments," *Physical Review D*, 2018.

[127] ATLAS collaboration, M. Aaboud et al., "Search for new high-mass phenomena in the dilepton final state using 36 fb^{-1} of proton-proton collision data at $\sqrt{s} = 13TeV$ with the ATLAS detector," *Journal of High Energy Physics*, vol. 10, 2017.

[128] ATLAS collaboration, M. Aaboud et al., "Search for new phenomena in dijet events using 37 fb^{-1} of pp collision data collected at $\sqrt{s} = 13TeV$ with the ATLAS detector," *Physical Review D*, vol. 96, 2017.

[129] ATLAS collaboration, "Search for dark matter and other new phenomena in events with an energetic jet and large missing transverse momentum using the ATLAS detector," ATLAS-CONF-2017-060.

[130] N. F. Bell, Y. Cai, and R. K. Leane, "Dark Forces in the Sky: Signals from Z' and the Dark Higgs," *Journal of Cosmology and Astroparticle Physics*, 2016.

[131] M. Duerr, A. Grohsjean, F. Kahlhoefer, B. Penning, K. Schmidt-Hoberg, and C. Schwanenberger, "Hunting the dark Higgs," *Journal of High Energy Physics*, vol. 2017, no. 4, 2017.

[132] J. Fan, M. Reece, and L. Wang, "Non-relativistic effective theory of dark matter direct detection," *Journal of Cosmology and Astroparticle Physics*, vol. 2010, no. 11, 2010.

[133] A. L. Fitzpatrick, W. Haxton, E. Katz, N. Lubbers, and Y. Xu, "The effective field theory of dark matter direct detection," *Journal of Cosmology and Astroparticle Physics*, vol. 2013, 2013.

[134] G. Branco, P. Ferreira, L. Lavoura, M. Rebelo, M. Sher, and J. P. Silva, "Theory and phenomenology of two-Higgs-doublet models," *Physics Reports*, vol. 516, no. 1-2, pp. 1–102, 2012.

[135] M. Bauer, U. Haisch, and F. Kahlhoefer, "Simplified dark matter models with two Higgs doublets: I. Pseudoscalar mediators," *Journal of High Energy Physics*, vol. 2017, no. 5, 2017.

[136] J. M. No, "Looking through the pseudoscalar portal into dark matter: Novel mono-Higgs and mono-Z signatures at the LHC," *Physical Review D*, vol. 93, no. 3, 2016.

[137] A. Angelescu and G. Arcadi, "Dark matter phenomenology of SM and enlarged Higgs sectors extended with vector-like leptons," *The European Physical Journal C*, 2017.

[138] S. Banerjee, D. Barducci, G. Bélanger, B. Fuks, A. Goudelis, and B. Zaldivar, "Cornering pseudoscalar-mediated dark matter with the LHC and cosmology," *Journal of High Energy Physics*, vol. 2017, no. 7, 2017.

[139] P. Tunney, J. M. No, and M. Fairbairn, "Probing the pseudoscalar portal to dark matter via $\bar{b}bZ(\longrightarrow \ell\ell) + \not{E}T$: From the LHC to the Galactic Center excess," *Physical Review D: Particles, Fields, Gravitation and Cosmology*, vol. 96, no. 9, 2017.

[140] N. F. Bell, G. Busoni, and I. W. Sanderson, "Two Higgs doublet dark matter portal," *Journal of Cosmology and Astroparticle Physics*, vol. 2018, no. 01, 2018.

[141] G. Arcadi, M. Lindner, F. S. Queiroz, W. Rodejohann, and S. Vogl, "Pseudoscalar mediators: a WIMP model at the neutrino floor," *Journal of Cosmology and Astroparticle Physics*, vol. 1803, no. 42, 2018.

[142] P. Pani and G. Polesello, "Dark matter production in association with a single top-quark at the LHC in a two-Higgs-doublet model with a pseudoscalar mediator," *High Energy Physics - Phenomenology*, vol. 21, pp. 8–15, 2018.

[143] S. Ipek, D. McKeen, and A. E. Nelson, "A Renormalizable Model for the Galactic Center Gamma Ray Excess from Dark Matter Annihilation," *Physical Review D: Particles, Fields, Gravitation and Cosmology*, vol. 90, 2014.

[144] L. L. Honorez, E. Nezri, J. F. Oliver, and M. H. G. Tytgat, "The inert doublet model: an archetype for dark matter," *Journal of Cosmology and Astroparticle Physics*, vol. 2007, 2007.

20

Thermodynamics of Modified Cosmic Chaplygin Gas

M. Sharif and Sara Ashraf

Department of Mathematics, University of the Punjab, Quaid-e-Azam Campus, Lahore 54590, Pakistan

Correspondence should be addressed to M. Sharif; msharif.math@pu.edu.pk

Academic Editor: Edward Sarkisyan-Grinbaum

We examine the thermodynamic features of an exotic fluid known as modified cosmic Chaplygin gas in the context of homogeneous isotropic universe model. For this purpose, the behavior of physical parameters is discussed that help to analyze nature of the universe. Using specific heat formalism, the validity of third law of thermodynamics is checked. Furthermore, with the help of thermodynamic entities, the thermal equation of state is also discussed. The thermodynamic stability is explored by means of adiabatic, specific heat and isothermal conditions from classical thermodynamics. It is concluded that the considered fluid configuration is thermodynamically stable and expands adiabatically for an appropriate choice of parameters.

1. Introduction

The discovery of accelerated expansion of the universe has unambiguously been proved by a diverse set of high-precision observational data accumulated from various astronomical sources [1–3]. Dark energy (DE) is considered as the root cause behind this tremendous change in cosmic history. It possesses negatively large pressure which violates the strong energy condition ($p+3\rho < 0$, where p and ρ are pressure and energy density, respectively) but its complete characteristics are still not known. Planck's observational data reveals that about 68.3% of our universe is filled with this mysterious form of energy while the other cosmic budget includes 4.9% ordinary matter and 26.8% dark matter (DM) [4, 5]. To explore the perplexing nature of DE, there began a search for different candidates that can play their role as an alternative for DE. The simplest candidate is the cosmological constant while other favorable approaches include quintessence, k-essence, and Chaplygin gas (CG) known as DE matter models [6–8].

Chaplygin gas is an intriguing model presented by a Russian physicist Chaplygin as a convenient soluble model to study the lifting force on the wing of an aeroplane in aerodynamics. It efficiently describes the cosmic expansion and elegantly discusses DM and DE in a unified form. The distinct feature of this model is its positive and bounded squared speed of sound (v_s^2) leading to stable results as compared to other fluids with negative pressure. Chaplygin gas acts as an alternative for dust dominated era with small value of the scale factor and tends to cosmic expansion for its large value while primordial universe cannot be discussed in this scenario [9–12]. Despite the fact that it does not meet the strong energy condition, it successfully shows consistency with the observational results accumulated from various cosmic probes such as Hubble space telescope, Wilkinson microwave anisotropy probe, and cosmic background explorer.

To discuss the cosmic history as well as get more accuracy with observational data, several modifications of CG have been presented which are obtained by introducing new parameters in its equation of state (EoS). Bento et al. [13] established the generalization of this model named generalized Chaplygin gas (GCG) which interprets the same evolutionary phases of the universe as CG. Benaoum [14] introduced the modified Chaplygin gas (MCG) which illustrates the radiation dominated era. González-Díaz [15] proposed the generalization of cosmic CG models known as generalized cosmic Chaplygin gas (GCCG) in such a way that it avoids big-rip (singularity at a finite time) which was previously presented in the DE models representing phantom era. This generalization provides stable and physical behavior models even when the vacuum matter configuration fulfills the phantom energy condition ($p + \rho < 0, \rho > 0, \omega < -1$, where ω is the EoS parameter). The other proposed CG models include variable Chaplygin gas (VCG), variable modified Chaplygin gas (VMCG), new variable MCG, extended CG, and modified cosmic Chaplygin gas (MCCG) [16–19].

Chaplygin gas models have stimulated many researchers to investigate their thermal stability. Santos et al. [20, 21] explored thermal stability of GCG as well as MCG and deduced that these fluid models verify the third law of thermodynamics along with the adiabatic expansion. Myung [22] proved the third law of thermodynamics for CG model and illustrated that it can represent a unified picture of DM and DE without any phase transition. Kahya and Pourhassan [23] analyzed extended CG model cosmologically as well as thermodynamically and found stable results without any phase transition against density perturbations. They also concluded that all laws of thermodynamics are satisfied for this exotic fluid which came out to be thermodynamically stable at all times. Panigrahi [24] studied thermodynamic behavior of VCG and observed that it is thermodynamically stable throughout the evolution. He also discussed thermal EoS which is an explicit function of temperature only and checked the validity of the third law. Panigrahi and Chatterjee [25] found that current accelerated expansion of the universe can be explained using VMCG model. Sharif and Sarwar [26] explored how GCCG can explain accelerated expansion of the universe by interpreting different physical parameters and showed that the fluid is adiabatically stable.

Here, we investigate thermodynamic stability of MCCG model in the background of isotropic and homogeneous universe model. In Section 2, we discuss the behavior of physical parameters such as pressure and EoS as well as deceleration parameters and analyze the stability using speed of sound. Section 3 deals with the thermodynamic stability of MCCG. The results are summarized in the last section.

2. Physical Parameters for MCCG

In this section, we discuss the behavior of MCCG in the background of FRW universe model for different physical parameters and examine its stability through squared speed of sound. The line element for FRW universe model is given by

$$ds^2 = dt^2 - a^2(t)\left(dr^2 + r^2 d\theta^2 + r^2 \sin^2\theta d\phi^2\right), \quad (1)$$

where $a(t)$ is the scale factor. The EoS for MCCG is defined as

$$P = A\rho - \rho^{-\alpha}\left[\left(\rho^{\alpha+1} - C\right)^{-\gamma} + C\right], \quad 0 < \alpha \leq 1, \; -b < \gamma < 0, \; b \neq 1, \quad (2)$$

where $C = Z/(\gamma + 1) - 1$ (Z is an arbitrary constant) and A is a positive constant. This EoS reduces to GCCG as $A \longrightarrow 0$ [26] while GCG is recovered in the limit $A \longrightarrow 0$ along with $\gamma \longrightarrow 0$ [13]. The energy density of the fluid configuration is given by

$$\rho = \frac{U}{V}, \quad (3)$$

where U and V represent the internal energy and volume, respectively. Classical thermodynamics provides a useful relationship among the quantities U, V, and P in the form

$$\left(\frac{dU}{dV}\right) = -P. \quad (4)$$

Using (2) in the above expression, we obtain

$$\frac{dU}{dV} + \frac{AU}{V} = \frac{V^{\alpha}}{U^{\alpha}}\left[C + \left(\frac{U^{\alpha+1}}{V^{\alpha+1}} - C\right)^{-\gamma}\right], \quad (5)$$

which is a nonlinear ordinary differential equation. Its solution is given by

$$U \approx V\left[\frac{(d/V)^{(\alpha+1)(A+1)}(A+1) + C + (-C)^{-\gamma}}{A + 1 + \gamma(-C)^{-\gamma-1}}\right]^{1/(\alpha+1)}, \quad (6)$$

where we have used the binomial expansion up to first order and d is an integration constant which is either universal constant or a function of entropy (S). The above equation can also be written as

$$U = V\left[\frac{(\varepsilon/V)^M + C + (-C)^{-\gamma}}{A + 1 + \gamma(-C)^{-\gamma-1}}\right]^{1/(\alpha+1)}, \quad (7)$$

where $\varepsilon = d(A+1)^{1/M}$, $M = (\alpha+1)(A+1)$, and $A+\gamma(-C)^{-\gamma-1} \neq -1$. It is clearly observed that internal energy of the fluid can only be discussed when γ is a whole number between the above-mentioned range for positive values of C. Using (3) and (7), the energy density of MCCG becomes

$$\rho = \left[\frac{(\varepsilon/V)^M + C + (-C)^{-\gamma}}{A + 1 + \gamma(-C)^{-\gamma-1}}\right]^{1/(\alpha+1)}. \quad (8)$$

In the following, we use this equation to discuss different physical parameters.

2.1. Pressure. The pressure of MCCG in terms of V can be obtained using (2) and (8) as

$$P = A\left(\frac{(\varepsilon/V)^M + C + (-C)^{-\gamma}}{A + 1 + \gamma(-C)^{-\gamma-1}}\right)^{1/(\alpha+1)} - \left(\frac{(\varepsilon/V)^M + C + (-C)^{-\gamma}}{A + 1 + \gamma(-C)^{-\gamma-1}}\right)^{-\alpha/(\alpha+1)} \times\left[C + \left(\frac{(\varepsilon/V)^M + C + (-C)^{-\gamma}}{A + 1 + \gamma(-C)^{-\gamma-1}} - C\right)^{-\gamma}\right]. \quad (9)$$

The graphical analysis of this equation is shown in Figure 1 for different values of Z with $A = 2$. The positive and negative behavior of pressure correspond to decelerated and accelerated eras of the universe, respectively. For $Z = 0.01$, it is observed that the accelerating universe at small volume tends to dust dominated universe at large volume. We note that the decelerating universe tends to accelerate as volume increases for $Z = -5, -7$ whereas accelerated phase is obtained only for $Z = -8$. The same behavior of pressure is observed for different values of the parameter A.

2.2. EoS Parameter. Here we discuss the effective EoS parameter of MCCG. Using (8) and (9), we have

$$\omega = \frac{P}{\rho} = A - \frac{C + \left(\left(\left((\varepsilon/V)^M + C + (-C)^{-\gamma}\right)/\left(A + 1 + \gamma(-C)^{-\gamma-1}\right)\right) - C\right)^{-\gamma}}{\left(\left((\varepsilon/V)^M + C + (-C)^{-\gamma}\right)/\left(A + 1 + \gamma(-C)^{-\gamma-1}\right)\right)}. \quad (10)$$

We study the following two extremal cases for volume to analyze the behavior of above equation.

(i) For small volume $V \ll \varepsilon$, the above equation reduces to

$$P \approx A\rho, \quad (11)$$

which is a barotropic EoS. In this case, ω will depend entirely on the value of A.

(ii) For large volume $V \gg \varepsilon$, (10) takes the form

$$\omega \approx \frac{P}{\rho} = A - \frac{C + \left(\left(\left(C + (-C)^{-\gamma}\right)/\left(A + 1 + \gamma(-C)^{-\gamma-1}\right)\right) - C\right)^{-\gamma}}{\left(\left(C + (-C)^{-\gamma}\right)/\left(A + 1 + \gamma(-C)^{-\gamma-1}\right)\right)}. \quad (12)$$

For $\omega = 0$, let volume be denoted by V_c, which is given by

$$V_c = \varepsilon \left[\frac{A + \gamma(-C)^{-\gamma-1}}{C + (-C)^{-\gamma}}\right]^{1/M}. \quad (13)$$

The EoS parameter discusses both accelerated and decelerated phases of the universe and successfully describes the phase transitions (dubbed as flip) at a critical value V_c between these cosmic phases. The proper flip occurs when $(\varepsilon/V)^M < 1$ while the inequality $C + (-C)^{-\gamma} \neq 0$ leads to real flip.

Figure 2 shows the behavior of ω for different values of Z. It is found that the value of V_c decreases as Z becomes negatively large. The negative values of Z show the decelerated cosmic phase at small volume undergoing acceleration at large volume while $Z = 1$ demonstrates only the decelerating phase. We observe that, at large volume, the considered values of $Z = 1, 0, -1.15$, and -2 correspond to stiff matter, dust dominated, ΛCDM, and phantom eras, respectively. Thus, the EoS parameter can interpret different evolutionary phases of the universe.

2.3. Deceleration Parameter. The deceleration parameter is given by

$$q = \frac{1}{2} + \frac{3P}{2\rho}. \quad (14)$$

Using (10), this parameter for MCCG takes the form

$$q = \frac{1}{2} + \frac{3}{2}\left[A - \frac{C + \left(\left(\left((\varepsilon/V)^M + C + (-C)^{-\gamma}\right)/\left(A + 1 + \gamma(-C)^{-\gamma-1}\right)\right) - C\right)^{-\gamma}}{\left(\left((\varepsilon/V)^M + C + (-C)^{-\gamma}\right)/\left(A + 1 + \gamma(-C)^{-\gamma-1}\right)\right)}\right]. \quad (15)$$

For small volume, it reduces to

$$q \approx \frac{1}{2} + \frac{3A}{2}, \quad (16)$$

which implies that the universe undergoes deceleration at its early stage since $A > 0$ while for large volume, (15) becomes

$$q \approx \frac{1}{2} + \frac{3}{2}\left[A - \frac{C + \left(\left(\left(C + (-C)^{-\gamma}\right)/\left(A + 1 + \gamma(-C)^{-\gamma-1}\right)\right) - C\right)^{-\gamma}}{\left(\left(C + (-C)^{-\gamma}\right)/\left(A + 1 + \gamma(-C)^{-\gamma-1}\right)\right)}\right]. \quad (17)$$

In this case, the flip occurs when deceleration parameter vanishes and the corresponding flip volume (V_f) is given by

$$V_f = \varepsilon \left[\frac{3A + 1 + 3\gamma(-C)^{-\gamma-1}}{C + (-C)^{-\gamma}}\right]^{1/M}, \quad (18)$$

provided that $C + (-C)^{-\gamma} \neq 0$ and the inequality $C + (-C)^{-\gamma} < 3(A + \gamma(-C)^{-\gamma-1}) + 1$ leads to proper flip. Figure 3 shows the evolution of deceleration parameter against volume for different values of Z. At small volume, the universe undergoes deceleration while accelerating behavior is observed at large volume for considered negative values of Z. The flip occurs at $V \approx 1.6$ and 3 for $Z = -2$ and -1.3, respectively. For

$Z = 0$, the deceleration parameter switches from acceleration to deceleration at $V_f \approx 2.2$ while no flip is observed for $Z = 2$.

2.4. Speed of Sound. Here we analyze the stability of MCCG using speed of sound as

$$v_s^2 = \left(\frac{\partial P}{\partial \rho}\right)_S = A + \frac{\gamma(\alpha+1)}{\left(\left(\left((\varepsilon/V)^M + C + (-C)^{-\gamma-1}\right)/\left(A+1+\gamma(-C)^{-\gamma-1}\right)\right) - C\right)^{\gamma+1}} + \frac{\alpha}{\left(\left((\varepsilon/V)^M + C + (-C)^{-\gamma-1}\right)/\left(A+1+\gamma(-C)^{-\gamma-1}\right)\right)} \times \left[C + \left(\frac{(\varepsilon/V)^M + C + (-C)^{-\gamma-1}}{A+1+\gamma(-C)^{-\gamma-1}} - C\right)^{-\gamma}\right], \quad (19)$$

whose feasible range is $0 < v_s^2 < 1$. This equation reduces to $v_s^2 = A$ at early universe while, for $V \gg \varepsilon$, we have

$$v_s^2 = A + \frac{\gamma(\alpha+1)}{\left(\left(\left(C + (-C)^{-\gamma}\right)/\left(A+1+\gamma(-C)^{-\gamma-1}\right)\right) - C\right)^{\gamma+1}} + \frac{\alpha}{\left(\left(C + (-C)^{-\gamma}\right)/\left(A+1+\gamma(-C)^{-\gamma-1}\right)\right)} \times \left[C + \left(\frac{C + (-C)^{-\gamma}}{A+1+\gamma(-C)^{-\gamma}} - C\right)^{-\gamma}\right]. \quad (20)$$

Figure 4 shows the behavior of squared speed of sound against the positive parameter A for different values of Z. It is observed that the viable ranges for A are $1.8 < A < 2.8$, $3.2 < A < 3.7$, and $5 < A < 5.6$ corresponding to $Z = -0.01$, 4, and 7, respectively. Thus, the stable results are found for the considered values of Z in a particular range of A.

3. Thermodynamic Stability

In this section, we discuss thermodynamic stability of MCCG during its evolution. The stability conditions are given by the following [27]:

(i) The pressure reduces for both adiabatic as well as isothermal expansions as

$$\left(\frac{\partial P}{\partial V}\right)_S < 0, \qquad \left(\frac{\partial P}{\partial V}\right)_T < 0, \quad (21)$$

where T represents temperature.

(ii) Specific heat at constant volume (c_V) is positive.

Differentiation of (9) with respect to volume yields

$$\left(\frac{\partial P}{\partial V}\right)_S = \frac{(A+1)\varepsilon^M}{V^{M+1}A+1+\gamma(-C)^{-\gamma-1}} \left(\frac{\varepsilon^M V^{-M} + C + (-C)^{-\gamma}}{A+1+\gamma(-C)^{-\gamma-1}}\right)^{-\alpha/(\alpha+1)} \times \left[-A(\alpha+1) + \alpha P \left(\frac{\varepsilon^M V^{-M} + C + (-C)^{-\gamma}}{A+1+\gamma(-C)^{-\gamma-1}}\right)^{-1/(\alpha+1)} - \gamma(\alpha+1)\left(\frac{\varepsilon^M V^{-M} + C + (-C)^{-\gamma}}{A+1+\gamma(-C)^{-\gamma-1}} - C\right)^{-\gamma-1}\right]. \quad (22)$$

When volume is very small, the above equation reduces to zero while, for large volume, we have the following expression:

$$\left(\frac{\partial P}{\partial V}\right)_S = \frac{(A+1)\varepsilon^M}{V^{M+1}\left(A+1+\gamma(-C)^{-\gamma-1}\right)} \left(\frac{C + (-C)^{-\gamma}}{A+1+\gamma(-C)^{-\gamma-1}}\right)^{-\alpha/(\alpha+1)} \times \left[\alpha P \left(\frac{C + (-C)^{-\gamma}}{A+1+\gamma(-C)^{-\gamma-1}}\right)^{-1/(\alpha+1)} - A(\alpha+1) - \gamma(\alpha+1) \times \left(\frac{C + (-C)^{-\gamma}}{A+1+\gamma(-C)^{-\gamma-1}} - C\right)^{-\gamma-1}\right]. \quad (23)$$

Figure 5 shows that the adiabatic condition is fulfilled for all the considered values of Z.

To investigate the positivity of specific heat at constant volume, we consider specific heat in terms of temperature and entropy as

$$c_v = T\left(\frac{\partial S}{\partial T}\right)_V, \quad (24)$$

where the temperature of MCCG is obtained from the following relation:

$$T = \frac{\partial U}{\partial S} = \left(\frac{\partial U}{\partial d}\right)\left(\frac{\partial d}{\partial S}\right). \quad (25)$$

Differentiating (6) with respect to d, we have

$$\frac{\partial U}{\partial d} = \frac{d^{M-1}(A+1)^2 U}{d^M(A+1) + CV^M + (-C)^{-\gamma}V^M}. \quad (26)$$

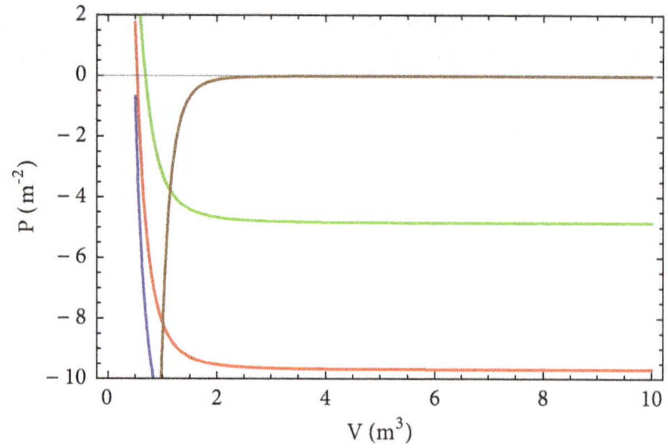

FIGURE 1: Plots of P versus V for $A = 2, \gamma = -2, \alpha = 0.1, d = 1$ with $Z = 0.01$ (brown), -5 (green), -7 (red), and -8 (blue).

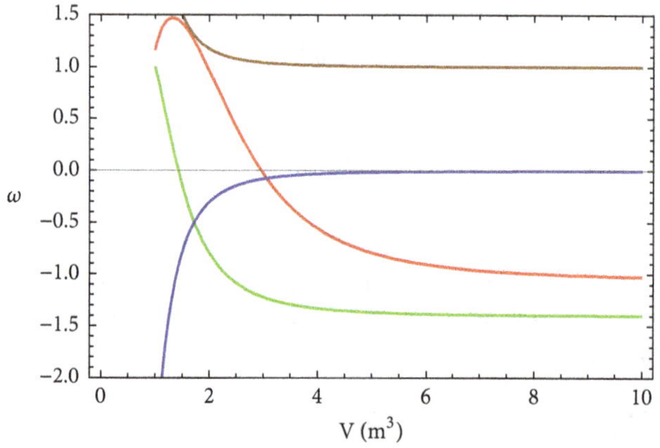

FIGURE 2: Plots of ω versus V for $\gamma = -2, A = 2, \alpha = 0.1, d = 1$ with $Z = -2$ (green), -1.15 (red), 0 (blue), and 1 (brown).

Substituting this relation in (25), the expression of T becomes

$$T = \frac{d^{M-1}(A+1)^2 U}{d^M (A+1) + CV^M + (-C)^{-\gamma} V^M} \left(\frac{\partial d}{\partial S}\right). \tag{27}$$

When d is a universal constant ($\partial d/\partial S = 0$), the temperature vanishes while it varies for CG expansion, so we consider $\partial d/\partial S \neq 0$. Here, we assume the case $\partial d/\partial S > 0$ to have a positive temperature which is cooled down through adiabatic expansion. Using the concept of dimensional analysis, (6) gives

$$[d]^{A+1} = [U, V]^A. \tag{28}$$

Using the relation $[U] = [T][S]$, the above equation becomes

$$[d] = [T]^{1/(A+1)} [S]^{1/(A+1)} [V]^{A/(A+1)}. \tag{29}$$

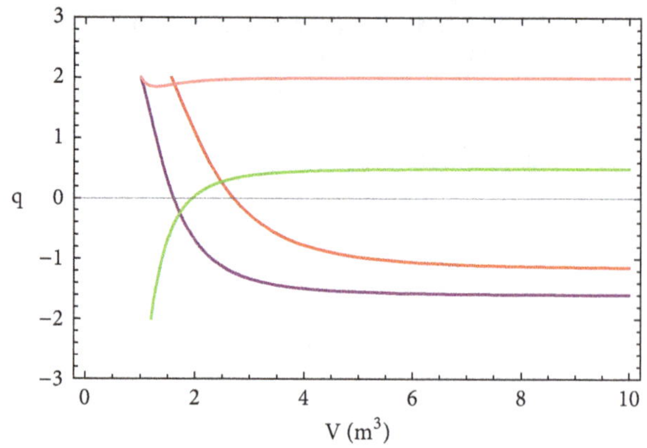

FIGURE 3: Plots of q versus V for $\gamma = -2, A = 2, \alpha = 0.1, d = 1$ with $Z = -2$ (purple), -1.3 (red), 0 (green), and 2 (pink).

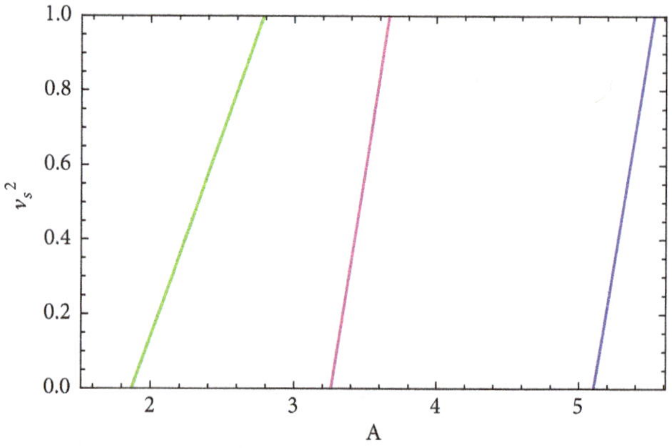

FIGURE 4: Plots of v_s^2 versus A for $\gamma = -2, \alpha = 0.1, d = 1$ with $Z = -0.01$ (green), 4 (magenta), and 7 (blue).

Taking $d = d(S)$, it follows that

$$d = \left(\tau \nu^A S\right)^{1/(A+1)}, \tag{30}$$

where τ and ν are constants having the dimensions of temperature and volume, respectively. Differentiating (30) with respect to S, we obtain

$$\frac{\partial d}{\partial S} = \frac{1}{A+1}\left(\frac{\tau \nu^A}{S^A}\right)^{1/(A+1)}. \tag{31}$$

Substituting this value in (27), the temperature of MCCG takes the form

$$T = \frac{d^{M-1}(A+1) U B^{1/(A+1)} S^{-A/(A+1)}}{d^M (A+1) + CV^M + (-C)^{-\gamma} V^M}, \tag{32}$$

where $B = \tau \nu^A$. Using (6) and (30) in the above equation, we have

$$T = \frac{S^\alpha B^{\alpha+1} (A+1) \left[(BS)^{\alpha+1} (A+1) + CV^M + (-C)^{-\gamma} V^M\right]^{-\alpha/(\alpha+1)}}{V^A \left(A + 1 + \gamma (-C)^{-\gamma-1}\right)^{1/(\alpha+1)}}. \tag{33}$$

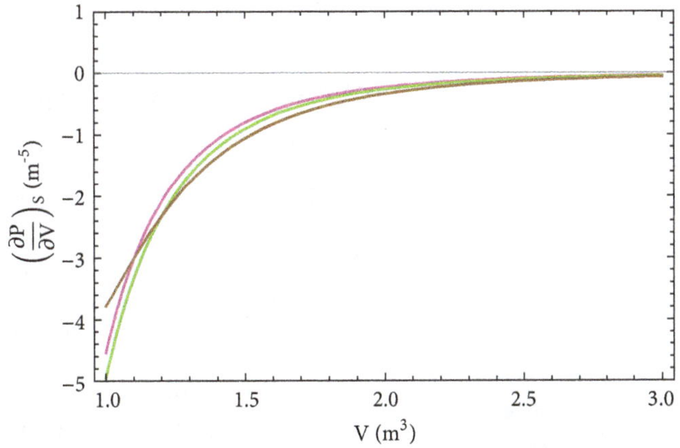

FIGURE 5: Plots of $(\partial P/\partial V)_S$ versus V for $\gamma = -2, A = 2, \alpha = 0.1, d = 1$ with $Z = -15$ (magenta), -7 (green), and -2 (brown).

For a positive definite entropy, we assume $0 < T < \tau$ and $0 < V < \nu$. It is worth mentioning here that when $T = 0$, the entropy vanishes which indicates that the considered fluid obeys third law of thermodynamics (if the temperature of a physical system approaches zero then entropy becomes zero). Differentiating (33) with respect to S, we obtain

$$\begin{aligned}\frac{\partial T}{\partial S} = \alpha Z S^{\alpha}\Big[&S^{-1}\left\{(A+1)(BS)^{\alpha+1}+CV^{M}\right.\\ &\left.+(-C)^{-\gamma}V^{M}\right\}^{-\alpha/(\alpha+1)} - B^{\alpha+1}S^{\alpha}(A+1)\\ &\cdot\left\{(A+1)(BS)^{\alpha+1}+CV^{M}\right.\\ &\left.+(-C)^{-\gamma}V^{M}\right\}^{(-2\alpha-1)/(\alpha+1)}\Big],\end{aligned} \quad (34)$$

where $Z = B^{\alpha+1}(A+1)/V^{A}(A+1+\gamma(-C)^{-\gamma-1})^{1/(\alpha+1)}$. Inserting (33) and (34) in (24), it follows that

$$c_V = \frac{S}{\alpha}\left[1 - \frac{(BS)^{\alpha+1}(A+1)}{\left[(BS)^{\alpha+1}(A+1)+CV^{M}+(-C)^{-\gamma}V^{M}\right]}\right]^{-1}. \quad (35)$$

For c_V to be real, $(BS)^{\alpha+1}(A+1) \neq (BS)^{\alpha+1}(A+1) + CV^{M} + (-C)^{-\gamma}V^{M}$ and to be positive, the inequality $(BS)^{\alpha+1}(A+1) < (BS)^{\alpha+1}(A+1) + CV^{M} + (-C)^{-\gamma}V^{M}$ must hold. The graphical analysis of (35) is shown in Figure 6 for different values of Z with $A = 2$. We observe that the positivity of specific heat is obtained for $Z = 6$ in the range $V > 1.5$ while MCCG is thermally stable for both values of $Z = 0$ and -0.5 throughout the evolution. It is also noted that when temperature is zero, thermal capacity vanishes which also assures the validity of third law of thermodynamics.

Finally, we analyze the behavior of considered model through isothermal condition. For this purpose, we assume $P = P(V,T)$ and by solving (9) and (30), we have

$$\begin{aligned}P = A&\left(\frac{(BS)^{\alpha+1}(A+1)V^{-M}+C+(-C)^{-\gamma}}{A+1+\gamma(-C)^{-\gamma-1}}\right)^{1/(\alpha+1)}\\ &-\left(\frac{(BS)^{\alpha+1}(A+1)V^{-M}+C+(-C)^{-\gamma}}{A+1+\gamma(-C)^{-\gamma-1}}\right)^{-\alpha/(\alpha+1)}\\ &\times\Bigg[C\\ &+\left(\frac{(BS)^{\alpha+1}(A+1)V^{-M}+C+(-C)^{-\gamma}}{A+1+\gamma(-C)^{-\gamma-1}}-C\right)^{-\gamma}\Bigg].\end{aligned} \quad (36)$$

The corresponding EoS parameter takes the form

$$\omega = \frac{P}{\rho} = A - \frac{C+\left(\left((BS)^{\alpha+1}(A+1)V^{-M}+C+(-C)^{-\gamma}\right)/\left(A+1+\gamma(-C)^{-\gamma-1}\right)-C\right)^{-\gamma}}{\left(\left((BS)^{\alpha+1}(A+1)V^{-M}+C+(-C)^{-\gamma}\right)/\left(A+1+\gamma(-C)^{-\gamma-1}\right)\right)}. \quad (37)$$

To check the isothermal condition, we should have $\rho = \rho(T)$ and $P = P(T)$. In our case, it is difficult to have a thermal EoS for MCCG as a function of temperature only since (33) is a complicated equation such that the explicit expression for S in terms of T cannot be extracted. For this reason, we are unable to analyze the isothermal condition in this scenario.

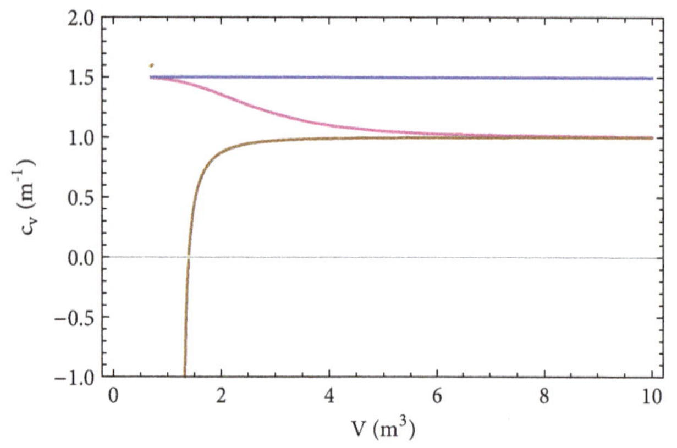

FIGURE 6: Plots of c_V versus V for $\gamma = -2, A = 2, \alpha = 0.1, \tau = 2.73, S = 1, \nu = 1, d = 1$ with $Z = -0.5$ (magenta), 0 (blue), and 6 (brown).

4. Conclusions

In this paper, we have analyzed thermodynamic stability of MCCG within the framework of FRW universe model. We have examined the expanding evolution of the universe through different physical parameters like pressure, effective EoS, and deceleration parameters as well as speed of sound. The results of these parameters can be summarized as follows:

(i) The consistent behavior of pressure with the evolutionary picture of the universe is obtained for the considered values of Z whereas inconsistent evolution is observed for its positive values (Figure 1).

(ii) The EoS parameter for MCCG depicts that decelerated and accelerated phases of our universe can be discussed for different values of parameter Z (Figure 2). We have also calculated the critical value at $\omega = 0$ and found that its value increases as Z increases from its negative values to zero.

(iii) The evolution of deceleration parameter against volume gives the decelerated universe when $V < V_f$ for negative values of Z while accelerating behavior is observed when $V > V_f$. For positive values of Z, we have found only deceleration while at $Z = 0$, acceleration occurs before the flip (Figure 3).

(iv) We have analyzed the stability of MCCG through speed of sound and obtained stable regions at large volume for the considered values of Z (Figure 4). For VMCG model, the squared speed of sound could be positive or negative [25] whereas for GCCG model, the stable regions do not exist in late universe [26].

Finally, we have investigated thermodynamic stability of considered fluid configuration using adiabatic, isothermal, and specific heat conditions. We have found the validity of adiabatic as well as positivity of specific heat for the considered values of Z (Figures 5 and 6). It is worth mentioning here that third law of thermodynamics is obeyed for MCCG. We conclude that MCCG expands adiabatically and the expansion is thermodynamically stable for a suitable choice of the parameters.

Conflicts of Interest

The authors declare that they have no conflicts of interest.

References

[1] A. G. Riess, A. V. Filippenko, and P. Challis, "Observational evidence from supernovae for an accelerating universe and a cosmological constant," *The Astronomical Journal*, vol. 116, no. 3, pp. 1009–1038, 1998.

[2] S. Perlmutter, G. Aldering, and G. Goldhaber, "Measurements of Ω and Λ from 42 high-redshift supernovae," *The Astrophysical Journal*, vol. 517, pp. 565–586, 1999.

[3] A. G. Riess, L.-G. Strolger, J. Tonry et al., "Type la supernova discoveries at $z > 1$ from the Hubble Space Telescope: evidence for past deceleration and constraints on dark energy evolution," *The Astrophysical Journal*, vol. 607, no. 2, p. 665, 2004.

[4] U. Alam, V. Sahni, A. A. Starobinsky, and J. Cosmol, "The case for dynamical dark energy revisited," *Journal of Cosmology and Astroparticle Physics*, vol. 2004, no. 06, p. 008, 2004.

[5] P. A. R. Ade, N. Aghanim, M. Arnaud et al., "Planck 2015 results XIII. Cosmological parameters," *Astronomy & Astrophysics*, vol. 594, article A13, 63 pages, 2016.

[6] A. Kamenshchik, U. Moschella, and V. Pasquier, "Chaplygin-like gas and branes in black hole bulks," *Physics Letters. B. Particle Physics, Nuclear Physics and Cosmology*, vol. 487, no. 1-2, pp. 7–13, 2000.

[7] E. J. Copeland, M. Sami, and S. Tsujikawa, "Dynamics of dark energy," *International Journal of Modern Physics D: Gravitation, Astrophysics, Cosmology*, vol. 15, no. 11, pp. 1753–1935, 2006.

[8] S. D. Bass, "Vacuum energy and the cosmological constant," *Modern Physics Letters A*, vol. 30, no. 22, Article ID 1540033, 2015.

[9] A. Kamenshchik, U. Moschella, and V. Pasquier, "An alternative to quintessence," *Physics Letters B*, vol. 511, no. 2-4, pp. 265–268, 2001.

[10] N. Bilic, G. B. Tupper, and R. D. Viollier, "Unification of dark matter and dark energy: the inhomogeneous Chaplygin gas," *Physics Letters B*, vol. 535, pp. 17–21, 2002.

[11] J. C. Fabris, S. V. Gonçalves, and P. E. de Souza, "Mass power spectrum in a universe dominated by the Chaplygin gas," *General Relativity and Gravitation*, vol. 34, no. 12, pp. 2111–2126, 2002.

[12] V. Gorini, A. Kamenshchik, U. Moschella, and V. Pasquier, "The Chaplygin gas as a model for dark energy," https://arxiv.org/abs/gr-qc/0403062.

[13] M. C. Bento, O. Bertolami, and A. A. Sen, "Generalized Chaplygin gas, accelerated expansion, and dark-energy-matter unification," *Physical Review D: Particles, Fields, Gravitation and Cosmology*, vol. 66, Article ID 043507, 2002.

[14] H. B. Benaoum, "Accelerated Universe from Modified Chaplygin Gas and Tachyonic Fluid," https://arxiv.org/abs/hep-th/0205140.

[15] P. F. González-Díaz, "You need not be afraid of phantom energy," *Physical Review D*, vol. 68, Article ID 021303, p. 68, 2003.

[16] Z. Guo and Y. Zhang, "Cosmology with a variable Chaplygin gas," *Physics Letters B*, vol. 645, p. 326, 2007.

[17] U. Debnath, "Variable modified Chaplygin gas and accelerating universe," *Astrophysics and Space Science*, vol. 312, p. 295, 2007.

[18] W. Chakraborty and U. Debnath, "A new variable modified Chaplygin gas model interacting with a scalar field," *Gravitation and Cosmology*, vol. 16, p. 223, 2010.

[19] B. Pourhassan, "Viscous modified cosmic Chaplygin gas cosmology," *International Journal of Modern Physics D*, vol. 22, 2013.

[20] F. Santos, M. Bedran, and V. Soares, "On the thermodynamic stability of the generalized Chaplygin gas," *Physics Letters B*, vol. 636, no. 2, pp. 86–90, 2006.

[21] F. C. Santos, M. L. Bedran, and V. Soares, "On the thermodynamic stability of the modified Chaplygin gas," *Physics Letters. B. Particle Physics, Nuclear Physics and Cosmology*, vol. 646, no. 5-6, pp. 215–221, 2007.

[22] Y. S. Myung, "Thermodynamics of Chaplygin gas," *Astrophysics and Space Science*, vol. 335, p. 561, 2011.

[23] E. O. Kahya and B. Pourhassan, "The universe dominated by the extended Chaplygin gas," *Modern Physics Letters A*, vol. 30, no. 13, Article ID 1550070, 2015.

[24] D. Panigrahi, "Thermodynamical behavior of the variable Chaplygin gas," *International Journal of Modern Physics D: Gravitation, Astrophysics, Cosmology*, vol. 24, no. 5, 1550030, 12 pages, 2015.

[25] D. Panigrahi and S. Chatterjee, "Thermodynamics of the variable modified Chaplygin gas," *Journal of Cosmology and Astroparticle Physic*, vol. 1605, p. 52, 2016.

[26] M. Sharif and A. Sarwar, "Thermodynamics of anisotropic emergent universe in nonlinear electrodynamics," *Modern Physics Letters A*, vol. 31, no. 22, 1650129, 12 pages, 2016.

[27] L. D. Landau and E. M. Lifshitz, *Statistical Physics Vol. 5 of Course of Theoretical Physics*, Addison Wesley, Boston, Massachusetts, US, 1958.

21

Novel Neutrino-Floor and Dark Matter Searches with Deformed Shell Model Calculations

D. K. Papoulias,[1] R. Sahu,[2] T. S. Kosmas,[3] V. K. B. Kota,[4] and B. Nayak[2]

[1] *Institute of Nuclear and Particle Physics, NCSR 'Demokritos', 15310 Agia Paraskevi, Greece*
[2] *National Institute of Science and Technology, Palur Hills, Berhampur, Odisha 761008, India*
[3] *Theoretical Physics Section, University of Ioannina, 45110 Ioannina, Greece*
[4] *Physical Research Laboratory, Ahmedabad 380 009, India*

Correspondence should be addressed to D. K. Papoulias; dimpap@cc.uoi.gr

Academic Editor: Enrico Lunghi

Event detection rates for WIMP-nucleus interactions are calculated for ^{71}Ga, ^{73}Ge, ^{75}As, and ^{127}I (direct dark matter detectors). The nuclear structure form factors, which are rather independent of the underlying beyond the Standard Model particle physics scenario assumed, are evaluated within the context of the deformed nuclear shell model (DSM) based on Hartree-Fock nuclear states. Along with the previously published DSM results for ^{73}Ge, the neutrino-floor due to coherent elastic neutrino-nucleus scattering (CEνNS), an important source of background to dark matter searches, is extensively calculated. The impact of new contributions to CEνNS due to neutrino magnetic moments and Z' mediators at direct dark matter detection experiments is also examined and discussed. The results show that the neutrino-floor constitutes a crucial source of background events for multi-ton scale detectors with sub-keV capabilities.

1. Introduction

In the last few decades, the measurements of the cosmic microwave background (CMB) radiation offered a remarkably powerful way of modelling the origin of cosmic-ray anisotropies and constraining the geometry, the evolution, and the matter content of our universe. Such observations have in general indicated the consistency of the standard cosmological model [1] and the fact that our universe hardly contains ~ 5% luminous matter, whereas the remainder consists of nonluminous dark matter (~ 23%) and dark energy (~ 72%) [2]. After the discovery of the CMB fluctuations by the Cosmic Background Explorer (COBE) satellite [3], the extremely high precision of the WMAP satellite and especially of the Planck third-generation space mission has helped us to produce maps for the CMB anisotropies and other cosmological parameters (see [4] for details). We also mention that high-resolution ground-based CMB data, like those of the Atacama Cosmology Telescope (ACT) [5] and the South Pole Telescope (SPT) [6] have recovered the underlying CMB spectra observed by the space missions.

Focusing on the topic we address in this work, it is worth noting that the CMB data, the Supernova Cosmology project [7], and so on suggest that most of the dark matter of the universe is cold. Furthermore, the baryonic cold dark matter (CDM) component can be considered to consist of either massive compact halo objects (MACHOs) like neutron stars, white dwarfs, Jupiters, etc. or Weakly Interacting Massive Particles (WIMPs) that constantly bombard Earth's atmosphere. Several results of experimental searches suggest that the MACHO fraction should not exceed a portion of about 20% [1]. On the theoretical side, within the framework of new physics beyond the Standard Model (SM), supersymmetric (SUSY) theories provide promising nonbaryonic candidates for dark matter [8] (for a review see [9]). In the simple picture, the dark matter in the galactic halo is assumed to be Weakly Interacting Massive Particles (WIMPs). The most appealing WIMP candidate for nonbaryonic CDM is the lightest supersymmetric particle (LSP) which is expected to be a neutral Majorana fermion traveling with nonrelativistic velocities [10].

In recent years, there have been considerable theoretical and experimental efforts towards WIMP detection through several nuclear probes [11–13]. Popular target nuclei include among others the ^{71}Ga, ^{73}Ge, ^{75}As, ^{127}I, ^{134}Xe, and ^{208}Pb isotopes [14, 15]. Towards the first ever dark matter detection, a great number of experimental efforts take place aiming at measuring the energy deposited after the galactic halo WIMPs scatter off the nuclear isotopes of the detection material. Because of the low count rates, due to the fact that the WIMP-nucleus interaction is remarkably weak, the choice of the detector plays very important role and for this reason spin-dependent interactions require the use of targets with nonzero spin. The Cryogenic Dark Matter Search (CDMS) experimental facility [16] has been designed to directly detect the dark matter by employing a ^{73}Ge as the target nucleus, setting the most sensitive limits on the interaction of WIMP with terrestrial materials to date. The development of upgrades is under way and will be located at SNOLAB. Another prominent dark matter experiment is the EDELWEISS facility in France [17] which uses high purity germanium cryogenic bolometers at milli-Kelvin temperatures. There are also other experimental attempts using detectors like ^{127}I, ^{129,131}Xe, ^{133}Cs, etc. (see [18–20]).

Inevitably, direct detection experiments are exposed to various neutrino emissions, such as those originating from astrophysical sources (e.g., Solar [21], Atmospheric [22, 23], and diffuse Supernova [24] neutrinos), Earth neutrinos (Geoneutrinos [25]), and in other cases even artificial terrestrial neutrinos (e.g., neutrinos from nearby reactors [26]). The subsequent neutrino interactions with the material of dark matter detectors, namely, the neutrino-floor, may perfectly mimic possible WIMP signals [27]. Thus, the impacts of the neutrino-floor on the relevant experiments looking for CDM as well as on the detector responses to neutrino interactions need be comprehensively investigated. Since Geoneutrino fluxes are relatively low, astrophysical neutrinos are recognised as the most significant background source that remains practically irreducible [28]. The recent advances of direct detection dark matter experiments, mainly due to the development of low threshold technology and high detection efficiency, are expected to reach the sensitivity frontiers in which astrophysical neutrino-induced backgrounds are expected to limit the observation potential of the WIMP signal [29].

In this work, we explore the impact of the most important neutrino background source on the relevant direct dark matter detection experiments by concentrating on the dominant neutrino-matter interaction channel, e.g., the coherent elastic neutrino-nucleus scattering (CEνNS) [30, 31]. It is worthwhile to mention that events of this process were recently measured for the first time by the COHERENT experiment at the Spallation Neutrino Source [32], completing the SM picture of electroweak interactions at low energies. Such a profound discovery motivated our present work and we will make an effort to shed light on the nuclear physics aspects. Neutrino nonstandard interactions (NSIs) [33] may constitute an important source of neutrino background and have been investigated recently in [34, 35]. Thus, apart from addressing the SM contributions to CEνNS [36], we also explore the impact of new physics contributions that arise in the context of electromagnetic (EM) neutrino properties [37, 38] as well as of those emerging in the framework of $U(1)'$ gauge interactions [39] due to the presence of new light Z' mediators [40, 41]. The aforementioned interaction channels may lead to a novel neutrino-floor as demonstrated by [42]. The latter could be detectable in view of the constantly increasing sensitivity of the upcoming direct detection experiments with multi-ton mass scale and sub-keV capabilities [43].

Direct detection dark matter experiments are currently entering a precision era, and nuclear structure effects are expected to become rather important and should be incorporated in astroparticle physics applications [45]. For this reason, our nuclear model is at first tested in its capabilities to adequately describe the nuclear properties before being applied to problems like dark matter detection. This work considers the deformed shell model (DSM), on the basis of Hartree-Fock (HF) deformed intrinsic states with angular momentum projection and band mixing [46], all with a realistic effective interaction and a set of single-particle states and single-particle energies, which is established to be rather successful in describing the properties of nuclei in the mass range $A = 60–90$ (see [47] for details regarding DSM and its applications). In particular, the DSM is employed for calculating the required nuclear structure factors entering the dark matter and neutrino-floor expected event rates by focusing on four interesting nuclei regarding dark matter investigations such as ^{71}Ga, ^{73}Ge, ^{75}As, and ^{127}I. Let us add that details of nuclear structure and dark matter event rates for ^{73}Ge obtained using DSM have been reported recently [48].

The paper has been organised as follows. Section 2 gives the main ingredients of WIMP-nucleus scattering, while Section 3 provides the formulation for neutrino-nucleus scattering (neutrino-floor) within and beyond the SM. Then in Section 4 we describe briefly the methodology of the DSM, and the main results of the present work are presented and discussed in Section 5. Finally, the concluding remarks are drawn in Section 6.

2. Searching WIMP Dark Matter

The Earth is exposed to a huge number of WIMPs originating from the galactic halo. Their direct detection through nuclear recoil measurements after scattering off the target nuclei at the relevant dark matter experiments is of fundamental interest in modern physics and is expected to have a direct impact on astroparticle physics and cosmology. In this section we discuss the mathematical formulation of WIMP-nucleus scattering. The formalism introduces an appropriate separation of the SUSY and nuclear parts entering the event rates of WIMP-nucleus interactions in our effort to emphasise the important role played by the nuclear physics aspects. In particular we perform reliable nuclear structure calculations within the context of DSM based on Hartree-Fock states.

2.1. WIMP-Nucleus Scattering. For direct detection dark matter experiments, the differential event rate of a WIMP with mass m_χ scattering off a nucleus (A, Z) with respect to the momentum transfer q can be cast in the form [1]

$$\frac{dR(u,v)}{dq^2} = N_t \phi \frac{d\sigma}{dq^2} f(v)\, d^3 v, \tag{1}$$

where $N_t = 1/(A m_p)$ denotes the number of target nuclei per unit mass, A stands for the mass number of the target nucleus, and m_p is the proton mass. In the above expression the WIMP flux is $\phi = \rho_0 v / m_\chi$, with ρ_0 being the local WIMP density. The distribution of WIMP velocity relative to the detector (or Earth) and also the motion of the Sun and Earth, $f(v)$, is taken into account and assumed to resemble a Maxwell-Boltzmann distribution to ensure consistency with the LSP velocity distribution. Note that, by neglecting the rotation of Earth in its own axis, $v = |\mathbf{v}|$ accounts for the relative velocity of WIMP with respect to the detector. For later convenience a dimensionless variable $u = q^2 b^2/2$ is introduced with b denoting the oscillator length parameter, and the corresponding WIMP-nucleus differential cross section in the laboratory frame reads [10, 15, 48–50]

$$\frac{d\sigma(u,v)}{du} = \frac{1}{2}\sigma_0 \left(\frac{1}{m_p b}\right)^2 \frac{c^2}{v^2} \frac{d\sigma_A(u)}{du}, \tag{2}$$

with

$$\begin{aligned}\frac{d\sigma_A}{du} &= \left[f_A^0 \Omega_0(0)\right]^2 F_{00}(u) \\ &+ 2 f_A^0 f_A^1 \Omega_0(0)\,\Omega_1(0)\,F_{01}(u) \\ &+ \left[f_A^1 \Omega_1(0)\right]^2 F_{11}(u) + \mathcal{M}^2(u).\end{aligned} \tag{3}$$

The first three terms account for the spin contribution due to the axial current, while the fourth term accounts for the coherent contribution arising from the scalar interaction. The coherent contribution is expressed in terms of the nuclear form factors given as

$$\begin{aligned}\mathcal{M}^2(u) = \Big(&f_S^0\left[Z F_Z(u) + N F_N(u)\right] \\ &+ f_S^1\left[Z F_Z(u) - N F_N(u)\right]\Big)^2.\end{aligned} \tag{4}$$

The coherent part in the approximation of nearly equal proton and neutron nuclear form factors $F_Z(u) \approx F_N(u)$ is given as

$$\mathcal{M}^2(u) = A^2 \left(f_S^0 - f_S^1 \frac{A-2Z}{A}\right)^2 |F(u)|^2. \tag{5}$$

The respective values of the nucleonic-current parameters f_V^0, f_V^1 for the isoscalar and isovector parts of the vector current (not shown here), f_A^0, f_A^1 for the isoscalar and isovector parts of the axial-vector current, and f_S^0, f_S^1 for the isoscalar and isovector parts of the scalar current depend on the specific SUSY model employed [51]. The spin structure functions $F_{\rho\rho'}(u)$ with $\rho, \rho' = 0, 1$ for the isoscalar and isovector contributions, respectively, take the form

$$F_{\rho\rho'}(u) = \sum_{\lambda,\kappa} \frac{\Omega_\rho^{(\lambda,\kappa)}(u)\,\Omega_{\rho'}^{(\lambda,\kappa)}(u)}{\Omega_\rho(0)\,\Omega_{\rho'}(0)}, \tag{6}$$

with

$$\begin{aligned}\Omega_\rho^{(\lambda,\kappa)}(u) &= \sqrt{\frac{4\pi}{2J_i+1}} \times \langle J_f|\,| \\ &\sum_{j=1}^{A}\left[Y_\lambda(\Omega_j)\otimes\sigma(j)\right]_\kappa j_\lambda\left(\sqrt{u}r_j\right)\omega_\rho(j)\,|\,|J_i\rangle.\end{aligned} \tag{7}$$

Here, $\omega_0(j) = 1$ and $\omega_1(j) = \tau(j)$ with $\tau = +1$ for protons and $\tau = -1$ for neutrons, while Ω_j represents the solid angle for the position vector of the j-th nucleon and j_λ stands for the well-known spherical Bessel function. The quantities $\Omega_\rho(0) = \Omega_\rho^{(0,1)}(0)$ are the static spin matrix elements (see, e.g., [8]). In this context, the WIMP-nucleus event rate per unit mass of the detector is conveniently written as

$$\begin{aligned}\langle R\rangle &= \left(f_A^0\right)^2 D_1 + 2 f_A^0 f_A^1 D_2 + \left(f_A^1\right)^2 D_3 \\ &+ A^2\left(f_S^0 - f_S^1\frac{A-2Z}{A}\right)^2 |F(u)|^2 D_4.\end{aligned} \tag{8}$$

The functions D_i enter the definition of the WIMP-nucleus event rate through the three-dimensional integrals, given by

$$D_i = \int_{-1}^{1} d\xi \int_{\psi_{\min}}^{\psi_{\max}} d\psi \int_{u_{\min}}^{u_{\max}} G(\psi,\xi)\, X_i du, \tag{9}$$

with

$$\begin{aligned}X_1 &= \left[\Omega_0(0)\right]^2 F_{00}(u), \\ X_2 &= \Omega_0(0)\,\Omega_1(0)\,F_{01}(u), \\ X_3 &= \left[\Omega_1(0)\right]^2 F_{11}(u), \\ X_4 &= |F(u)|^2.\end{aligned} \tag{10}$$

In the latter expression, D_1, D_2, and D_3 account for the spin-dependent parts of (3), while D_4 is associated with the coherent contribution.

In this work, the nuclear wave functions $\langle J_f|$ and $|J_i\rangle$ entering (7) are calculated within the nuclear DSM of [46, 47]. For a comprehensive discussion on the explicit form of the function $G(\psi)$, the integration limits of (9) and the various parameters entering into these, the reader is referred to [48].

3. Neutrino-Nucleus Scattering

The neutrino-floor stands out as an important source of irreducible background to WIMP searches at a direct detection experiment. In this work we explore the neutrino-floor due to neutrino-nucleus scattering since the corresponding floor

TABLE 1: Solar neutrino fluxes and uncertainties in the framework of the employed high metallicity SSM (for details, see the text).

type	$E_{\nu_{\max}}$ [MeV]	flux [$cm^{-2}s^{-1}$]
pp	0.423	$(5.98 \pm 0.006) \times 10^{10}$
pep	1.440	$(1.44 \pm 0.012) \times 10^{8}$
hep	18.784	$(8.04 \pm 1.30) \times 10^{3}$
$^7Be_{low}$	0.3843	$(4.84 \pm 0.48) \times 10^{8}$
$^7Be_{high}$	0.8613	$(4.35 \pm 0.35) \times 10^{9}$
^8B	16.360	$(5.58 \pm 0.14) \times 10^{6}$
^{13}N	1.199	$(2.97 \pm 0.14) \times 10^{8}$
^{15}O	1.732	$(2.23 \pm 0.15) \times 10^{8}$
^{17}F	1.740	$(5.52 \pm 0.17) \times 10^{6}$

coming from neutrino-electron scattering is relatively low [52]. Motivated by the novel neutrino interaction searches using reactor neutrinos of [42], here we consider various astrophysical neutrino sources in our calculations that involve the conventional and beyond the SM interactions channels (see below).

3.1. Differential Event Rate at Dark Matter Detectors. For a given interaction channel $x = \mathrm{SM}, \mathrm{EM}, Z'$, the differential event rate dR_ν/dT_N of CEνNS processes at a dark matter detector is obtained through the convolution of the normalised neutrino energy distribution $\lambda_\nu(E_\nu)$ of the background neutrino source in question (i.e., Solar, Atmospheric and Diffuse Supernova Neutrinos, as seen below) with the CEνNS cross section, as follows [53]:

$$\left(\frac{dR_\nu}{dT_N}\right)_x = \mathscr{K}\int_{E_\nu^{\min}}^{E_\nu^{\max}} \lambda_\nu(E_\nu)\,\frac{d\sigma_x}{dT_N}(E_\nu, T_N)\,dE_\nu, \qquad (11)$$

where $E_\nu^{\max}$ is the maximum neutrino energy of the source in question (for the case of Solar neutrinos see, e.g., Table 1) and $E_\nu^{\min} = \sqrt{MT_N/2}$ is the minimum neutrino energy that is required to yield a nuclear recoil with energy T_N. In the latter expression $\mathscr{K} = t_{\mathrm{run}} N_{\mathrm{targ}} \Phi_\nu$ with t_{run} being the exposure time, N_{targ} is the number of target nuclei and Φ_ν is the assumed neutrino flux.

3.1.1. Standard Model Interactions. Assuming SM interactions only, at low and intermediate neutrino energies $E_\nu \ll M_W$, the weak neutral-current CEνNS process is adequately described by the four-fermion effective interaction Lagrangian [33, 36]

$$\mathscr{L}_{\mathrm{SM}} = -2\sqrt{2}G_F \sum_{\substack{f=u,d\\ \alpha=e,\mu,\tau}} g_{\alpha\alpha}^{f,P}\left[\bar{\nu}_\alpha \gamma_\rho L \nu_\alpha\right]\left[\bar{f}\gamma^\rho P f\right], \qquad (12)$$

where $P = \{L, R\}$ denote the chiral projectors, $\alpha = \{e, \mu, \tau\}$ represents the neutrino flavour, and $f = \{u, d\}$ is a first generation quark. By including the radiative corrections of [54], the P-handed couplings of the f quarks to the Z-boson are expressed as

$$\begin{aligned} g_{\alpha\alpha}^{u,L} &= \rho_{\nu N}^{NC}\left(\frac{1}{2} - \frac{2}{3}\widehat{\kappa}_{\nu N}\widehat{s}_Z^2\right) + \lambda^{u,L}, \\ g_{\alpha\alpha}^{d,L} &= \rho_{\nu N}^{NC}\left(-\frac{1}{2} + \frac{1}{3}\widehat{\kappa}_{\nu N}\widehat{s}_Z^2\right) + \lambda^{d,L}, \\ g_{\alpha\alpha}^{u,R} &= \rho_{\nu N}^{NC}\left(-\frac{2}{3}\widehat{\kappa}_{\nu N}\widehat{s}_Z^2\right) + \lambda^{u,R}, \\ g_{\alpha\alpha}^{d,R} &= \rho_{\nu N}^{NC}\left(\frac{1}{3}\widehat{\kappa}_{\nu N}\widehat{s}_Z^2\right) + \lambda^{d,R}, \end{aligned} \qquad (13)$$

with $\widehat{s}_Z^2 = \sin^2\theta_W = 0.2312$, $\rho_{\nu N}^{NC} = 1.0086$, $\widehat{\kappa}_{\nu N} = 0.9978$, $\lambda^{u,L} = -0.0031$, $\lambda^{d,L} = -0.0025$, and $\lambda^{d,R} = 2\lambda^{u,R} = 7.5\times10^{-5}$.

In this work we restrict our study only to low momentum transfer in order to satisfy the coherent condition $|\mathbf{q}| \le 1/R_A$, where R_A is the nuclear size and $|\mathbf{q}|$ is the magnitude of the three-momentum transfer [31]. Focusing on the dominant CEνNS channel, the relevant SM differential cross section with respect to the nuclear recoil energy T_N takes the form [41]

$$\begin{aligned} \frac{d\sigma_{\mathrm{SM}}}{(dT_N)}(E_\nu, T_N) = \frac{G_F^2 M}{\pi}\Bigg[&\left(\mathscr{Q}_W^V\right)^2\left(1 - \frac{MT_N}{2E_\nu^2}\right) \\ &+ \left(\mathscr{Q}_W^A\right)^2\left(1 + \frac{MT_N}{2E_\nu^2}\right)\Bigg], \end{aligned} \qquad (14)$$

with E_ν denoting the neutrino energy and M the mass of the target nucleus. The relevant vector ($\mathscr{Q}_W^V$) and axial-vector ($\mathscr{Q}_W^A$) weak charges entering the CEνNS cross section are given by the relations [55]

$$\begin{aligned} \mathscr{Q}_W^V(Q^2) &= \left[g_p^V Z F_Z^V(Q^2) + g_n^V N F_N^V(Q^2)\right], \\ \mathscr{Q}_W^A(Q^2) &= \left[g_p^A(Z_+ - Z_-) + g_n^A(N_+ - N_-)\right] F_A(Q^2). \end{aligned} \qquad (15)$$

Here, $Z_\pm$ ($N_\pm$) stands for the number of protons (neutrons) with spin up (+) and spin down (−), respectively, while g_p^A

(g_n^A) represent the axial-vector couplings of protons (neutrons) to the Z^0 boson. At the nuclear level, the relevant vector (axial-vector) couplings of protons g_p^V (g_p^A) and neutrons g_n^V (g_n^A) take the form

$$\begin{aligned} g_p^V &= 2\left(g_{\alpha\alpha}^{u,L} + g_{\alpha\alpha}^{u,R}\right) + \left(g_{\alpha\alpha}^{d,L} + g_{\alpha\alpha}^{d,R}\right), \\ g_n^V &= \left(g_{\alpha\alpha}^{u,L} + g_{\alpha\alpha}^{u,R}\right) + 2\left(g_{\alpha\alpha}^{d,L} + g_{\alpha\alpha}^{d,R}\right), \\ g_p^A &= 2\left(g_{\alpha\alpha}^{u,L} - g_{\alpha\alpha}^{u,R}\right) + \left(g_{\alpha\alpha}^{d,L} - g_{\alpha\alpha}^{d,R}\right), \\ g_n^A &= \left(g_{\alpha\alpha}^{u,L} - g_{\alpha\alpha}^{u,R}\right) + 2\left(g_{\alpha\alpha}^{d,L} - g_{\alpha\alpha}^{d,R}\right). \end{aligned} \quad (16)$$

The axial-vector nucleon form factor takes into account the spin structure of the nucleon and is defined as [56]

$$F_A\left(Q^2\right) = g_A\left(1 + \frac{Q^2}{M_A^2}\right)^{-2}, \quad (17)$$

where $g_A = 1.267$ is the free axial-vector coupling constant and the axial mass is taken to be $M_A = 1$ GeV, while strange quark effects have been neglected.

We note that for spin-zero nuclei the axial-vector contribution vanishes, while for the odd-A nuclei considered in the present study $\mathcal{Q}_W^A$ it is negligible and of the order of $\mathcal{Q}_W^A/\mathcal{Q}_W^V \sim 1/A$. The weak charges in (15) encode crucial information regarding the finite nuclear size through the proton $F_Z^V(Q^2)$ and neutron $F_N^V(Q^2)$ nuclear form factors, which in our work are obtained within the context of the DSM (see below), as functions of the momentum transfer $-q^\mu q_\mu = Q^2 = 2MT_N$. Contrary to similar studies assuming the conventional Helm-type form factors, the present work also takes into account the nuclear effects due to the nonspherical symmetric nuclei employed in dark matter searches.

3.1.2. Electromagnetic Neutrino Contributions. Turning our attention to new physics phenomena we now address potential contributions to CEνNS in the framework of nontrivial neutrino EM interactions that may lead to a new neutrino-floor at low detector thresholds. In this framework, the presence of an effective neutrino magnetic moment μ_ν leads to an EM contribution of the differential cross section that has been written as [41]

$$\left(\frac{d\sigma}{dT_N}\right)_{\text{SM+EM}} = \mathcal{G}_{\text{EM}}\left(E_\nu, T_N\right)\frac{d\sigma_{\text{SM}}}{dT_N}. \quad (18)$$

Neglecting axial effects, the EM contribution to CEνNS at a direct detection dark matter is encoded in the factor

$$\mathcal{G}_{\text{EM}} = 1 + \frac{1}{G_F^2 M}\left(\frac{\mathcal{Q}_{\text{EM}}}{\mathcal{Q}_W^V}\right)^2 \frac{\left(1 - T_N\right)/E_\nu/T_N}{1 - MT_N/2E_\nu^2}, \quad (19)$$

where the relevant EM charge $\mathcal{Q}_{\text{EM}}$ is written in terms of the electron mass m_e, the fine-structure constant a_{EM}, and the effective neutrino magnetic moment as [57]

$$\mathcal{Q}_{\text{EM}} = \frac{\pi a_{\text{EM}}\mu_\nu}{m_e} Z. \quad (20)$$

In contrast to the $\sim N^2$ dependence of the SM case, (19) and (20) imply the existence of a Z^2 coherence along with a characteristic $\sim 1/T_N$ enhancement of the total cross section. This implies a potential distortion of the expected recoil spectrum at very low recoil energies that may be detectable at future direct dark matter detection with sub-keV operation thresholds.

For the sake of completeness, we stress that the effective neutrino magnetic moment μ_ν is expressed through neutrino amplitudes of positive and negative helicity states, e.g., the 3-vectors a_+ and a_- and the neutrino transition magnetic moment matrix, λ, in flavour basis, as [37, 58]

$$\mu_\nu^2 = a_+^\dagger \lambda \lambda^\dagger a_+ + a_-^\dagger \lambda \lambda^\dagger a_-. \quad (21)$$

Then, the effective neutrino magnetic moment is written in mass basis through a proper rotation; for a detailed description of this formalism see [59].

3.1.3. Novel Mediator Contribution. We now explore novel mediator fields that could be accommodated in the context of simplified $U(1)'$ scenarios [60, 61] predicting the existence of a new Z' vector mediator with mass $M_{Z'}$ [62]. Such beyond the SM interactions may constitute a new neutrino-floor at direct detection dark matter experiments [39].

The presence of a Z' mediator gives rise to subleading contributions to the SM CEνNS rate, described by the Lagrangian [63]

$$\mathcal{L}_{\text{vec}} = Z'_\mu\left(g_{Z'}^{qV}\bar{q}\gamma^\mu q + g_{Z'}^{\nu V}\bar{\nu}_L\gamma^\mu\nu_L\right) + \frac{1}{2}M_{Z'}^2 Z'_\mu Z'^\mu, \quad (22)$$

where only left-handed neutrinos are assumed (right-handed neutrinos in the theory would lead to vector-axial-vector cancellations). The resulting cross section reads [41]

$$\left(\frac{d\sigma}{dT_N}\right)_{\text{SM}+Z'} = \mathcal{G}_{Z'}^2(\text{Q})\frac{d\sigma_{\text{SM}}}{dT_N}, \quad (23)$$

with the factor $\mathcal{G}_{Z'}$ being written in terms of the neutrino-vector coupling $g_{Z'}^{\nu V}$, as

$$\mathcal{G}_{Z'}(\text{Q}) = 1 - \frac{1}{2\sqrt{2}G_F}\frac{\mathcal{Q}_{Z'}}{\mathcal{Q}_W^V}\frac{g_{Z'}^{\nu V}}{Q^2 + M_{Z'}^2}. \quad (24)$$

The relevant charge in this case is expressed through the vector quark couplings $^{qV}_{Z'}$ to the Z' boson, as [39]

$$\mathcal{Q}_{Z'} = \left(2g_{Z'}^{uV} + g_{Z'}^{dV}\right)Z + \left(g_{Z'}^{uV} + 2g_{Z'}^{dV}\right)N. \quad (25)$$

Let us mention that emerging degeneracies can be either reduced through multidetector measurements [61] or broken in the framework of NSIs [64]. For completeness we note that, despite being not present for the low energies considered here, these couplings could be changed by currently unknown in-medium effects (see, e.g., [23] and references therein).

3.2. Neutrino Sources

3.2.1. Solar Neutrinos. In terrestrial searches for dark matter candidates at low energies, the Solar neutrinos emanating from the interior of the Sun generated through various fusion reactions produce a dominant background for direct CDM detection experiments. Assuming WIMP masses less than 10 GeV, an estimated total Solar neutrino flux of about $6.5 \times 10^{11} \mathrm{cm}^{-2}\mathrm{s}^{-1}$ [65] hitting the Earth is expected to appreciably limit the sensitivity of such experiments [27]. On the other hand, the theoretical uncertainties of Solar neutrinos are presently quite large and depend strongly on the assumed Solar neutrino model. To maintain consistency with existing Solar data, in this work we consider the high metallicity Standard Solar Model (SSM) [21]. We note however that the dominant Solar neutrino component coming from the primary proton-proton channel (*pp* neutrinos) that accounts for about 86% the Solar neutrinos flux has been recently measured by the Borexino experiment with an uncertainty of 1% [66]. Through CEνNS, the direct detection dark matter experiments are mainly sensitive to two sources of Solar neutrinos, namely, the ^{8}B and the *hep* neutrinos which cover the highest energy range of the Solar neutrino spectrum. Since ^{8}B neutrinos are generated from the decay $^8\mathrm{B} \longrightarrow {}^7\mathrm{Be}^* + e^+ + \nu_e$, while *hep* neutrinos from $^3\mathrm{He} + p \longrightarrow {}^4\mathrm{He} + e^+ + \nu_e$, both sources occur in the aftermath of the *pp* chain. Following previous similar studies [28], in this work, we explore the neutrino-floor extending our analysis to the lowest neutrino energies, by considering the *pep* neutrino line which belongs to the *pp* chain and the e^--capture reaction on ^{7}Be that leads to two monochromatic beams at 384.3 and 861.3 keV as well as the well-known CNO cycle. The latter neutrinos appear as three continuous spectra (^{13}N, ^{15}O, and ^{17}F) with end point energies close to the *pep* neutrinos.

3.2.2. Atmospheric Neutrinos. Atmospheric neutrinos are decay products of the particles (mostly pions and kaons) produced as a result of cosmic-ray scattering in the Earth's atmosphere. The generated secondary particles decay to ν_e, $\overline{\nu}_e$, ν_μ, and $\overline{\nu}_\mu$ constituting a significant background to dark matter searches especially for WIMP masses above 100 GeV. In particular, the effect is crucial on the discovery potential of WIMPs with spin-independent cross section of the order of $10^{-48}\mathrm{cm}^2$. The direct detection dark matter experiments, due to the lack of directional sensitivity, are in principle sensitive to the lowest energy (less than ~100 MeV) atmospheric neutrinos. For this reason, in our present work atmospheric neutrinos are considered by employing the low-energy flux coming out of the FLUKA code simulations [22].

3.2.3. Diffuse Supernova Neutrinos. The weak glow of MeV neutrinos emitted from the total number of core-collapse supernovae, known as the Diffuse Supernova Neutrino Background (DSNB), creates an important source of neutrino background specifically for the WIMPs mass range 10–30 GeV [24]. Despite the appreciably lower flux compared to Solar neutrinos, DSNB neutrino energies are higher than those of the Solar neutrino spectrum. In our simulations, the adopted DSNB distributions (usually of Fermi-Dirac or power-law type) correspond to temperatures 3 MeV for ν_e, 5 MeV for $\overline{\nu}_e$, and 8 MeV for the other neutrino flavours denoted as ν_x or $\overline{\nu}_x$, $x = \mu, \tau$.

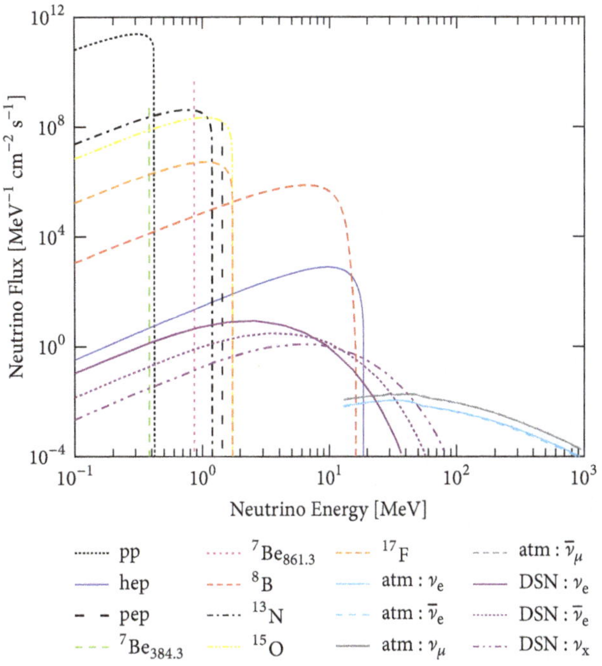

FIGURE 1: Unoscillate neutrino flux considered in the present study, including the Solar, Atmospheric, and DSNB spectra.

Figure 1 shows the unoscillate neutrino flux considered in the present study, illustrating the Solar neutrino spectra of the dominant neutrino sources assuming the high metallicity Standard Solar Model (SSM) as defined in [21]. Also shown is the low-energy atmospheric neutrino flux as obtained from the FLUKA simulation [22] as well as the DSNB spectrum [67]. The corresponding neutrino types, maximum energies, and fluxes are listed in Table 1.

4. Deformed Shell Model

In the formalism of the WIMP-nucleus or neutrino-nucleus event rates of Sections 2 and 3, both for the case of elastic or inelastic interaction channels, the nuclear physics and particle physics (SUSY model) parts appear almost completely separated. In the present work our main focus drops on the nuclear physics aspects which are contained in the nuclear structure factors discussed in Section 2. Special attention is paid on the factors D_i of (9) that depend on the spin structure functions and the nuclear form factors. These quantities have been calculated using the DSM method [48] (for a comprehensive discussion of DSM see [47]) given the kinematics and the assumptions describing the WIMP particles.

The construction of the many-body wave functions for the initial $|J_i^\pi\rangle$ and final $|J_f^\pi\rangle$ nuclear states in the framework of DSM involves performance of the following steps. (i) At first, one chooses a model space consisting of a given set of spherical single-particle (sp) orbits, sp energies, and the

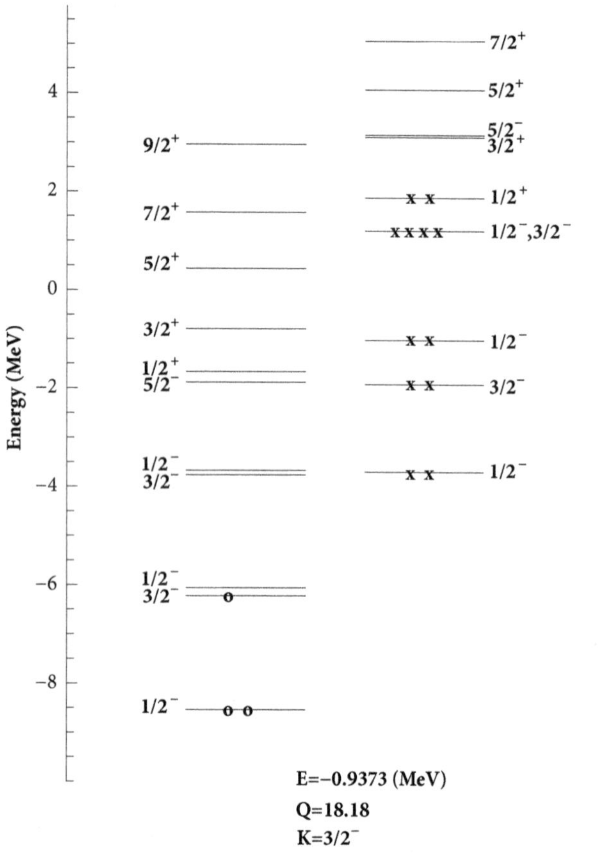

FIGURE 2: HF single-particle spectra for ^{71}Ga corresponding to the lowest prolate configuration. In the figure, circles represent protons and crosses represent neutrons. The HF energy E in MeV, the mass quadrupole moment Q in units of the square of the oscillator length parameter, and the total azimuthal quantum number K are given in the figure.

appropriate two-body effective interaction matrix elements. For ^{71}Ga and ^{75}As, the spherical sp orbits are $1p_{3/2}$, $0f_{5/2}$, $1p_{1/2}$, and $0g_{9/2}$ with energies 0.0, 2.20, 2.28, and 5.40 MeV and 0.0, 0.78, 1.08, and 3.20 MeV, respectively, while the assumed effective interaction is the modified Kuo interaction [68]. Similarly for ^{127}I, the sp orbits, their energies, and the effective interaction are taken from a recent paper [69]. (ii) Assuming axial symmetry and solving the HF single-particle equations self-consistently, the lowest-energy prolate (or oblate) intrinsic state for the nucleus in question is obtained. An example is shown in Figure 2 for ^{71}Ga. (iii) The various excited intrinsic states then are obtained by making particle-hole (p-h) excitations over the lowest-energy intrinsic state (lowest configuration). (iv) Then, because the HF intrinsic nuclear states $|\chi_K(\eta)\rangle$ (K is azimuthal quantum number and η distinguishes states with the same K) do not have definite angular momentum, angular momentum projected states $|\phi^J_{MK}(\mu)\rangle$ are constructed as

$$|\phi^J_{MK}(\eta)\rangle = \frac{2J+1}{8\pi^2\sqrt{N_{JK}}}\int d\Omega D^{J^*}_{MK}(\Omega)\, R(\Omega)\, |\chi_K(\eta)\rangle . \quad (26)$$

In the previous expression, $\Omega = (\alpha, \beta, \gamma)$ represents the Euler angles, $R(\Omega)$ denotes the known general rotation operator, and the Wigner D-matrices are defined as $D^J_{MK}(\Omega) = \langle JM|R(\Omega)|JK\rangle$. Here, N_{JK} is the normalisation constant which by assuming axial symmetry is defined as

$$N_{JK} = \frac{2J+1}{2}\int_0^{\pi} d\beta \sin\beta\, d^J_{KK}(\beta)\, \langle\chi_K(\eta)| \cdot e^{-i\beta J_y}\, |\chi_K(\eta)\rangle , \quad (27)$$

where the functions $d^J_{KK}(\beta)$ are the diagonal elements of the matrix $d^J_{MK}(\beta) = \langle JM|e^{-i\beta J_y}|JK\rangle$. ($v$) Finally, the good angular momentum states ϕ^J_{MK} are orthonormalised by band mixing calculations and then, in terms of the index η, it is possible to distinguish between different states having the same angular momentum J,

$$|\Phi^J_M(\eta)\rangle = \sum_{K,\alpha} S^J_{K\eta}(\alpha)\, |\phi^J_{MK}(\alpha)\rangle . \quad (28)$$

Within the DSM method, for the evaluation of the reduced nuclear matrix element entering (6) and (7), we first calculate the single-particle matrix elements of the relevant operators $t^{(l,s)J}_{\nu}$, as

$$\langle n_i l_i j_i \| \tilde{t}^{(l,s)J} \| n_k l_k j_k \rangle = \sqrt{(2j_k+1)(2j_i+1)(2J+1)(s+1)(s+2)} \times \begin{Bmatrix} l_i & \frac{1}{2} & j_i \\ l_k & \frac{1}{2} & j_k \\ l & s & J \end{Bmatrix} \langle l_i \| \sqrt{4\pi} Y^l \| l_k \rangle \langle n_i l_i | j_l(kr) | n_l l_k \rangle , \quad (29)$$

where $\{--\}$ is the 9-j symbol. For more details, the reader is referred to [70–72]. It should be noted that in the DSM method one considers an adequate number of intrinsic states in the band mixing calculations.

DSM calculations are performed in the same spirit as in spherical shell model where one takes a model space and a suitable effective interaction (single-particle orbitals, single-particle energies, and a two-body effective interaction). This procedure has been found to be quite successful in describing the spectroscopic properties and electromagnetic properties of many nuclei in the mass region A = 60–90 and has also been applied to double beta decay nuclear transition matrix elements [47]. In addition, this model has been used recently in calculating the event rates for dark matter detection [48]. With the proper choice of effective interaction, one will not be considering core excitations. This is a standard prescription in shell model as well as in DSM. To go beyond this, one has to use no-core shell model or DSM with much larger set of single-particle orbitals (inclusion of core orbitals), such refinements are planned to be employed in future calculations.

We note that the many-body nuclear calculations performed take into account in the usual way the inert core orbits

Table 2: $2k$ values of the occupied proton and neutron single particle deformed orbits of the HF intrinsic states used in the calculation for each nucleus. The second column gives the serial no. of the HF intrinsic states used. All the $2k$ values are of negative parity unless explicitly shown. The (+), (−), or (±) sign before the $2k$ values implies that either the time-like, time-reversed, or both orbits are occupied. In columns 3 and 4, 3_1 means the first $3/2^-$ HF deformed sp orbit, 3_2 means the second $3/2^-$ deformed HF orbit, and so on (see also Figure 2). Detailed information regarding the structure each of the deformed HF sp orbits, their energies, and the parentage of each of the HF intrinsic state in the Φ^J states (e.g., the linear combination of ϕ^J_{MK} obtained in the band mixing diagonalisation) can be obtained from the authors.

Nucleus	Serial No.	proton orbits			neutron orbits						
^{71}Ga	1	$\pm1_1$	$+3_1$		$\pm1_1$	$\pm1_2$	$\pm3_1$	$\pm3_2$	$\pm1_3$	$\pm1_1^+$	
	2	$\pm1_1$	$+1_2$		$\pm1_1$	$\pm1_2$	$\pm3_1$	$\pm3_2$	$\pm1_3$	$\pm1_1^+$	
	3	$\pm1_1$	-3_1		$\pm1_1$	$\pm1_2$	$\pm3_1$	$\pm3_2$	$\pm1_3$	$+1^+$	$+3^+$
	4	$\pm1_1$	$+3_1$		$\pm1_1$	$\pm1_2$	$\pm3_1$	$\pm3_2$	$\pm1_3$	$\pm5_1$	
^{73}Ge	1	$\pm1_1$	$\pm1_2$		$\pm1_1$	$\pm1_2$	$\pm3_1$	$\pm3_2$	$\pm1_3$	$\pm1_1^+$	$+3_1^+$
	2	$\pm1_1$	$\pm1_2$		$\pm1_1$	$\pm1_2$	$\pm3_1$	$\pm3_2$	$\pm1_3$	$\pm3_1^+$	$+1_1^+$
	3	$\pm1_1$	$\pm3_1$		$\pm1_1$	$\pm1_2$	$\pm3_1$	$\pm3_2$	$\pm1_3$	$\pm1_1^+$	$+3_1^+$
^{75}As	1	$\pm1_1$	$\pm1_2$	$+3_1$	$\pm1_1$	$\pm1_2$	$\pm3_1$	$\pm1_1^+$	$\pm3_1^+$	$\pm3_2$	$\pm1_3$
	2	$\pm1_1$	$\pm3_1$	$+1_2$	$\pm1_1$	$\pm1_2$	$\pm3_1$	$\pm1_1^+$	$\pm3_1^+$	$\pm3_2$	$\pm1_3$
	3	$\pm1_1$	$\pm1_2$	$+3_1$	$\pm1_1$	$\pm1_2$	$\pm3_1$	$\pm1_1^+$	$\pm3_1^+$	$\pm3_2$	$\pm5_1^+$
	4	$\pm1_1$	$\pm3_1$	$+1_3$	$\pm1_1$	$\pm1_2$	$\pm3_1$	$\pm1_1^+$	$\pm3_1^+$	$\pm3_2$	$\pm5_1^+$
	5	$\pm1_1$	$\pm1_2$	$+3_1$	$\pm1_1$	$\pm1_2$	$\pm3_1$	$\pm1_1^+$	$\pm3_1^+$	$\pm3_2$	$\pm5_1$
	6	$\pm1_1$	$\pm3_1$	$+1_2$	$\pm1_1$	$\pm1_2$	$\pm3_1$	$\pm1_1^+$	$\pm3_1^+$	$\pm3_2$	$\pm5_1$
^{127}I	1	$\pm7_1^+$	$+5_1^+$		$\pm7_1^+$	$\pm5_1^+$	$\pm3_1^+$	$\pm11_1$	$\pm1_1^+$	$\pm5_2^+$	$\pm9_1$
					$\pm3_2^+$	$\pm1_2^+$	$\pm7_1$	$\pm5_1$	$\pm3_1$		
	2	$\pm7_1^+$	$+5_1^+$		$\pm7_1^+$	$\pm5_1^+$	$\pm3_1^+$	$\pm11_1$	$\pm1_1^+$	$\pm5_2^+$	$\pm9_1$
					$\pm3_2^+$	$\pm1_2^+$	$\pm7_1$	$\pm5_1$	$\pm3_3^+$		
	3	$\pm7_1^+$	$+3_1^+$		$\pm7_1^+$	$\pm5_1^+$	$\pm3_1^+$	$\pm11_1$	$\pm1_1^+$	$\pm5_2^+$	$\pm9_1$
					$\pm3_2^+$	$\pm1_2^+$	$\pm7_1$	$\pm5_1$	$\pm3_1$		
	4	$\pm7_1^+$	$+3_1^+$		$\pm7_1^+$	$\pm5_1^+$	$\pm3_1^+$	$\pm11_1$	$\pm1_1^+$	$\pm5_2^+$	$\pm9_1$
					$\pm3_2^+$	$\pm1_2^+$	$\pm7_1$	$\pm5_1$	$\pm3_3^+$		
	5	$\pm7_1^+$	$+1_1^+$		$\pm7_1^+$	$\pm5_1^+$	$\pm3_1^+$	$\pm11_1$	$\pm1_1^+$	$\pm5_2^+$	$\pm9_1$
					$\pm3_2^+$	$\pm1_2^+$	$\pm7_1$	$\pm5_1$	$\pm3_1$		
	6	$\pm7_1^+$	$+1_1^+$		$\pm7_1^+$	$\pm5_1^+$	$\pm3_1^+$	$\pm11_1$	$\pm1_1^+$	$\pm5_2^+$	$\pm9_1$
					$\pm3_2^+$	$\pm1_2^+$	$\pm7_1$	$\pm5_1$	$\pm3_3^+$		

(completely filled by the protons and neutrons) and the extra-core nucleons moving in the assumed model space under the influence of an effective interaction. The explicit $2k$ values of the occupied nucleon single-particle deformed orbits of the HF intrinsic states considered in our calculations are listed in Table 2.

5. Results and Discussion

5.1. Nuclear Physics Aspects. To maximise the significance of our WIMP-nucleus and neutrino-floor calculations, the reliability of the obtained nuclear wave functions is tested by comparing the extracted energy level spectrum and magnetic moments with available experimental data. The consistency of this method obtained for ^{73}Ge has been already presented in [48]. Furthermore, in the DSM calculations for ^{71}Ga and ^{75}As, we restrict ourselves to prolate solutions only, since the oblate solution does not reproduce the energy spectra and electromagnetic properties of these nuclei. It also does not mix with the prolate solution. Hence, we neglect the oblate solutions in the calculations. For each of these nuclei, we consider only four intrinsic prolate states which should be sufficient to explain the systematics of the ground state and close lying excited state. Due to size restrictions, in Figure 3 we illustrate only the calculated spectrum for ^{71}Ga.

For ^{75}As, the ground state is $3/2^-$ and there are also two $1/2^-$ and $3/2^-$ levels around 0.12 MeV. In addition, there is a collective band consisting of $5/2^-$, $9/2^-$, and $13/2^-$ $17/2^-$ levels at 0.279, 1.095, 2.150, and 3.091 MeV, respectively. All these levels are well reproduced by the DSM method. Turning to the ^{127}I spectrum, there are four observed collective bands with band heads $5/2^+$, $(7/2^+)_{1,2}$, and $9/2^+$. There are evidences suggesting that low-lying states in ^{127}I have oblate deformation [73]. Hence, for this nucleus, we consider only oblate configurations and take the six lowest oblate intrinsic states in the band mixing calculation. These intrinsic states are found to provide adequate description of the energy spectrum and electromagnetic properties for this nucleus. The calculations for this nucleus utilise a new effective interaction developed by an Italian group very recently [69]. The new effective interaction is seen to reproduce well the ^{127}I spectrum; details will be presented elsewhere.

We thus conclude that concerning the evaluation of the WIMP-nucleus and CEνNS event rates we are interested in

TABLE 3: List of potential dark matter detectors considered in the present study. The calculated magnetic moments for the ground states of ^{71}Ga, ^{73}Ge, ^{75}As, and ^{127}I are shown. The results involve the bare gyromagnetic ratios and experimental data are from [44]. The ground state J^π and the harmonic oscillator size b are also shown.

Nucleus	A	Z	J^π	$< l_p >$	$< S_p >$	$< l_n >$	$< S_n >$	μ (nm)	Exp	b [fm^{-1}]
Ga	71	31	$3/2^-$	0.863	0.257	0.369	0.011	2.259	2.562	1.90
Ge	73	32	$9/2^+$	0.581	−0.001	3.558	0.362	−0.811	−0.879	1.91
As	75	33	$(3/2^-)_1$	0.667	0.164	0.626	0.042	1.422	1.439	1.92
I	127	53	$5/2^+$	2.395	−0.211	0.313	2.343	1.207	2.813	2.09

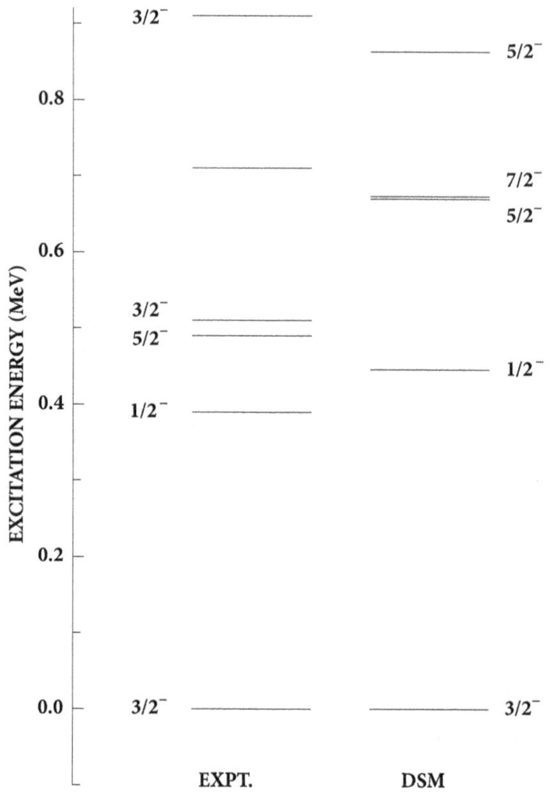

FIGURE 3: Comparison of deformed shell model results with experimental data for ^{71}Ga for low-lying states. The experimental values are taken from [44].

this work, the required ground state wave functions obtained through the DSM method are reliable and the intrinsic states used in the subsequent analysis are considered sufficient. From the perspective of nuclear physics, spin contributions constitute significant ingredients in the evaluation of WIMP-nucleus event rates. For this reason, the first stage of our work involves the calculation of the magnetic moment, which is decomposed into an orbital and spin part. The relevant results for the proton and neutron contributions to the orbital and spin parts concerning the ground states of the four nuclear isotopes studied in this paper are given in Table 3. A comparison between the obtained magnetic moments and the respective experimental data is also provided. Despite the fact that these calculations adopt bare values of g-factors neglecting quenching effects, the obtained DSM results of the ground state magnetic moments are consistent with the experimental values.

Having successfully reproduced the energy spectrum and the magnetic moments within the context of the DSM wave functions, we evaluate important nuclear physics inputs entering the WIMP-nucleus and CEνNS cross sections. Figures 4 and 5 present a comparison between the DSM nuclear form factors and the effective Helm-type ones employed in various similar studies, where, as can be seen, the DSM results differ from the Helm-type ones. The behaviour of the proton form factor for ^{71}Ga is found to be different from those of the other nuclei and this may be due to the nearby proton shell closure and the neutron subshell closure. Calculations with several different effective interactions are under way to rule out the possibility of any deficiency of the effective two-body interaction used. We furthermore illustrate the spin structure functions of WIMP-^{71}Ga elastic scattering calculated using (6) and (7). The variation of F_{00}, F_{01}, and F_{11} with respect to the parameter u is shown in Figure 6, while similar results are obtained for ^{75}As and ^{127}I (for the ^{73}Ge case see [48]).

The consistency of our nuclear physics DSM calculations has been extensively explored in this work and compared with existing experimental data (see Figures 2 and 3 and Table 3) making the considered form factors reliable. Specifically we have tested the reliability of this model to describe nuclear structure properties such as excitation spectra and nuclear magnetic moments. We mention that DSM has been tested in the past in many nuclei in the A = 60–90 region [47] (see above).

5.2. WIMP-Nucleus Rates and the Neutrino-Floor. The WIMP-nucleus event rates and the neutrino-floor due to neutrino-nucleus scattering are calculated for a set of interesting nuclear targets such as ^{71}Ga, ^{73}Ge, ^{75}As, and ^{127}I (see Table 3). In evaluating the neutrino-induced backgrounds, we consider only the dominant CEνNS channel, since neutrino-electron events are expected to produce less events by about one order of magnitude [27]. For the case of a ^{71}Ga target, in Figure 7 we provide the coefficients D_i associated with the spin dependent and coherent interactions given in (9) as functions of the WIMP mass m_χ by assuming three typical values of the detector threshold energy T_N= 0, 5, 10 keV. For the special case of T_N = 0, all plots peak at m_χ ~35 GeV, while for higher threshold energies D_i are shifted towards higher values of the WIMP mass. The calculations take also into account the annual modulation which is represented by the curve thickness. As can be seen from the figure, the modulation signal varies with respect to the WIMP mass, being larger for $m_\chi \leq 50$ GeV, while its magnitude is slightly different for the spin dependent and coherent channels.

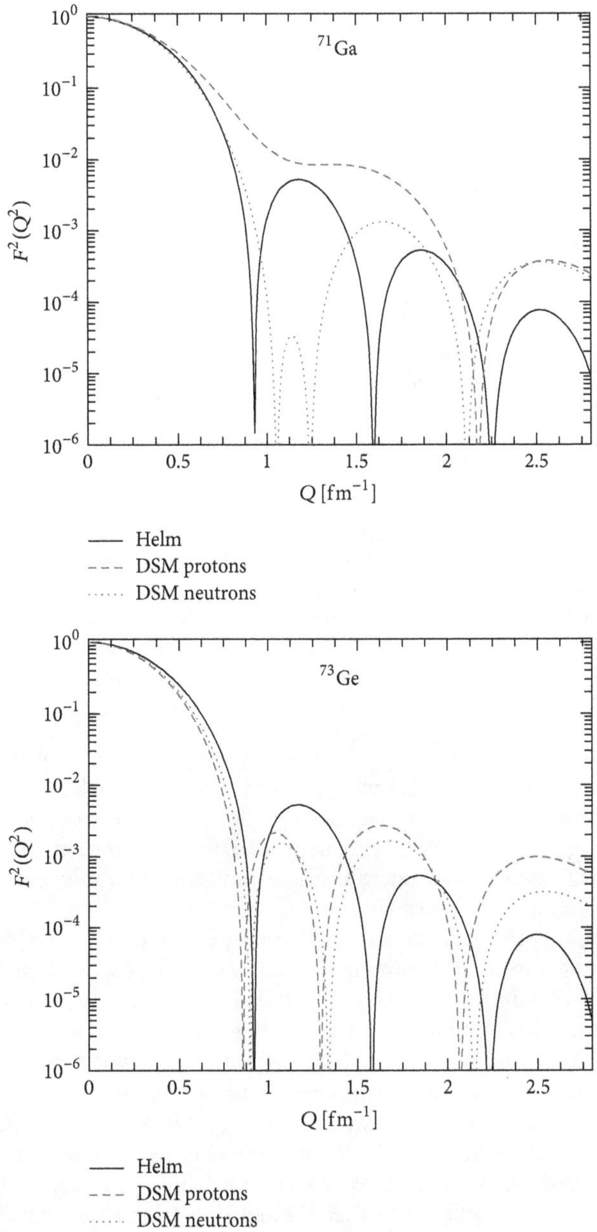

FIGURE 4: Comparison of the DSM nuclear form factors of the ^{71}Ga and ^{73}Ge isotopes, obtained in the present work with the corresponding effective Helm form factors.

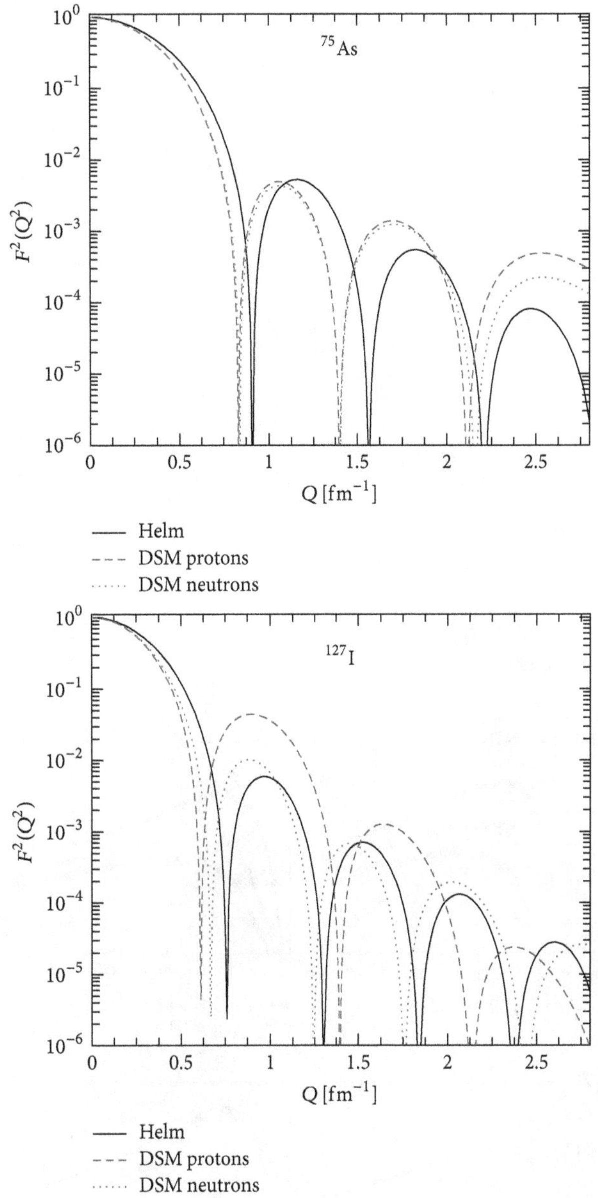

FIGURE 5: Same as in Figure 4 but for the ^{75}As and ^{127}I isotopes.

Proceeding further, in Figure 8 we evaluate the expected event rates for the four target nuclei assuming elastic WIMP scattering for WIMP candidates with mass $m_\chi = 110$ GeV, by adopting the nucleonic-current parameters $f_A^0 = 3.55 \times 10^{-2}$, $f_A^1 = 5.31 \times 10^{-2}$, $f_S^0 = 8.02 \times 10^{-4}$, and $f_S^1 = -0.15 \times f_S^0$. As in the previous discussion, the thickness of the graph accounts for the annual modulation. We find that there is a strong dependence of the event rate on the studied nuclear isotope. Again the modulation is found to decrease for heavier mass. Among the four studied nuclei, we come out with a larger event rate for the case of a ^{71}Ga nuclear detector, since D_1, D_2, and D_3 are all positive and have similar values. For ^{73}Ge, D_2 is negative and its magnitude is comparable to D_1 and D_3, while for ^{75}As, D_3 is positive but small, and finally for ^{127}I, D_2 and D_3 are relatively smaller and D_1 is large. The coherent contribution D_4 has more or less similar values for all nuclei considered.

For each component of the Solar, Atmospheric, and DSNB neutrino distributions we calculate the expected neutrino-floor due to CEνNS, by considering the target nuclei presented in Table 3. In our calculations, we neglect possible recoil events arising from Geoneutrinos as they are expected to be at least one order of magnitude less that the aforementioned neutrino sources (see, e.g., [25, 26]). In order to make a quantitative estimate of the neutrino-floor, here we do not consider neutrino oscillations and we assume that CEνNS is a flavour blind process in the SM. The

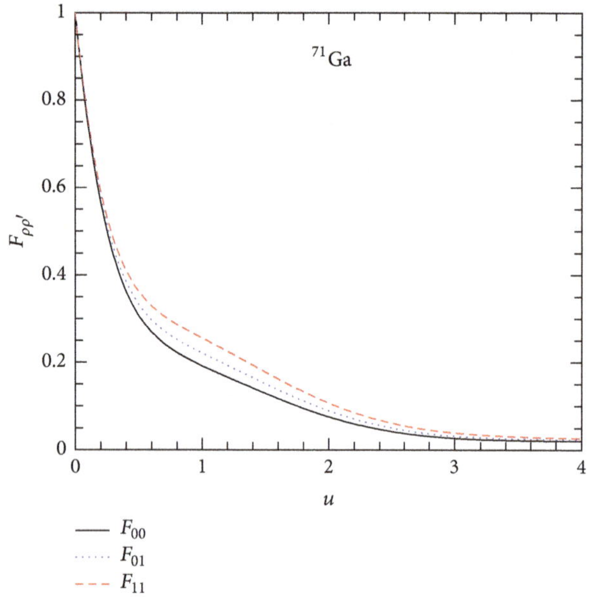

FIGURE 6: Normalised spin structure functions of ^{71}Ga for the ground state.

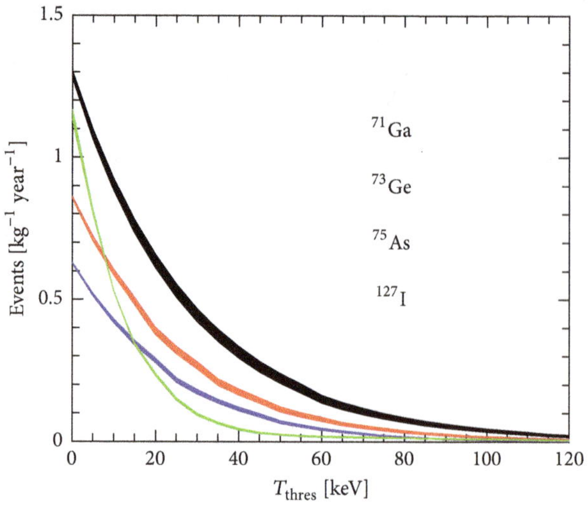

FIGURE 8: The WIMP event rates for ^{71}Ga, ^{73}Ge, ^{75}As, and ^{127}I detectors in units of kg^{-1} year^{-1} as a function of the detector threshold T_N. The nuclear threshold T_N energy through the limit of the integration in (9). The thickness of the curve represents the annual modulation which decreases with increasing nuclear mass.

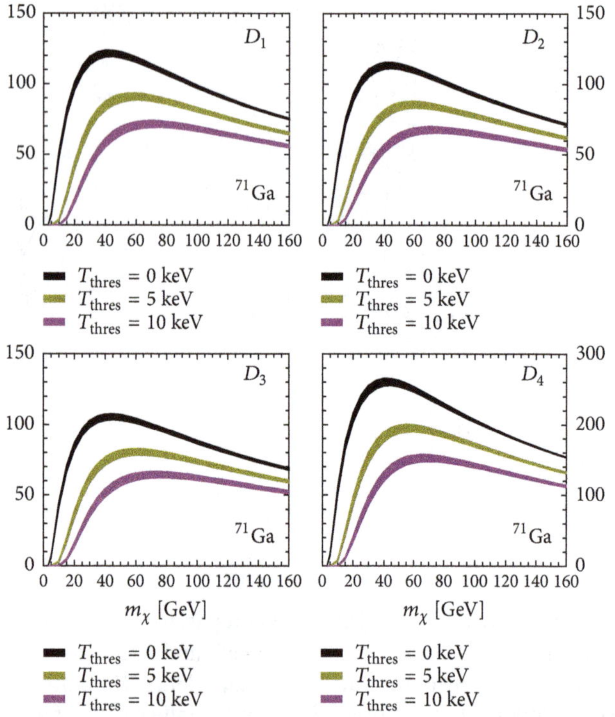

FIGURE 7: Nuclear structure coefficients D_i for ^{71}Ga plotted as a function of the WIMP mass. The graphs are plotted for three values of the detector threshold 0, 5, and 10 keV. The thickness of the graphs represents annual modulation.

differential event rate due to CEνNS, for the various dark matter detectors considered in the present study, is presented in Figure 9. It can be noticed that the neutrino background is dominated by Solar neutrinos at very low recoil energies. We stress that, for the typical keV-recoil thresholds of the current direct detection dark matter experiments only the *hep* and ^{8}B sources constitute a possibly detectable background. From our results we conclude that, for recoil energies above about 10 keV, Atmospheric neutrinos dominate the neutrino background event rates, having a tiny contribution coming from the DSNB spectrum.

The number of expected background events due to CEνNS for each component of the Solar, Atmospheric, and DSNB neutrino fluxes is illustrated in Figure 10. Similar to the differential event case, at low energies the neutrino background is dominated by the Solar neutrino spectrum with the dominant components being the *hep* and ^{8}B neutrino sources. The results imply that future multi-ton scale detectors with sub-keV sensitivities may be also sensitive to ^{7}Be and *pp* neutrinos. We comment however that such sensitivities will be further limited due to the quenching effect of the nuclear recoil spectrum which is not taken into account here. Moreover, it is worth mentioning that neutrino-induced and WIMP-nucleus scattering processes provide similar recoil spectra; e.g., the recoil spectrum of ^{8}B neutrinos may mimic that of a WIMP with mass 6 GeV (100 MeV) [28].

At this point, we consider additional interactions in the context of new physics beyond the SM that may enhance the CEνNS rate at a direct detection dark matter experiment. Specifically we study the impact of neutrino EM properties as well as the impact of new interactions due to a Z' mediator, on the neutrino floor. In our calculations we assume the existence of a neutrino magnetic moment $\mu_\nu = 4.3 \times 10^{-9}\mu_B$, extracted from CE$\nu$NS data in [41] as well as the corresponding limit from $\overline{\nu}_e - e^-$ scattering data of the GEMMA experiment, e.g., $\mu_{\overline{\nu}_e} = 2.9{\times}10^{-11}\mu_B$ [74]. Regarding the Z' interaction we consider typical values such as M'_Z = 10 MeV, $g^2_{Z'} = 10^{-6}$, and M'_Z = 1GeV, $g^2_{Z'} = 10^{-6}$

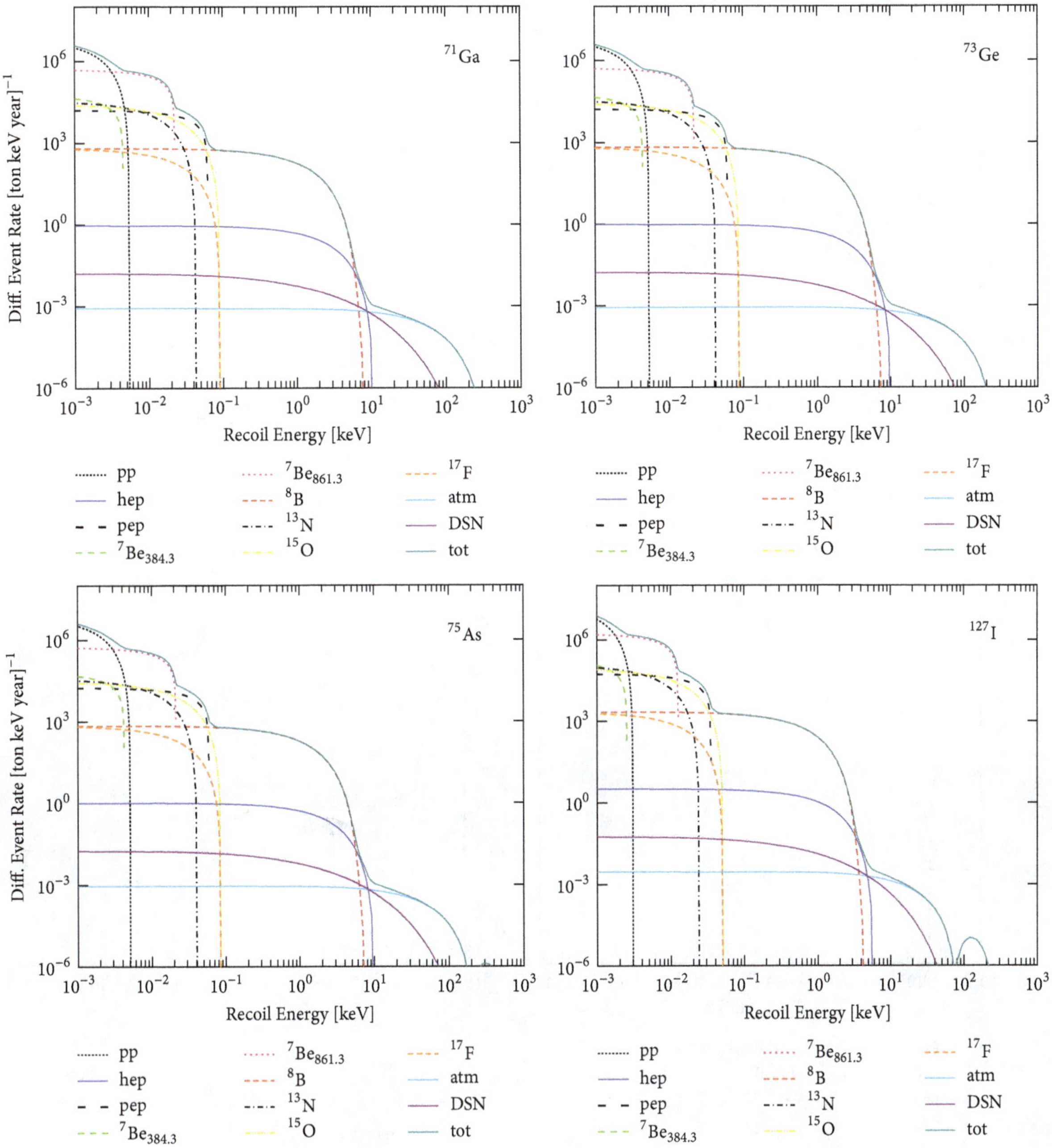

FIGURE 9: Differential event rate of the neutrino-floor assuming ^{71}Ga, ^{73}Ge, ^{75}As, and ^{127}I as cold dark matter detectors. The individual components coming from the Solar, Atmospheric, and DSNB flux are also shown.

[75]. Following [64], by assuming universal couplings, our calculations involve the product of neutrino and quark Z' couplings defined as (for a comprehensive study involving the flavour dependence of the Z' couplings the reader is refereed to [76])

$$g_{Z'}^2 = \frac{g_{Z'}^{\nu V} \mathcal{Q}_{Z'}}{3A}. \quad (30)$$

The corresponding results are presented in Figure 11, indicating that such new physics phenomena may constitute a crucial source of background even for multi-ton scale detectors with sub-keV capabilities. We stress, however, that the latter conclusion depends largely on the assumed parameters, which currently are unknown.

Before closing, we estimate the difference in the calculated number of neutrino-floor events between the conventional Helm-type and DSM predictions by defining the ratio

$$\mathscr{R} = \frac{\mathrm{DSM}_{\mathrm{events}}}{\mathrm{Helm}_{\mathrm{events}}}. \quad (31)$$

For each nuclear system, the corresponding results are presented in Figure 12 indicating that the differences can become significant, especially in the high energy tail of the detected recoil spectrum.

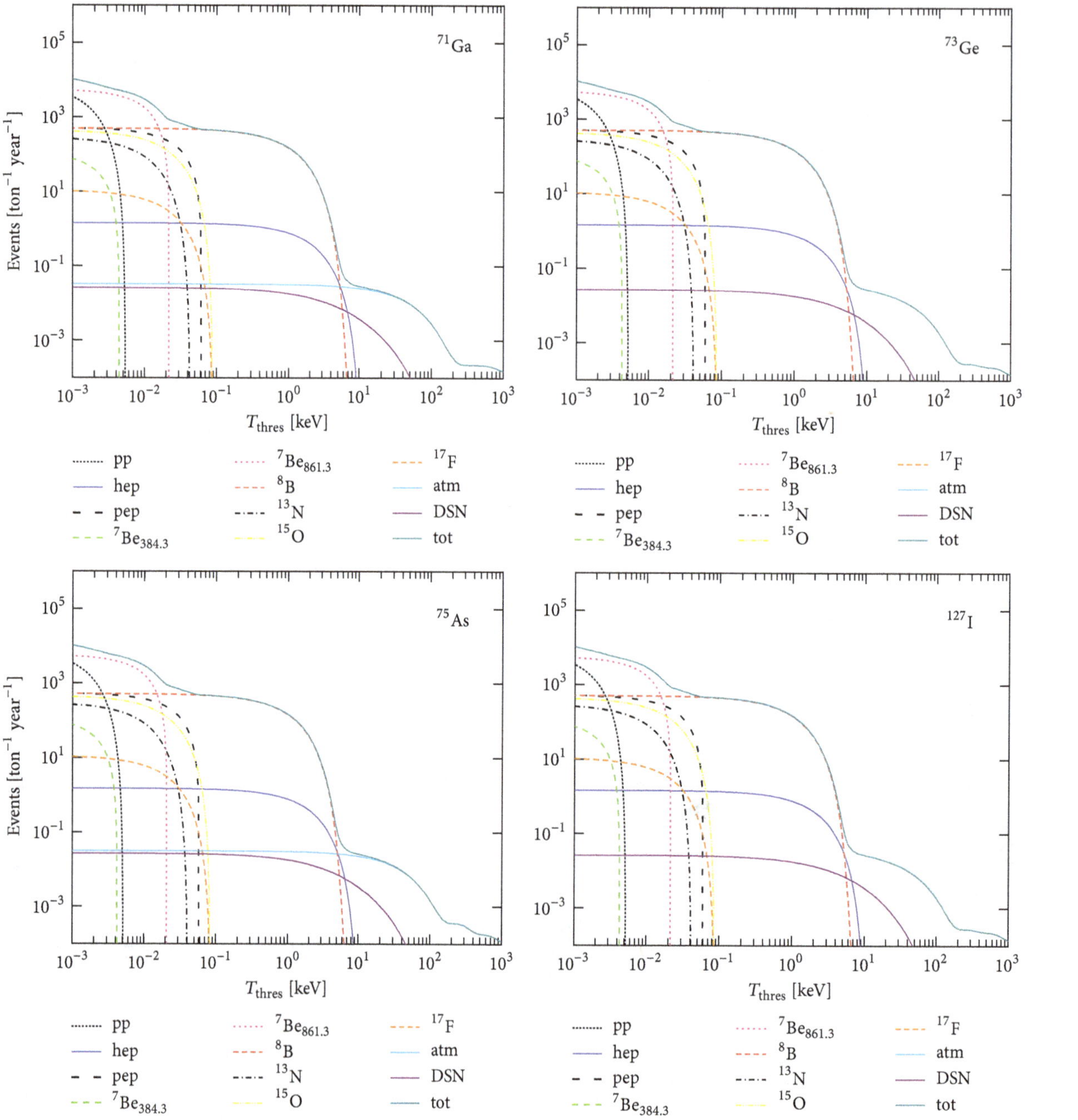

FIGURE 10: Same as in Figure 9 but for the number of events above the detector threshold.

6. Conclusions

In this work, we studied comprehensively the expected event rates in WIMP-nucleus and neutrino-floor processes by performing reliable calculations for a set of prominent nuclear materials of direct dark matter detection experiments. The detailed calculations involve crucial nuclear physics inputs in the framework of the deformed shell model based on Hartree-Fock nuclear states. This way, the nuclear deformation and the spin structure effects of odd-A isotopes that play significant role in searching for dark matter candidates are incorporated. The chosen nuclear detectors involve popular nuclear isotopes in dark matter investigations such as the ^{71}Ga, ^{73}Ge, ^{75}As, and ^{127}I isotopes. The DSM results indicate that ^{71}Ga needs further investigation by employing another effective two-body interaction than the one used in the chosen set of nuclear isotopes.

The deformed shell model (DSM) employed for the nuclear structure calculations in this work is very well tested in many examples in the past [47] for nuclei with A=60–90. Therefore, in our study we have chosen the dark matter candidates ^{71}Ga, ^{73}Ge, and ^{75}As. In addition, to extend DSM to heavier nuclei of interest in dark matter detection, we have considered ^{127}I and the results, reported in the present paper, are quite encouraging. In the near future we will consider Xe isotopes that are also of current interest. For lighter candidate

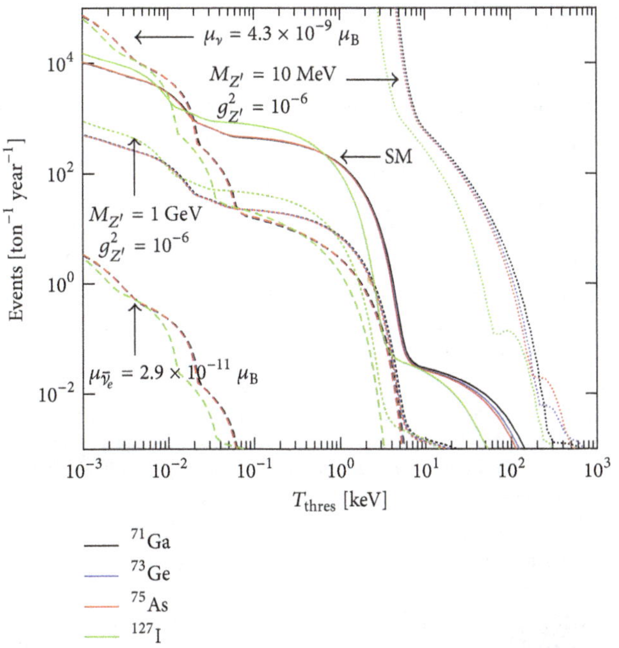

FIGURE 11: The neutrino-floor for various interaction channels. Solid, dashed, and dotted lines correspond to SM, EM, and Z' contributions, respectively.

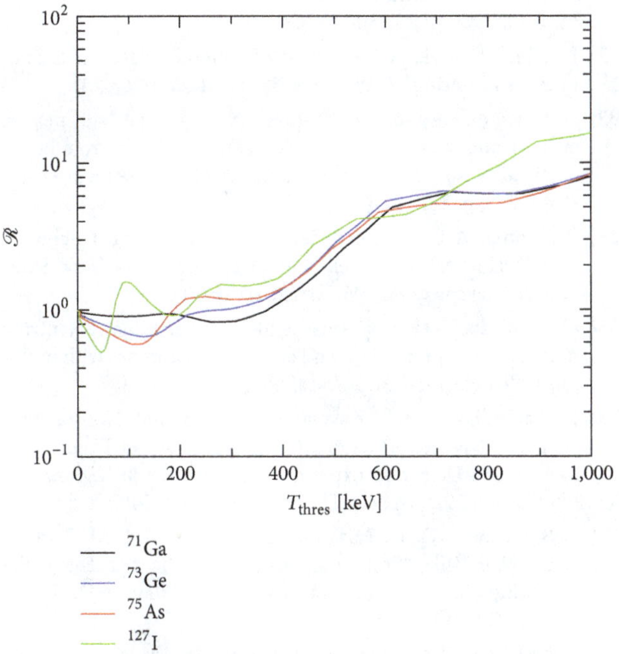

FIGURE 12: The ratio $\mathcal{R}$ as a function of the detector threshold.

nuclei, such as Na, Si, and Ar, clearly shell model will be better choice and DSM may also be tested for these isotopes.

More importantly, by exploiting the expected neutrino-floor due to Solar, Atmospheric, and DSNB neutrinos, which constitute an important source of background to dark matter searches, the impacts of new physics CEνNS contributions based on novel electromagnetic neutrino properties and Z' mediator bosons have been estimated and discussed. Our results also indicate that the addressed novel contributions may lead to a distortion of the expected recoil spectrum that could limit the sensitivity of upcoming WIMP searches. Such aspects could also provide key information concerning existing anomalies in B-meson decay at the LHCb experiment [77] and offer new insights into the LMA-Dark solution [78, 79].

Finally, the present results indicate that the addressed nuclear effects may become significant, leading to alterations especially in the high energy tail of the expected neutrino-floor as described by effective nuclear calculations, thus motivating further studies in the context of advanced nuclear physics methods such as the deformed shell model or the Quasiparticle Random Phase approximations and others. Such a comprehensive study using available data of the COHERENT experiment is under way and will be presented elsewhere.

Conflicts of Interest

The authors declare that they have no conflicts of interest.

Acknowledgments

R. Sahu is thankful to SERB of Department of Science and Technology (Government of India) for financial support. D. K. Papoulias is grateful to Professor Naumov for stimulating discussions.

References

[1] G. Jungman, M. Kamionkowski, and K. Griest, "Supersymmetric dark matter," *Physics Reports*, vol. 267, no. 5-6, pp. 195–373, 1996.

[2] G. Hinshaw, D. Larson, E. Komatsu et al., "Nine-Year Wilkinson Microwave Anisotropy Probe (WMAP) Observations: Cosmological Parameter Results," *The Astrophysical Journal Supplement Series*, vol. 208, p. 19, 2013.

[3] G. F. Smoot, C. L. Bennett, and A. Kogut, "Structure in the COBE differential microwave radiometer first-year maps," *The Astrophysical Journal*, vol. 396, no. 1, pp. L1–L5, 1992.

[4] P. A. R. Ade, N. Aghanim, M. Arnaud et al., "Planck 2015 results. XIII. Cosmological parameters," *Astronomy & Astrophysics*, vol. A13, p. 594, 2016.

[5] S. Das, T. Louis, M. R. Nolta et al., "The Atacama Cosmology Telescope: Temperature and Gravitational Lensing Power Spectrum Measurements from Three Seasons of Data," *Journal of Cosmology and Astroparticle Physics*, vol. 2014, no. 04, p. 14, 2014.

[6] E. M. George, C. L. Reichardt, K. A. Aird et al., "A measurement of secondary cosmic microwave background anisotropies from the 2500-square-degree SPT-SZ survey," *The Astrophysical Journal*, vol. 799, p. 177, 2015, https://arxiv.org/abs/1408.3161.

[7] E. Gawiser, "Extracting Primordial Density Fluctuations," *Science*, vol. 280, no. 5368, pp. 1405–1411.

[8] T. S. Kosmas and J. D. Vergados, "Cold dark matter in SUSY theories: the role of nuclear form factors and the folding with the LSP velocity," *Physical Review D: Particles, Fields, Gravitation and Cosmology*, vol. 55, no. 4, pp. 1752–1764, 1997.

[9] G. Bertone, D. Hooper, and J. Silk, "Particle dark matter: evidence, candidates and constraints," *Physics Reports*, vol. 405, no. 5-6, pp. 279–390, 2005.

[10] M. Kortelainen, T. Kosmas, J. Suhonen, and J. Toivanen, "Event rates for CDM detectors from large-scale shell-model calculations," *Physics Letters B*, vol. 632, no. 2-3, pp. 226–232, 2006.

[11] J. L. Feng, "Dark Matter Candidates from Particle Physics and Methods of Detection," *Annual Review of Astronomy and Astrophysics*, vol. 48, p. 495, 2010, https://arxiv.org/abs/1003.0904.

[12] A. L. Fitzpatrick, W. Haxton, E. Katz, N. Lubbers, and Y. Xu, "The effective field theory of dark matter direct detection," *Journal of Cosmology and Astroparticle Physics*, vol. 2013, no. 02, pp. 004-004, 2013.

[13] J. Vergados, F. Avignone, P. Pirinen, P. Srivastava, M. Kortelainen, and J. Suhonen, " Theoretical direct WIMP detection rates for transitions to the first excited state in ," *Physical Review D: Particles, Fields, Gravitation and Cosmology*, vol. 92, no. 1, 2015.

[14] P. C. Divari, T. S. Kosmas, J. D. Vergados, and L. D. Skouras, "Shell model calculations for light supersymmetric particle scattering off light nuclei," *Physical Review C: Nuclear Physics*, vol. 61, no. 5, 2000.

[15] E. Holmlund, M. Kortelainen, T. S. Kosmas, J. Suhonen, and J. Toivanen, "Microscopic calculation of the LSP detection rates for the ^{71}Ga, ^{73}Ge and ^{127}I dark-matter detectors," *Physics Letters B*, vol. 584, no. 1-2, pp. 31–39, 2004.

[16] D. S. Akerib and CDMS Collaboration, "Exclusion Limits on the WIMP-Nucleon Cross-Section from the First Run of the Cryogenic Dark Matter Search in the Soudan Underground Lab," *Physical Review D*, vol. 72, Article ID 052009, 2005, https://arxiv.org/abs/astro-ph/0507190.

[17] A. Broniatowski and EDELWEISS Collaboration, "A new high-background-rejection dark matter Ge cryogenic detector," *Physics Letters B*, vol. 681, no. 4, pp. 305–309, 2009.

[18] K. Freese, M. Lisanti, and C. Savage, "Annual Modulation of Dark Matter: A Review," *Reviews of Modern Physics*, vol. 85, no. 4, pp. 1561–1581, 2013.

[19] D. S. Akerib, S. Alsum, H. M. Araújo et al., "Results from a search for dark matter in the complete LUX exposure," *Physical Review Letters*, vol. 118, Article ID 021303, 2017, https://arxiv.org/abs/1608.07648.

[20] C. Fu and PandaX-II Collaboration, "Spin-Dependent Weakly-Interacting-Massive-Particle–Nucleon Cross Section Limits from First Data of PandaX-II Experiment," *Physical Review Letters*, vol. 118, Article ID 071301, 2017, [Erratum: *Physical Review Letters*, vol. 120, no.4, Article ID 049902, 2018].

[21] W. C. Haxton, R. G. Hamish Robertson, and A. M. Serenelli, "Solar Neutrinos: Status and Prospects," *Annual Review of Astronomy and Astrophysics*, vol. 51, p. 21, 2013, https://arxiv.org/abs/1208.5723.

[22] G. Battistoni, A. Ferrari, T. Montaruli, and P. R. Sala, "The atmospheric neutrino flux below 100 MeV: The FLUKA results," *Astroparticle Physics*, vol. 23, no. 5, pp. 526–534, 2005.

[23] K. C. Y. Ng, J. F. Beacom, A. H. G. Peter, and C. Rott, "Solar atmospheric neutrinos: A new neutrino floor for dark matter searches," *Physical Review D: Particles, Fields, Gravitation and Cosmology*, vol. 96, Article ID 103006, 2017.

[24] J. F. Beacom, "The Diffuse Supernova Neutrino Background," *Annual Review of Nuclear and Particle Science*, vol. 60, pp. 439–462, 2010.

[25] J. Monroe and P. Fisher, "Neutrino backgrounds to dark matter searches," *Physical Review D*, vol. 76, Article ID 033007, 2007.

[26] G. B. Gelmini, V. Takhistov, and S. J. Witte, "Casting a wide signal net with future direct dark matter detection experiments," *Journal of Cosmology and Astroparticle Physics*, 2018.

[27] J. Billard, E. Figueroa-Feliciano, and L. Strigari, "Implication of neutrino backgrounds on the reach of next generation dark matter direct detection experiments," *Physical Review D: Particles, Fields, Gravitation and Cosmology*, vol. 89, no. 2, Article ID 023524, 2014.

[28] C. A. O'Hare, "Dark matter astrophysical uncertainties and the neutrino floor," *Physical Review D*, vol. 94, Article ID 063527, 2016.

[29] J. B. R. Battat et al., "Readout technologies for directional WIMP Dark Matter detection," *Physics Reports*, vol. 662, pp. 1–46, 2016.

[30] D. Z. Freedman, "Coherent effects of a weak neutral current," *Physical Review D: Particles, Fields, Gravitation and Cosmology*, vol. 9, no. 5, pp. 1389–1392, 1974.

[31] V. A. Bednyakov and D. V. Naumov, "Coherency and incoherency in neutrino-nucleus elastic and inelastic scattering," https://arxiv.org/abs/1806.08768.

[32] D. Akimov et al., "Observation of coherent elastic neutrino-nucleus scattering," *Science*, vol. 357, pp. 1123–1126, 2017.

[33] D. K. Papoulias and T. S. Kosmas, "Nuclear aspects of neutral current non-standard ν-nucleus reactions and the role of the exotic $\mu^- \longrightarrow e^-$ transitions experimental limits," *Physics Letters B*, vol. 728, p. 482, 2014.

[34] P. Coloma, M. C. Gonzalez-Garcia, M. Maltoni, and T. Schwetz, "A COHERENT enlightenment of the neutrino Dark Side," *Physical Review D*, vol. 96, Article ID 115007, 2017.

[35] D. Aristizabal Sierra, N. Rojas, and M. H. G. Tytgat, "Neutrino non-standard interactions and dark matter searches with multi-ton scale detectors," https://arxiv.org/abs/1712.09667.

[36] D. Papoulias and T. Kosmas, "Standard and Nonstandard Neutrino-Nucleus Reactions Cross Sections and Event Rates to Neutrino Detection Experiments," *Advances in High Energy Physics*, vol. 2015, Article ID 763648, 17 pages, 2015.

[37] T. S. Kosmas, O. G. Miranda, D. K. Papoulias, M. Tortola, and J. W. F. Valle, "Probing neutrino magnetic moments at the Spallation Neutron Source facility," *Physical Review D*, vol. 92, Article ID 013011, 2015.

[38] T. S. Kosmas, O. G. Miranda, D. K. Papoulias, M. Tortola, and J. W. F. Valle, "Sensitivities to neutrino electromagnetic properties at the TEXONO experiment," *Physics Letters B*, vol. 750, p. 459, 2015.

[39] E. Bertuzzo, F. F. Deppisch, S. Kulkarni, Y. F. Perez Gonzalez, and R. Z. Funchal, "Dark matter and exotic neutrino interactions in direct detection searches," *Journal of High Energy Physics*, vol. 2017, no. 4, 2017.

[40] J. B. Dent, B. Dutta, J. L. Newstead, and L. E. Strigari, "Dark matter, light mediators, and the neutrino floor," *Physical Review D: Particles, Fields, Gravitation and Cosmology*, vol. 95, no. 5, 2017.

[41] D. Papoulias and T. Kosmas, "COHERENT constraints to conventional and exotic neutrino physics," *Physical Review D: Particles, Fields, Gravitation and Cosmology*, vol. 97, no. 3, 2018.

[42] A. Cisterna, T. Delsate, L. Ducobu, and M. Rinaldi, "Sensitivity to Z-prime and nonstandard neutrino interactions from ultralow threshold neutrino-nucleus coherent scattering," *Physical Review D*, vol. 93, Article ID 013015, 2016.

[43] L. E. Strigari, "Neutrino floor at ultralow threshold," *Physical Review D*, vol. 93, Article ID 103534, 2016.

[44] http://www.nndc.bnl.gov/ensdf.

[45] S. Gardner and G. Fuller, "Dark matter studies entrain nuclear physics," *Progress in Particle and Nuclear Physics*, vol. 71, pp. 167–184, 2013.

[46] T. S. Kosmas, A. Faessler, and R. Sahu, " Transition matrix elements for ," *Physical Review C: Nuclear Physics*, vol. 68, no. 5, 2003.

[47] V. K. B. Kota and R. Sahu, *Structure of Medium Mass Nuclei: Deformed Shell Model and Spin-Isospin Interact-ing Boson Model*, CRC Press, 2016.

[48] R. Sahu and V. K. Kota, "Deformed shell model study of event rates for WIMP-73Ge scattering," *Modern Physics Letters A*, vol. 32, no. 38, p. 1750210, 2017.

[49] P. Pirinen, P. Srivastava, J. Suhonen, and M. Kortelainen, " Shell-model study on event rates of lightest supersymmetric particles scattering off ," *Physical Review D: Particles, Fields, Gravitation and Cosmology*, vol. 93, no. 9, 2016.

[50] P. Toivanen, M. Kortelainen, J. Suhonen, and J. Toivanen, "Large-scale shell-model calculations of elastic and inelastic scattering rates of lightest supersymmetric particles (LSP) on I-127, Xe-129, Xe-131, and Cs-133 nuclei," *Physical Review C*, vol. 79, Article ID 044302, 2009.

[51] J. D. Vergados and T. S. Kosmas, "Searching for Cold Dark Matter. A case of coexistence of Supersymmetry and Nuclear Physics," *Physics of Atomic Nuclei*, vol. 61, p. 1066, 1998, https://arxiv.org/abs/hep-ph/9802270.

[52] J. Wyenberg and I. M. Shoemaker, "Mapping The Neutrino Floor For Dark Matter-Electron Direct Detection Experiments," *Physical Review D*, vol. 79, Article ID 115026, 2018.

[53] D. K. Papoulias and T. S. Kosmas, "Neutrino transition magnetic moments within the non-standard neutrino-nucleus interactions," *Physics Letters B*, vol. 747, pp. 454–459, 2015.

[54] J. Beringer, J. F. Arguin, and R. M. Barnett, "Review of particle physics," *Physical Review D: Particles, Fields, Gravitation and Cosmology*, vol. 86, no. 1, Article ID 010001, 2012.

[55] J. Barranco, O. G. Miranda, and T. I. Rashba, "Probing new physics with coherent neutrino scattering off nuclei," *Journal of High Energy Physics*, vol. 2005, article 021, 14 pages, 2005.

[56] W. M. Alberico, S. M. Bilenky, and C. Maieron, "Strangeness in the nucleon: neutrino—nucleon and polarized electron—nucleon scattering," *Physics Reports*, vol. 358, no. 4, pp. 227–308, 2002.

[57] K. Scholberg, "Prospects for measuring coherent neutrino-nucleus elastic scattering at a stopped-pion neutrino source," *Physical Review D: Particles, Fields, Gravitation and Cosmology*, vol. 73, Article ID 033005, 2006.

[58] W. Grimus, M. Maltoni, T. Schwetz, M. A. Tortola, and J. W. F. Valle, "Constraining Majorana neutrino electromagnetic properties from the LMA-MSW solution of the solar neutrino problem," *Nuclear Physics B*, vol. 648, no. 1-2, pp. 376–396, 2003.

[59] B. Cañas, O. Miranda, A. Parada, M. Tórtola, and J. Valle, "Addendum to "Updating neutrino magnetic moment constraints," *Physics Letters B*, vol. 757, p. 568, 2016.

[60] J. B. Dent, B. Dutta, S. Liao, J. L. Newstead, L. E. Strigari, and J. W. Walker, "Probing light mediators at ultralow threshold energies with coherent elastic neutrino-nucleus scattering," *Physical Review D: Particles, Fields, Gravitation and Cosmology*, vol. 96, no. 9, 2017.

[61] I. M. Shoemaker, "A COHERENT Search Strategy for Beyond Standard Model Neutrino Interactions," *Physical Review D: Particles, Fields, Gravitation and Cosmology*, vol. 95, Article ID 115028, 2017.

[62] M. Lindner, W. Rodejohann, and X.-J. Xu, "Coherent Neutrino-Nucleus Scattering and new Neutrino Interactions," *Journal of High Energy Physics*, vol. 3, no. 97, 2017.

[63] D. G. Cerdeño, M. Fairbairn, T. Jubb, P. A. Machado, A. C. Vincent, and C. Bœhm, "Erratum to: Physics from solar neutrinos in dark matter direct detection experiments," *Journal of High Energy Physics*, vol. 2016, no. 9, 2016.

[64] J. Liao and D. Marfatia, "COHERENT constraints on nonstandard neutrino interactions," *Physics Letters B*, vol. 775, pp. 54–57, 2017.

[65] V. Antonelli, L. Miramonti, C. Pena Garay, and A. Serenelli, "Solar Neutrinos," https://arxiv.org/abs/1208.1356.

[66] G. Bellini, J. Benziger, and D. Bick, "Neutrinos from the primary proton-proton fusion process in the Sun," *Nature*, vol. 512, no. 7515, pp. 383–386, 2014.

[67] S. Horiuchi, J. F. Beacom, and E. Dwek, "Diffuse supernova neutrino background is detectable in Super-Kamiokande," *Physical Review D: Particles, Fields, Gravitation and Cosmology*, vol. 79, no. 8, 2009.

[68] D. P. Ahalpara, K. H. Bhatt, and R. Sahu, "Collective bands in 81Sr," *Journal of Physics G: Nuclear and Particle Physics*, vol. 11, no. 6, pp. 735–743, 1985.

[69] L. Coraggio, L. De Angelis, T. Fukui, A. Gargano, and N. Itaco, "Calculation of Gamow-Teller and two-neutrino double-β decay properties for 130Te and 136Xe with a realistic nucleon-nucleon potential," *Physical Review C*, vol. 59, Article ID 064324, 2018.

[70] R. Sahu and V. K. B. Kota, "Deformed shell model for T=0 and T=1 bands in 62Ga and 66As," *Physical Review C*, vol. 66, Article ID 024301, 2002.

[71] R. Sahu and V. K. B. Kota, "Deformed shell model for collective T=0 and T=1 bands in 46V and 50Mn," *Physical Review C*, vol. 67, Article ID 054323, 2003.

[72] P. C. Srivastava, R. Sahu, and V. K. B. Kota, "Shell model and deformed shell model spectroscopy of 62Ga," *The European Physical Journal A*, vol. 51, no. 3, 2015.

[73] B. Ding et al., "High-spin states in ^{127}I," *Physical Review C*, vol. 85, Article ID 044306, 2012.

[74] A. Beda, V. Brudanin, V. Egorov et al., "The results of search for the neutrino magnetic moment in GEMMA experiment," *Advances in High Energy Physics*, vol. 2012, Article ID 350150, 12 pages, 2012.

[75] J. Billard, J. Johnston, and B. J. Kavanagh, "Prospects for exploring New Physics in Coherent Elastic Neutrino-Nucleus Scattering," https://arxiv.org/abs/1805.01798.

[76] M. Abdullah, J. B. Dent, B. Dutta, G. L. Kane, S. Liao, and L. E. Strigari, "Coherent Elastic Neutrino Nucleus Scattering (CEνNS) as a probe of $Z\prime$ through kinetic and mass mixing effects," *Physical Review D*, vol. 98, Article ID 015005, 2018.

[77] M. Abdullah, M. Dalchenko, B. Dutta et al., "Bottom-quark Fusion Processes at the LHC for Probing $Z\prime$ Models and B-meson Decay Anomalies," *Physical Review D*, vol. 97, Article ID 075035, 2018.

[78] Y. Farzan, "A model for large non-standard interactions of neutrinos leading to the LMA-Dark solution," *Physics Letters B*, vol. 748, pp. 311–315, 2015.

[79] P. Coloma, P. B. Denton, M. C. Gonzalez-Garcia, M. Maltoni, and T. Schwetz, "Curtailing the dark side in non-standard neutrino interactions," *Journal of High Energy Physics*, vol. 2017, no. 4, 2017.

22

Simulations of Gamma-Ray Emission from Magnetized Microquasar Jets

Odysseas Kosmas[1] **and Theodoros Smponias**[2]

[1] *Modelling and Simulation Centre, MACE, University of Manchester, Sackville Street, Manchester, UK*
[2] *Division of Theoretical Physics, University of Ioannina, 45110 Ioannina, Greece*

Correspondence should be addressed to Odysseas Kosmas; odysseas.kosmas@manchester.ac.uk

Academic Editor: Athanasios Hatzikoutelis

In this work, we simulate γ-rays created in the hadronic jets of the compact object in binary stellar systems known as microquasars. We utilize as the main computational tool the 3D relativistic magnetohydrodynamical code PLUTO combined with in-house derived codes. Our simulated experiments refer to the SS433 X-ray binary, a stellar system in which hadronic jets have been observed. We examine two new model configurations that employ hadron-based emission mechanisms. The simulations aim to explore the dependence of the γ-ray emissions on the dynamical as well as the radiative properties of the jet (hydrodynamic parameters of the mass-flow density, gas-pressure, temperature of the ejected matter, high energy proton population inside the jet plasma, etc.). The results of the two new scenarios of initial conditions for the microquasar stellar system studied are compared to those of previously considered scenarios.

1. Introduction

The emissions of γ-rays, neutrinos, etc. within the jets of microquasars (MQs) have recently gained great interest among researchers seeking to understand the structure properties and evolution of X-ray binary systems [1–3].

Special interests appeared on the γ-ray emission mechanisms inside the hadronic jets, as the photon-hadron interactions [4, 5] and the hadron-hadron interactions [6, 7] as well as the γ-ray absorption that help to deepen our knowledge on microquasars evolution [8].

On the other hand, the strong magnetic field in the jets may significantly affect the total internal γ-ray and neutrino emissions by tuning several processes determining the high energy proton population (synchrotron radio emission, etc.). Therefore, magnetic field effects should be appropriately incorporated and treated in jet models [9, 10].

Recently, neutrinos from galactic microquasars, even though not being detected so far, have been modelled and several simulations have been performed towards this aim. Such modelling may support future attempts to detect them (see, e.g., [7, 11]).

Invariably, the jets of microquasars as well as in general the astrophysical jets may be described as fluid flow emanating from the vicinity of the compact object. Such a microquasar system is the SS433 X-ray binary consisted of a donor (companion) star and a compact stellar object which emits relativistic jets in various wavelength bands. Currently it is the only microquasar observed with a definite hadronic content in its jets, as verified from observations of spectral lines [1, 2, 9].

Radiative transfer calculations may be performed at every point in the jet (for a range of frequencies/energies, at every location) [12], providing the relevant emission and absorption coefficients. In such cases, finally a line of sight integration may derive synthetic images of jet γ-ray emission, at the energy-window of interest [12, 13].

The relativistic treatment of jets takes into account various energy loss mechanisms that occur through several hadronic processes [4–7]. In the known fluid approximation, macroscopically the jet matter behaves as a fluid collimated by the magnetic field. At a smaller scale, consideration of the kinematics of the jet plasma becomes necessary for treating shock acceleration effects.

Many authors consider that the proton-proton (p-p) collisions between fast high energy protons (nonthermal protons) and bulk-flow slow (thermal) protons constitute the dominant cooling process of the high energy proton population of the jet. This mechanism explains the main part of the γ-rays and neutrinos produced in the binary SS433 system. The acceleration of thermal protons (diffusive first-order shock acceleration) occurs above a minimum threshold proton energy [7].

Assuming a Maxwellian energy distribution for the "slow" protons, only a tiny portion of the total bulk proton jet flow, i.e., the fastest of them, may undergo diffusive shock acceleration and may jump to the fast proton population. Hence, the fast protons constitute a small fraction of the total jet proton density which subsequently produce γ-rays, neutrinos, etc. In this work, we assume that this is the dominant mechanism generating high energy γ-rays in the SS433 microquasar jets.

For the sake of completeness, we mention that another rather important mechanism has been suggested based on the hadronic interactions occurring within the jet-wind interaction zone [7]. In this scenario, the γ-rays are generated from the decay of neutral pions, as $\pi^0 \longrightarrow \gamma + \gamma$. Pions are created via inelastic collisions of jet protons, ejected from the compact object, and ions of the stellar wind (such a process may also occur in the vicinity of the extended disk of the binary system) [6]. The latter emission mechanism is rather weak in SS433 [7].

So far microquasar γ-ray emissions have been observed through Cherenkov telescopes (HESS, MAGIC, and CTA) and orbital telescopes (INTEGRAL, Fermi) [15–20]. We also mention that, for low energy γ-rays, ongoing and future or next generation measurements with INTEGRAL (ESA satellite) and Fermi (NASA orbital telescope) may provide new data. Furthermore, very high energy γ-rays, in general above about 30GeV, can be studied with ground-based Cherenkov telescopes [21].

Phenomenologically, estimations of high energy γ-ray emission from MQs have extensively been carried out [9, 12]. In this work, using the 3D relativistic hydrocode PLUTO [22] and some in-house (mainly radiative transfer) code (now written both in Mathematica and in C) [23–25], we model γ-ray emissions from hadronic microquasar jets in the E_γ-energy range 1.2 GeV $\leq E_\gamma \leq 10^2 - 10^3$ TeV.

The emission/absorption coefficients are computed on the basis of Monte Carlo simulations of terrestrial particle-particle collision experimental data [12, 26, 27] that describe γ-ray emission in MQs. Such simulations provide analytical parametrization for emission and absorption coefficients in a wide range of γ-ray energies (frequencies) produced in microquasar jets [12, 28].

Furthermore, by exploiting the hydrodynamic variable values supplied by PLUTO, our line of sight code may provide emission/absorption coefficients for every location in the jet. The results produced this way depend on the initial high energy proton distribution inserted in the hydrodynamical model jet [12, 29].

In the rest of the paper, at first (Section 2) the main MQs emissions mechanisms are briefly summarized. Then (Section 2.4), the radiative transfer method and the calculational procedure for obtaining gamma-ray emission are briefly described. The results of the 3D relativistic hydrocode PLUTO for the emission/absorption coefficients are presented in Section 4. Finally (Section 5), the main conclusions extracted in this work are summarized.

2. Outline of MQ Jet Emission Mechanisms

The SS433 microquasar, an eclipsing X-ray binary system with a compact object most likely a black hole, comprises two oppositely directed precessing hadronic jets. The spectrum of the companion (donor) star suggests that it is rather a late A-type MQ. In modelling γ-ray emission from SS433 in our present work, we assume that they are created mainly through p-p interactions between fast (relativistic) and slow (cold) protons within its hadronic jets.

Other production mechanisms, though not excluded, are considered less important. For example, some authors considered that the high energy γ-rays in hadronic MQ jets, are produced from p-p collisions taking place in the jets due to the interaction of relativistic protons with target protons of the rather weak stellar wind created in the companion star [6].

The main reaction chain that produces γ-rays starting from p-p interaction through the pion decay channel is written as

$$p + p \longrightarrow p + p + \pi^0 \longrightarrow p + p + 2\gamma \qquad (1)$$

($m_p = 1.67 \times 10^{-24}$ g and $m_\pi = 2.38 \times 10^{-25}$ g). References [6, 7, 26] present an analytical description of the evolution of reaction chain within the jet. Here we assume that a very energetic but small proton population, N_{fp} (formed due to shock fronts in the jet), interacts with the bulk-flow jet protons.

From the latter protons, high energy protons are produced through first-order Fermi acceleration that occurs at shocks within the jet [7]. Such shocks are considered rather homogeneously distributed throughout the jet. The jet matter density is closely related to the density of the aforementioned shocks; thus, the internal shocks convert a portion of the bulk kinetic energy of cold protons, K, to the fast protons energy E_p of the multidirectional motion. The rate, t_{acc}^{-1}, at which some slow protons are transferred to the high energy distribution is described in [1] by

$$r = t_{acc}^{-1} = E^{-1}\frac{dE}{dt} \simeq \beta^2 \frac{ceB}{E_p}, \qquad (2)$$

where e denotes the proton charge, $\beta = u_{jet}/c$ with u_{jet} being the jet matter's local velocity, and B denotes the magnetic field.

Concerning the magnetic field B, we assume that this is either constant or it decreases with the distance from the jet base as $B \sim z^{-1}$ [30, 31]. We stress that such a variation leads to a decrease of proton acceleration not more than two orders

TABLE 1: Scenario C (run 3) has artificially accelerated precession, while scenario D (run 4) has all the densities of the system increased by a few orders of magnitude, in order to account for a higher jet-mass flow-rate (jet's kinetic luminosity). The parameter n refers to a "normalization" process that equates the results of two different methods of γ-ray emission calculations, one applied for energies above E_γ=100 GeV and the other below this limit.

Parameter/Scenario	C (run3)	D (run4)	Comments
cell size ($\times 10^{10} cm$)	0.40	0.40	PLUTO's computational cell
ρ_{jet} (cm^{-3})	1.0×10^{11}	1.0×10^{14}	jet's matter density
ρ_{sw} (cm^{-3})	1.0×10^{11}	1.0×10^{12}	stellar wind density
ρ_{adw} (cm^{-3})	1.0×10^{11}	1.0×10^{13}	accretion disk wind density
t_{run}^{max} (s)	1.5×10^3	1.5×10^3	model execution time
Method	P. L.	P. L.	Piecewise Linear
Integrator	Ch. Tr.	Ch. Tr.	Characteristic Tracing
EOS	Ideal	Ideal	Equation of state
n	0.1005	0.1005	E_γ = 100 GeV normalisation
BinSep (cm)	4.0×10^{12}	4.0×10^{12}	Binary star separation
$M_{BH}/M_\odot$	3-10	3-10	Mass range of collapsed star
$M_*/M_\odot$	10-30	10-30	Mass range of Main Seq. star
$\beta = v_0/c$	0.26	0.26	Initial jet speed
L_k^p	10^{36}	10^{39}	Jet kinetic luminosity
grid resolution	$300 * 500 * 300$	$300 * 500 * 300$	PLUTO grid resolution (xyz)

of magnitude compared to its value around the jet's base $z = z_0$. We further stress that the acceleration rate of fast protons, $r = t_{acc}^{-1}$, depends on the magnetic field B as indicated in (2).

Alternatively, (2) gives the production rate of fast protons at every "location" in the jet, though the production of γ-rays from these fast protons occurs at a next stage (by "location" we mean a hydrodynamical grid-cell which, microscopically, is very large); of the order of 10^{10} cm [7]. We note that the presence of β^2 in (2) cuts off γ-ray emission from slow-moving matter into the jet (acceleration sites are much less in slow matter).

Moreover, in jet emission calculations the β^2 is incorporated into the jet density. Also, the proton acceleration rate cannot affect the γ-ray emission rate, unless β drops below some value. As we will see below, in this work, instead of the proton density ρ, we also use the product of jet matter density times the velocity squared ρu^2 [7, 13].

Regarding the particles ejected from a hadronic jet, we consider that they are mostly slow (thermal) protons of density n_{sp} and a small portion of fast (nonthermal) protons of density n_{fp}. The energy distribution of the fast protons in the jet's frame (see below) is described by [7, 13]

$$n_{fp}(E') = K_0 (E')^{-\alpha} \quad (3)$$

($E' \equiv E'_{fp}$) which is a power law type distribution. The parameter α takes the value $\alpha = 2$ and K_0 denotes a normalization constant [13].

In Table 1, we tabulate the values of some model jet parameters (together with explanation of their symbols) relevant to γ-ray emissions from the SS433 binary system (The scenarios C and D are described below).

2.1. Flux in Observation and Jet's Frame. In our γ-ray flux calculations, we denote the flux density of the fast (slow) proton populations as J_{fp} (J_{sp}), in the observation frame, and as J'_{fp} (J'_{sp}), in the jet's frame. Moreover, at a given jet point, in the jet (moving) frame, the fast protons energy-spectrum is described by (3). For the corresponding fast protons spatial density we adopt the relation

$$n_{fp}(E') \propto w \frac{dN'_{fp}}{dE'} = w K_0 (E')^{-\alpha} \quad (4)$$

where

$$w = n_{sp}\beta^2 \quad (5)$$

($\beta \equiv u/c$, with $u = |\mathbf{u}|$ being the magnitude of the total velocity vector) where n_{sp} is in protons/cm^3 [29]. From (5) one can conclude that the emission from the fast-moving matter of the jet is larger compared to the emission from slow-moving matter which is because in (2), t_{acc} is proportional to β^2 and creates fast proton jet density which subsequently allows for p-p collisions to occur and for γ-rays to be produced. [6].

2.2. The Model of Jet's Dynamics. The jet is assumed to travel along the y axis (we consider particles of mass m_p). Then, for the flux densities, we can write (steady state)

$$J_{sp} = m_p w = m_p n_{sp} \beta^2 \quad (6)$$

In general, u is not necessarily parallel to the y-axis, but it may point almost anywhere which in turn means that emission may occur from jet matter moving in any direction. Furthermore, the emission mechanism is based on "randomly oriented turbulent shocks", so the emission is considered multidirectional (no secondary emissions from scattering are assumed, since more shocks exist wherever the jet matter moves faster).

In our simulations, large turbulences of the jet flow may appear which favor shocks existence. This is due to the assumed strong dependence on the local velocity of the jet or ambient matter (acceleration rate is proportional to u^2 and further $J = \rho u^2$). Here, instead of the simple ρ dependence, we adopt, in addition, the ρu^2 dependence to distinguish the moving matter of the jet from that of the surrounding medium. This way, the calculation of γ-rays and neutrino emissions from the jet are decoupled from the influence of the surrounding matter. Then, the jet's contribution to high energy γ-ray emission is mostly dependent on its internal turbulence (turbulence here means spatial number density of proton accelerating shocks randomly oriented) [7].

From the above discussion, we note in short that the proton acceleration efficiency of the model jet is proportional to the square of the local velocity of the flow (the fast proton density is considered proportional to the square of the local velocity). Thus, the fast protons spatial density, n_{fp}, is also taken as proportional to the slow protons spatial density, n_{sp} as well as to the square of the local velocity.

Furthermore, for hydrodynamical jets [12, 13] the fast proton current density, $J'_{fp}(E')$, as a function of their energy, is given by

$$J'_{fp}\left(E'\right) = \frac{c}{4\pi}K_1 n_j \beta_j^2 \left(E'\right)^{-\alpha} \tag{7}$$

In the latter expression, n_j denotes the slow bulk jet protons local density hydrodynamical model (PLUTO code) [12].

2.3. The Current Density in the Observer's Frame. Regarding the transformation, to the observer frame, we write [32]

$$J_{fp}\left(E_p, t\right) = \frac{c}{4\pi}K_1 n_{sp}\beta^2 F. \tag{8}$$

F represents a function of stationary frame energy E_p written as

$$F = \frac{\gamma^{-\alpha+1}\left(E_p - \beta_b\sqrt{E_p^2 - m_p^2c^4}\cos i_j\right)^{-\alpha}}{\left[\sin i_j{}^2 + \gamma^2\left(\cos i_j - \left(\beta_b E_p\right)/\sqrt{E_p^2 - m_p^2c^4}\right)^2\right]^{1/2}} \tag{9}$$

In the latter equation, $i_j(t)$ denotes the angle between the jet axis and the line of sight (for SS433 microquasar $\beta_b = v_b/c = 0.26$), and

$$\gamma = \left[1 - \beta_b^2\right]^{-1/2} \tag{10}$$

is the jet Lorentz factor.

Thus, F provides the relation of J_{fp} (for laboratory frame) that depends on the γ-ray energy E_γ as measured in laboratory frame (see (7)). In conclusion, one can work with laboratory frame quantities only, which are also the jet model quantities (for other symbols the reader is referred to [12, 13, 28, 29]

2.4. The 3D Radiative Transfer in Time-Dependent Jet. The propagation of γ-rays along a one-dimensional line of sight (without scattering) we address here is based on the relation

$$\frac{dI_\nu}{dl} = -I_\nu\kappa_\nu + \epsilon_\nu \tag{11}$$

where κ_ν is the absorption coefficient at a given frequency (energy) ν and ϵ_ν is the relevant emission coefficient. I denotes the intensity and l is the length along the line of sight.

By considering the model jet artificially imaged in γ-rays, we calculate the emission from a small jet element corresponding to a computational cell (for simulated emissions). To this aim, we first define the quantity

$$dI_\gamma = J_\gamma dV = \rho dV\frac{dN_\gamma}{dE_\gamma} = dm_{cell}\frac{dN_\gamma}{dE_\gamma} \tag{12}$$

to represent the intensity created from a cell of volume dV, at a given frequency, or γ-ray energy E_γ, while ρ is the hydrodensity of the cell ($\rho = m_p n_{sp}$) and dN_γ stands for the emission coefficient of the cell at the same frequency. An alternative version of the above quantity is

$$dI_\gamma = \rho dV\frac{dN_\gamma}{dE_\gamma}u^2, \tag{13}$$

where u is the local jet matter velocity.

3. Use of PLUTO Code for Gamma-Ray Emission Calculations

Our calculation of the emission coefficient dN_γ proceeds directly starting from the hydrodynamical properties of the model jet (supplied by a check point of the PLUTO code). These quantities enter the calculation of the emission coefficients, ϵ_γ, at every computational cell of the 3D hydrodynamical model grid.

We mention that the production of synthetic images from the data is carried out by using the line of sight code constructed in [12] (here, instead of the radio emission and absorption coefficients we require their γ-ray equivalents, ϵ_γ and k_γ; in the case of neutrino production we need only emission coefficients).

In using the hydrodynamical code, PLUTO, the energy E_γ (in GeV) refers to the observed γ-ray energy. The quantity n_{sp}, i.e., the bulk-flow slow jet proton number density of ejected particles, is the dynamically important. This number density is taken to represent the hydrodynamic number density of the PLUTO code, namely,

$$n_{sp} = n_{j(\text{PLUTO})} \tag{14}$$

The fast proton density, n_{fp}, though not-important dynamically, is radiatively important. For SS433, in the fast proton power law energy distribution, the index α takes the value $\alpha = 2$, and the ratio of the initial jet beam speed u_b divided by the speed of light is $u_b/c = \beta_b = 0.26$ [13].

In the hydrodynamic (HD) simulations with PLUTO code, as we have done previously [12, 13], the magnetic field lines are assumed to follow the matter flow. Their tangling with the jet material makes applicable the fluid approximation within the jet [33]. In this case, the magnetic field could not affect the flow dynamics which is however possible in the magnetohydrodynamic (MHD) treatment of PLUTO. In the relativistic hydrodynamical version of PLUTO, the magnetic field within the jet's medium is assumed rather strong so as the coupling effects permit the fluid approximation to be applicable. At the same time, the dynamical effects of the magnetic field on the relativistic flow are not permitted [33].

We mention that, in modelling the microquasar SS433 system with PLUTO, only one of the twin jets is considered. The counterjet is presumed to exist outside the model space (at the bottom of $x - z$ plane), but its interference with our model system is considered very small [34]. The computational grid is 3D Cartesian (x, y, z), homogeneous and the boundary conditions are adopted to be reflective at the jet's base ($x - z$ plane) and outflow at all other planes of the computational domain (box).

The grid spans $120 \times 200 \times 120$ (for x, y, z, respectively), in model length units (equal to 10^{10}cm) and the resolution used is $300 \times 500 \times 300$ (for x, y, z, respectively. The jet emanates from the middle of the $x - z$ plane, at the point (60, 0, 60)$\times 10^{10}$cm and then advances while precessing around the (60, y, 60) line (parallel to the y-axis). The precession angle for the SS433 jet used (δ = 0.2 radians), is slightly smaller than the value of 21 degrees of [34], in order to allow the use of finer resolution.

The centre of the companion star is supposed to be outside of the box, at the point (400, 0, 400) while compact object is situated at the point (60, 0, 60), i.e., at the jet's base. We remind that, because the exact orbital separation in SS433 is not well known, this estimation is within an order of magnitude (the objects are orbiting around their centre-of-mass). We also mention that we assume that the companion star is not included in the model [12]. However, its wind is included through its density which is taken as decreasing with distance r (as $1/r^2$), away from its centre.

Furthermore, we also include a simplified accretion disk wind through the jet's dynamic interaction with both winds. This means that our model is less realistic as we approach the companion star and accretion disk locations but the results are reliable in the vicinity of the jet. For a detailed discussion related to important phenomena of the jet's interaction zone with nearby winds, the reader is referred to the Refs. [12, 13, 28, 29] and references therein.

4. Results of Simulations for the New Jet Model Scenarios

In this section, we present the results for two new scenarios of initial conditions (referred to the microquasar stellar system SS433) obtained as follows: (i) Hydrodynamical simulation carried out by utilizing as main computational tool the 3D relativistic hydrocode PLUTO and (ii) Gamma-ray emission synthetic images obtained with the line of sight integration.

In scenario C, the jet precesses faster than reality, therefore precession effects are enhanced. In this case, the jet involves artificially accelerated precession, in order to better investigate the effects of precession on the surrounding winds, within the limited time-frame of the model run.

In scenario D, the jet is quite heavier than both winds, in order to consider the possibility of a dense jet beam, containing the estimated jet mass flow of SS433 while remaining more focused and more both narrow. The heavier jet of this scenario crosses the winds with greater ease. Also the effects of its interaction with the winds appear decreased.

We note that another characteristic scenario would have been a jet much lighter than both winds, but this would have taken longer simulation time, and practically more difficult.

In both cases, the jet begins to expand into the accretion disk wind, but at a more limited pace, due to the increased density of that wind in the model. As soon as the jet head reaches the stellar wind region, however, the jet's expansion rate increases greatly (especially sideways), in the form of a side shock that accumulates ambient matter. At the same time, the accretion disk matter is expelled outwards, from the vicinity of the jet base, forming a "ring" around the jet. The accretion disk wind is swept in a prominent way, being denser than the stellar wind, leading to the creation of a halo around the jet base (see below).

The structure develops throughout the model run, therefore suggesting the possibility of its persistence later on, when the jet reaches its lobe in the W50 nebula. This is similar (to a certain extent) in structure to that discussed in [14]. The above scenarios are applied to the SS433 microquasar as described below.

4.1. Description of Runs for Scenarios C and D

4.1.1. Simulations of Scenario C . Scenario C (medium resolution, see Figures 1, 2, 3, and 4) has been chosen to cover the case of artificially fast precession of the jet. This way we may investigate the effects of precession on the system observables. The precessing jet sweeps across more of the ambient matter in a given period of time, as compared to an otherwise same but nonprecessing jet (run2-scenario), with both jet examples considered to be moving through identical surroundings. Consequently, the effects of the precessing jet on its surrounding environment are, in terms of affected volume, more prominent than when precession is absent. More ambient matter is displaced and part of it ends up being dragged along by the jet, albeit at a pace clearly slower than when precession is slower.

The precessing jet model was therefore run, at an accelerated precession rate, and that showed an enhanced effect of sweeping the accretion disk wind matter from the jet cone. This, in effect, caused the formation of an enhanced, outward moving, "halo", around the jet base (Figures 1 and 2). The precessing jet advances, first through the accretion disk wind, and then through the stellar wind, opening its path at an accelerating pace, due to meeting with progressively lower resistance, due to the falling density of the winds. The latter originated from both the accretion disk and the companion star. The gradient of the stellar wind, within the computational grid, is, however, not big, due to the increased distance from its origin, the companion star.

The jet precession pronounces the sweeping of the inner, denser part of the accretion disk wind, leading, later on, to the formation of an expanding approximately torus-shaped halo surrounding the jet whose axis roughly coincides with the axis around which the jet precesses. The torus

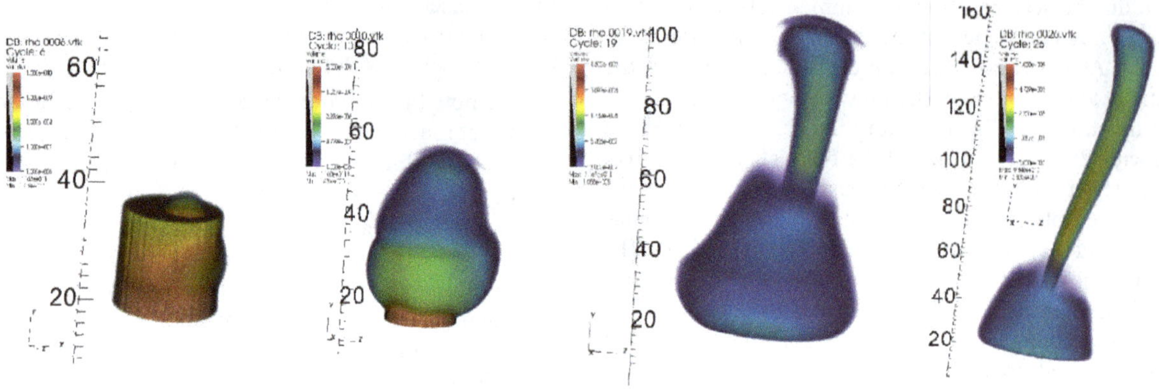

(a) (b) (c) (d)

FIGURE 1: Scenario C: three-dimensional density snapshots obtained as in scenario A of [13], but now we added an accelerated precession of the jet that leads to the formation of a revolving jet flow. See the rightmost panel of Table 1, where the jet is the most evolved among the plots (parameters of run3, snapshot intervals 50/count). The jet's behaviour resembles to that of scenario A (run1) of [13]. Also, now the jet sweeps a larger volume of ambient matter and propagates slower than in the nonprecession case, though now more ambient matter is further activated for emission. Matter piles up around the jet base, hinting on a halo of slower moving (outbound) material there, crudely reminiscent of the suggestion of [14] for a halo formed around SS433.

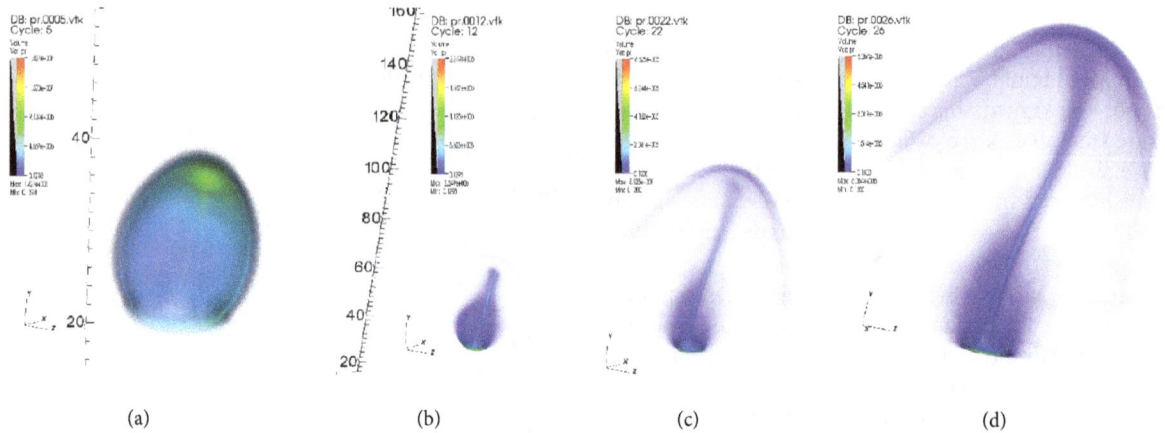

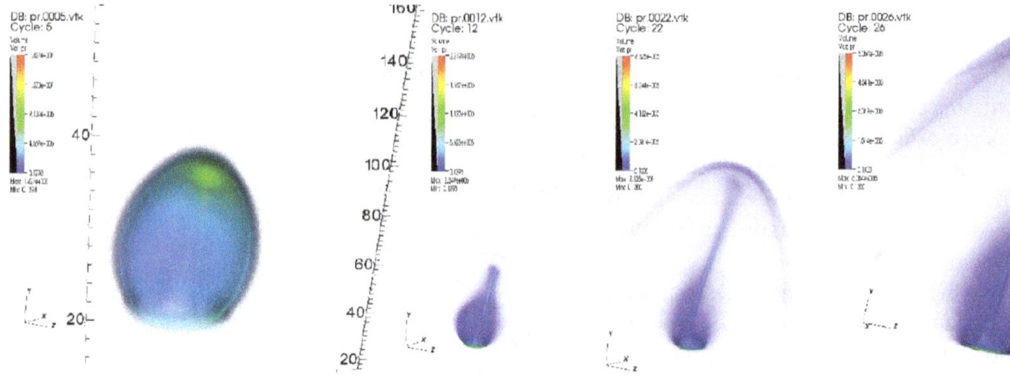

(a) (b) (c) (d)

FIGURE 2: Scenario C: 3D illustrations of pressure evolution in the vicinity of the jet, as well as in the jet itself (in this case linear plots are produced, in order to better display the periphery of the system; see run3 parameters in Table 1). The three rightmost plots (b, c, d) share the same spatial size scale, whereas (a) has been magnified. The jet precession is visible, especially in (c) and (d), while a pressurized mass concentration can be found in the vicinity of the jet base, persisting throughout the model run and slowly advancing outwards. The front of the jet itself, due to precession, advances in an asymmetrical way.

consists mainly of accretion disk wind matter, forming a loose "barrel", or hollow cylinder (Figures 1 and 2) around the jet, having a velocity component (clearly slower than the jet, i.e., subrelativistic) parallel to the jet's precession axis, and a sideways expansion velocity component as well.

In the scale of the simulation, the companion star has a nonnegligible distance from the jet base, therefore in the immediate vicinity of the jet base, it is the more localized accretion disk wind that dominates over the stellar wind, in terms of density. It is, therefore, the accretion disk wind matter that is mainly expelled from the incoming jet and rushed outwards, swept over by the precessing jet. This finding might suggest a behaviour that over a much longer timescale leads to the formation of a "ruff" of material, around the cone swept by the precessing jet, perhaps along the lines of the "bow-tie" structure recently observed in SS433 [14].

The jet precession makes it possible to drag an increased quantity of surrounding matter along the jet which is in contrast to a nonprecessing jet where the swept matter is significantly less. This happens because the precessing jet covers a cone with an opening angle much larger than the jet's own. Therefore, more ambient matter is displaced than from a straight jet. Furthermore, the partially sideways motion of the precessing jet further disturbs the surrounding winds, pushing and dragging them in an outwards direction. However, the time scale for precession is much bigger than the jet crossing time of the model space. Therefore, a longer term simulation, perhaps including replenishment of the

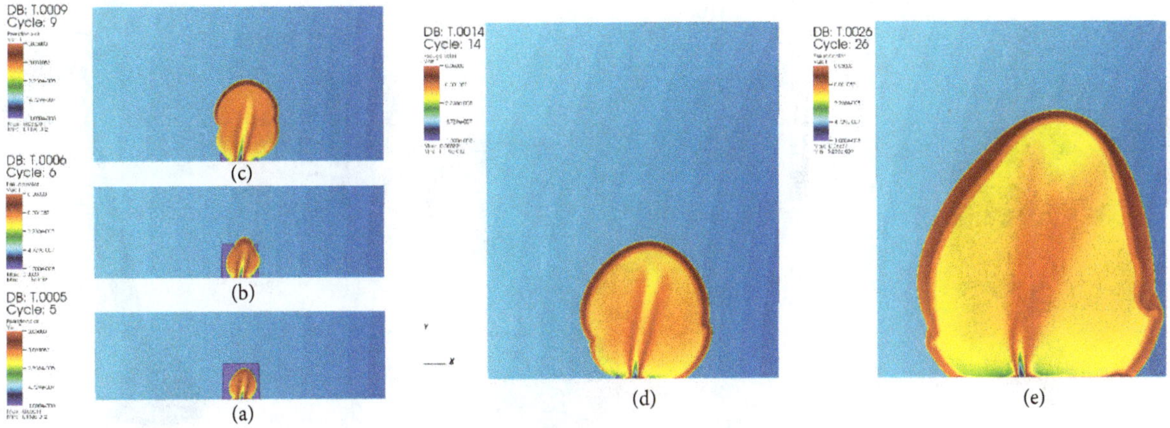

FIGURE 3: Scenario C: a series of slices representing the time evolution of temperature, as in scenario A of [13], but now assuming increased precession rate (see text). The temperature image depicts 2D slices cut along the jet, at a specific time instant. The precessing jet can be seen (d,e) to sweep across a larger (compared to scenario A [13]) portion of the surrounding wind volume, piling up additional matter around its head and sides (see run3 parameters in Table 1). The expansion rate is lower in the accretion disk wind region and increases in the stellar wind region, but afterwards it stays relatively constant, as the density gradient of the wind (visible in the images) is not too large along the jet path.

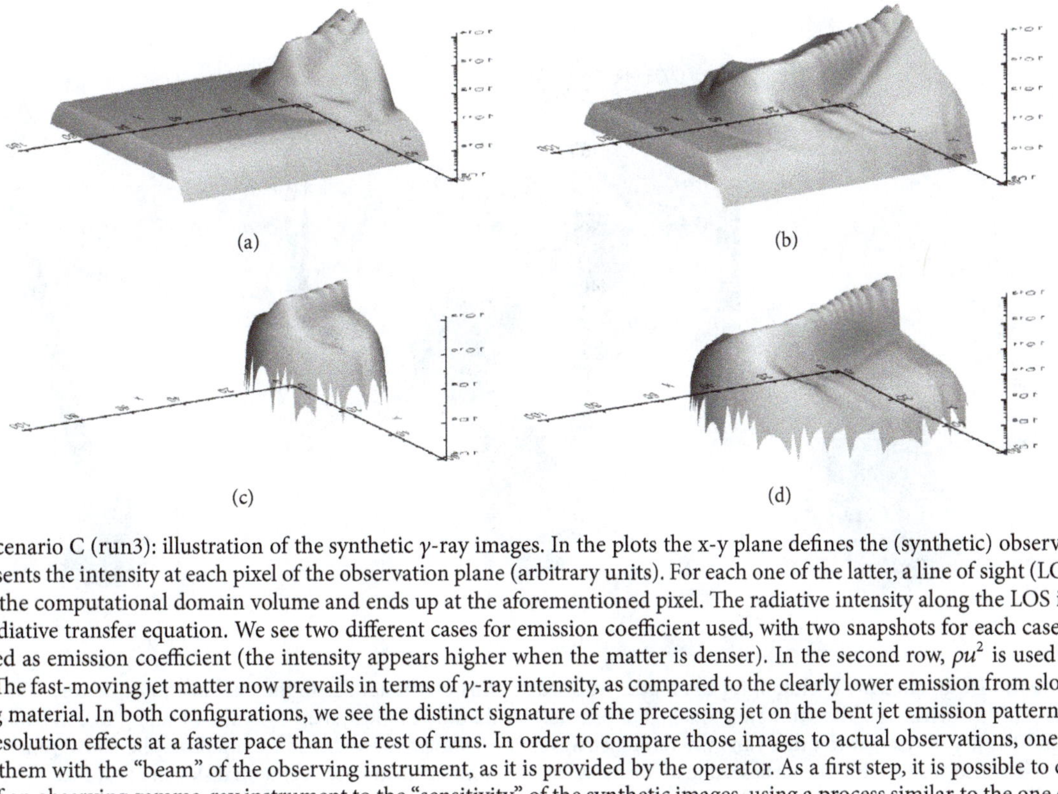

FIGURE 4: Scenario C (run3): illustration of the synthetic γ-ray images. In the plots the x-y plane defines the (synthetic) observation plane. z-axis represents the intensity at each pixel of the observation plane (arbitrary units). For each one of the latter, a line of sight (LOS) is drawn that crosses the computational domain volume and ends up at the aforementioned pixel. The radiative intensity along the LOS is calculated using the radiative transfer equation. We see two different cases for emission coefficient used, with two snapshots for each case. In the first row, ρ is used as emission coefficient (the intensity appears higher when the matter is denser). In the second row, ρu^2 is used as emission coefficient. The fast-moving jet matter now prevails in terms of γ-ray intensity, as compared to the clearly lower emission from slower moving surrounding material. In both configurations, we see the distinct signature of the precessing jet on the bent jet emission patterns formed, as well as on resolution effects at a faster pace than the rest of runs. In order to compare those images to actual observations, one would have to convolve them with the "beam" of the observing instrument, as it is provided by the operator. As a first step, it is possible to compare the sensitivity of an observing gamma-ray instrument to the "sensitivity" of the synthetic images, using a process similar to the one described in SK14.

winds as well, would be needed in order to study the effects of precession on the jet's environment.

4.1.2. Simulations of Scenario D. In scenario D (medium resolution) discussed in this work (see Figures 5, 6, 7, and 8), the jet is assumed heavier than its surrounding winds, which in turn are also somewhat heavier than those of the other cases, leading to a faster crossing of the computational domain (Figure 5). Sideways expansion is also swift as the increased jet mass density allows for a faster "sweep" of the wind matter. Leftovers from the displaced accretion disk wind matter can be seen piling up around the jet base (Figure 6).

The jet forms a funnel that transfers mass outwards at an increased flow rate. The properties of the jet's surroundings

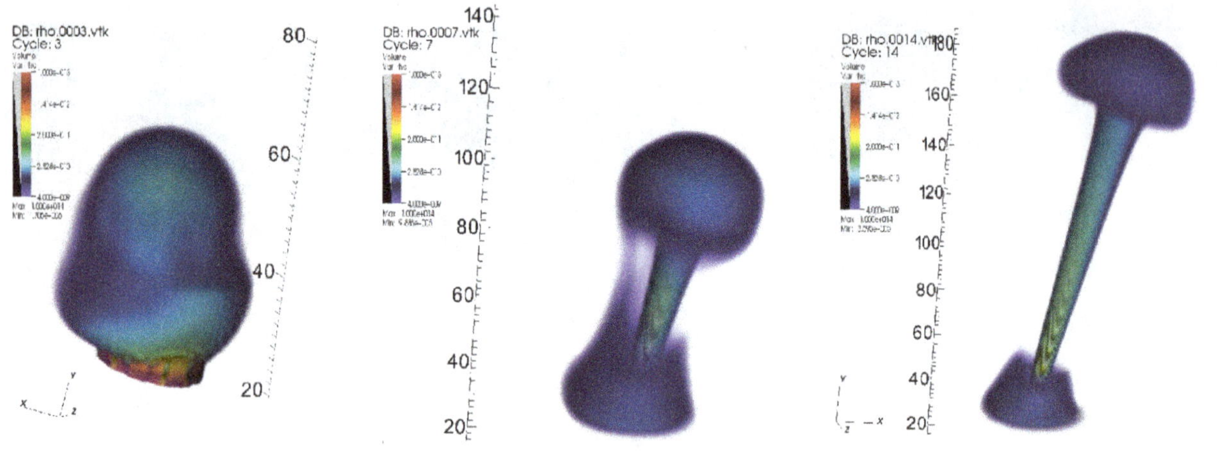

(a) (b) (c)

FIGURE 5: Scenario D: a series of snapshots depicting the density evolution (logarithmic plots) for the run4. Here we can see 3D plots of the jet proton density in space, at specific time instants of the simulation run. This time the jet is quite much heavier than both surrounding winds, leading to a decreased jet crossing time of the model space. The jet mass flow rate now rests closer to the estimates for SS433. The jet breaks through the accretion disk wind construct and soon crosses the stellar wind at a rapid pace (run4 parameters in Table 1, snapshot intervals of 50/count).

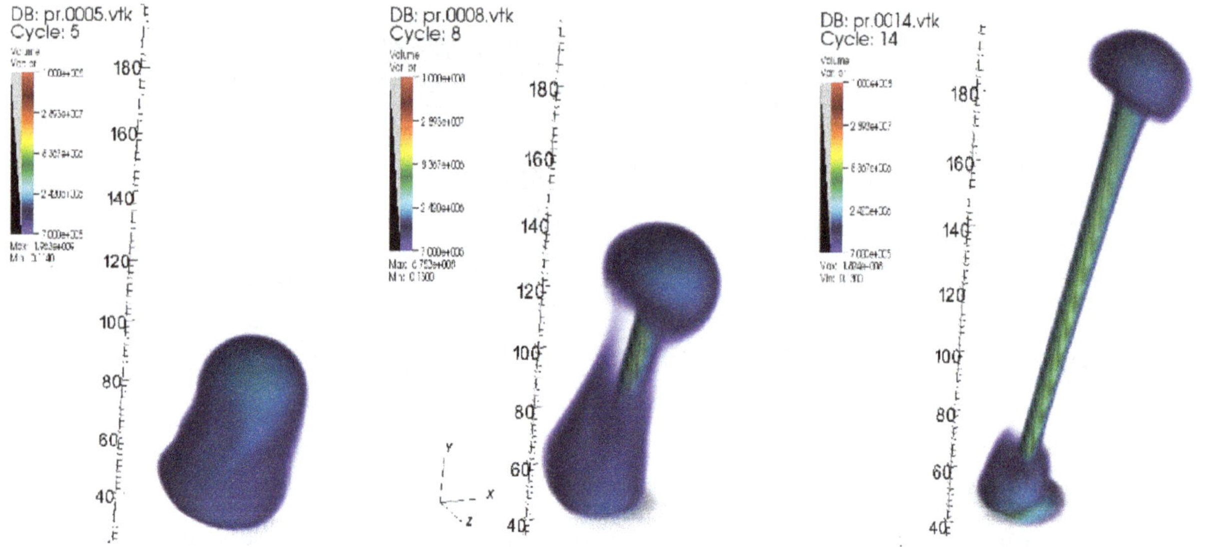

(a) (b) (c)

FIGURE 6: Scenario D: the pressure evolution for the model jet of run4. This plot is similar to the one for the jet density. The patterns are similar to those of the density for the same run. The higher jet's density dominates the system dynamics, leading to comparatively poorer features around the jet, in relation to other runs where the jet was lighter. The pressure distribution follows the above general pattern. A pressure increase is also seen around the jet base, due to the mass leftover from the disturbed accretion disk wind that used to be there (run4 parameters in Table 1).

are now less pronounced as the expansion meets with reduced resistance from ambient matter. The dynamic behaviour of the jet dominates the hydrosimulation, with the wind's matter giving way to the jet (Figures 7 and 8). This case covers the possibility of jet production as a very dense inflow at the source, thus enriching nearby interstellar matter with important mass outflow per time interval.

All of the hydrocode snapshots, of hydrodynamic parameters, have been created with the VisIt visualization code (https://wci.llnl.gov/simulation/computer-codes/visit), whereas the synthetic γ-ray images have been produced using IDL (https://www.harrisgeospatial.com/SoftwareTechnology/IDL.aspx).

4.2. Gamma-Ray Emission Synthetic Images. The above discussed runs with the hydrocode PLUTO have been performed for a precessing jet model of SS433. As the simulation

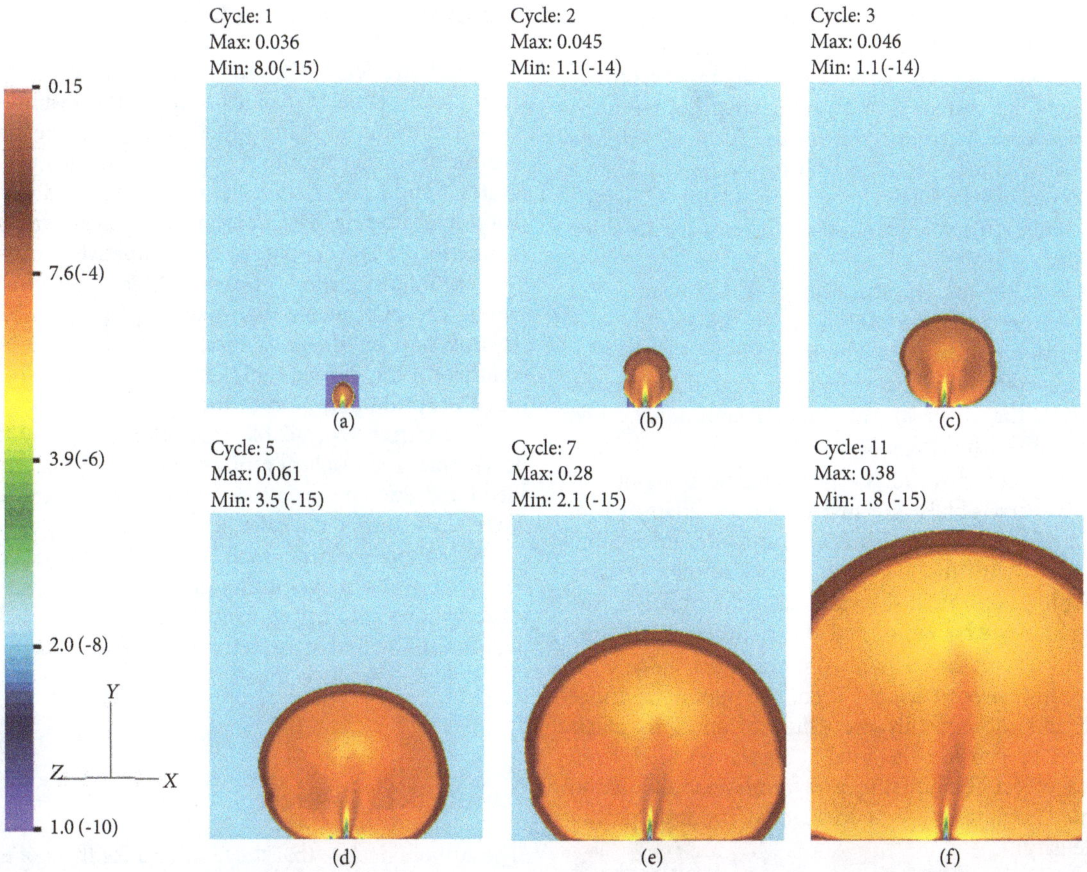

FIGURE 7: Scenario D: a series of plots for the jet temperature of run4 (run4 parameters in Table 1). This plot shows 2D slices cut along the jet, depicting the temperature on the slice, at a specific time instant. The, faster than previous runs, advance of the jet, through surrounding wind matter, is characterized by an expanding shock front. Said expansion is initially slower, till the jet crosses the simplified accretion disk wind construct. Then, a higher rate of expansion occurs, especially to the sides, as the stellar wind gradient is less pronounced, since we have $\Delta x < \Delta r$, where x is the jet crossing distance and r is the distance to the binary companion star.

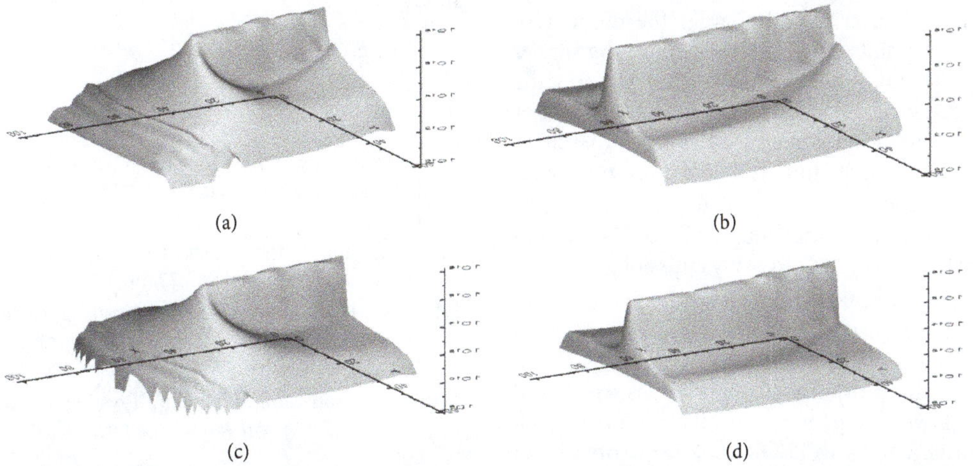

FIGURE 8: Scenario D: the top row shows two snapshots of the "heavy jet" run4 (run4 parameters in Table 1), using the hydrodynamical density ρ as the emission coefficient (arbitrary units). The jet makes its way through the surrounding winds, piling up matter ahead of its head and around the jet shock front, as well as around the jet base (remnants of the accretion disk wind). The bottom row shows the line of sight integration images of the same scenario (D), where now ρu^2 is the emission coefficient. This time only matter that is both fast and dense, simultaneously, does contribute to the emission. The jet base, therefore, now emits much less, since, only the jet flow there is both fast and dense, whereas the remnants of the wind constitute dense but slow matter. The impact of resolution effects, from near the jet base, can also be seen as humps on the back of the jet intensity ridge.

proceeds, at some point the computational model space data is transferred to an output file to be processed (with the available line of sight (LOS) code) for producing a synthetic γ-ray image of the system. The data of a snapshot from the PLUTO hydrocode is transferred, in the form of 3D data arrays (density, velocity vector, pressure: ρ, u_x, u_y, u_z, P) to a routine that performs the LOS integration [12, 13]. Along the LOS, γ-ray emission and absorption coefficients are provided at each point.

At this level, the orbital separation of the compact object and the donor star is taken to be about 4×10^{12} cm so the stellar wind origin does not coincide with the jet base. Also, the jet is taken to travel through a halo produced from the accretion disk (centred at the jet's base). This matter is commonly called the accretion disk wind [12, 13].

With our method we follow two separate steps of calculations. We first obtain the hydrodynamic quantities of density and velocity with the PLUTO hydrocode and second, the integration (with the LOS code) provides the intensity. This decoupling allows, in the calculation of emission, the LOS integration to be performed using either ρ (in CGS units) or ρv^2 (in speed of light dimensionless units, $c = 1$) as the emission coefficient. Then, the synthetic image is formed, and subsequently the values of all of its pixels are summed to provide the total intensity released from the studied object. Finally, the γ-ray emission calculation is performed separately in Mathematica (for unit proton number density).

In each of the above scenarios (runs), for γ-rays a synthetic image was produced, at a suitable model time, using the relevant radiative transfer code of [12] (only the emission coefficient was employed). As can be seen, in all the synthetic γ-ray images (Figures 4 and 8), at each point of the computational volume, the denser the matter, the higher the emission at that point is.

Furthermore, it is clear that the larger the number of significant emission points along a line of sight, the higher the total emission of the whole line of sight is. By adding the dependence on the local velocity, denser but slower matter cannot emit any significant amount of γ-rays. Therefore, emission from the jet body and its interaction zone with surrounding media can be seen to be stronger in the $\rho\beta^2$ maps. In addition, the rest of the (roughly inert) medium in the system contributes very little to γ-ray emission.

Before closing we note that currently we apply the above-mentioned method to carry out jet emission simulations for other microquasar systems like the Cygnus X-1 and Cygnus X-3. From the viewpoint of observations we should mention that, for lower energy γ-rays, orbital platforms, such as NASA's Fermi and ESA's INTEGRAL, already offered an important relevant body of observations for various systems [21, 35]. Very high energy γ-rays can also be studied using data provided by ground-based Cherenkov telescopes such as HESS, MAGIC and HEGRA. Hydrodynamical jet models, combined with artificial imaging, offer realistic estimates of the conditions in the jet and surrounding environments, adding insight to open questions about the γ-ray jet emission from a microquasar.

5. Summary and Conclusions

A precessing jet was modelled using relativistic hydrodynamic code (PLUTO). Furthermore, the results were processed using line of sight code, assuming that the flow velocity u is much smaller than the speed of light. The LOS code integrates along lines of sight the equation of radiative transfer without scattering. The emission coefficients are, in general, a function of hydrodynamical and radiative parameters.

The intensity result of each LOS is assigned to the pixel where the LOS meets the imaging plane of "observation". In this way, an image is formed which could be called a synthetic γ-ray image. Only emission is used for this paper, but absorption may both be incorporated.

The currently available resolution for γ-ray imaging is lower than the resolution of synthetic imaging. Yet the connections between the emission properties and the underlying system dynamics do offer useful constrains on a variety of system parameters, such as the jet kinetic luminosity energy L_k. The above occurs in the light of potential future observations, originating from orbital γ-ray telescopes, from terrestrial Cherenkov detector arrays and even from underground neutrino detectors.

Conflicts of Interest

The authors declare that they have no conflicts of interest.

Acknowledgments

Dr. Odysseas Kosmas wishes to acknowledge the support of EPSRC via Grant EP/N026136/1 "Geometric Mechanics of Solids".

References

[1] M. C. Begelman, S. P. Hatchett, C. F. McKee, C. L. Sarazin, and J. Arons, "Beam models for SS 433," *The Astrophysical Journal*, vol. 238, pp. 722–730, 1980.

[2] B. Margon, "Observations of SS 433," *Annual Review of Astronomy and Astrophysics*, vol. 22, no. 1, pp. 507–536, 1984.

[3] R. M. Hjellming and K. J. Johnston, "Radio emission from conical jets associated with X-ray binaries," *The Astrophysical Journal*, vol. 328, pp. 600–609, 1988.

[4] A. Levinson and E. Waxman, "Probing microquasars with TeV neutrinos," *Physical Review Letters*, vol. 87, Article ID 171101, 2001.

[5] C. Distefano, D. Guetta, E. Waxman, and A. Levinson, "Neutrino flux predictions for known galactic microquasars," *The Astrophysical Journal*, vol. 575, no. 1 I, pp. 378–383, 2002.

[6] G. E. Romero, D. F. Torres, M. M. K. Bernadó, and I. F. Mirabel, "Hadronic gamma-ray emission from windy microquasars," *Astronomy & Astrophysics*, vol. 410, no. 2, pp. L1–L4, 2003.

[7] M. M. Reynoso, G. E. Romero, and H. R. Christiansen, "Production of gamma rays and neutrinos in the dark jets of the

microquasar SS433," *Monthly Notices of the Royal Astronomical Society*, vol. 387, no. 4, pp. 1745–1754, 2008.

[8] B. Cerutti, G. Dubus, J. Malzac et al., "Absorption of high-energy gamma rays in Cygnus X-3," *Astronomy & Astrophysics*, vol. 529, 2011.

[9] I. F. Mirabel and L. F. Rodrguez, "Sources of relativistic jets in the galaxy," *Annual Review of Astronomy and Astrophysics*, vol. 37, pp. 409–443, 1999.

[10] M. M. Reynoso and G. E. Romero, "Magnetic field effects on neutrino production in microquasars," *Astronomy & Astrophysics*, vol. 493, no. 1, pp. 1–11, 2009.

[11] P. Lipari, M. Lusignoli, and D. Meloni, "Flavor composition and energy spectrum of astrophysical neutrinos," *Physical Review D: Particles, Fields, Gravitation and Cosmology*, vol. 75, Article ID 123005, 2007.

[12] T. Smponias and T. S. Kosmas, "Modelling the equatorial emission in a microquasar," *Monthly Notices of the Royal Astronomical Society*, vol. 412, no. 2, pp. 1320–1330, 2011.

[13] T. Smponias and T. S. Kosmas, "Dynamical and radiative simulations of γ-ray jets in microquasars," *Monthly Notices of the Royal Astronomical Society*, vol. 438, no. 2, pp. 1014–1026, 2014.

[14] K. M. Blundell and P. Hirst, "Jet propulsion of wind ejecta from a major flare in the black hole microquasar SS433," *The Astrophysical Journal Letters*, vol. 735, no. 1, 2011.

[15] J. A. Hinton, "The status of the HESS project," *New Astronomy Reviews*, vol. 48, no. 5-6, pp. 331–337, 2004.

[16] C. Baixeras, D. Bastieribo, C. Bigongiari et al., "Commissioning and first tests of the MAGIC telescope," *Nuclear Instruments and Methods in Physics Research Section A*, vol. 518, pp. 188–192, 2004.

[17] M. Actis, G. Agnetta, and F. Aharonian, "Design concepts for the Cherenkov Telescope Array CTA: an advanced facility for ground-based high-energy gamma-ray astronomy," *Experimental Astronomy*, vol. 32, pp. 193–316, 2011.

[18] F. Aharonian, A. G. Akhperjanian, and K.-M. Aye, "Discovery of very high energy gamma rays associated with an X-ray binary," *Science*, vol. 309, no. 5735, pp. 746–749, 2005.

[19] T. Y. Saito, R. Zanin, P. Bordas et al., "Microquasar observations with the MAGIC telescope," https://arxiv.org/abs/0907.1017.

[20] A. A. Abdo, M. Ackermann et al., "Modulated high-energy gamma-ray emission from the microquasar Cygnus X-3," *Science*, vol. 326, pp. 1512–1516, 2009.

[21] S. Hayashi, F. Kajino, T. Naito et al., "Search for VHE gamma rays from SS433/W50 with the CANGAROO-II telescope," *Astroparticle Physics*, vol. 32, pp. 112–119, 2009.

[22] A. Mignone, G. Bodo, S. Massaglia et al., "PLUTO: a numerical code for computational astrophysics," *The Astrophysical Journal Supplement Series*, vol. 170, no. 1, pp. 228–242, 2007.

[23] O. T. Kosmas and D. S. Vlachos, "Local path fitting: a new approach to variational integrators," *Journal of Computational and Applied Mathematics*, vol. 236, no. 10, pp. 2632–2642, 2012.

[24] O. Kosmas and S. Leyendecker, "Analysis of higher order phase fitted variational integrators," *Advances in Computational Mathematics*, vol. 42, no. 3, pp. 605–619, 2016.

[25] O. T. Kosmas, "Charged particle in an electromagnetic field using variational integrators," in *Proceedings of the International Conference on Numerical Analysis and Applied Mathematics: Numerical Analysis and Applied Mathematics, ICNAAM 2011*, pp. 1927–1931, Greece, September 2011.

[26] S. R. Kelner, F. A. Aharonian, and V. V. Bugayov, "Energy spectra of gamma rays, electrons, and neutrinos produced at proton-proton interactions in the very high energy regime," *Physical Review D*, vol. 79, Article ID 039901, 2009.

[27] D. F. Torres and A. Reimer, "Hadronic beam models for quasars and microquasars," *Astronomy & Astrophysics* , vol. 528, article L2, 2011.

[28] T. Smponias and O. T. Kosmas, "High energy neutrino emission from astrophysical jets in the Galaxy," *Advances in High Energy Physics*, vol. 2015, Article ID 921757, 7 pages, 2015.

[29] T. Smponias and O. Kosmas, "Neutrino Emission from Magnetized Microquasar Jets," *Advances in High Energy Physics*, vol. 2017, Article ID 4962741, 7 pages, 2017.

[30] F. M. Rieger, V. Bosch-Ramon, and P. Duffy, "Fermi acceleration in astrophysical jets," *Astrophysics and Space Science*, vol. 309, no. 1-4, pp. 119–125, 2007.

[31] V. Bosch-Ramon, G. E. Romero, and J. M. Paredes, "A broadband leptonic model for gamma-ray emitting microquasars," *Astronomy & Astrophysics* , vol. 447, no. 1, pp. 263–276, 2006.

[32] D. Purmohammad and J. Samimi, "On the hadronic beam model of TeV γ-ray flares from blazars," *Astronomy & Astrophysics* , vol. 371, no. 1, pp. 61–67, 2001.

[33] A. Ferrari, "Modeling extragalactic jets," *Annual Review of Astronomy and Astrophysics*, vol. 36, no. 1, pp. 539–598, 1998.

[34] S. Fabrika, "The jets and supercritical accretion disk," *Astrophysics and Space Physics Reviews*, vol. 12, pp. 1–152, 2004.

[35] M. G. Aartsen, M. Ackermann, and J. Adams, "Search for prompt neutrino emission from gamma-ray bursts with IceCube," *The Astrophysical Journal Letters*, vol. 805, no. 1, article L5, 2015.

23

Inflation in $f(R, \phi)$ Gravity with Exponential Model

Farzana Kousar, Rabia Saleem, and M. Zubair

Department of Mathematics, COMSATS University Islamabad, Lahore-Campus, Pakistan

Correspondence should be addressed to M. Zubair; mzubairkk@gmail.com

Academic Editor: Torsten Asselmeyer-Maluga

We are taking action of $f(R)$ gravity with a nonminimal coupling to a massive inflaton field. A $f(R, \phi)$ model is chosen which leads to the *scalar-tensor theory* which can be transformed to Einstein frame by conformal transformation. To avoid the vagueness of the frame dependence, we evaluate the exact analytical solutions for inflationary era in Jordan frame and find a condition for graceful exit from inflation. Furthermore, we calculate the perturbed parameters (i.e., number of e-folds, slow-roll parameters, scalar and tensor power spectra, corresponding spectral indices, and tensor to scalar ratio). It is showed that the tensor power spectra lead to blue tilt for this model. The trajectories of the perturbed parameters are plotted to compare the results with recent observations.

1. Introduction

The late-time cosmic acceleration was discovered in 1998 [1, 2], based on the observations of type Ia supernovae (SN Ia) that opened up a new door of research in the field of cosmology. The main ingredient of this acceleration, named as dark energy (DE) [3], has been still a paradigm in spite of tremendous efforts to understand its origin over the last decade [4, 5]. Dark energy is different from ordinary matter in the respect that it has huge negative pressure whose equation of state (EoS) is close to -1. Independent observational data like SN Ia [6, 7], cosmic microwave background radiation (CMBR) [8, 9], and baryon acoustic oscillations (BAO) [10–12] proved that about 70% of the energy density content of the recent universe comprises DE.

Currently, there are various approaches to construct the models for the explanation of the behavior of DE. One way is to modify the dynamical field equations by taking negative pressure in the form of energy momentum tensor $T_{\mu\nu}$. This class of models includes inflation [13], quintessence [14, 15], k-essence [16, 17], and perfect fluid models. The perfect fluid models are solved with the combination of EoS like Chaplygin gas model and the generalization of this model [18, 19]. On the basis of particle physics, there have been several efforts to find the scalar field models for the explanation of DE [20–22]. The alternative way for the construction of DE model is to modify the "Einstein-Hilbert action" to attain the modified gravity theories like $f(R)$ gravity, $f(T)$ (where T is torsion scalar) gravity, $f(R, T)$ and Gauss-Bonnet gravity [23–31].

The compelling research phenomena of "cosmological inflation" are introduced by Guth (1981) [32]. In hot big-bang (HBB) theory, inflation is a notion implemented on a very initial cosmic stage of expansion. The scale of inflation is assumed to be long enough since over and the standard evolution is rebuilt to hold the prominent triumphs, such as CMBR and nucleosynthesis. In spite of all of its successes, there are some puzzles with HBB theory, which generate inflation [33].

The "flatness issue" [34]$\Omega = 1$ on the time scale, where $\Omega = \rho/\rho_c$, ρ_c being critical density. The curvature term, $(aH)^2$ (a, H is the scale factor and the Hubble parameter) in standard big-bang model, decreases with respect to time lead to $\Omega(t)$ varying from unity. However, recent observations suggested that the value of $\Omega(t)$ is closed to unity, thus it must be same in the early-time such that its value is $|\Omega(t_{Pl})| < \mathcal{O}(10^{-64})$ at *Planck time* while during nucleosynthesis $|\Omega(t_{nucleo})| < \mathcal{O}(10^{-16})$. The difference in numeric values suggests that initial conditions should be fine-tuned. An inappropriate choice generates a cosmos, which either soon expands before the formation of structure or quickly collapses.

The "horizon problem [34]" illustrates "why the temperature of CMBR is the same all over the sky?" The exactly same temperature of CMBR in east and west directions is detected through antenna whereas the radiation coming from opposite directions is separated by $28BLY$. It is well known that travel speed of information is always less than the speed of light, hence neither the radiation nor the regions ever have been in thermal contact. Any two cosmic regions could be in thermal equilibrium if and only if they are closed enough to communicate with each other. So, question arises that without any causal connection, how was thermal equilibrium between two regions developed?

Inflationary mechanism is basically introduced to solve the classical shortcomings attached with HBB model. More precisely, during inflationary phase the factor $a(t)$ grows exponentially ($\ddot{a} < 0$) and evolution equation immediately yields $\rho + 3P < 0$; since ρ is a positive quantity, therefore to hold the mentioned inequality, P must be negative (*i.e.*, $P < -\rho/3$). Symmetry breaking is a technique which helps in achieving this negative pressure. The cosmic model with cosmological constant (Λ) satisfying EoS $P = -\rho$ is the usual example of cosmic inflationary model. The quantity ρ_λ decayed into ordinary matter with passage of time, leading to graceful exit from inflation and sustained the HBB model. Unluckily, Λ is known to be very ad hoc mechanism. An outstanding inflationary model should follow a reasonable hypothesis for the origin of Λ and a graceful exit from the phase of inflation [35].

The *phase transition* is a successful mechanism to achieve inflation, especially a dramatic stage in time-line of the universe where universe really alters its properties. In fact, the present cosmos have undergone a chain of phase transitions as its temperature cooled down. Scalar field, an unusual form of matter with negative pressure, is assumed to be responsible for these transitions in cosmic phases. The inflaton decayed at the end of evolutionary phase and inflation terminates, hopefully expanding the universe by a factor of 10^{27} or more. Moreover, modified gravity theories (MGT) [36–39] provide a new way to get inflation. In these MGT, higher-derivative curvature corrections in Einstein's theory lead to early-time acceleration (see [40, 41] for review and [33, 42–49] for applications).

Liddle and Samuel [50] discussed the effects of nonstandard expansion between two cosmic phases, end of inflation, and the current cosmic stage, resulting that the expected number of e-folding (N) can be reformed and significantly increased in some cases. Walliser [51] solved general scalar-tensor theories of gravity and found the differential equations which successfully inflate the universe. Garcia-Bellido and Quiros [52] solved the problem of inflation, based on a general scalar-tensor theory of gravity. They determined a particular class of models with a Brans-Dicke like behavior during inflation. The result converted continuously to general relativity during the radiation and matter-dominated eras. They solved numerical equations of motion and found a subclass of models. Lahiri and Bhattacharya [53] formulated a general mechanism to analyze the linear perturbations during inflation based on the gauge-ready approach. They solved the first order slow-roll equations for scalar and tensor perturbations and obtained the super-horizon solutions for different perturbations after inflation.

Myrzakulov et al. [54] described the inflation with the reference to $f(R,\phi)$-theories and generated a class of models which support early-time acceleration. Sharif and Saleem [55] studied the warm inflation in the framework of locally rotationally symmetric Bianchi type I universe model. They presented the graphical analysis of the perturbed parameters to check the comparability of the considered model with recent data. In Jordan frame, Mathew et al. [56] constructed exact solution with nonminimal coupled action of $f(R)$ gravity to a massive inflaton field. They proved that the solutions were the same as in scalar-tensor theory. They also explained the dynamics of *tensor power spectrum* associated to this model.

Inspiring by the technique used in [56], we build a cosmic inflationary model with a massive inflaton field that has fundamental place in the standard model of particle physics. The *Einstein-Hilbert (EH) action* is considered as a constrained case of a generalized action with higher order curvature invariants; $f(R)$ gravity is the example of such an action [36, 57, 58]. General theory of relativity (GR) cannot be renormalizable, so it is not possible to quantize it conventionally. However, the modified EH action containing higher order curvature terms can be renormalizable [59, 60], due to which $f(R)$ gravity is taken to be an interesting alternative to GR. The associated $f(R)$ field equations are nontrivial due to its fourth order. In addition, these theories do not experience *Oströgradsky instability* [61].

A conformal transformation can be applied to $f(R,\phi)$ action to convert it to EH action with an additional (canonical) inflaton field [62]. The *scalar-tensor gravity theories* suffer from a long-lasting controversy about the choice of physical frame either Einstein or Jordan [63]. Although the two: Jordan frame (original) as well as the Einstein frame are under conformal transformation, it is unclear how the observable quantities are related to the physical quantities computed in the two frames [35, 64]. To get rid of these controversies and the vagueness in the selection of frame, here we take the action without implementing any transformation to frame, any other theory, or variables [63]. Since the EH action does not possess any nonminimal coupling term, so there is no motivation of performing conformal transformation. Generally, we cannot trust in these techniques presented in literature, and we work with a new analytical method developed in [56].

The manuscript is arranged as follows. In Section 2, firstly, we consider a model and obtain the exact analytical solutions in de-Sitter case and secondly we find inflationary solutions numerically with an exit for different initial conditions. In Section 3, we discuss the scalar and tensor power spectra and prove that the solution of Hubble parameter represents a saddle point. The compatibility of the model with recent data is checked through graphical analysis of the perturbed parameters. In the last section, we conclude the results.

2. Model and Background Solution

The $f(R,\phi)$ theory is described by the action given as

$$S = \int d^4x\sqrt{-g}\left[\frac{1}{2}f(R,\phi) - \frac{1}{2}\omega g^{ab}\nabla_a\phi\nabla_b\phi - V(\phi)\right], \quad (1)$$

where $V(\phi)$ denotes the effective potential related to inflaton field. We are taking the following $f(R,\phi)$ model

$$f(R,\phi) = \frac{1}{\kappa}\left[R + h(\phi)R^2 e^{\alpha R}\right], \quad (2)$$

with a coupling function denoted by $h(\phi)$. By expanding the exponential in terms of Ricci scalar up to first order, we have

$$f(R,\phi) \simeq \frac{1}{\kappa}\left[R + h(\phi)\left(R^2 + \alpha R^3\right)\right]. \quad (3)$$

In Jordan frame, the corresponding $f(R,\phi)$ field equations are as follows [65]:

$$\Box\phi + \frac{1}{2\omega}\left(\omega_{,\phi}\phi^{;a}_{;a} + f_{,\phi} - 2V_{,\phi}\right) = 0, \quad (4)$$

$$FG^p_q = \omega\left(\phi^{;p}\phi_{;q} - \frac{1}{2}\delta^p_q\phi^{;c}\phi_{;c}\right) - \frac{1}{2}\delta^p_q\left(RF - f + 2V\right) + F^{;p}_{;q} - \delta^p_q\Box F, \quad (5)$$

where $F = \partial f(R,\phi)/\partial R$. In case of scalar field and modified gravity, the corresponding stress-tensors are defined, respectively, as

$$T^{\phi}_{\mu\nu} = \omega\left(\partial_\mu\phi\partial_\nu\phi - \frac{1}{2}g_{\mu\nu}\partial_c\phi\partial^c\phi\right) - g_{\mu\nu}V(\phi), \quad (6)$$

$$T^{MG}_{\mu\nu} = \frac{1}{2}g_{\mu\nu}(f - FR) + \nabla_\mu\nabla_\mu F - g_{\mu\nu}\Box F. \quad (7)$$

Now, we will find exact solution analytically in de-Sitter case.

2.1. Background Inflationary Solution. The line element of flat FRW space time is

$$ds^2 = -dt^2 + a(t)^2\left(dx^2 + dy^2 + dz^2\right), \quad (8)$$

where $a(t)$ denotes the scale factor. Using FRW space time, we have equation of motion for ϕ (4) and field equations (5) of the form, respectively,

$$-\omega\dot\phi^2 - 6F\left(\dot H + H^2\right) + 6\dot F H + 2V - f = 0, \quad (9)$$

$$4FH^2 + 2F\left(\dot H + H^2\right) - \omega\dot\phi^2 - 2\ddot F - 4FH - 2V + f = 0, \quad (10)$$

$$2\omega\ddot\phi + 6\omega\dot\phi H - \frac{h_{,\phi}f}{h} + \frac{12h_{,\phi}H^2}{h\kappa} + \frac{6h_{,\phi}\dot H}{h\kappa} + 2V_{,\phi} = 0, \quad (11)$$

where $h = h(\phi)$ is the coupling function. Here, we are considering the following assumptions

$$a(t) = a_0 e^{H_D t}, \qquad \phi = \phi_0 e^{-nH_D t}, \quad (12)$$

where n, ϕ_0, and H_D are constants. Substituting (12) in (9)-(11) and solving these equations, we get the coupling function of the form

$$h(\phi) = \lambda_0 + \lambda_1\phi^2 + \lambda_n\phi^{-1/n}, \quad (13)$$

where $\lambda_1 = -n\omega\kappa/48H_D^2(2n+1)(18\alpha H_D^2 + 1)$ and $\lambda_n = nc_1$. The scalar field potential is

$$V(\phi) = V_0 + V_1\phi^2 + V_n\phi^{-1/n}, \quad (14)$$

where

$$V_0 = \frac{3H_D^2\lambda_0}{\kappa},$$
$$V_1 = 24\lambda_1 H_D^4\left\{-18\alpha H_D^2 + n^2\left(36\alpha H_D^2 + 2\right) - 5n\left(18\alpha H_D^2 + 1\right)\right\}, \quad (15)$$
$$V_n = -\frac{72\lambda_n H_D^4\left(12\alpha H_D^2 + 1\right)}{\kappa}.$$

Here λ_0 is the constant of integration. It can be seen that this is an exact solution obtained from the background equations. We are mentioning here some important points: first, we have obtained the solution without using conformal transformation. According to the best of our knowledge, in Jordan frame no exact solution exists. Second, in case of exact analytical de-Sitter solution, the inflaton field (12) decreases as time increases. Third, coupling function $h(\phi)$ directly depends on $V(\phi)$.

From (13) and (14), it can be seen that V_0 depends on λ_0, and similarly V_n depends on λ_n and λ_1 is related to V_1. If we choose $\lambda_0 = \lambda_n = 0$, then it is obvious that V_0 and V_n also vanished.

3. Special Case: $\lambda_0 = \lambda_n = 0$

Here, we consider $\lambda_0 = \lambda_1 = 0$. The coupling function and potential are reduced to

$$h(\phi) = \lambda_1\phi^2, \qquad V(\phi) = V_1\phi^2. \quad (16)$$

This leads to conclude the following points. It can be seen that λ_1 is a positive definite implying that $18\alpha H_D^2 + 1 < 0$ or $\alpha < -1/(18H_D^2)$. Since during inflation the parameter H is large, this leads to small negative value of α. From the stress-tensor (6) and (7), we can calculate $\rho + 3P$ as

$$\rho + 3P \equiv -T^0_0 + T^\alpha_\alpha = 2\phi_0^2 H_D^2 e^{-2nH_D t}\left[n^2 + V1 - \lambda_1\left\{72H_D^2\left(1 + 2n + 2n^2\right) + 648\alpha H_D^4 \times\left(3 + 2n + 4n^2\right)\right\}\right]. \quad (17)$$

In (17), the first and second terms are related to the canonical inflaton field while the last quantity represents the modifications to the gravity. The numeric value $n < 1/2$ yields the negativity of the third term while it is positive for $n > 1/2$. However, $n \ll 1$ leads to $\rho + 3P < 0$ and for $n \gg 1$, we have $\rho + 3P > 0$. In further analysis, we are taking $n \gg 1$. We can say from the above discussion that either it corresponds to exit for large values of n or by varying the initial condition of $\dot{\phi}$. If $\dot{\phi}_{t=0} \neq \dot{\phi}^{dS}_{t=0}$, then we will check what kind of inflation exists using the relation $\dot{\phi} \propto n$.

4. First Order Scalar and Tensor Model Perturbations

Now we will discuss the scalar/tensor power spectra for considered inflationary model (3). We have used the same notation as used in [66] then it will be easy to compare. For our model, analysis of [66] is not applicable.

4.1. Perturbations. For FRW space time first order perturbations are

$$ds^2 = -(1+2\theta)\,dt^2 - a\left(\beta_{,\alpha} + B_\alpha\right) dt\, dx^\alpha + a^2\left[g^{(3)}_{\alpha\beta}\left(1-2\psi\right) + 2\gamma_{,\alpha|\beta} + 2C_{\alpha|\beta} + 2C_{\alpha\beta}\right]. \quad (18)$$

In the above expression, the scalar perturbations are represented by $dt \equiv ad\eta$ and $\theta(x,t)$, $\beta(x,t)$, $\psi(x,t)$, $\gamma(x,t)$. The quantities $B_\alpha(x,t)$ as well as $C_\alpha(x,t)$ denote the trace-free vector perturbation and $C_{\alpha\beta}(x,t)$ presents the trace-free and transverse tensor perturbations. The decomposed form of inflaton field is $\phi(x,t) = \overline{\phi}(t) + \delta\phi(x,t)$.

In Fourier space and scalar perturbation of Newtonian gauge are as follows [65, 66]:

$$-F\psi + F\theta + \delta F = 0, \quad (19)$$

$$-2F\dot{\psi} - 2FH\theta - \dot{F}\theta + \dot{\phi}\delta\phi + \dot{\delta F} - H\delta F = 0, \quad (20)$$

$$6FH\dot{\psi} + 6FH^2\theta + 2F\frac{k^2}{a^2}\psi - \dot{\phi}^2\theta + 3\dot{F}\dot{\psi} + 6\dot{F}H\theta + \dot{\phi}\dot{\delta\phi} - \ddot{\phi}\delta\phi - 3H\dot{\phi}\delta\phi - 3H\dot{\delta F} + 3\dot{H}\delta F + 3H^2\delta F - \frac{k^2}{a^2}\delta F = 0, \quad (21)$$

$$6F\ddot{\psi} + 12F\dot{H}\theta + 6FH\dot{\theta} + 12FH\dot{\psi} + 12FH^2\theta - 2F\frac{k^2}{a^2}\theta + 3\dot{F}\dot{\psi} + 6\dot{F}H\theta + \dot{F}\dot{\theta} + 4\dot{\phi}^2\theta + 6\theta\ddot{F} - 4\dot{\phi}\dot{\delta\phi} - 2\ddot{\phi}\delta\phi - 6H\dot{\phi}\delta\phi - 3\ddot{\delta F} - 3H\dot{\delta F} + 6H^2\delta F - \frac{k^2}{a^2}\delta F = 0, \quad (22)$$

$$\ddot{\delta\phi} + 3H\dot{\delta\phi} - \frac{1}{2}f_{\phi\phi} + V_{\phi\phi}\delta\phi + \frac{k^2}{a^2}\delta\phi - 3\dot{\phi}\dot{\psi} - 6H\dot{\phi}\theta - \dot{\phi}\dot{\theta} - 2\ddot{\phi}\theta + 3F_\phi\ddot{\psi} + 6F_\phi\dot{H}\theta + 3HF_\phi\dot{\theta} + 12F_\phi H\dot{\psi} + 12F_\phi H^2\theta + 2F_\phi\frac{k^2}{a^2}\psi - F_\phi\frac{k^2}{a^2}\theta = 0, \quad (23)$$

$$\delta F - F_\phi\delta\phi + F_R\delta R = 0, \quad (24)$$

where $\delta R = -6\ddot{\psi} - 12\dot{H}\theta - 6H\dot{\theta} - 24H\dot{\psi} - 24H^2\theta - 4(k^2/a^2)\psi + 2(k^2/a^2)\theta$. While the tensor perturbations are as follows [65]:

$$\ddot{C}^\alpha_\beta + \left(\frac{\dot{F}}{F} + 3H\right)\dot{C}^\alpha_\beta + \frac{k^2}{a^2}C^\alpha_\beta = 0. \quad (25)$$

4.2. Scalar Power Spectrum. Here we calculate the equation, which satisfies the 3-curvature perturbation $\mathscr{R}$ and derives the related power spectrum. We are using the technique followed in [67]. The $\mathscr{R}$ in Jordan frame is stated as

$$\mathscr{R} = \psi + \frac{H}{\dot{\phi}}\delta\phi. \quad (26)$$

Equations (19)-(24) are highly ordered and nonlinear, so we use different techniques to calculate the 3-curvature perturbation equation. First, we will use $\theta + \psi = \Theta$ and find the solution of differential equation. Physically, in Einstein frame, Θ represents the Bardeen potential. Using (20)-(22), we have differential equation in Θ as

$$F\ddot{\Theta} + \left(3\dot{F} + FH - \frac{2F\ddot{\phi}}{\dot{\phi}}\right)\dot{\Theta} + \left(\frac{Fk^2}{a^2} + \frac{2\dot{F}\ddot{\phi}}{\dot{\phi}} - \frac{2FH\ddot{\phi}}{\dot{\phi}} - \ddot{F} + \dot{F}H + 4F\dot{H}\right)\Theta + \left(-\dot{\phi}^2 + 3\ddot{F} + 3\dot{F}H - 6F\dot{H} - \frac{6\dot{F}\ddot{\phi}}{\dot{\phi}}\right)\theta = 0. \quad (27)$$

Using background quantities for de-Sitter, assuming $n \ll 1$ and large values of k, we have differential equation in terms of Θ as

$$\ddot{\Theta} + (1-4n)H_D\dot{\Theta} + \frac{k^2}{a^2}\Theta - 4n(1-n)H_D{}^2\theta = 0. \quad (28)$$

Applying the small wavelength limit $k/a \gg 1$, the last two terms of the left hand side can be written as

$$\left(\frac{k^2}{a^2} - 4n(1-n)H_D{}^2\right)\theta + \frac{k^2}{a^2}\psi \simeq \frac{k^2}{a^2}\Theta, \quad (29)$$

which can further be written as

$$\ddot{\Theta} + (1-4n)H_D\dot{\Theta} + \frac{k^2}{a^2}\Theta = 0. \quad (30)$$

In terms of Θ, $\delta\phi$ is as follows

$$\delta\phi = -\frac{\phi_0 e^{-nH_Dt}}{nH_D}\left(\dot{\Theta} + H_D\Theta\right). \quad (31)$$

Rewriting the perturbation equations in terms of $\mathscr{R}$ and using (26) and (30), we have Θ in terms of $\mathscr{R}$ as

$$\Theta = -\left(3a^2\ddot{R} + 12H_D a^2\dot{R} + 2k^2R\right)\frac{nH_D}{\phi_0}e^{nH_Dt}, \quad (32)$$

substituting it into the perturbation equations, we get the differential equation in terms of $\mathscr{R}$:

$$\ddot{\mathscr{R}} + 3H_D\dot{\mathscr{R}} + \frac{k^2}{a^2}\mathscr{R} = 0. \quad (33)$$

It is an important result which shows that higher order differential equation can be reduced to second order. For the power spectrum, we can evaluate solution to the above differential equation for the short wavelength limit and usethe Bunch-Davies vacuum at the initial epoch of inflation as

$$R_< = \frac{H_D}{2a\sqrt{k}}e^{-ik\eta}. \quad (34)$$

In the long wavelength limit, we get $R_> = C$. By matching $R_<$ and $R_>$ at horizon crossing ($|k\eta| = 2\pi$), we get

$$C = \frac{\sqrt{2}H_D\pi}{k^{3/2}}, \quad (35)$$

leading to the following scale-invariant scalar power spectrum

$$\mathscr{P}_R = H_D{}^2. \quad (36)$$

For $n \ll 1$, above analysis is an analytical expression. It is not possible to obtain the semianalytical expression for other cases, which shows the tilted spectrum. Next, we derive the tensor power spectrum without using $n \ll 1$ limit which leads to blue tilt.

4.3. Tensor Power Spectrum. Following [65], the required equation of motion for tensor perturbation of exact de-Sitter solution can be obtained as

$$\ddot{C}^\alpha_\beta + (3-2n)H_D\dot{C}^\alpha_\beta + \frac{k^2}{a^2}C^\alpha_\beta = 0. \quad (37)$$

Defining $C^\alpha_\beta = v_g/z_g$ and $z_g = ae^{-nH_Dt}$, we have

$$v_g{}'' + \left(k^2 - \frac{z_g{}''}{z_g}\right) = 0. \quad (38)$$

The above-mentioned differential equation yields a result as a combination of Hankel function:

$$v_g = \sqrt{-\eta}\left(\widetilde{C}_1 H^{(1)}_{3/2-n}(-k\eta) + \widetilde{C}_2 H^{(2)}_{3/2-n}(-k\eta)\right). \quad (39)$$

Fixing Bunch-Davies vacuum to the initial state, we obtained $\widetilde{C}_1 = \sqrt{\pi/4}$; $\widetilde{C}_2 = 0$. Thus tensor perturbation C^α_β provides

$$v_g = \sqrt{\frac{\pi}{4}}\sqrt{-\eta}H^{(1)}_{3/2-n}(-k\eta). \quad (40)$$

The corresponding power spectrum ($\mathscr{P}_g$) can be evaluated in the following form $\mathscr{P}_g = 8(k^3/2\pi^2)|\mathscr{C}^\alpha_\beta|^2$ and modified as

$$\mathscr{P}_g = 8\left(\frac{k}{k_*}\right)^{2n}\frac{2^{-2n}}{4\pi^2}H_D{}^2\left(\frac{\Gamma(3/2-n)}{\Gamma(3/2)}\right)^2 e^{2nH_Dt_*}. \quad (41)$$

The tensor spectral index is calculated as $n_T = 2n$, implying a blue-tilted spectrum for a decaying inflaton field, i.e., blue tilted for $n > 1/2$ and red tilted for $n < 1/2$.

5. Stability of Inflationary Solution

Now we will discuss the stability of de-Sitter solution and will examine the variations of initial values whether they show inflationary phase, super inflation, or smooth exit. Hence, to show the inflationary solution many initial values exist and a mess of models have been discussed in literature which show the saddle point [68, 69]. The field equations (9)-(11) in terms of the variable $\Delta = \dot{\phi}/\phi$ can be written as

$$\begin{aligned}\dot{\Delta} &= 2V_1 + 2592\alpha\lambda_1\dot{H}H^4 + 1296\alpha\lambda_1\dot{H}^2H^2 \\ &\quad + 1728\alpha\lambda_1H^6 + 216\alpha\lambda_1\dot{H}^3 - 144\lambda_1\dot{H}H^2 \\ &\quad - 144\lambda_1H^4 - 36\lambda_1\dot{H}^2 - 3H\Delta - \Delta^2,\end{aligned} \quad (42)$$

$$\begin{aligned}\ddot{H} &= \frac{1}{36\lambda_1H - 648\alpha\lambda_1H\dot{H} - 1296\alpha\lambda_1H^3}\left[\frac{3H^2}{\phi_0{}^2}\right. \\ &\quad \cdot e^{2nH_Dt} - 180\lambda_1\dot{H}H^2 + 972\alpha\lambda_1H^2\dot{H}^2 \\ &\quad - 3024\alpha\lambda_1H^6 + 1296\alpha\lambda_1\dot{H}H^4 + \frac{1}{2}\Delta^2 - 18\lambda_1\dot{H}^2 \\ &\quad + 216\alpha\lambda_1\dot{H}^3 - V_1 - 72\lambda_1H\dot{H}\Delta - 144\lambda_1H^3\Delta \\ &\quad + 648\alpha\lambda_1H\dot{H}^2\Delta + 2592\alpha\lambda_1H^5\Delta \\ &\quad \left. + 2592\alpha\lambda_1H^3\dot{H}\Delta\right].\end{aligned} \quad (43)$$

In terms of Δ, the above-mentioned equations show that the evolution of H and N, etc., does not involve ϕ or $\dot{\phi}$ and only depends on Δ.

A vector v is defined as

$$v = \begin{pmatrix} H \\ \dot{H} \\ \Delta \end{pmatrix}. \quad (44)$$

It is worth noticed that the solution of de-Sitter model ($H = H_D$) behaves as an equilibrium point ($\dot{v}_{eq} = 0$) where

$$\{v\}_{eq} = \begin{pmatrix} H_D \\ 0 \\ -nH_D \end{pmatrix}, \quad (45)$$

and the expression $\dot{v} = f(v)$ is written as

$$\dot{v} = \begin{pmatrix} \dot{H} \\ \ddot{H} \\ \dot{\Delta} \end{pmatrix}. \quad (46)$$

As we have mentioned above, perturbing $v = v_{eq} + \delta v$ and expansion of $f(v)$ for δv around the equilibrium point provide

$$\delta v_i = \{\partial_j f_i\}_{eq} \delta v_j = J_{ij} \delta v_j, \quad (47)$$

where

$$J_{ij} = \begin{pmatrix} \frac{\partial \dot{H}}{\partial H} & \frac{\partial \dot{H}}{\partial \dot{H}} & \frac{\partial \dot{H}}{\partial \Delta} \\ \frac{\partial \ddot{H}}{\partial H} & \frac{\partial \ddot{H}}{\partial \dot{H}} & \frac{\partial \ddot{H}}{\partial \Delta} \\ \frac{\partial \dot{\Delta}}{\partial H} & \frac{\partial \dot{\Delta}}{\partial \dot{H}} & \frac{\partial \dot{\Delta}}{\partial \Delta} \end{pmatrix}. \quad (48)$$

Let us consider the eigen value λ_i and eigen vector μ_i of the Jacobian. Hence trajectory of phase space is introduced by

$$\delta v_i = \sum_{i=1}^{i=3} c_i \mu_i e^{(\lambda_i t)}. \quad (49)$$

where the values of constants c_i's have to be constraint from initial values of Hi and $\dot{\phi}/\phi$. In our case, we have one real and two complex eigen values which are too lengthy in expression due to which we did not mention them in the paper. For large values of (λ), N is equivalent to

$$N \approx \frac{H_D}{\lambda} \ln\left(\frac{H_D{}^2}{\lambda\left(H_D - H_j\right)}\right). \quad (50)$$

Figure 1 shows the behavior of slow-roll parameter $\epsilon = -\dot{H}/H^2$ versus N for various values of $\dot{\phi}$. The parameter ϵ can be evaluated as follows:

$$\epsilon = \frac{24H^2\left(18\alpha H^2 + 1\right)\left(2H\lambda_1\phi\dot{\phi} - 2\lambda_1\dot{\phi}^2 - 2\lambda_1\ddot{\phi}\phi\right)}{2H^2\left\{1 - 12H^2\lambda_1\phi^2\left(36\alpha H^2 + 2\right)\right\}}. \quad (51)$$

Figure 1 is plotted for $\epsilon - N$ trajectories taking some initial values. It can be observed that for $\dot{\phi} < 1.4\phi_D$, the inflationary phase sustained as $\epsilon < 1$ and ϵ attains a constant value less than unity for $\dot{\phi} \geq 1.4\phi_D$, fixing the other parameters as $\omega_0 = -0.005$, $\phi_0 = -0.3$, $n = 200$, $H_D = 4 \times 10^4$, $H_j = 2$, $\alpha = -10^{-7}$. It can be seen that in the space $\dot{\phi} < \phi_D$, the inflationary phase exists without an exit which represents ϵ diverging to $-\infty$ while $\dot{\phi} > \phi_D$ leads to the inflationary era with an exit. The initial condition $\dot{\phi} = \phi_D$ generates $\epsilon = 0$. In this case, the results are obtained for standard number of e-folds, i.e., $N \simeq 50, 60$, which is in good agreement with observational data. Further, it is observed that as the value of $\dot{\phi}/\phi_D$ is directly proportional to N, as increment in initial value produced an increase in N. Hence rate of inflation increases as $\dot{\phi}$ increases.

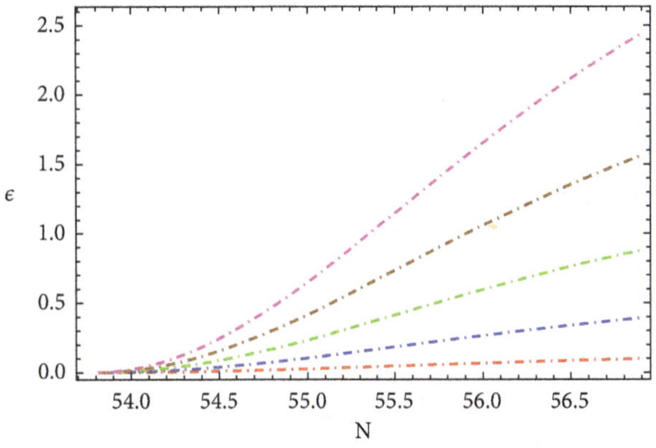

FIGURE 1: Evolution of ϵ versus N for various values of $\dot{\phi}$.

The trajectories are attracted toward its origin with increasing initial values. This shows that divergence of initial constraints form de-Sitter values yields either inflation with graceful exit or super inflation. Hence de-Sitter analytical solution represents saddle point.

Figure 2 is plotted for ϕ versus N (left plot) and versus time t (right plot) for standard number of e-folds. It can be seen that scalar field is decaying with the evolution of time. The trajectories of $\phi - N$ show the same behavior as $\epsilon - N$. The scalar field ϕ is expressed as

$$\phi = \frac{6\kappa\omega H\dot{\phi}}{4\kappa v_1 - 288\lambda_1 - 3456\alpha\lambda_1 H^6}. \quad (52)$$

The scalar spectral index is defined as [70, 71]

$$n_s = 1 - 4\epsilon_1 - 2\epsilon_2 + 2\epsilon_3 - 2\epsilon_4, \quad (53)$$

where

$$\begin{aligned} \epsilon_1 &= -\frac{\dot{H}}{H^2}, \\ \epsilon_2 &= \frac{\ddot{\phi}}{H\dot{\phi}}, \\ \epsilon_3 &= \frac{\dot{F}}{2HF}, \\ \epsilon_4 &= \frac{\dot{E}}{2HE}, \\ E &= \omega F + \frac{3\dot{F}^2}{2\dot{\phi}^2}. \end{aligned} \quad (54)$$

The tensor to scalar ratio is defined as $r = P_g/P_r$. For better understanding of the inflationary model's compatibility with recent data, we have plotted parametric plots (Figure 3) in which scalar spectral index is plotted versus tensor to scalar ratio for $n > 1$. It is observed that for $n = 1.61$, we have

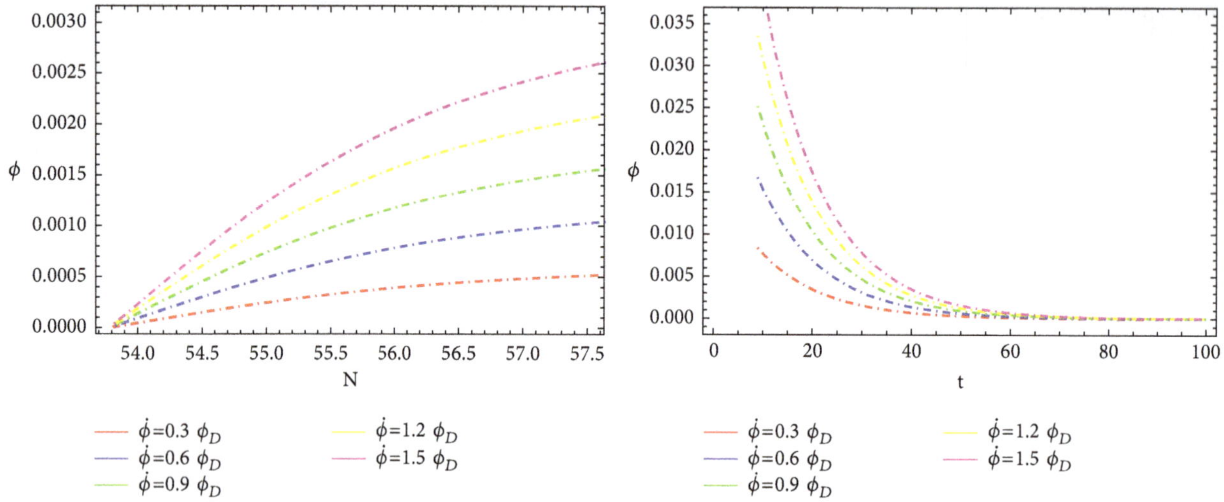

Figure 2: Evolution of ϕ versus N for different initial values of $\dot{\phi}$.

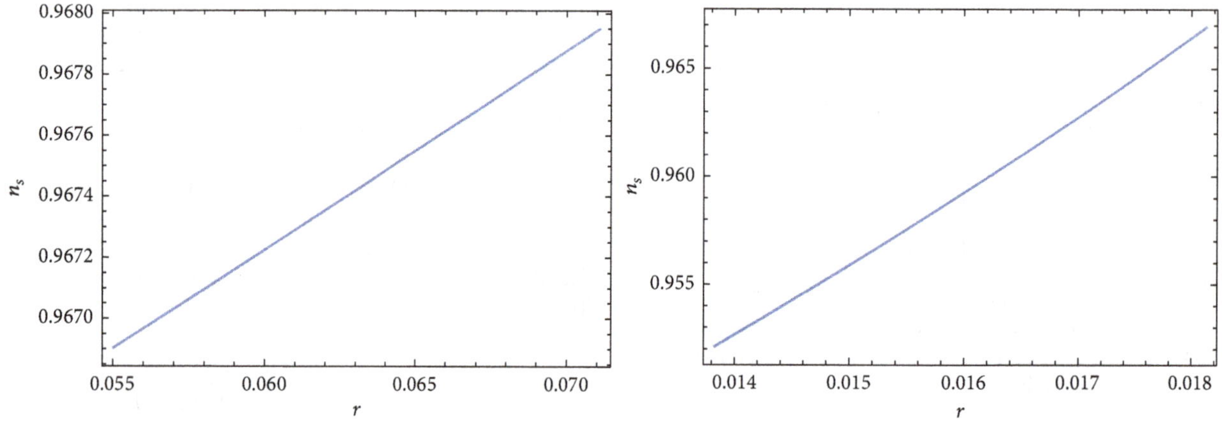

Figure 3: Left plot shows the behavior of r versus n_s for $\dot{\phi} = 0.2\phi_D$ with $n = 1.6$ and right plot for $\dot{\phi} = 0.2\phi_D$ with $n = 1.8$.

the standard value of spectral index $n_s = 0.96$ and an upper bound of tensor to scalar ratio is obtained as $r < 0.36$, which is compatible with WMAP7 [72], while for $n = 1.8$, we get $r < 0.11$ in accordance with Plank 2015 [73].

6. Discussion

In this manuscript, we derived the exact inflationary model in Jordan frame for $f(R,\phi)$ model. Even though it is common to study the $f(R)$ models in Einstein frame, we showed explicitly that in the Einstein frame, action contains noncanonical kinetic term. Thus, the advantage of the conformal transformation is negated. Thus, we performed the background and the first order perturbation analysis in the Jordan frame. A nonminimally coupling of massive inflaton field with $f(R)$ action is considered and has ignored the self-interacting inflaton potential. We have evaluated the expression $\rho + 3P$, which led to conclude that for the values $n \ll 1, \rho+3P < 0$ and for $n \gg 1$, we have $\rho+3P > 0$. In [56], authors also calculated the term $\rho + 3P$ and showed that for $n \gg 1 \Longrightarrow \rho + 3P = 0$ which did not lead to inflation while the inequality $\rho+3P < 0$ is satisfied for $n \ll 1$.

It is explicitly shown that the model supports inflationary solution with graceful exit and the number of e-folds depends on deviation of initial values from de-Sitter scenario. Further, we have discussed the scalar and tensor perturbations for the chosen $f(R,\phi)$ model. We have used the new analytical method devised in [56] to reduce the higher order scalar perturbation equations to second order in 3-curvature perturbation. We analytically obtained the scalar power spectrum in $n \gg 1$ limit and proved the scale invariance of scalar power spectrum. We obtained the tensor power spectrum for $n > 1/2$ and showed that the spectrum is blue tilted.

To get insight, we plotted the graphs of parameter ϵ and inflaton field for various physical initial conditions of $\dot{\phi}$, i.e., $\dot{\phi} = 0.3\phi_D, 0.6\phi_D, 0.9\phi_D, 1.2\phi_D, 1.5\phi_D$ and $\dot{\phi} = \phi_D$. The $\epsilon - N$ trajectories showed that in the parametric space $\dot{\phi} < 1.4\phi_D$, the inflationary phase sustained as $\epsilon < 1$ and ϵ attains a constant value less than unity for $\dot{\phi} \geq 1.4\phi_D$. It can be seen that in the space $\dot{\phi} < \phi_D$, the inflationary phase exists without an exit which represents ϵ diverging to $-\infty$ while $\dot{\phi} > \phi_D$ leads to the inflationary era with an exit. The initial condition $\dot{\phi} = \phi_D$ generates $\epsilon = 0$. In this case, the results are obtained for standard number of e-folds, i.e., $N \simeq 50, 60$, which is in good agreement with observational data. Further, it is observed that as the value of $\dot{\phi}/\phi_D$ is directly proportional to N, as increment in initial value produced an increase in N. Hence the rate of inflation increases as $\dot{\phi}$ increases. The trajectories are attracted toward its origin with increasing initial values. This shows that change in initial conditions forms value of de-Sitter implying either inflation with an exit or super inflation and proves that the obtained de-Sitter solution behaves as a saddle point. The plot $\phi - N$ represents the decaying behavior with evolution of time. The trajectories of ϵ and ϕ versus N for various initial conditions of $\dot{\phi}$ and $n = 0.01, 0.1$ are also plotted in [56, 67]. These plots show inflationary era and then an exit from inflation for $\dot{\phi} < \phi_D$ (which is opposite to our case) and number of e-folding lies between 80 and 90. Our results are compatible with [56, 67] for standard values of perturbed parameters.

For better understanding of the inflationary model's compatibility with recent data, we have plotted parametric plots (Figure 3) in which scalar spectral index is plotted versus tensor to scalar ratio for $n > 1$. It is observed that for $n = 1.61$, we have the standard value of spectral index $n_s = 0.96$ and an upper bound of tensor to scalar ratio is obtained as $r < 0.36$ which is compatible with WMAP7 [72], while for $n = 1.8$, we get $r < 0.11$ in accordance with Plank 2015 [73].

One of the main points is that the inflationary models in general relativity go to red-tilt [74, 75]. Our discussion and the references [56, 67] led to a conclusion that the modified theories of gravity and general relativity can be distinguished by the fact that, in modified theories, the tensor spectrum is blue tilted. It is worth mentioning here that our results reduced to [67] by choosing $\alpha = 0$.

Conflicts of Interest

The authors declare that they have no conflicts of interest.

References

[1] A. G. Riess, A. V. Filippenko, and P. Challis, "Observational evidence from supernovae for an accelerating universe and a cosmological constant," *The Astronomical Journal*, vol. 116, pp. 1009–1038, 1998.

[2] S. Perlmutter, G. Aldering, and G. Goldhaber, "Measurements of Ω and Λ from 42 High-Redshift Supernovae," *The Astrophysical Journal*, vol. 517, no. 2, pp. 565–586, 1999.

[3] D. Huterer and M. S. Turner, "Prospects of probing the dark energy via supernova distance measurements," *Physical Review D*, vol. 60, Article ID 081301, 1999.

[4] V. Sahni and A. A. Starobinsky, "The case for a positive cosmological Λ-term," *International Journal of Modern Physics D*, vol. 9, no. 4, pp. 373–443, 2000.

[5] S. M. Carroll, "The cosmological constant," *Living Reviews in Relativity*, vol. 4, article 1, 2001.

[6] P. Astier, J. Guy, and N. Regnault, "The Supernova Legacy Survey: measurement of Ω_M, Ω_λ and w from the first year data set," *Astronomy & Astrophysics*, vol. 447, no. 1, pp. 31–48, 2006.

[7] A. G. Riess and Supernova Search Team Collaboration, "Type la supernova discoveries at z>1 from the *Hubble Space Telescope*: evidence for past deceleration and constraints on dark energy evolution," *The Astrophysical Journal*, vol. 607, no. 2, p. 665, 2004.

[8] D. N. Spergel, L. Verde, and H. V. Peiris, "First year *Wilkinson Microwave Anisotropy Probe (WMAP)* observations: determination of cosmological parameters," *The Astrophysical Journal Supplement Series*, vol. 148, no. 1, pp. 175–194, 2003.

[9] D. N. Spergel, R. Bean, and O. Doré, "Three-year Wilkinson Microwave Anisotropy Probe (WMAP) observations: implications for cosmology," *The Astrophysical Journal Supplement Series*, vol. 170, no. 2, pp. 377–408, 2007.

[10] D. J. Eisenstein, I. Zehavi, and D. W. Hogg, "Detection of the Baryon acoustic peak in the large-scale correlation function of SDSS luminous red Galaxies," *The Astrophysical Journal*, vol. 633, no. 2, pp. 560–574, 2005.

[11] W. J. Percival, S. Cole, D. J. Eisenstein et al., "Measuring the baryon acoustic oscillation scale using the sloan digital sky survey and 2dF galaxy redshift survey," *Monthly Notices of the Royal Astronomical Society*, vol. 381, no. 3, pp. 1053–1066, 2007.

[12] W. J. Percival, "Baryon acoustic oscillations in the sloan digital sky survey data release 7 galaxy sample, MNRAS," *Astronomical Society of India*, vol. 401, p. 2148, 2010.

[13] A. A. Starobinsky, "Relict gravitation radiation spectrum and initial state of the universe," *JETP Letters*, vol. 30, p. 682, 1979.

[14] Y. Fujii, "Origin of the gravitational constant and particle masses in a scale-invariant scalar-tensor theory," *Physical Review D: Particles, Fields, Gravitation and Cosmology*, vol. 26, no. 10, pp. 2580–2588, 1982.

[15] S. M. Carroll, "Quintessence and the rest of the world: suppressing long-range interactions," *Physical Review Letters*, vol. 81, p. 3067, 1998.

[16] T. Chiba, T. Okabe, and M. Yamaguchi, "Kinetically driven quintessence," *Physical Review D: Particles, Fields, Gravitation and Cosmology*, vol. 62, Article ID 023511, 2000.

[17] C. Armendariz-Picon, V. Mukhanov, and P. J. Steinbardt, "Essentials of *k*-essence," *Physical Review D: Particles, Fields, Gravitation and Cosmology*, vol. 63, Article ID 103510, 2001.

[18] A. Y. Kamenshchik et al., "An alternative to quintessence," *Physics Letters B*, vol. 511, p. 265, 2001.

[19] M. C. Bento et al., "Generalized Chaplygin gas, accelerated expansion and dark-energy-matter unification," *Physical Review D*, vol. 66, Article ID 043507, 2002.

[20] J. A. Frieman et al., "Cosmology with ultralight pseudo nambu-goldstone bosons," *Physical Review Letters*, vol. 75, p. 2077, 1995.

[21] P. Brax and J. Martin, "Quintessence and supergravity," *Physics Letters. B. Particle Physics, Nuclear Physics and Cosmology*, vol. 468, no. 1-2, pp. 40–45, 1999.

[22] P. Brax and J. Martin, "Robustness of quintessence," *Physical Review D: Particles, Fields, Gravitation and Cosmology*, vol. 61, no. 10, Article ID 103502, 2000.

[23] H. A. Buchdahl, *Astronomical Society of India*, vol. 150, p. 1, 1970.

[24] R. Ferraro and F. Fiorini, "Modified teleparallel gravity: inflation without an inflaton," *Physical Review D: Particles, Fields, Gravitation and Cosmology*, vol. 75, no. 8, Article ID 084031, 2007.

[25] G. Cognola, "Dark energy in modified Gauss-Bonnet gravity: Late-time acceleration and the hierarchy problem," *Physical Review D*, vol. 73, Article ID 084007, 2006.

[26] T. Harko, F. S. N. Lobo, S. Nojiri, and S. D. Odintsov, "*f(R,T)* gravity," *Physical Review D: Particles, Fields, Gravitation and Cosmology*, vol. 84, Article ID 024020, 2011.

[27] M. Zubair, H. Azmat, and I. Noureen, *International Journal of Modern Physics D*, vol. 27, no. 4, Article ID 1850047, 2018.

[28] H. Azmat, M. Zubair, and I. Noureen, *International Journal of Modern Physics D*, vol. 27, no. 1, 2018.

[29] M. Zubair, G. Mustafa, S. Waheed, and G. Abbas, "Existence of stable wormholes on a non-commutative-geometric background in modified gravity," *The European Physical Journal C*, vol. 77, no. 10, 2017.

[30] M. Sharif and M. Zubair, "Anisotropic universe models with perfect fluid and scalar field in $f(R,T)$ gravity," *Journal of the Physical Society of Japan*, vol. 81, no. 11, Article ID 114005, 2012.

[31] M. Sharif and M. J. Zubair, "Analysis of *f(R)* theory corresponding to NADE and NHDE," *Journal of the Physical Society of Japan*, vol. 82, Article ID 014002, 2013.

[32] A. H. Guth, "Inflationary universe: a possible solution to the horizon and flatness problems," *Physical Review D: Particles, Fields, Gravitation and Cosmology*, vol. 23, pp. 347–356, 1981.

[33] A. A. Starobinsky, "A new type of isotropic cosmological models without singularity," *Physics Letters B*, vol. 91, no. 1, pp. 99–102, 1980.

[34] A. R. Liddle and D. H. Lyth, *Cosmological Inflation and Large-Scale Structure*, Cambridge University Press, Cambridge, UK, 2000.

[35] A. Linde, *Particle Physics and Inflationary Cosmology*, Harwood Academic, Chur, Switzerland, 1990.

[36] S. Nojiri and S. D. Odintsov, "Unified cosmic history in modified gravity: from *F(R)* theory to Lorentz non-invariant models," *Physics Reports*, vol. 505, pp. 59–144, 2011.

[37] S. Nojiri and S. D. Odintsov, *International Journal of Geometric Methods in Modern Physics*, vol. 4, p. 115, 2007.

[38] S. Capozziello and M. de Laurentis, "Extended Theories of Gravity," *Physics Reports*, vol. 509, no. 4-5, pp. 167–321, 2011.

[39] R. Myrzakulov, L. Sebastiani, and S. Zerbini, "Some aspects of generalized modified gravity models," *International Journal of Modern Physics D*, vol. 22, no. 8, Article ID 1330017, 2013.

[40] K. Bamba and S. D. Odintsov, "Inflationary cosmology in modified gravity theories," *Symmetry*, vol. 7, no. 1, pp. 220–240, 2015.

[41] L. Sebastiani and R. Myrzakulov, "*F(R)* gravity and Inflation," *International Journal of Geometric Methods in Modern Physics*, vol. 12, Article ID 1530003, 2015.

[42] S. D. Odintsov and V. K. Oikonomou, "Viable mimetic F(R) gravity compatible with Planck observations," *Annals of Physics*, vol. 363, pp. 503–514, 2015.

[43] K. Bamba, S. D. Odintsov, and P. V. Tretyakov, *The European Physical Journal C*, vol. 75, p. 344, 2015.

[44] M. Rinaldi et al., "Reconstructing the inflationary *f(R)* from observations," *Journal of Cosmology and Astroparticle Physics*, vol. 2014, no. 8, p. 15, 2014.

[45] M. Rinaldi, "Inflation in scale-invariant theories of gravity," *Physical Review D*, vol. 91, Article ID 123527, 2015.

[46] L. Sebastiani, "Nearly starobinsky inflation from modified gravity," *Physical Review D*, vol. 89, Article ID 023518, 2014.

[47] K. Bamba, "Trace-anomaly driven inflation in modified gravity and the BICEP2 result," *Physical Review D*, vol. 90, Article ID 043505, 2014.

[48] R. Myrzakulov, S. D. Odintsov, and L. Sebastiani, "Inflationary universe from higher-derivative quantum gravity," *Physical Review D: Particles, Fields, Gravitation and Cosmology*, vol. 91, no. 8, Article ID 083529, 15 pages, 2015.

[49] R. Myrzakulov, L. Sebastiani, and S. Zerbini, "Some aspects of generalized modified gravity models," *The European Physical Journal C*, vol. 75, p. 215, 2015.

[50] A. R. Liddle and M. Samuel, "How long before the end of inflation were observable perturbations produced?" *Physical Review D*, vol. 68, Article ID 103503, 2003.

[51] D. Walliser, "Successful inflation in scalar-tensor theories of gravity," *Nuclear Physics B*, vol. 378, p. 150, 1992.

[52] J. Garcia-Bellido and M. Quiros, "Extended inflation in scalar-tensor theories of gravity," *Physics Letters B*, vol. 243, p. 45, 1990.

[53] J. Lahiri and G. Bhattacharya, "Perturbative analysis of multiple-field cosmological inflation," *Annals of Physics*, vol. 321, no. 4, pp. 999–1023, 2006.

[54] R. Myrzakulov, L. Sebastiani, and S. Vagnozzi, "Inflation in $f(R, \phi)$-theories and mimetic gravity scenario," *The European Physical Journal C*, vol. 75, no. 9, 2015.

[55] M. Sharif and R. Saleem, "Warm anisotropic inflationary universe model," *The European Physical Journal C*, vol. 74, no. 2, 2014.

[56] J. Mathew, J. P. Johnson, and S. Shankaranarayanan, "Inflation with *f(R,φ)* in Jordan frame," *General Relativity and Quantum Cosmology*, vol. 1, 2017.

[57] T. Clifton, P. G. Ferreira, A. Padilla, and C. Skordis, "Modified gravity and cosmology," *Cosmology and Nongalactic Astrophysics*, vol. 1, 2011.

[58] A. De Felice and S. Tsujikawa, "f(R) theories," *Living Reviews in Relativity*, vol. 13, p. 3, 2010.

[59] D. Benedetti, P. F. Machado, and F. Saueressig, "Asymptotic safety in higher-derivative gravity," *Modern Physics Letters A*, vol. 24, no. 28, pp. 2233–2241, 2009.

[60] T. P. Sotiriou and V. Faraoni, "f(R) theories of gravity," *Reviews of Modern Physics*, vol. 82, p. 451, 2010.

[61] R. P. Woodard, "Avoiding dark energy with *1/R* modifications of gravity," *The Invisible Universe-Dark Matter and Dark Energy*, vol. 720, p. 403, 2007.

[62] V. F. Mukhanov, H. A. Feldman, and R. H. Brandenberger, "Theory of cosmological perturbations," *Physics Reports*, vol. 215, no. 5-6, pp. 203–333, 1992.

[63] T. Multamäki and I. Vilja, "Cosmological expansion and the uniqueness of the gravitational action," *Physical Review D: Particles, Fields, Gravitation and Cosmology*, vol. 73, no. 2, 2006.

[64] J. E. Lidsey, A. R. Liddle, E. W. Kolb, E. J. Copeland, T. Barreiro, and M. Abney, "Reconstructing the inflaton potential—an overview," *Reviews of Modern Physics*, vol. 69, no. 2, pp. 373–410, 1997.

[65] J.-C. Hwang and H. Noh, "Conserved cosmological structures in the one-loop superstring effective action," *Physical Review D: Particles, Fields, Gravitation and Cosmology*, vol. 61, no. 4, Article ID 043511, 5 pages, 2000.

[66] J. Hwang and H. Noh, "Cosmological perturbations in generalized gravity theories," *Physical Review D: Particles, Fields, Gravitation and Cosmology*, vol. 54, no. 2, pp. 1460–1473, 1996.

[67] J. P. Johnson, J. Mathew, and S. Shankaranarayanan, "Inflation driven by exponential non-minimal coupling of inflaton with gravity," 2017, https://arxiv.org/abs/1706.10150.

[68] G. Tambalo and M. Rinaldi, "Inflation and reheating in scale-invariant scalar-tensor gravity," *General Relativity and Quantum Cosmology*, vol. 1, pp. 49–52, 2016.

[69] P. Kanti, R. Gannouji, and N. Dadhich, "Early-time cosmological solutions in Einstein-scalar-Gauss-Bonnet theory," *Physical Review D: Particles, Fields, Gravitation and Cosmology*, vol. 92, no. 8, 083524, 15 pages, 2015.

[70] J. C. Hwang and H. Noh, "*f(R)* gravity theory and CMBR constraints," *Physics Letters B*, vol. 506, p. 13, 2001.

[71] H. Noh and J. C. Hwang, "Constraints on inflation models from supersymmetry breakings," *Physics Letters B*, vol. 515, p. 231, 2001.

[72] D. Larson and etal., "Seven-year wilkinson microwave anisotropy probe (WMAP) observations: power spectra and WMAP-derived parameters," *The Astrophysical Journal Supplement Series*, vol. 192, p. 16, 2011.

[73] P. A. R. Ade and Planck Collaboration, "Planck 2015 results. XIII. Cosmological parameters," *Astronomy & Astrophysics*, vol. 594, 63 pages, 2015.

[74] V. F. Mukhanov, *Physical Foundations of Cosmology*, Cambridge University Press, Cambridge, UK, 2005.

[75] S. Weinberg, *Cosmology*, Oxford University Press, New York, NY, USA, 2008.

24

Constraints on Gravitation from Causality and Quantum Consistency

Mark P. Hertzberg

Institute of Cosmology & Department of Physics and Astronomy, Tufts University, Medford, MA 02155, USA

Correspondence should be addressed to Mark P. Hertzberg; mark.hertzberg@tufts.edu

Academic Editor: Diego Saez-Chillon Gomez

We examine the role of consistency with causality and quantum mechanics in determining the properties of gravitation. We begin by examining two different classes of interacting theories of massless spin 2 particles—gravitons. One involves coupling the graviton with the lowest number of derivatives to matter, the other involves coupling the graviton with higher derivatives to matter, making use of the linearized Riemann tensor. The first class requires an infinite tower of terms for consistency, which is known to lead uniquely to general relativity. The second class only requires a finite number of terms for consistency, which appears as another class of theories of massless spin 2. We recap the causal consistency of general relativity and show how this fails in the second class for the special case of coupling to photons, exploiting related calculations in the literature. In a companion paper Hertzberg and Sandora (2017), this result is generalized to a much broader set of theories. Then, as a causal modification of general relativity, we add light scalar particles and recap the generic violation of universal free-fall they introduce and its quantum resolution. This leads to a discussion of a special type of scalar-tensor theory: the $F(\mathscr{R})$ models. We show that, unlike general relativity, these models do not possess the requisite counterterms to be consistent quantum effective field theories. Together this helps to remove some of the central assumptions made in deriving general relativity.

1. Introduction

General relativity is consistent with observations over a vast range of length scales. The $1/r^2$ force law has been tested down to fractions of a millimeter, while precisions tests of the relativistic theory have occurred on solar system scales, binary pulsars, and even the recent gravitational wave observations of merging binary black holes. On galactic and cosmological scales there is also agreement, though it does require the introduction of some as yet undiscovered form of dark matter and a small but nonzero amount of dark energy [1].

The latter has provided some of the central motivations for considering alternatives to general relativity. It is difficult to understand why the vacuum energy is so small, despite there being known large contributions from massive particles running in loops, such as top quarks. Furthermore, the coincidence problem (why there is a comparable amount of matter and dark energy today), as well as the cosmological horizon, homogeneity, and flatness problems are also sometimes invoked as motivations. Also, there are a suite of difficulties in understanding general relativity as a quantum theory, including nonrenormalizability, trans-Planckian unitarity violation, black hole information paradox, and global issues associated with de Sitter space and eternal inflation.

This range of primarily conceptual challenges, leads one to enquire just how inevitable general relativity is; whether theoretically consistent alternatives exist. It is sometimes thought that indeed general relativity follows inevitably as the *unique* consistent theory of massless spin 2 particles at low energies. Following uniquely if one only assumes the Lorentz symmetry applied to the spin 2 degrees of freedom. That to deviate from general relativity requires either a violation of Lorentz symmetry, or the introduction of additional degrees of freedom.

In this letter we clarify some aspects of this basic idea. Firstly, we point out that in fact classes of theories of massless spin 2 particles exist, which are Lorentz invariant and do not propagate new degrees of freedom. The most basic one involving the least number of derivatives, so-called minimal coupling, and the others involving higher derivatives (the latter can be organized to not propagate any additional degrees of freedom). While the first leads to general relativity, the seconds appears as another class of theories of spin 2, and was earlier introduced in Ref. [2]. We examine this second class in the special case of coupling to photons. We examine the propagation of photons in the second theory, showing there is superluminality (by exploiting related results in the literature) and forbidding a possible UV completion. In a companion paper [3] this idea is developed further and generalized to a much larger set of theories, including couplings to fermions and scalars, using a more systematic analysis.

Secondly, we study conventional ways to modify general relativity by the addition of new light scalars. We emphasize that in the Standard Model of particle physics only the Lorentz symmetry is postulated and most couplings compatible with it are observed. Similarly, new scalars should generically come with many parameters leading to violation of the observed universality of free-fall. However, quantum effects typically remove the problem by making the scalars heavy. This leads to an examination of the so-called $F(\mathscr{R})$ models, which do have the property of immediately ensuring the universality of free-fall. We show that such theories, unlike general relativity, fail to have the appropriate set of counterterms in the domain of applications of these theories, and so they fail to be consistent quantum effective field theories.

2. Massless Spin 2 Particles

We are interested in constructing theories of massless spin 2 particles from the ground up. As is well known, the massless spin 2 unitary representation of the Lorentz group involves two helicities in 3+1 dimensions. It is useful to embed these two degrees of freedom into a symmetric tensor field in order to build a local theory. We shall denote this $h_{\mu\nu}$ and we will shortly discuss to what extent this may be interpreted as a metric field. Since $h_{\mu\nu}$ is a 10 component object, we need to remove 8 of the 10 degrees of freedom; 4 are removed by the introduction of constraints on $h_{\mu 0}$, while another 4 can be removed by the introduction of an identification $h_{\mu\nu} \equiv h_{\mu\nu}+\partial_\mu\alpha_\nu+\partial_\nu\alpha_\mu$, where α_μ, the gauge function, parameterizes a family of physical equivalent representations of the same state $|\psi\rangle$.

If one fixes the gauge, then under a Lorentz transformation $\Lambda^\rho{}_\mu$, the field $h_{\mu\nu}$ transforms as

$$h_{\mu\nu} \longrightarrow \Lambda^\rho{}_\mu \Lambda^\sigma{}_\nu h_{\rho\sigma} + \partial_\mu \Omega_\nu + \partial_\nu \Omega_\mu \quad (1)$$

where Ω_μ depends on the gauge choice and $\Lambda^\rho{}_\mu$. Since $h_{\mu\nu}$ is evidently *not* a Lorentz tensor, it is generally very difficult to construct a Lorentz invariant interacting theory when we couple $h_{\mu\nu}$ to matter. There does exist, however, a manifestly gauge invariant and indeed Lorentz covariant 4-tensor we can construct out of derivatives of $h_{\mu\nu}$, the linearized Riemann tensor

$$R^{(L)}_{\mu\nu\rho\sigma} \equiv \frac{1}{2}\left(\partial_\rho\partial_\nu h_{\mu\sigma} + \partial_\sigma\partial_\mu h_{\nu\rho} - \partial_\sigma\partial_\nu h_{\mu\rho} - \partial_\rho\partial_\mu h_{\nu\sigma}\right) \quad (2)$$

which we will explicitly make use of in the upcoming "Type II" theories.

The free theory of these spin 2 particles is associated with terms of the form $\sim (\partial h)^2$. By demanding Lorentz/gauge invariance, only a unique set of terms is allowed, up to boundary terms, which is

$$\mathscr{L}^h_{kin} = \frac{1}{2}(\partial h)^2 - \frac{1}{2}\partial h_{\mu\nu}\partial h^{\mu\nu} + \partial_\mu h^{\mu\nu}\partial_\nu h - \partial_\mu h^{\rho\sigma}\partial_\rho h^\mu_\sigma \quad (3)$$

where $h \equiv h^\mu{}_\mu$ and we have scaled the coefficient of the first term to $1/2$ without loss of generality.

3. Type I: Lowest Number of Derivatives

The interaction that involves the least number of derivatives, and hence would be most relevant at large distances, is to attempt to couple $h_{\mu\nu}$ directly to matter as follows:

$$\mathscr{L}_{int} = h_{\mu\nu}\tau^{\mu\nu}_M \quad (4)$$

where $\tau^{\mu\nu}_M$ is some symmetric tensor built out of the matter fields, whose properties we shall shortly identify. Evidently this term is not gauge invariant for a generic $\tau^{\mu\nu}_M$, which means the theory is not unitary as it would propagate the wrong number of degrees of freedom. Alternatively, one could try to define this term in a particular gauge to avoid additional degrees of freedom, but then one would find the term is not Lorentz invariant as $h_{\mu\nu}$ is not a proper Lorentz tensor.

It is easy to see that under a gauge transformation and an integration by parts, the problem is fixed by taking $\tau^{\mu\nu}_M$ to be conserved $\partial_\mu\tau^{\mu\nu}_M = 0$. So $\tau^{\mu\nu}_M$ should be proportional to the matter energy-momentum tensor $T^{\mu\nu}_M$ as follows $\tau^{\mu\nu}_M = -(\kappa/2)T^{\mu\nu}_M$, where κ is a coupling. We are assured that this is conserved due to translation invariance (at least to leading order; more on this shortly). This immediately implies that all matter particles ϕ_i must couple universally to $h_{\mu\nu}$, with strength $\kappa_1 = \kappa_2 = \ldots = \kappa$, as only the *total* energy-momentum tensor is conserved for interacting particles. This implies the (weak) equivalence principle.

This ensures the theory is gauge invariant on-shell. To also be gauge invariant off-shell one must endow the matter fields with a gauge transformation rule, such as $\phi_i \longrightarrow \phi_i + \kappa\alpha_\mu\partial^\mu\phi_i$ for scalars [4]. However, the gauge invariance is only ensured to order κ, since the presence of this interaction means that energy and momentum will in general be exchanged between matter and gravitons, so $T^{\mu\nu}_M$ is no longer exactly conserved. This requires several fixes: (i) one must include the coupling of $h_{\mu\nu}$ to itself as gravitons carry energy and momentum, (ii) at higher order in κ one must modify the gauge transformation rule for $h_{\mu\nu}$ to involve higher order corrections, (iii) the gauge transformation rule for matter must also involve higher order corrections, such as $\phi_i \longrightarrow \phi_i + \kappa\alpha_\mu\partial^\mu\phi_i + (1/2)\kappa^2\alpha_\mu\alpha_\nu\partial^\mu\partial^\nu\phi_i + \ldots$, and (iv) an infinite tower of interaction terms, in powers of κ, must be included with the schematic " " form

$$\mathcal{L}_{int} = \kappa\left(-\frac{h_{\mu\nu}T_M^{\mu\nu}}{2} + \text{“}h\,(\partial h)^2\text{”}\right) + \kappa^2\left(\text{“}h^2\widehat{T}_M\text{”} + \text{“}h^2\,(\partial h)^2\text{”}\right) + \ldots \tag{5}$$

where every term is determined uniquely in terms of κ, up to boundary terms and field redefinitions. Amazingly, this infinite series can be resummed for any matter Lagrangian $\mathcal{L}_M(\phi_i, \eta_{\mu\nu})$ [5], giving the Einstein-Hilbert action

$$S_I = \int d^4x\sqrt{-g}\left[\frac{\mathcal{R}}{16\pi G} + \mathcal{L}_M\left(\phi_i, g_{\mu\nu}\right)\right] \tag{6}$$

where $G \equiv \kappa^2/(16\pi)$, $g_{\mu\nu} \equiv \eta_{\mu\nu} + \kappa h_{\mu\nu}$, and the matter Lagrangian involves the lift to "minimal coupling" $\eta_{\mu\nu} \longrightarrow g_{\mu\nu}$ and $\partial_\mu \longrightarrow \nabla_\nu$. The gauge invariance is lifted to the full diffeomorphism invariance $\phi_i(x^\mu) \longrightarrow \phi_i(x^\mu + \kappa\alpha^\mu)$ and $\mathcal{R}$ is the fully nonlinear Ricci scalar. Now $g_{\mu\nu}$ inevitably has a geometric interpretation.

Hence Type I coupling leads uniquely to general relativity and all of its successes. This theory can even be quantized, in the low energy regime, with concrete quantum gravity predictions such as [6, 7]; we shall return to this issue later. Furthermore, by including some small, but nonzero, vacuum energy in the matter Lagrangian this can even account for cosmic acceleration. This theory does lead to conceptual puzzles, such as the cosmological constant problem and black hole information paradox, as mentioned in the introduction, but is in great agreement with observations.

4. Type II: Higher Number of Derivatives

Here we would like to describe a much bigger class of theories of massless spin 2 particles; this was introduced earlier in Ref. [2] and some other work includes Ref. [8]. By exploiting the gauge invariant object $R^{(L)}_{\mu\nu\rho\sigma}$ defined in eq. (2) we can immediately write down a manifestly gauge/Lorentz invariant class of theories by coupling it into essentially any four index object $\widetilde{\tau}_M^{\mu\nu\rho\sigma}$ with interaction $\mathcal{L}_{int} = R^{(L)}_{\mu\nu\rho\sigma}\widetilde{\tau}_M^{\mu\nu\rho\sigma}$ giving

$$S_{II} = \int d^4x\left[\mathcal{L}^h_{kin} + \mathcal{L}_M\left(\phi_i, \eta_{\mu\nu}\right) + R^{(L)}_{\mu\nu\rho\sigma}\widetilde{\tau}_M^{\mu\nu\rho\sigma}\right] \tag{7}$$

Note that this object involves only a *finite* number of terms in powers of $h_{\mu\nu}$, unlike the Type I theory that involves an infinite tower of terms. (Both types are of course nonrenormalizable in 3+1 dimensions so higher order terms will be generated by quantum mechanics, which we shall return to later). Also note that for any matter Lagrangian $\mathcal{L}_M$, we are essentially free to choose any new independent object $\widetilde{\tau}_M^{\mu\nu\rho\sigma}$ which may be built out of many new parameters, unlike in Type I where the interaction term comes uniquely specified by the matter Lagrangian through the minimal coupling procedure. We further note that in Type II we can have many different flavors of massless spin 2 particles, without any contradiction, while in Type I there can only be a single flavor.

As a concrete example of the interaction, if the matter involves N vector fields with field strengths $F_{\mu\nu,i}$ we could choose $\widetilde{\tau}_M^{\mu\nu\rho\sigma}$ to be

$$\widetilde{\tau}_M^{\mu\nu\rho\sigma} = \sum_{i=1}^{N}\left[a_i F_i^{\mu\nu}F_i^{\rho\sigma} + b_i\eta^{\mu\rho}F_i^{\nu\beta}F^{\sigma}_{i\beta} + c_i\eta^{\mu\rho}\eta^{\nu\sigma}F_i^2\right] \tag{8}$$

where a_i, b_i, c_i are couplings. In general some constraints may be placed on the relative sizes of a_i, b_i, c_i to avoid higher time derivatives and ghosts, however there is no requirement for the couplings to be universal. So $a_1 \neq a_2 \neq a_3 \neq \ldots$ is permitted by the Lorentz symmetry. Hence such a theory does not imply the (weak) equivalence principle.

5. Principles in Physics

If we considered the equivalence principle to be another fundamental postulate, then this would suffice to reject this entire Type II class in favor of the very special Type I. However, in this work we only take the Lorentz symmetry as a fundamental postulate, and the equivalence principle is to be derived rather than assumed.

In fact it is useful to put this point of view in a broader perspective. It is often suggested that modern particle physics is built out of various additional postulates, such as the "gauge principle" or "principle of minimal coupling". However, if we examine the structure of the Standard Model, in particular its symmetries, a different picture emerges. (i) Exact symmetries: CPT derives from locality and unitarity, while $SU(3)\times SU(2)\times U(1)$ gauge is derived as an identification to remove the unphysical components of fields associated with twelve spin 1 particles. (ii) Approximate symmetries: $U(1)_B$, $U(1)_{B-L}$ are derived as accidental. (iii) Asymmetries: C, P, T, chiral, scale, etc., are not derivable from Lorentz symmetry and are not realized in nature. So in the Standard Model of particle physics, only that which follows from the Lorentz symmetry (applied to unitary representations) is realized, and no additional postulates appear to be required.

Furthermore, the global $U(1)$, associated with electric charge, derives from considering the analogous Type I theory of massless photons coupled to charged matter with the lowest number of derivatives as $A_\mu J_M^\mu$, which requires coupling to a very special conserved J_M^μ (just as $\tau_M^{\mu\nu}$ above). On the other hand, we can also consider an analogous Type II theory of massless photons coupled to neutral matter with a higher number of derivatives as

$$\mathcal{L}_{int} = F_{\mu\nu}\widetilde{J}_M^{\mu\nu} \tag{9}$$

which does not require anything special about $\widetilde{J}_M^{\mu\nu}$ (just as $\widetilde{\tau}_M^{\mu\nu\rho\sigma}$ above). In fact this describes the low energy effective theory of photons coupled to neutrinos with $\widetilde{J}_M^{\mu\nu} \propto \overline{\psi}\sigma^{\mu\nu}\psi$, etc. So at the level of the effective theory, both Type I and Type II theories are realized in nature for photons coupled to matter. This begs a question for gravitation: why has nature chosen Type I and not the much larger class Type II of gravitons coupled to matter?

There is reason to think that causality provides a possible answer to this question. We examine this in greater detail and for a much broader class of models in a companion paper [3]. For now we illustrate the idea in the special case of coupling to photons.

6. Causality in Type I

It is well known that in general relativity with standard matter sources there is no problem with causality [9, 10]. This can be seen as follows. Consider photons minimally coupled to gravity in Type I. In the geometrics optics limit, photons obey the null geodesic equation $k_\mu k_\nu g^{\mu\nu} = 0$, where k_μ is the photon's 4-momentum. The leading deflection from null propagation $k_\mu^{(0)}$ on the Minkowski cone comes from expanding in powers of the gravitational coupling as $k_\mu = k_\mu^{(0)} + \kappa k_\mu^{(1)} + \ldots$ and $g^{\mu\nu} = \eta^{\mu\nu} - \kappa h^{\mu\nu} + \ldots$, giving $k_\mu k_\mu \eta^{\mu\nu} \approx \kappa k_\mu^{(0)} k_\nu^{(0)} h^{\mu\nu}$. Then using the linearized Einstein equations in the Lorenz gauge $\Box \bar{h}^{\mu\nu} = -\kappa T_M^{\mu\nu}$ and solving for a particular solution gives [9, 10]

$$k_\mu k_\nu \eta^{\mu\nu} \approx 4G \int d^3x' \frac{k_\mu^{(0)} k_\nu^{(0)} T_M^{\mu\nu}(\mathbf{x}', t_R)}{|\mathbf{x} - \mathbf{x}'|} \tag{10}$$

which clearly satisfies $k_\mu k_\mu \eta^{\mu\nu} \geq 0$ for any matter $T_M^{\mu\nu}$ that satisfies null-energy condition. Hence light stays inside the Minkowski cone and slows down in accord with the Shapiro time delay.

7. Causality in Type II

Let us focus on the case of the four index object $\tilde{\tau}_M^{\mu\nu\rho\sigma}$ given in eq. (8) and focus on a single species, the photon. The modified Maxwell equation for a region of space-time that is Ricci flat is

$$\partial_\mu F^\mu{}_\nu - 4aR^{(L)}_{\mu\nu\rho\sigma}\partial^\mu F^{\rho\sigma} = 0 \tag{11}$$

Note that this equation is *exact*, and the derivatives in Type II are just ordinary, not covariant, derivatives.

In the geometric optics limit, the leading deflection from null propagation on the Minkowski cone is [11]

$$k_\mu k_\nu \eta^{\mu\nu} \approx -8aR^{(L)}_{\mu\nu\rho\sigma} k^\mu k^\rho \epsilon_p^\nu \epsilon_p^\sigma \Big|_{(0)} \tag{12}$$

where all terms on the right hand side are evaluated for free null propagation. Here ϵ_p^ν is the photon's polarization unit vector with $p = 1, 2$ for each of the two modes. For an appropriate choice of polarization and direction of propagation, relative to the gravitational field, one can arrange for $k_\mu k_\nu \eta^{\mu\nu} < 0$ going outside the Minkowski cone. Since these theories are manifestly Lorentz invariant, this leads to problems with causality. This idea is greatly generalized in a companion paper [3].

In a related context of QED, minimally coupled to gravity, it is known that one can integrate out the electron and generate terms of the form (8) (with the nonlinear Riemann tensor) with coefficients $a, b, c \sim \alpha\kappa/m_e^2$ [12]. However this does not produce superluminality since the leading Shapiro time delay dominates over this effect in its domain of applicability. Related ideas appear in the context of string theory [13]. However, in the context of this new class of spin 2 theories, where this is the leading interaction, superluminality appears unavoidable.

8. Additional Fields

There does exist a manifestly causal way to modify gravitation. This involves the introduction of additional degrees of freedom. Since fermions do not mediate long range forces, and vectors (with minimal coupling) have sources that tend to neutralize, we focus on the remaining case of adding light scalars.

Usually in the literature a scalar χ is added that is taken to couple universally to the trace of the matter energy-momentum tensor at leading order as

$$S = S_I + \int d^4x\sqrt{-g}\left[\frac{1}{2}(\partial\chi)^2 + \gamma\chi g_{\mu\nu}T_M^{\mu\nu} + \ldots\right] \tag{13}$$

where γ is a coupling. Here the universal coupling is inserted to be compatible with the (weak) equivalence principle. But as described earlier in this letter, the Standard Model does not give any reason to utilize any principle beyond that of just the Lorentz symmetry. Since χ is a gauge singlet scalar, we could take any gauge invariant term in the Standard Model $\mathscr{L}_{SM,j}$ and multiply it by χ

$$S = S_I + \int d^4x\sqrt{-g}\left[\frac{1}{2}(\partial\chi)^2 + \chi\sum_j \gamma_j \mathscr{L}_{SM,j} + \ldots\right] \tag{14}$$

where γ_j are arbitrary couplings and obtain a Lorentz invariant theory.

So, to assume the form (13) is to tune the theory to be compatible with tests of the universality of free-fall, which have constrained $|\Delta a/a| < 10^{-13}$ [14]. One may appeal to technical naturalness to justify such universal couplings [15], or to link χ to the dilaton of a spontaneously broken scale symmetry [16], or to special fields associated with extra dimensions. However, generic scalars beyond the Standard Model do not have this feature of universal coupling. Instead Lorentz symmetry suggests the nonuniversal (14) is much more generic.

This poses a challenge to deriving the (weak) equivalence principle. However, if we take quantum effects into account, then a generic scalar χ will pick up a mass from Standard Model particles running in a loop of the form $\Delta m_\chi \sim \gamma\Lambda_{UV}^2/(4\pi)$ (or $\Delta m_\chi \sim \gamma m_{SM}\Lambda_{UV}/(4\pi)$), using a hard UV cutoff on the loop integral Λ_{UV}. For $\gamma \gtrsim \kappa$, and unless the cutoff is extremely low, the scalar χ will be typically heavy and unable to mediate long range forces, so general relativity is recovered at large distances.

9. $F(\mathscr{R})$ Gravity: Classical Treatment

A popular framework that is both causal and enforces the universality of free-fall is the so-called $F(\mathscr{R})$ models. Here the Einstein-Hilbert action eq. (6) is modified as $\mathscr{R} \longrightarrow$

$F(\mathcal{R})$. Note that inside any nonlinear function $F(\mathcal{R})$ are higher derivatives. This can be seen by expanding around a Minkowski background obtaining $\mathcal{R} = \kappa(\partial_\mu\partial_\nu h^{\mu\nu} - \Box h) + \mathcal{O}(\kappa^2 h(\partial h)^2)$. The $\sim \kappa\partial^2 h$ terms here only lead to a total derivative in the action if $F(\mathcal{R})$ is linear in $\mathcal{R}$, but for nonlinear $F(\mathcal{R})$ these higher derivatives have consequences. At the classical level, these consequences can be captured by the introduction of a scalar χ with action

$$S_\chi = \int d^4x\sqrt{-g}\left[\frac{\mathcal{R}}{16\pi G} + \mathcal{L}_M\left(\phi_i, g_{\mu\nu} f(\chi)\right) + \frac{1}{2}(\partial\chi)^2 - V(\chi)\right] \tag{15}$$

where $f(\chi)$ and $V(\chi)$ are functions that depend on the choice of F. Note that χ couples to matter in a universal way through the single function $g_{\mu\nu} f(\chi)$, satisfying the (weak) equivalence principle.

A popular example is $F(\mathcal{R}) = \mathcal{R} + \zeta\mathcal{R}^2/M_{\rm Pl}^2$ (with $M_{Pl} \equiv 1/\sqrt{16\pi G}$ and $\zeta \gg 1$) which is a model of inflation [17]. Here the classically equivalent scalar χ plays the role of the inflaton. Its potential turns out to be

$$V(\chi) = \frac{M_{\rm Pl}^4}{4\zeta}\left(1 - \exp\left(-\frac{\chi}{\sqrt{3}M_{\rm Pl}}\right)\right)^2 \tag{16}$$

For large field values $\chi \gg M_{\rm Pl}$ the potential is exponentially flat and inflation takes place. One computes correlation functions of the scalar mode of the form $\langle {\rm BD}|\chi(x)\chi(y)|{\rm BD}\rangle$ (where $|{\rm BD}\rangle$ is the Bunch-Davies vacuum) to obtain an approximately scale invariant spectrum of density perturbations with small red tilt $n_s - 1 \approx -2/N_e \sim -0.04$ and tensor-to-scalar ratio $r \approx 12/N_e^2 \sim 0.004$. These predictions are compatible with recent data [18, 19]. Similarly there exist many popular models of dark energy associated with various choices of $F(\mathcal{R})$ [20].

10. $F(\mathcal{R})$ Gravity: Quantum Treatment

Here we would like to examine $F(\mathcal{R})$ gravity as an effective field theory. To begin, let us return to the Einstein-Hilbert action eq. (6) and try to study it as a quantum theory. The quantum partition function is

$$Z_I = \int \prod_p \mathcal{D}h_p \prod_i \mathcal{D}\phi_i e^{iS_I[h_p,\phi_i]/\hbar} \tag{17}$$

where the first measure of the path integral is over the *two* modes of the graviton labelled $p = 1, 2$ and the second measure is over the matter fields. In practice there are various complications associated with gauge fixing and constraints, but this is the formal structure. In principle this allows one to compute various correlation functions such as $\langle h_p(x)h_{p'}(y)\rangle$, $\langle\phi_i(x)\phi_j(y)\rangle$. If we perform the path integral partially by integrating down to some scale Λ, we will generate a new Wilsonian effective action including corrections such as

$$S_{I,eff} = S_I(\Lambda) + \int d^4x\sqrt{-g}\left[c_1(\Lambda)\mathcal{R}^2 + c_2(\Lambda)\mathcal{R}_{\mu\nu}\mathcal{R}^{\mu\nu} + c_3(\Lambda)\mathcal{R}_{\mu\nu\rho\sigma}\mathcal{R}^{\mu\nu\rho\sigma} + \ldots\right] \tag{18}$$

(plus some nonlocal terms, etc.). These additional terms are required as counterterms to cancel divergences associated with graviton loops. As long as we focus on only the physical degrees of freedom, such as the two modes of the graviton, we can use this effective theory to compute quantum effects. (In fact quantization of the leading Einstein-Hilbert term S_I already gives rise to long range corrections to gravitation; see Refs. [6, 7]).

Note that the effective Lagrangian in eq. (18) formally involves higher derivatives due to the presence of the terms $\mathcal{R}^2$, etc. Furthermore the presence of these types of terms might, at first sight, seem to justify the kind of $F(\mathcal{R})$ actions that we wrote above. However, it is essential to not use these higher derivative terms incorrectly; the original path integral is only defined with a measure for the two modes of the graviton and the matter fields. The measure does *not* include integration over additional degrees of freedom, such as a scalar χ. Instead the path integral forces these to be spurious additional degrees of freedom; they can never be external and on-shell.

By contrast, the $F(\mathcal{R})$ models, as applied to inflation and dark energy, etc., explicitly make use of the additional scalar χ. This scalar is given its own dynamics, its own phase space, and its own independent set of fluctuations that are used as the source of density perturbations. This means the $F(\mathcal{R})$ models are disconnected from a rigorous quantum treatment of the Einstein-Hilbert action.

One might attempt to quantize $F(\mathcal{R})$ using the path integral. However, the path integral requires one to integrate over all the physical fields in the theory. So we would need to explicitly use the scalar χ and form

$$Z_\chi = \int \prod_p \mathcal{D}h_p \mathcal{D}\chi \prod_i \mathcal{D}\phi_i e^{iS_\chi[h_p,\chi,\phi_i]/\hbar} \tag{19}$$

Again by integrating over high energy modes, we form a new type of Wilsonian effective action

$$S_{\chi,eff} = S_\chi(\Lambda) + \int d^4x\sqrt{-g}\left[d_1(\Lambda)(\partial\chi)^4 + \ldots\right] \tag{20}$$

(plus generating $\mathcal{R}^2$ terms, etc., again). We emphasize the presence of new counterterms, such as $(\partial\chi)^4$. These additional counterterms are required to cancel new divergences that arise from the ability to put the χ external and on-shell. It is very important to note that such additional terms *cannot* be put in the $F(\mathcal{R})$ form, and more generally, cannot put in the form of some function only of the metric when generic matter is included. Hence the $F(\mathcal{R})$ models, which exploit the dynamics of the additional degree of freedom, contain cut-off dependence in the quantum theory without the required counterterms to be a consistent quantum effective theory.

We note that in other formulations of gravity, such as Palatini, similar conclusions hold. Namely, in constructions in which there are no new degrees of freedom, then the theory can always be recast into the general relativity form with a collections of appropriate counter-terms, while, in constructions in which there is a new degree of freedom, it can only be self-consistently quantized by reorganization into the scalar-tensor form.

11. Outlook

The above arguments help toward deriving general relativity as the only consistent theory involving massless spin 2 at low energies. (i) Type II theories that utilize higher derivative couplings (but can avoid extra degrees of freedom) can lead to problems with causality; see [3] for an extended analysis. (ii) Additional scalars, which would generically lead to nonuniversal free-fall, are typically expected to be heavy due to quantum effects. (iii) $F(\mathscr{R})$ models are not consistent quantum effective field theories.

However, important puzzles remain, including understanding dark energy. The smallness of the vacuum energy within the framework of general relativity does have a candidate explanation by introducing many (heavy) scalars, leading to a potential with an exponentially large number of vacua. Though it is unclear how to formulate probabilities in this context. While the behavior of gravitation at the Planck scale requires further new physics.

Disclosure

An earlier version of this manuscript was presented at the 13th International Symposium on Cosmology and Particle Astrophysics (CosPA 2016).

Conflicts of Interest

The author declares that they have no conflicts of interest.

Acknowledgments

I would like to thank Raphael Flauger, Jaume Garriga, Alan Guth, David Kaiser, Juan Maldacena, McCullen Sandora, and Mark Trodden for helpful conversations. I would like to thank the Tufts Institute of Cosmology for support.

References

[1] A. G. Riess, A. V. Filippenko, and P. Challis, "Observational evidence from supernovae for an accelerating universe and a cosmological constant," *The Astronomical Journal*, vol. 116, no. 3, p. 1009, 1998.

[2] R. M. Wald, "Spin-two fields and general covariance," *Physical Review D*, vol. 33, 1986.

[3] M. P. Hertzberg and M. Sandora, "General relativity from causality," *Journal of High Energy Physics*, vol. 119, 2017.

[4] M. D. Schwartz, *Quantum Field Theory and the Standard Model*, Cambridge University Press, Cambridge, 2014.

[5] S. Deser, "Self-interaction and gauge invariance," *General Relativity and Gravitation*, vol. 1, no. 1, pp. 9–18, 1970.

[6] J. F. Donoghue, "Leading quantum correction to the Newtonian potential," *Physical Review Letters*, vol. 72, 1994.

[7] L. H. Ford, M. P. Hertzberg, and J. Karouby, "Quantum gravitational force between polarizable objects," *Physical Review Letters*, vol. 116, no. 15, Article ID 151301, 2016.

[8] D. Bai and Y. H. Xing, "Higher derivative theories for interacting massless gravitons in Minkowski spacetime," *Nuclear Physics B*, vol. 932, 2018.

[9] M. Visser, B. Bassett, and S. Liberati, "Superluminal censorship," *Nuclear Physics B. Proceedings Supplement*, vol. 88, pp. 267–270, 2000.

[10] A. Adams, N. Arkani-Hamed, S. Dubovsky, A. Nicolis, and R. Rattazzi, "Causality, analyticity and an IR obstruction to UV completion," *Journal of High Energy Physics*, vol. 2006, 2006.

[11] G. M. Shore, "Quantum gravitational optics," *Contemporary Physics*, vol. 44, 2003.

[12] I. T. Drummond and S. J. Hathrell, "QED vacuum polarization in a background gravitational field and its effect on the velocity of photons," *Physical Review D: Particles, Fields, Gravitation and Cosmology*, vol. 22, no. 2, pp. 343–355, 1980.

[13] X. O. Camanho, J. D. Edelstein, J. Maldacena, and A. Zhiboedov, "Causality constraints on corrections to the graviton three-point coupling," *Journal of High Energy Physics*, vol. 2016, no. 2, 2016.

[14] S. Schlamminger, K.-Y. Choi, T. A. Wagner, J. H. Gundlach, and E. G. Adelberger, "Test of the equivalence principle using a rotating torsion balance," *Physical Review Letters*, vol. 100, no. 4, Article ID 041101, 2008.

[15] L. Hui and A. Nicolis, "Equivalence principle for scalar forces," *Physical Review Letters*, vol. 105, no. 23, 2010.

[16] C. Armendariz-Picon and R. Penco, "Quantum equivalence principle violations in scalar-tensor theories," *Physical Review D: Particles, Fields, Gravitation and Cosmology*, vol. 85, no. 4, 2012.

[17] A. A. Starobinsky, "A new type of isotropic cosmological models without singularity," *Physics Letters B*, vol. 91, no. 1, pp. 99–102, 1980.

[18] G. Hinshaw, "WMAP Collaboration," arXiv:1212.5226.

[19] P. A. R. Ade, "Planck Collaboration," arXiv:1303.5082.

[20] T. P. Sotiriou and V. Faraoni, "$f(R)$ theories of gravity," *Reviews of Modern Physics*, vol. 82, no. 1, article 451, 2010.

25

A Covariant Canonical Quantization of General Relativity

Stuart Marongwe

Department of Physics and Astronomy, Botswana International University of Science and Technology, P. Bag 16, Palapye, Botswana

Correspondence should be addressed to Stuart Marongwe; stuartmarongwe@gmail.com

Guest Editor: Farook Rahaman

A Hamiltonian formulation of General Relativity within the context of the Nexus Paradigm of quantum gravity is presented. We show that the Ricci flow in a compact matter free manifold serves as the Hamiltonian density of the vacuum as well as a time evolution operator for the vacuum energy density. The metric tensor of GR is expressed in terms of the Bloch energy eigenstate functions of the quantum vacuum allowing an interpretation of GR in terms of the fundamental concepts of quantum mechanics.

1. Introduction

The gravitational field which is elegantly described by Einstein's field equations has so far eluded a quantum description. Much effort has been placed into formulating General Relativity (GR) in terms of Hamilton's equations since a Hamiltonian formulation of a classical field theory leads naturally to its quantization. The earliest such attempt is the ADM formalism [1], named for its authors Richard Arnowitt, Stanley Deser, and Charles W. Misner first published in 1959. This formalism starts from the assumption that space is foliated into a family of time slices Σ_t, labeled by their time coordinate t and with space coordinates on each slice given by x^k. The dynamic variables of this theory are then taken to be the metric tensor of three-dimensional spatial slices $\gamma_{ij}(t.x^k)$ and their conjugate momenta $\pi^{ij}(t.x^k)$. Using these variables it is possible to define a Hamiltonian and thereby write the equations of motion for GR in Hamilton's form. The time slices are then welded together using four Lagrange multipliers and components of a shift vector field. An extensive review of this formalism can be found in the literature notably in [2–5].

The ADM formalism was first applied by Bryce De Witt in 1967 [6] to quantize gravity which resulted in the Wheeler–De Witt equation of quantum gravity. It is a functional differential equation in which the three-dimensional spatial metrics have the form of an operator acting on a wave function. This wave function contains all of the information about the geometry and matter content of the universe of each time slice. However, the Hamiltonian no longer determines the evolution of the system and leads to the problem of timelessness [7, 8]. Hawking rightly points out that the very act of splitting space-time into space and time destroys the spirit of GR (general covariance) and therefore not much can be gained from this approach to quantization of gravity. Perturbative covariant approaches to the problem of quantum gravity have an inherent weakness in that they depend on a fine classical background. It is therefore difficult to obtain a self-consistent quantum theory of gravity with a classical background space-time. Steven Carlip [9, 10] and Claus Kiefer [11, 12] have made excellent and extensive reviews of the problems faced by current approaches to the problem of quantum gravity. Carlip [9] in particular singles out the lack of a firm conceptual understanding of the foundational concepts of quantum gravity as the source of much of the difficulty in understanding quantum gravity. These deep conceptual issues result in technical problems in the attempt to develop a consistent quantum theory of gravity.

In this paper we report a successful covariant canonical quantization of the gravitational field which preserves the success of GR while simultaneously explaining Dark Energy (DE) and Dark Matter (DM). This approach to quantization takes place in 4-space of metric signature (-1,1,1,1) in which the quanta are excitations of the quantum vacuum called Nexus gravitons. Though the Nexus Paradigm has been introduced in the following papers [13–15], the aim of this study is to explicitly express the Hamiltonian formulation of the theory using the Bloch eigenstate functions of the quantum vacuum. These wave functions contain information about the energy state of the quantum vacuum which in turn dictates the geometry of space-time.

2. Methods

Our first step towards a covariant canonical quantization begins with defining a quantized space-time and its quanta. We then modify Einstein's vacuum equations to be consistent with the quantized space-time followed by the defining of Hamilton's equations of the quantized space-time. This step is then followed by the Poisson brackets which provide the bridge between classical and quantum mechanics (QM). The covariant canonical quantization procedure is carried out within the context of the Nexus Paradigm of quantum gravity.

2.1. Quantization of 4-Space and the Nexus Graviton. The primary objective of physics is the study of functional relationships amongst measurable physical quantities. In particular, a unifying paradigm of physical phenomena should reveal the functional relationship between the fundamental physical quantities of 4-space and 4-momentum. Currently GR and QM offer the best predictions of the results of measurement of physical phenomena in their respective domains using different languages. GR describes gravitation in the language of geometry and thus far, it has been difficult to apply the the language of wave functions used in QM to give it a QM description. The problem of quantum gravity is therefore to interpret GR in terms of the wavefunctions of QM. Translating the language of measurement of GR into that of QM becomes the primary objective of this present attempt to resolve the problem of quantum gravity.

Measurements in GR take place in a local patch of a Reimannian manifold. This local patch can be considered as a flat Minkowski space. The line element in Minkowski space which is the subject of measurement can be computed through the inner product of the local coordinates as

$$\begin{aligned}\Delta x^{\mu}\Delta x_{\mu} &= \Delta x^2 + \Delta y^2 + \Delta z^2 - c^2\Delta t^2 \\ &= (A\Delta x + B\Delta y + C\Delta z + iDc\Delta t) \\ &\quad \cdot (A\Delta x + B\Delta y + C\Delta z + iDc\Delta t)\end{aligned} \quad (1)$$

On multiplying the right hand side we see that to get all the cross terms such as $\Delta x\Delta y$ to cancel out we must assume

$$\begin{aligned}AB + BA &= 0 \\ A^2 = B^2 &= \cdots = 1\end{aligned} \quad (2)$$

The above conditions therefore imply that the coefficients (A, B, C, D) generate a Clifford algebra and therefore must be matrices. We rewrite these coefficients in the 4-tuple form as $(\gamma^1, \gamma^2, \gamma^3, \gamma^0)$ which may be summarized using the Minkowski metric on space-time as follows:

$$\{\gamma^{\mu}, \gamma^{\nu}\} = 2\eta^{\mu\nu} \quad (3)$$

The gammas are of course the Dirac matrices. Thus in order to satisfy (1) we can express a displacement 4-vector as

$$\Delta x^{\mu} = a\gamma^{\mu} \quad (4)$$

where a is the amplitude of a displacement vector in Minkowski space. If we consider the Hubble diameter as the maximum dimension of the local patch of space then $a = r_{HS}$ where r_{HS} is the Hubble radius. This choice of a maximum amplitude is justified from the fact that we cannot physically interact with objects beyond the Hubble 4-radius. It is important to note that the line element is the square of the amplitude of the displacement 4-vector. Thus if we are to express GR in terms of the language of QM we must make the radical assumption that the displacement vectors in Minkowski space are pulses of 4-space which can be expressed in terms of Fourier functions as follows:

$$\begin{aligned}\Delta x_n^{\mu} &= \frac{2r_{HS}}{n\pi}\gamma^{\mu}\int_{-\infty}^{\infty}\operatorname{sinc}(kx)\, e^{ikx}dk \\ &= \gamma^{\mu}\int_{-\infty}^{\infty} a_{nk}\varphi_{(nkx)}dk\end{aligned} \quad (5)$$

$$\text{Where } \frac{2r_{HS}}{n\pi} = \sum_{k=-\infty}^{k=+\infty} a_{nk} \quad (6)$$

Here $\varphi_{(nkx)} = \operatorname{sinc}(kx)e^{ikx}$ are Bloch energy eigenstate functions. The Bloch functions can only allow the four wave vector to assume the following quantized values:

$$k^{\mu} = \frac{n\pi}{r_{HS}^{\mu}} \qquad n = \pm 1, \pm 2 \ldots 10^{60} \quad (7)$$

The minimum 4-radius in Minkowski space is the Planck 4-length since it is impossible to measure this length without forming a black hole. The 10^{60} states arise from the ratio of Hubble 4-radius to the Planck 4-length. The displacement 4-vectors in each eigenstate of space-time generate an infinite Bravais 4-lattice. Also, condition (7) transforms (5) to

$$\Delta x_n^{\mu} = \gamma^{\mu}\int_{-nk_1}^{nk_1} a_{nk}\varphi_{(n,k,x)}dk \quad (8)$$

The second assumption we make is that each displacement 4-vector is associated with a conjugate pulse of four-momentum which can also be expressed as a Fourier integral

$$\begin{aligned}\Delta p_{(n)\mu} &= \frac{2np_1}{\pi}\gamma_{\mu}\int_{-nk_1}^{nk_1}\varphi_{(n,k,x)}dk \\ &= \gamma_{\mu}\int_{-nk_1}^{nk_1} c_{nk}\varphi_{(n,k,x)}dk\end{aligned} \quad (9)$$

where $p_{(1)\mu}$ is the four-momentum of the ground state.

A displacement 4-vector and its conjugate 4-momentum satisfy the Heisenberg uncertainty relation

$$\Delta x_n \Delta p_n \geq \frac{\hbar}{2} \quad (10)$$

The Uncertainty Principle plays the important role of generating a vector bundle, out of the total uncertainty space E of trivial displacement 4-vectors from which a closed compact manifold X is formed, i.e., $(\pi : E \longrightarrow X)$. Each point on the manifold is associated with a vector which is along a normal to the manifold.

The wave packet described by (8) is essentially a particle of four-space. The spin of this particle can be determined from the fact that each component of the four-displacement vector transforms according to the law

$$\Delta x'^{\mu}_{n} = \exp\left(\frac{1}{8}\omega_{\mu\upsilon}\left[\gamma_{\mu}, \gamma_{\upsilon}\right]\right)\Delta x^{\mu}_{n} \tag{11}$$

where $\omega_{\mu\upsilon}$ is an antisymmetric 4x4 matrix parameterizing the transformation.

Therefore, each component of the 4-vector has a spin half. A summation of all the four half spins yields a total spin of *2*. We give the name the Nexus graviton to this particle of 4-space since the primary objective of quantum gravity is to find the nexus between the concepts of GR and QM.

From (7) the norm squared of the 4-momentum of the n-th state graviton is

$$(\hbar)^2 k^{\mu}k_{\mu} = \frac{E_n^2}{c^2} - \frac{3\left(nhH_0\right)^2}{c^2} = 0 \tag{12}$$

where H_0 is the Hubble constant (2.2 x 10^{-18} s^{-1}) and can be expressed in terms of the cosmological constant, Λ, as

$$\Lambda_n = \frac{E_n^2}{(hc)^2} = \frac{3k_n^2}{(2\pi)^2} = n^2\Lambda \tag{13}$$

We infer from (13) that the Nexus graviton (or displacement 4-vector) in the n-th quantum state forms a trivial vector bundle via the Uncertainty Principle which generates a compact Riemannian manifold of positive Ricci curvature that can be expressed in the form

$$G_{(nk)\mu\upsilon} = n^2\Lambda g_{(n,k)\mu\upsilon} \tag{14}$$

where $G_{(nk)\mu\upsilon}$ is the Einstein tensor of space-time in the n-th state. Equation (14) depicts a contracting geodesic ball and as explained in [13–15] this is DM which is an intrinsic compactification of the elements space-time in the n-th quantum state. This compactification is a result of the superposition of several plane waves as described by (8) to form an increasingly localized wave packet as more waves are added. Similary the converse is also true. The loss of harmonic waves expands the elements of space-time which gives rise to DE. Thus the DE arises from the emission of a ground state graviton such that (14) becomes

$$G_{(nk)\mu\upsilon} = \left(n^2 - 1\right)\Lambda g_{(n,k)\mu\upsilon} \tag{15}$$

These are Einstein's vacuum field equations in the quantized space-time. If the graviton field is perturbed by the presence of baryonic matter then (15) becomes

$$\begin{aligned} G_{(nk)\mu\upsilon} &= kT_{\mu\nu} + \left(n^2 - 1\right)\Lambda g_{(n,k)\mu\upsilon} \\ &= kT_{\mu\nu} + \left(n^2 - 1\right)k\rho_{DE}g_{(n,k)\mu\upsilon} \end{aligned} \tag{16}$$

where ρ_{DE} is the density of DE.

From [14] the static solution to (14) for a spherically symmetric Nexus graviton is computed as

$$\begin{aligned} ds^2 = &-\left(1 - \left(\frac{2}{n^2}\right)\right)c^2dt^2 + \left(1 - \left(\frac{2}{n^2}\right)\right)^{-1}dr^2 \\ &+ r^2\left(d\theta^2 + \sin^2\theta d\varphi^2\right) \end{aligned} \tag{17}$$

There are no singularities in (17). At high energies, characterized by microcosmic scale wavelengths of the Nexus graviton and high values of n, space-time is flat and highly compact. This implies a continuous length contraction of the local coordinates via the addition of more waves resulting in an increase in localization. Also, from a world line perspective of a test particle, it implies that the deviation from a rectilinear trajectory due to uncertainities in its location in 4-space is small. Thus at microcosmic scales, in the realm of subatomic particles space-time is extremely flat. The world line begins to substantially deviate from a well defined rectilinear path at low energies as the uncertainities in its location are increased. That is, at low values of n, the world line becomes degenerate allowing a multiplicity of trajectories which in GR are averaged by the Ricci curvature tensor. Thus gravity is a low energy phenomenon which vanishes asymptotically at high energies.

The gravitational effects of the Nexus graviton manifest at large scales and for galaxies, these effects begin to manifest when the density of DE is equal to the density of baryonic matter as described by (16) or when the acceleration due to baryonic matter is equal and opposite to the acceleration due to the emission of the ground state graviton as described in [14]. A solution to eqn (16) from [14] in the weak field at galactic scale radii is

$$\frac{d^2r}{dt^2} = \frac{GM(r)}{r^2} + H_0v_n - H_0c \tag{18}$$

Here c is the speed of light.

The first term on the right is the Newtonian gravitational acceleration, the second term is a radial acceleration due to space-time in the n-th quantum state, and the final term is acceleration due to DE. The dynamics becomes strongly non-Newtonian when

$$\frac{GM(r)}{r^2} = H_0c = \frac{v_n^2}{r} \tag{19}$$

These are conditions in which the acceleration due to baryonic matter is annulled by that due to the DE. Under such conditions

$$r = \frac{v_n^2}{H_0c} \tag{20}$$

Substituting for r in (19) yields

$$v_n^4 = GM(r)H_0c \tag{21}$$

This is the Baryonic Tully–Fisher relation. The conditions permitting the DE to cancel out the acceleration due to baryonic matter leave quantum gravity as the unique source of gravity. Thus condition (19) reduces (18) to

$$\frac{d^2r}{dt^2} = \frac{dv_n}{dt} = H_0v_n \tag{22}$$

from which we obtain the following equations of galactic and cosmic evolution:

$$r_n = \frac{1}{H_0}e^{(H_0t)}\left(GM(r)H_0c\right)^{1/4} \tag{23}$$

$$v_n = e^{(H_0 t)} \left(GM(r) H_0 c\right)^{1/4} \tag{24}$$

$$a_n = H_0 e^{(H_0 t)} \left(GM(r) H_0 c\right)^{1/4} \tag{25}$$

Here r_n is the radius of curvature of space-time in the n-th quantum state (which is also the radius of the n-th state nexus graviton), v_n the radial velocity of objects embedded in that space-time, and a_n their radial acceleration within it. The amplification of the radius of curvature with time explains the existence of ultra-diffuse galaxies and the spiral shapes of most galaxies. The increase in radial velocity with time explains why early type galaxies composed of population II stars are fast rotators. Equation (25) explains late time cosmic acceleration which began once condition (19) was satisfied or equivalently from (16), when the density of baryonic matter was at the same value as that of DE. Thus condition (19) also explains the Coincidence Problem.

2.2. Canonical Transformations in the Nexus Paradigm. In classical mechanics, a system is described by n independent coordinates $(q_1, q_2, \dots q_n)$ together with their conjugate momenta $(p_1, p_2, \dots p_n)$. In the Nexus Paradigm, the labeling q_n refers to a creation of a Nexus graviton in the n-th quantum state associated with a conjugate momentum p_n. The Hamiltonian equation

$$\dot{q}_n = \frac{\partial H}{\partial p_n} \tag{26}$$

refers to the rate of expansion or contraction of space-time generated by the addition or emission of harmonic waves to the Nexus graviton and

$$\dot{p}_n = -\frac{\partial H}{\partial q_n} \tag{27}$$

refers to the force field associated with the addition or emission of harmonic waves to the Nexus graviton. It is important to note that this force field generates an isotropic expansion or contraction of space-time within the spatiotemporal dimensions of the graviton.

We can also rewrite the Hamiltonian equations in terms of Poisson brackets which are invariant under canonical transformations as

$$\begin{aligned} \dot{q}_n &= \{q_n, H\}, \\ \dot{p}_n &= \{p_n, H\} \end{aligned} \tag{28}$$

The Poisson brackets provide the bridge between classical and QM and in QM, these brackets are written as

$$\begin{aligned} \dot{\hat{q}}_n &= \left[\hat{q}_n, \hat{H}\right], \\ \dot{\hat{p}}_n &= \left[\hat{p}_n, \hat{H}\right] \end{aligned} \tag{29}$$

and obey the following commutation rules

$$\begin{aligned} \left[\hat{q}_n, \hat{q}_s\right] &= 0, \\ \left[\hat{p}_n, \hat{p}_s\right] &= 0, \\ \left[\hat{q}_n, \hat{p}_s\right] &= \delta_{ns} \end{aligned} \tag{30}$$

2.3. The Hamiltonian Formulation for the Quantum Vacuum. The Nexus graviton is a pulse of 4-space which can only expand or contract and does not execute translational motion implying that the Hamiltonian density of the system is equal to the Lagrangian density.

$$H = L \tag{31}$$

GR is a metric field in which the energy density in four-space determines its value. Since the Bloch energy eigenstate functions determine the energy of space-time, we must seek to express the metric in terms of the Bloch wave functions. To this end we shall express the eigenstate four-space components of the Nexus graviton in the k-th band as

$$\Delta x^{\mu}_{n,k} = z^{\mu}_{n,k} = a_{nk} \gamma^{\mu} \operatorname{sinc}(kx)\, e^{ikx} \tag{32}$$

such that an infintesimal four-radius within the k-th band is computed as

$$dr_{n,k} = \frac{\partial z^{\mu}_{n,k}}{\partial k^{\mu}} dk^{\mu} = ix a_{nk} \gamma^{\mu} \operatorname{sinc}(kx)\, e^{ikx} dk \tag{33}$$

In (33) the first-order derivative of the periodic sinc function is equal to zero for all integral values of n. The interval within the band is then computed as

$$\begin{aligned} ds^2 = dr^2_{n,k} &= \frac{\partial z^{\mu}_{n,k}}{\partial k^{\mu}} \frac{\partial z^{\mu}_{n,k}}{\partial k^{\nu}} dk^{\mu} dk^{\nu} \\ &= \alpha\beta \gamma^{\mu} \gamma^{\nu} \varphi_{(n,k,x)} \varphi_{(n,k,x)} dk^{\mu} dk^{\nu} \end{aligned} \tag{34}$$

Here the interval is described in terms of the reciprocal lattice with $\alpha = ix_{\mu} a_{nk}$ and $\beta = ix_{\nu} a_{nk}$. The metric tensor of four-space in the k-th band is therefore associated with the Bloch energy eigenstate functions of the quantum vacuum as follows:

$$g^{\mu\nu}_{(n,k)} = \gamma^{\mu} \gamma^{\nu} \varphi_{(n,k,x)} \varphi_{(n,k,x)} = \eta^{\mu\nu} \varphi_{(n,k,x)} \varphi_{(n,k,x)} \tag{35}$$

Equation (35) translates the geometric language of GR into the wave function language of QM.

We initiate the translation procedure of GR into QM by first finding the Lagrange density for the quantized vacuum from (15) which following Einstein and Hilbert is found to be

$$L_{EH} = k\left(R - 2\left(n^2 - 1\right)\Lambda\right) \tag{36}$$

Given that the Einstein tensor in a compact manifold is equal to the Ricci flow

$$-\partial_t g_{\mu\nu} = \Delta g_{\mu\nu} = R_{\mu\nu} - \frac{1}{2} R g_{\mu\nu} = G_{\mu\nu} \tag{37}$$

therefore the equations of motion of the quantum vacuum obtained from (36) yield the following quantized field equations:

$$\begin{aligned} &-\partial_t \left(\gamma^{\mu} \varphi_{(n,k,x)} \gamma^{\nu} \varphi_{(n.k.x)}\right) \\ &\quad = \left(n^2 - 1\right) \Lambda \gamma^{\mu} \varphi_{(n,k,x)} \gamma^{\nu} \varphi_{(n,k,x)} \end{aligned} \tag{38}$$

which can be written as

$$\begin{aligned} &\partial_t \left(\gamma^{\mu} \varphi_{(n-1,k,x)} \gamma^{\nu} \varphi_{(n+1,k,x)}\right) \\ &\quad = \frac{-i^2}{(2\pi)^2} \nabla_{\mu} \nabla_{\nu} \gamma^{\mu} \varphi_{(n-1,k,x)} \gamma^{\nu} \varphi_{(n+1,k,x)} \\ &\quad = \frac{1}{4\pi^2} \nabla_{\mu} \nabla_{\nu} \gamma^{\mu} \varphi_{(n-1,k,x)} \gamma^{\nu} \varphi_{(n+1,k,x)} \end{aligned} \tag{39}$$

where

$$\varphi_{(n-1,k,x)} = \operatorname{sinc}\left((n-1) k_1 x\right) e^{i(n-1)k_1 x} \tag{40}$$

$$\varphi_{(n+1,k,x)} = \text{sinc}\left((n+1)\,k_1 x\right) e^{i(n+1)k_1 x} \tag{41}$$

$$\frac{3k_1^2}{(2\pi)^2} = \Lambda \tag{42}$$

For large values of n the Bloch functions statisfy the condition

$$\varphi_{(n-1,k,x)} \approx \varphi_{(n,k,x)} \approx \varphi_{(n+1,k,x)} \tag{43}$$

The quantum vacuum can therefore be interpreted as a system in which there are a constant annihilation and creation of quanta as implied by (40) and (41) which causes the Nexus graviton to either expand or contract.

2.4. The Hamiltonian Formulation in the Presence of Matter Fields. We now seek to introduce matter fields into the quantum vacuum. If we compare the quantized metric of (17) describing the gravity within a Nexus graviton with the Schwarzschild metric describing the gravitational field around baryonic matter, we notice that we can describe the gravitational field around baryonic matter in terms of the quantum state of space-time through the relation

$$\frac{2}{n^2} = \frac{2GM}{c^2 r} \tag{44}$$

This yields a relationship between the quantum state of space-time and the amount of baryonic matter embedded within it as follows:

$$n^2 = \frac{c^2 r}{GM} = \frac{c^2}{v^2} \tag{45}$$

Equation (45) reveals a family of concentric stable circular orbits $r_n = n^2 GM/c^2$ with corresponding orbital speeds of $v_n = c/n$. Thus in the Nexus Paradigm, unlike in GR, the innermost stable circular orbit ocurrs at $n = 1$ or at half the Schwarzchild radius which implies that the event horizon predicted by the Nexus Paradigm is half the size predicted in GR. Also (45) reveals how the Nexus graviton in the n-th quantum state imitates DM if M is considered as the apparent mass of the DM. Through this comparision, we can also deduce that the deflection of light through gravitational lensing by space-time in the n-th quantum state is

$$\alpha = \frac{4}{n^2} \tag{46}$$

Thus gravitational lensing can be used to constrain the value of the quantum state n of space-time within a lensing system.

Having obtained the relationship between the quantum state of space-time and the amount of baryonic matter embedded within it, the result of (45) is then added to (39) to yield the time evolution of the quantum vacuum in the presence of baryonic matter as

$$\begin{aligned} &\partial_t \left(\gamma^\mu \varphi_{(n-1,n,k,x)} \gamma^\nu \varphi_{(n+1,n,k,x)}\right) \\ &= \frac{1}{4\pi^2} \nabla_\mu \nabla_\nu \gamma^\mu \varphi_{(n-1,k,x)} \gamma^\nu \varphi_{(n+1,k,x)} \\ &- n^2 \Lambda \gamma^\mu \varphi_{(n,k.x)} \gamma^\nu \varphi_{(n,k,x)} \end{aligned} \tag{47}$$

Thus the time evolution of the quantum vacuum in the presence of matter resembles thermal flow in the presence of a heat sink. The second term on the R.H.S. of (47) is an 8-cell or 4-cube that operates as a sinc filter with a four-wave cut-off of

$$k_c = nk_1 = 2\pi \sqrt{\frac{\Lambda}{3} \bullet \frac{c^2 r}{GM}} \tag{48}$$

The filtration of high frequencies from the vacuum lowers the quantum vacuum state and generates a gravitational field in much the same way as the Casimir Effect is generated. The physical process of filtration occurs as follows: A 4-cube with baryonic matter embedded with in it acquires inertia. The inertia gives the cell inductive impedance and becomes less reponsive to high frequency vibrations of the quantum vacuum. Thus the heavier the cell, the lower the cut-off frequency. Equation (48) shows that when the radius of the cell is equal to half the Schwarzschild radius, the only permited frequency is that of the ground state. Thus the inside of a black hole is in the ground state having a negative metric signature.

We now introduce a test particle of mass m, into the quantum vacuum perturbed by matter fields. The particle will flow along with the Ricci flow and the Hamiltonian of the system becomes

$$\begin{aligned} &\widehat{H} \left(\gamma^\mu \varphi_{(n-1,n,k,x)} \gamma^\nu \varphi_{(n+1,n,k,x)}\right) \\ &= \frac{\widehat{\mathrm{P}}_\mu \widehat{\mathrm{P}}_\nu}{2m} \left(\gamma^\mu \varphi_{(n-1,k,x)} \gamma^\nu \varphi_{(n+1,k,x)}\right) \\ &- V \left(\gamma^\mu \varphi_{(n,k,x)} \gamma^\nu \varphi_{(n,k,x)}\right) \end{aligned} \tag{49}$$

Here

$$V = n^2 \frac{\hbar^2}{2m} \Lambda = \frac{rc^2}{GM} \cdot \frac{3\hbar^2 k_1^2}{2m} \tag{50}$$

$$\widehat{P}_\mu = -i\hbar \nabla_\mu \tag{51}$$

$$\widehat{H} = -i\hbar \partial_t \tag{52}$$

Equation (49) is equivalent to (31) in which the Hamiltonian is equal to the Lagrangian.

$$\begin{aligned} &\widehat{H} \left(\gamma^\mu \varphi_{(n-1,n,k,x)} \gamma^\nu \varphi_{(n+1,n,k,x)}\right) \\ &= \mathrm{L} \left(\gamma^\mu \varphi_{(n-1,n,k,x)} \gamma^\nu \varphi_{(n+1,n,k,x)}\right) \end{aligned} \tag{53}$$

The Lagrangian is not relativistic because the energy reference frames or space-time states involved in the transition dynamics $n-1$, n, and $n+1$ are separated by a small energy gap, $E = \sqrt{3} \bullet hH_0$. The exchanged quantum of energy between states is responsible for "welding" them together.

Equation (50) is the gravitational interaction in reciprocal space-time. The weakness of the gravitational interaction is due to the small value of the cosmological constant.

3. Line Elements and Information in K-Space

A close inspection of (34) reveals the relationship between a line element in 4-space and its conjugate line in K-space

$$ds^2 = \alpha\beta d\kappa^2 \tag{54}$$

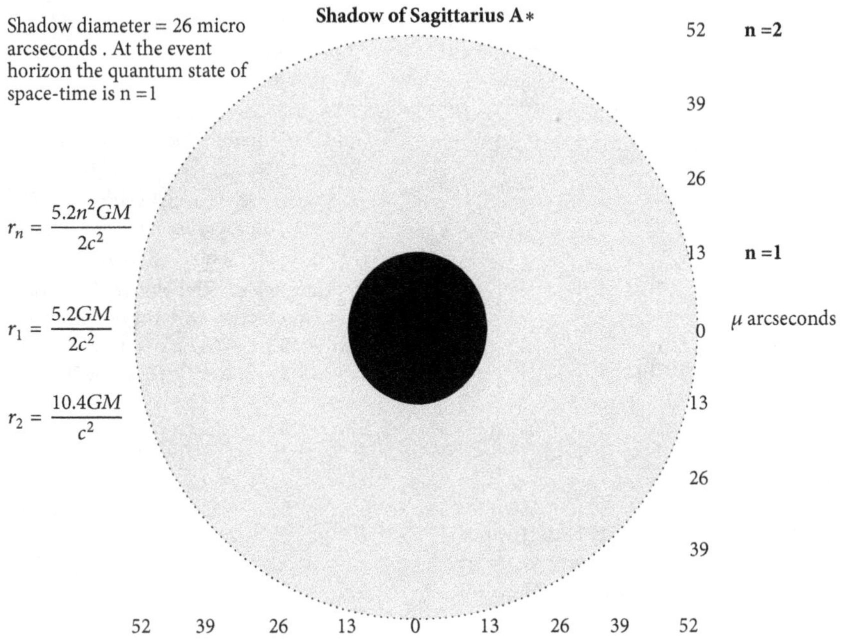

FIGURE 1: Shadow of Sagittarius A*.

where

$$d\kappa^2 = \gamma^{\mu}\varphi_{(n,k,x)}\gamma^{\nu}\varphi_{(n,k,x)}dk^{\mu}dk^{\nu} \quad (55)$$

The term $\gamma^{\mu}\varphi_{(n,k,x)}\gamma^{\nu}\varphi_{(n,k,x)}$ refers to the normalized information flux density passing through an elementary surface $dk^{\mu}dk^{\nu}$ in K-space. The gradient of the flux density yields the baseband bandwidth $k_{(n)\mu}$.

$$\partial_{\mu}\gamma^{\mu}\varphi_{(n,k,x)}\gamma^{\nu}\varphi_{(n,k,x)} = k_{(n)\mu}\gamma^{\mu}\varphi_{(n,k,x)}\gamma^{\nu}\varphi_{(n,k,x)} \quad (56)$$

This gradient is also a measure of the acutance or sharpness of the information. A large bandwidth provides detailed information while a small bandwidth provides diffuse information. Equation (49) describes the diffusion of information as it flows into a gravitational well. Unitarity is preserved if the flux is summed over the total diffusion surface.

$$\iint_0^{\Sigma} \gamma^{\mu}\varphi_{(n,k,x)}\gamma^{\nu}\varphi_{(n,k,x)}dk_{\mu}dk_{\nu} = 1 \quad (57)$$

Here the integral implies that one bit of information is found on a surface $\Sigma = k_n^2$ in K-space which corresponds to an area $A_n = 4\pi^2/k_n^2$ in 4-space. The k_1 bandwidth permitted within a black hole generates the smallest elementary surface or pixel in K-space of k_1^2 suggesting that information inside a blackhole is diluted to one bit per area equivalent to the square of the Hubble 4-radius in 4-space or from (48) the information surface density is $k_1^2 = 4\pi^2\Lambda/3$. Thus the cosmological constant can also be interpreted as a unit of information surface density. The ratio of the Planck surface in K-space to the smallest elementary surface in K-space yields 10^{120}. This huge number represents the maximum number of pixels that can fit on the largest surface in K-space. Hence the expression $W = k_{\text{p}}^2/k_{\text{n}}^2$ represents the number of empty pixel slots available in the n-th quantum state or the number of degenerate energy levels.

$$|\varphi\rangle_n = c_1|\psi_1\rangle_n + c_2|\psi_2\rangle_n \cdots c_W|\psi_W\rangle_n. \quad (58)$$

The entropy of the n-state therefore becomes

$$S = k_B \ln W = k_B \ln \frac{k_{\text{p}}^2}{k_{\text{n}}^2} = k_B \ln \frac{A_n}{l_{\text{p}}^2} = k_B \ln \frac{c^3 A_n}{G\hbar} \quad (59)$$

Here we observe that objects fall into a gravitational field because it leads to an increase in entropy. In strong fields the information is delocalized. A black hole does not annihilate information; it simply diffuses or dilutes it by increasing the entropy. It can also be interpreted as a low pass filter that permits one bit of information to pass per Hubble time. Since information is delocalized close to a black hole, then we do not expect the Event Horizon Telescope to observe a silhouette surrounded by a well defined accretion disc but a circular silhouette surrounded by a cloud of degenerate (delocalized) matter (grey area) as depicted in Figure 1.

The radii are calculated from (45) and the magnification factor, 5.2/2, arises from gravitational lensing. The silhouette is half the size predicted in GR.

4. Discussion

A successful covariant canonical quantization of the gravitational field has been presented in which we find that gravity is akin to a thermal flow of space-time and that space-time can also be described in terms of a reciprocal lattice. The presence of matter creates an impure lattice and a potential well arises in the region of perturbation through filtration of high frequencies from the quantum vacuum. A test particle flows along with the quantum vacuum and the presence of a potential well will cause it to flow towards the sink. This formulation will find important applications in high energy lattice gauge field theories where it may help expand the standard model of particle physics by eliminating divergent terms in the current theory and predicting hitherto unknown phenomena.

Conflicts of Interest

The author declares no conflicts of interest.

Acknowledgments

The author gratefully appreciates the funding and support from the Department of Physics and Astronomy at the Botswana International University of Science and Technology (BIUST).

References

[1] R. Arnowitt, S. Deser, and C. W. Misner, "Dynamical structure and definition of energy in general relativity," *Physical Review A: Atomic, Molecular and Optical Physics*, vol. 116, no. 5, pp. 1322–1330, 1959.

[2] Y. Sendouda, N. Deruelle, M. Sasaki, and D. Yamauchi, "Higher curvature theories of gravity in the adm canonical formalism," *International Journal of Modern Physics: Conference Series*, vol. 01, pp. 297–302, 2011.

[3] E. Gourgoulhon, "3+1 formalism and bases for numerical relativity," 2007, https://arxiv.org/abs/gr-qc/0703035.

[4] N. Kiriushcheva and S. Kuzmin, "The Hamiltonian formulation of general relativity: myths and reality," *Open Physics*, vol. 9, no. 3, pp. 576–615, 2011.

[5] E. Anderson, "Problem of time in quantum gravity," *Annalen der Physik*, vol. 524, no. 12, pp. 757–786, 2012.

[6] B. S. DeWitt, "Quantum theory of gravity. I. The canonical theory," *Physical Review A: Atomic, Molecular and Optical Physics*, vol. 160, article 1113, 1967.

[7] M. Bojowald and P. A. Hohn, "An Effective approach to the problem of time," *Classical and Quantum Gravity*, vol. 28, no. 3, Article ID 035006, 2011.

[8] C. Kiefer, "Does time exist in Quantum Gravity?" in *Towards a Theory of Spacetime Theories. Einstein Studies*, D. Lehmkuhl, G. Schiemamm, and E. Scholz, Eds., vol. 13, pp. 287–295, Birkhauser, New York, NY, USA, 2017.

[9] S. Carlip, "Quantum gravity: a progress report," *Reports on Progress in Physics*, vol. 64, no. 8, pp. 885–942, 2001.

[10] S. Carlip, D.-W. Chiou, W.-T. Ni, and R. Woodard, "Quantum gravity: a brief history of ideas and some prospects," *International Journal of Modern Physics D: Gravitation, Astrophysics, Cosmology*, vol. 24, no. 11, Article ID 1530028, 23 pages, 2015.

[11] C. Kiefer, "Quantum gravity: general introduction and recent developments," *Annalen der Physik*, vol. 15, no. 1-2, pp. 129–148, 2006.

[12] C. Kiefer, "Conceptual problems in quantum gravity and quantum cosmology," *ISRN Mathematical Physics*, vol. 2013, Article ID 509316, 17 pages, 2013.

[13] S. Marongwe, "The Nexus graviton: A quantum of Dark Energy and Dark Matter," *International Journal of Geometric Methods in Modern Physics*, vol. 11, no. 06, Article ID 1450059, 2014.

[14] S. Marongwe, "The Schwarzschild solution to the Nexus graviton field," *International Journal of Geometric Methods in Modern Physics*, vol. 12, no. 4, Article ID 1550042, 10 pages, 2015.

[15] S. Marongwe, "The electromagnetic signature of gravitational wave interaction with the quantum vacuum," *International Journal of Modern Physics D: Gravitation, Astrophysics, Cosmology*, vol. 26, no. 3, Article ID 1750020, 15 pages, 2017.

Permissions

All chapters in this book were first published in AHEP, by Hindawi Publishing Corporation; hereby published with permission under the Creative Commons Attribution License or equivalent. Every chapter published in this book has been scrutinized by our experts. Their significance has been extensively debated. The topics covered herein carry significant findings which will fuel the growth of the discipline. They may even be implemented as practical applications or may be referred to as a beginning point for another development.

The contributors of this book come from diverse backgrounds, making this book a truly international effort. This book will bring forth new frontiers with its revolutionizing research information and detailed analysis of the nascent developments around the world.

We would like to thank all the contributing authors for lending their expertise to make the book truly unique. They have played a crucial role in the development of this book. Without their invaluable contributions this book wouldn't have been possible. They have made vital efforts to compile up to date information on the varied aspects of this subject to make this book a valuable addition to the collection of many professionals and students.

This book was conceptualized with the vision of imparting up-to-date information and advanced data in this field. To ensure the same, a matchless editorial board was set up. Every individual on the board went through rigorous rounds of assessment to prove their worth. After which they invested a large part of their time researching and compiling the most relevant data for our readers.

The editorial board has been involved in producing this book since its inception. They have spent rigorous hours researching and exploring the diverse topics which have resulted in the successful publishing of this book. They have passed on their knowledge of decades through this book. To expedite this challenging task, the publisher supported the team at every step. A small team of assistant editors was also appointed to further simplify the editing procedure and attain best results for the readers.

Apart from the editorial board, the designing team has also invested a significant amount of their time in understanding the subject and creating the most relevant covers. They scrutinized every image to scout for the most suitable representation of the subject and create an appropriate cover for the book.

The publishing team has been an ardent support to the editorial, designing and production team. Their endless efforts to recruit the best for this project, has resulted in the accomplishment of this book. They are a veteran in the field of academics and their pool of knowledge is as vast as their experience in printing. Their expertise and guidance has proved useful at every step. Their uncompromising quality standards have made this book an exceptional effort. Their encouragement from time to time has been an inspiration for everyone.

The publisher and the editorial board hope that this book will prove to be a valuable piece of knowledge for researchers, students, practitioners and scholars across the globe.

List of Contributors

Ping Li, Miao He, Jia-Cheng Ding, Xian-Ru Hu and Jian-Bo Deng
Institute of Theoretical Physics, Lanzhou University, Lanzhou 730000, China

Tanwi Bandyopadhyay
Adani Institute of Infrastructure Engineering, Ahmedabad-382421, India

M. Abdollahi Zadeh
Physics Department and Biruni Observatory, College of Sciences, Shiraz University, Shiraz 71454, Iran

A. Sheykhi
Physics Department and Biruni Observatory, College of Sciences, Shiraz University, Shiraz 71454, Iran
Research Institute for Astronomy and Astrophysics of Maragha (RIAAM), Maragha, Iran

Farzana Kousar, Rabia Saleem and M. Zubair
Department of Mathematics, COMSATS University Islamabad, Lahore-Campus, Pakistan

Youngsub Yoon
Department of Physics and Astronomy, Seoul National University, Seoul 08826, Republic of Korea

Yan-Song Liu
Department of Physics, Shanxi Datong University, Datong 037009, China

Meng-Sen Ma
Institute of Theoretical Physics, Shanxi Datong University, Datong 037009, China
Department of Physics, Shanxi Datong University, Datong 037009, China

Pameli Saha and Ujjal Debnath
Department of Mathematics, Indian Institute of Engineering Science and Technology, Shibpur, Howrah 711 103, India

Alexander Y. Yosifov
Department of Physics and Astronomy, Shumen University, Bulgaria

Lachezar G. Filipov
Space Research and Technology Institute, Bulgarian Academy of Sciences, Bulgaria

Ovidiu Cristinel Stoica
Department of Theoretical Physics, National Institute of Physics and Nuclear Engineering - Horia Hulubei, Bucharest, Romania

A. S. Sefiedgar and M. Mirzazadeh
Department of Physics, Faculty of Basic Sciences, University of Mazandaran, Babolsar 47416-95447, Iran

Shan-Quan Lan
Department of Physics, Lingnan Normal University, Zhanjiang, 524048, Guangdong, China

Mansoureh Hosseinpour and Hassan Hassanabadi
Faculty of Physics, Shahrood University of Technology, Shahrood, Iran

S. Davood Sadatian and S. M. Hosseini
Department of Physics, Faculty of Basic Sciences, University of Neyshabur, Neyshabur, Iran

Babak Vakili
Research Institute for Astronomy and Astrophysics of Maragha (RIAAM), Maragha, Iran
Department of Physics, Central Tehran Branch, Islamic Azad University, Tehran, Iran

R. Shojaee and F. Darabi
Department of Physics, Azarbaijan Shahid Madani University, Tabriz, Iran

K. Nozari
Department of Physics, Faculty of Basic Sciences, University of Mazandaran, Babolsar, Iran
Research Institute for Astronomy and Astrophysics of Maragha (RIAAM), Maragha, Iran

Aleksander Stachowski
Astronomical Observatory, Jagiellonian University, Orla 171, 30-244 Kraków, Poland

Marek Szydłowski
Astronomical Observatory, Jagiellonian University, Orla 171, 30-244 Kraków, Poland
Mark Kac Complex Systems Research Centre, Jagiellonian University, Łojasiewicza 11, 30-348 Kraków, Poland

Krzysztof Urbanowski
Institute of Physics, University of Zielona Góra, Prof. Z. Szafrana 4a, 65-516 Zielona Góra, Poland

Irina Radinschi and Marius Mihai Cazacu
Department of Physics "Gh. Asachi" Technical University, Iasi 700050, Romania

Theophanes Grammenos
Department of Civil Engineering, University of Thessaly, 383 34 Volos, Greece

Farook Rahaman
Department of Mathematics, Jadavpur University, Kolkata 700 032,West Bengal, India

Andromahi Spanou
School of Applied Mathematics and Physical Sciences, National Technical University of Athens, 157 80 Athens, Greece

Surajit Chattopadhyay
Department of Mathematics, Amity University, Kolkata 700135, India

Antonio Pasqua
Department of Physics, University of Trieste, 34127 Trieste, Italy

Susmita Sarkar and Farook Rahaman
Department of Mathematics, Jadavpur University, Kolkata 700 032,West Bengal, India

Irina Radinschi
Department of Physics, Gheorghe Asachi Technical University, 700050 Iasi, Romania

Theophanes Grammenos
Department of Civil Engineering, University of Thessaly, 383 34 Volos, Greece

Joydeep Chakraborty
Department of Mathematics, Nagar College, West Bengal, India

G. J. M. Zilioti
Universidade Federal do ABC (UFABC), Santo André, 09210-580 São Paulo, Brazil

R. C. Santos
Departamento de Física, Universidade Federal de São Paulo (UNIFESP), 09972-270 Diadema, SP, Brazil

J. A. S. Lima
Departamento de Astronomia, Universidade de São Paulo (IAGUSP), Rua do Matão 1226, 05508-900 São Paulo, Brazil

Enrico Morgante
Deutsches Elektronen-Synchrotron DESY, Notkestraße 85, 22607 Hamburg, Germany

M. Sharif and Sara Ashraf
Department of Mathematics, University of the Punjab, Quaid-e-Azam Campus, Lahore 54590, Pakistan

D. K. Papoulias
Institute of Nuclear and Particle Physics, NCSR 'Demokritos', 15310 Agia Paraskevi, Greece

R. Sahu and B. Nayak
National Institute of Science and Technology, Palur Hills, Berhampur, Odisha 761008, India

T. S. Kosmas
Theoretical Physics Section, University of Ioannina, 45110 Ioannina, Greece

V. K. B. Kota
Physical Research Laboratory, Ahmedabad 380 009, India

Odysseas Kosmas
Modelling and Simulation Centre, MACE, University of Manchester, Sackville Street, Manchester, UK

Theodoros Smponias
Division of Theoretical Physics, University of Ioannina, 45110 Ioannina, Greece

Mark P. Hertzberg
Institute of Cosmology & Department of Physics and Astronomy, Tufs University, Medford, MA 02155, USA

Stuart Marongwe
Department of Physics and Astronomy, Botswana International University of Science and Technology, Palapye, Botswana

Index

www.ingramcontent.com/pod-product-compliance
Lightning Source LLC
Chambersburg PA
CBHW080529200326
41458CB00012B/4382
9781632387332